1991
NATIONAL REPAIR &
REMODELING ESTIMATOR

Fourteenth Edition

by
Albert S. Paxton

Edited by
J.A. O'Grady

Craftsman

Craftsman Book Company
6058 Corte del Cedro, P.O. Box 6500, Carlsbad, CA 92008

Preface

The author has corresponded with manufacturers and wholesalers of building material supplies and surveyed retail pricing services. From these sources he developed average material prices which should apply in most parts of the country.

Wherever possible, labor and material costs have been given in conjunction with our Average Daily Production figures. However, the reader still has the option of developing local labor figures. Simply divide your total local crew manhours per day by your crew's average daily production, then multiply by the average local crew rate per hour. Using your local labor figures will make your estimates more accurate. What is a realistic labor figure to one reader may well be low to another reader because of variations in labor efficiency. Many of these output figures were developed by time studies on various job sites.

The topics in this book are arranged in alphabetical order, A to Z. To assist the reader in finding specific items, there is a complete index at the end of the book, and a main subject index at the beginning.

For each construction topic with unit price information, the reader is provided with a range of cost data: *Costs Based On Large Volume* and *Costs Based On Small Volume*. It's up to you to decide whether your jobs constitute large or small volume jobs. Factors that enter into this judgment are:

- Wage scales (union vs. non-union)
- Population density (metropolitan vs. suburban)
- Local economic conditions
- Availability of competitive bidding

The data in this book has been carefully researched, typeset, and proofread. The author and editorial staff have made every reasonable effort to ensure no errors remain in the finished product. However, the author and the publisher make no guarantees and assume no liability for the accuracy or usefulness of the data. The book represents an estimating *guideline* that should be refined by the user's expertise and knowledge of his own local conditions.

© 1991   Craftsman Book Company

IBSN 0-934041-63-6

Main Subject Index

How To Use This Book

						Costs Based On Large Volume							Costs Based On Small Volume			
1	2	3	4	5	6	7	8	9	10	11	12	13	14	15	16	17
Description	Oper	Unit	Crew Size	Avg Day Prod	Avg Mat'l Unit Cost	Avg Equip Unit Cost	Avg Labor Unit Cost	Avg Total Unit Cost	Avg Price Incl O&P	Crew Size	Avg Day Prod	Avg Mat'l Unit Cost	Avg Equip Unit Cost	Avg Labor Unit Cost	Avg Total Unit Cost	Avg Price Incl O&P

The descriptions and cost data in this book are arranged in a series of columns, which are described below. The cost data is divided into two categories: **Costs Based On Large Volume** and **Costs Based On Small Volume**. These two categories provide the estimator with a pricing range for each construction topic.

The **Description** column (1) contains the pertinent, specific information necessary to make the pricing information relevant and accurate.

The **Operation** column (2) contains a description of the construction repair or remodeling operation being performed. Generally the operations are Demolition, Install, and Reset.

The **Unit** column (3) contains the unit of measurement or quantity which applies to the item described.

The **Crew Size** columns (4 and 11) contain a description of the trade that usually installs or labors on the specified item. They include information on the labor trade that installs the material and the typical crew size. Alpha-numeral abbreviations are used in the crew size column. Full descriptions of these abbreviations are in the **Crew Compositions** and **Wage Rates** table.

The **Average Daily Production** columns (5 and 12) include the average rate of the Crew (Col. 4) per day.

The units per day in this book don't take into consideration unusually large or small quantities. But items such as travel, accessibility to work, experience of workmen, and protection of undamaged property, which can favorably or adversely affect productivity, have been considered in developing average output. For further information about labor, see "Notes — Labor" in the Notes Section of some specific items.

The **Average Material Unit** Cost columns (6 and 13) contain an average material cost for products (including, in many cases, the by-products used in installing the products). No allowance for sales tax, delivery charges, overhead and profit, has been included. Percentages for waste, shrinkage, or coverage have been taken into consideration unless indicated. For other information, see "Dimensions" or "Installation" in the Notes Section.

If the item described has many or very unusual by-products which are essential to determining the Average Material Unit Cost, the author has provided examples of material pricing. These examples are placed throughout the book in the Notes Section.

Average Daily Production and **Average Material Unit Cost** should assist the estimator in:

1. Checking prices quoted by others.

2. Developing local prices.

The estimator should verify labor rates and material prices locally. Though the prices in this book are average material prices, prices vary from locality to locality. A local hourly wage rate should normally include taxes, benefits, and insurance. Some contractors may also include overhead and profit in the hourly rate.

The **Average Equipment Unit Cost** columns (7 and 14) contain an average equipment cost based on both the average daily rental and the cost per day if owned and depreciated. The costs of daily maintenance and the operator are included.

The **Average Labor Unit Cost** columns (8 and 15) contain an average labor cost based on the average output, using data from the Average Daily Production columns and the Crew Compositions and Wage Rates table. The average labor unit cost is determined by dividing the Manhours Per Day by the Average Daily Production, then multiplying by the Average Crew Rate per hour. The rates do include fringe benefits, taxes, and insurance. Examples that show how to determine the average labor unit cost are provided in the Notes Section.

The **Average Total Unit Cost** columns (9 and 16) contain the sum of the Material, Equipment, and Labor Unit Cost columns. It does not include an allowance for overhead and profit. Total Cost + O&P = Total Price.

The **Average Total Price Including Overhead and Profit** columns (10 and 17) result from adding an O&P allowance to Total Cost. This allowance reflects the author's interpretation of average fixed and variable overhead expenses and the labor intensiveness of the operation vs. the costs of materials for the operation. This allowance factor varies throughout the book, depending on the operation. Each contractor interprets O&P differently. The range can be from 15% to 80% of the Average Total Unit Cost.

Estimating Techniques

Estimating Repair/Remodeling Jobs: The unforeseen, unpredictable, or unexpected can ruin you.

Each year, the residential repair and remodeling industry grows, and is currently outpacing residential new construction due to increases in land costs, labor wage rates, interest rates, material costs, and economic uncertainty. When people can't afford a new home, they tend to remodel their old one. And there are always houses that need repair, from natural disasters or accidents like fire. The professional repair and remodeling contractor is moving to the forefront of the industry.

Repair and remodeling spawns three occupations: the contractor and his workers, the insurance company property claims adjuster, and the property damage appraiser. Each of these professionals shares common functions, including estimating the cost of the repair or remodeling work.

Estimating isn't an exact science. Yet the estimate determines the profit or loss for the contractor, the fairness of the claim payout by the adjustor, and the amount of grant or loan by the appraiser. Quality estimating must be uppermost in the mind of each of these professionals. And accurate estimates are possible only when you know exactly what materials are needed and the number of manhours required for demolition, removal, and installation. Remember, profits follow the professional. To be profitable you must control costs — and cost control is directly related to accurate, professional estimates.

There are four general types of estimates, each with a different purpose and a corresponding degree of accuracy:

- The *guess* method: "All bathrooms cost $5,000." or "It looks like an $8,000 job to me."

- The *per measure* method: (I like to call it the *surprise package*.) "Remodeling costs $60 per SF and the job is 500 SF, so the price is $30,000."

These two methods are the least accurate and accomplish little for the adjustor or the appraiser. The contractor might use the methods for qualifying customers (e.g., "I thought a bathroom would only cost $2,000."), but never as the basis for bidding or negotiating a price.

- The *piece estimate* or *stick-by-stick* method.

- The *unit cost estimate* method.

These two methods yield a detailed estimate itemizing all of the material quantities and costs, the labor manhours and wage rates, the subcontract costs, and the allowance for overhead and profit.

Though time-consuming, the detailed estimate provides the most accuracy and predictability. It's a very satisfactory tool for negotiating either the contract price or the adjustment of a building loss. The piece estimate and the unit cost estimate rely on historical data, such as manhours per specific job operation and recent material costs. The successful repair and remodeling contractor, or insurance/appraisal company, maintains records of previous jobs detailing allocation of crew manhours per day and materials expended.

While new estimators don't have historical data records, they can rely on reference books, magazines, and newsletters to determine manhours and material costs. It is important to remember that **the reference must pertain to repair and remodeling.** This book is designed *specifically* to meet this requirement.

The reference material must specialize in repair and remodeling work because there's a large cost difference between new construction and repair and remodeling. Material and labor construction costs vary radically with the size of the job or project. Economies of scale come in to play. The larger the quantity of materials, the better the purchase price should be. The larger the number of units to be installed, the greater the labor efficiency.

Repair and remodeling work, compared to new construction, is more expensive due to a normally smaller volume of work. Typical repair work involves only two or three rooms of a house, or one roof. In new construction, the job size may be three to five complete homes or an entire development. Additionally, a lot of repair and remodeling is done with the house occupied,

working around the normal, daily activities of the occupants. In new construction, the approach is systematic and logical — work proceeds from the ground up to the roof and to the inside of the structure.

Since the jobs are small, the repair and remodeling contractor doesn't employ trade specialists. Repairers employ the "jack-of-all-trades" who is less specialized and therefore less efficient. This isn't to say the repairer is less professional than the trade specialist. On the contrary, the repairer must know about many more facets of construction: not just framing, but painting, finish carpentry, roofing, and electrical as well. But because the repairer has to spread his expertise over a greater area, he will be less efficient than the specialist who repeats the same operation all day long.

Another factor reducing worker efficiency is poor access to the work area. With new construction, where building is an orderly "from the ground up" approach, workers have easy access to the work area for any given operation. The workers can spread out as much as needed, which facilitates efficiency and minimizes the manhours required to perform a given operation.

The opposite situation exists with repair and remodeling construction. Consider an example where the work area involves fire damage on the second floor. Materials either go up through the interior stairs or through a second story window. Neither is easy when the exterior and interior walls have a finished covering such as siding and sheetrock. That results in greater labor costs with repair and remodeling because it takes more manhours to perform many of the same tasks.

If, as a professional estimator, you want to start collecting historical data, the place to begin is with *daily worker time sheets* that detail:

(1) total hours worked by each worker per day

(2) what specific operations each worker performed that day

(3) how many hours (to the nearest tenth) each worker used in each operation performed that day.

Second, you must catalog all material invoices daily, being sure that quantities and unit costs per item are clearly indicated.

Third, maintain a record of overhead expenses attributable to the particular project. Then after a number of jobs, you'll be able to calculate an average percentage of the job's gross amount that's attributable to overhead. Many contractors add 45% for overhead and profit to their total direct costs (direct labor, direct material, and direct subcontract costs). But that figure may not be right for *your* jobs.

Finally, each week you should reconcile in a job summary file the actual costs versus the estimated costs, *and determine why there is any difference*. This information can't immediately help you on this job since the contract has been signed, but it will be invaluable to you on your next job.

Up to now I've been talking about general estimating theory. Now let's be more specific. On page 9 is a *Building Repair Estimate* form. Each line is keyed to an explanation. A filled-out copy of the form is also provided, and a blank, full-size copy that you can reproduce for your own use.

The Building Repair Estimate form is adaptable for use by contractor, adjustor, or appraiser. Use of the form will yield a *detailed estimate* that will identify:

- The room or area involved, including sizes, dimensions and measurements.

- The kind and quality of material to be used.

- The type of work to be performed (demolish, remove, install, remove and reset) by what type of crew.

- The quantities of materials to be used and verification of their prices.

- The crew manhours per job operation and verification of the hourly wage scale.

- All arithmetical calculations that can be verified.

- Areas of difference between your estimate and others.

- Areas that will be a basis for negotiation and discussion of details. Each job estimate begins with a *visual inspection* of the work site. If it's a repair job, you've got to see the damage. Without a visual inspection, you can't select a method of repair and you can't properly evaluate the opinions of others regarding repair or replacement. With either repair or remodeling work, the visual inspection is essential to uncover the "hiders" — the unpredictable, unforeseen, and unexpected problems that can turn profit into loss, or simplicity into nightmare. You're looking

for the many variables and unknowns that exist behind an exterior or interior wall covering.

Along with the Building Repair Estimate form, use this checklist to make sure you're not forgetting anything.

Checklist:

- Site accessibility: Will you store materials and tools in the garage? Is it secure? You can save one-half hour to an hour each day by storing tools in the garage. Will the landscaping prevent trucks from reaching the work site? Are wheelbarrows or concrete pumpers going to be required?

- Soil: What type and how much water content? Will the soil change your excavation estimate?

- Utility lines: What's under the soil and where? Should you schedule the utilities to stake their lines?

- Soundness of the structure: If you're going to remodel, repair or add on, how sound is that portion of the house that you're going to have to work around? Where are the load-bearing walls? Are you going to remove and reset any walls? Do the floor joists sag?

- Roof strength: Can the roof support the weight of another layer of shingles. (Is four layers of composition shingles already too much)?

- Electrical: Is another breaker box required for the additional load?

This checklist is by no means complete, but it is a start. Take pictures! A Polaroid camera will quickly pay for itself. When you're back at the office, the picture helps reconstruct the scene. Before and after pictures are also a sales tool representing your professional expertise.

During the visual inspection always be asking yourself "what if" this or that happened. Be looking for potential problem areas that would be extremely labor intensive or expensive in material to repair or replace.

Also spend some time getting to know your clients and their attitudes. Most of repair and remodeling work occurs while the house is occupied. If the work will be messy, let the homeowners know in advance. Their satisfaction is your ultimate goal — and their satisfaction will provide you a pleasant working atmosphere. You're there to communicate with them. At the end of an estimate and visual inspection, the homeowner should have a clear idea of what you can or can't do, how it will be done, and approximately how long it will take. Don't discuss costs now! Save the estimating for your quiet office with a print-out calculator and your cost files or reference books.

What you create on your estimate form during a visual inspection is a set of rough notes and diagrams that make the estimate speak. To avoid duplications and omissions, estimate in a *systematic sequence of inspection*. There are two questions to consider. First, where do you start the estimate? Second, in what order will you list the damaged or replaced items? It's customary to start in the room having either the most damage or requiring the most extensive remodeling. The sequence of listing is important. Start with either the floor or the ceiling. When starting with the floor, you might list items in the following sequence: Joists, subfloor, finish floor, base — listing from bottom to top. When starting with the ceiling, you reverse, and list from top to bottom. The important thing is to *be consistent* as you go from room to room! It's a good idea to figure the roof and foundation separately, instead of by the room.

After completing your visual inspection, go back to your office to cost out the items. Talk to your material supply houses and get unit costs for the quantity involved. Consult your job files or reference books and assign crew manhours to the different job operations.

There's one more reason for creating detailed estimates. Besides an estimate, what else have your notes given you? A material take-off sheet, a lumber list, a plan and specification sheet — the basis for writing a job summary for comparing estimated costs and profit versus actual costs and profit — and a project schedule that minimizes down time.

Here's the last step: Enter an amount for overhead and profit. No matter how small or large your work volume is, be realistic — everyone has overhead.

An office, even in your home, costs money to operate. If family members help out, pay them. Everyone's time is valuable!

Don't forget to charge for performing your estimate. A professional expects to be paid. You'll render a better product if you know you're being paid for your time. If you want to soften the blow to the owner, say the first hour is free or that the cost of the estimate will be deducted from the job price if you get the job.

In conclusion, whether you're a contractor, adjustor, or appraiser, you're selling your personal service, your ideas, and your reputation. To be successful you must:

- Know yourself and your capabilities.

- Know what the job will require by ferreting out the "hiders."

- Know your products and your work crew.

- Know your productivity and be able to deliver in a reasonable manner and within a reasonable allotment of time.

- Know your client and make it clear that all change orders, no matter how large or small, will cost money.

BUILDING REPAIR ESTIMATE

Blank keyed form:

DATE (33)		
INSURED (1)	CLAIM OR POLICY NO. (2)	PAGE (3) OF
LOSS ADDRESS (4)	HOME PH.	CAUSE OF LOSS (5)
CITY (7)	BUS. PH.	OTHER INS. Y N (6)
BLDG R.C.V. (7)	BLDG A.C.V (8)	INSURANCE AMOUNT (9)
INSURANCE REQUIRED R.C.V. (10) % A.C.V. (11) %		

DESCRIPTION OF ITEM (20)	UNIT COST OR MATERIAL PRICE ONLY (12)			LABOR PRICE ONLY (16)		
	UNIT (13)	UNIT PRICE (14)	TOTAL (COL A) (15)	HOURS (17)	RATE (18)	TOTAL (COL B) (19)

THIS IS NOT AN ORDER TO REPAIR   TOTALS (21) (22)

The undersigned agrees to complete and guarantee repairs at a total of $ (29)   TOTAL COLUMN A

REPAIRER (30) — (23)
STREET (31) — (24)
CITY   PHONE — (25)
BY (32) — (26)
ADJUSTER (34)   DATE OF A/P (35) — (27)
ADJ. LICENSE NO. (IF ANY)   (28) GRAND TOTAL
SERVICE OFFICE NAME (36)

NOTE: THIS FORM DOES NOT REPLACE THE NEED FOR FIELD NOTES, SKETCHES AND MEASUREMENTS

BUILDING REPAIR ESTIMATE

Filled example form:

DATE		
INSURED John Q. Smith	CLAIM OR POLICY NO. DP 0029	PAGE 1 OF 2
LOSS ADDRESS 123 So. Main St.	HOME PH. 555-1241	CAUSE OF LOSS Fire
CITY Anywhere, Anystate 00010	BUS. PH. 555-1438	OTHER INS. Y (N)
BLDG R.C.V. 100,000 BLDG A.C.V 90,000	INSURANCE AMOUNT $100,000	
INSURANCE REQUIRED R.C.V.(80%) A.C.V.(80%)		

DESCRIPTION OF ITEM	UNIT COST OR MATERIAL PRICE ONLY			LABOR PRICE ONLY		
	UNIT	UNIT PRICE	TOTAL (COL A)	HOURS	RATE	TOTAL (COL B)
Install 1/2" sheetrock (standard)						
on walls, including tape						
and finish 400/736	400	0.26	104.00	4.4	20.79	91.48
Paint walls, roller						
1 coat primer 600/1840	600	0.05	30.00	2.6	19.27	50.10
2 coats latex flat 600/1040	600	0.09	54.00	4.6	19.27	88.64

THIS IS NOT AN ORDER TO REPAIR   TOTALS   188.00   230.22

The undersigned agrees to complete and guarantee repairs at a total of $515.40   TOTAL COLUMN A   188.00

REPAIRER   ABC Construction — 418.22
STREET   316 E. 2nd Street   Tax — 11.28
CITY   Anywhere   PHONE — 429.50
BY   Jack Williams   10% Overhead — 42.95
ADJUSTER Stan Jones   DATE OF A/P N/A   10% Profit — 42.95
ADJ. LICENSE NO. (IF ANY)   561-84   GRAND TOTAL — 515.40
SERVICE OFFICE NAME   Phoenix

NOTE: THIS FORM DOES NOT REPLACE THE NEED FOR FIELD NOTES, SKETCHES AND MEASUREMENTS

Keyed Explanations Of The Building Repair Estimate Form

1. Insert name of insured(s).

2. Insert claim number or, if claim number is not available, insert policy number or binder number.

3. Insert the page number and the total number of pages.

4. Insert street address, city and state where loss or damage occurred.

5. Insert type of loss (wind, hail, fire, water, etc.)

6. Check YES if there is other insurance, whether collectible or not. Check NO if there's only one insurer.

7. Insert the present replacement cost of the building. What would it cost to build the structure today?

8. Insert present actual cash value of the building.

9. Insert the amount of insurance applicable. If there is more than one insurer, insert the total amount of applicable insurance provided by all insurers.

10. If the amount of insurance required is based on replacement cost value, circle RCV and insert the percent required by the policy, if any.

11. If the amount of insurance required is based on actual cash value, circle AV and insert the percent required by the policy, if any.

Note: (regarding 10 and 11) if there is a non-concurrency, i.e., one insurer requires insurance to 90% of value while another requires insurance to 80% of value, make a note here. Comment on the non-concurrency in the settlement report.

12. The installed price and/or material price only, as expressed in columns 13 through 15, may include any of the following (expressed in units and unit prices):

   - Material only (no labor)

   - Material and labor to replace

   - Material and labor to remove and replace

   Unit Cost is determined by dividing dollar cost by quantity. The term cost, as used in unit cost, is not intended to include any allowance, percentage or otherwise, for overhead or profit. Usually, overhead and profit are expressed as a percentage of cost. Cost must be determined first. Insert a line or dash in a space(s) in columns 13, 14, 15, 17, 18 or 19 if the space is not to be used.

13. The units column includes both the quantity and the unit of measure, i.e., 100 SF, 100 BF, 200 CF, 100 CY, 20 ea., etc.

14. The unit price may be expressed in dollars, cents or both. If the units column has 100 SF and if the unit price column has $.10, this would indicate the price to be $.10 per SF.

15. The total column is merely the dollar product of the quantity (in column 13) times the price per unit measure (in column 14).

16-19. These columns are normally used to express labor as follows: hours times rate per hour. However, it is possible to express labor as a unit price, i.e., 100 SF in column 13, a dash in column 17, $.05 in column 18 and $5.00 in column 19.

20. Under description of item, the following may be included:

   - Description of item to be repaired or replaced (studs 2" x 4" 8'0" #2 Fir, Sheetrock 1/2", etc.)

   - Quantities or dimensions (20 pcs., 8'0" x 14'0", etc.)

   - Location within a room or area (north wall, ceiling, etc.)

   - Method of correcting damage (paint - 1 coat; sand, fill and finish; R&R; remove only; replace; resize; etc.)

21-22. Dollar totals of columns A and B respectively.

23-27. Spaces provided for items not included in the body of the estimate (subtotals, overhead, profit, sales tax, etc.)

28. Total cost of repair.

29. Insert the agreed amount here. The agreement may be between the claim representative and the insured or between the claim rep and the repairer. If the agreed price is different from the grand total, the reason(s) for the difference should be itemized on the estimate sheet. If there is no room, attach an additional estimate sheet.

30. PRINT the name of the insured or the repairer so that it is legible.

31. PRINT the address of the insured or repairer legibly. Include phone number.

32. The insured or a representative of the repairer should sign here indicating agreement with the claim rep's estimate.

33. Insured or representative of the repairer should insert date here.

34. Claim rep should sign here.

35. Claim rep should insert date here.

36. Insert name of service office here.

BUILDING REPAIR ESTIMATE

INSURED

LOSS ADDRESS

CITY

BLDG R.C.V.   BLDG A.C.V.

INSURANCE REQUIRED   R.C.V.(%)   A.C.V.(%)

DATE

CLAIM OR POLICY NO.

PAGE   OF

HOME PH.

CAUSE OF LOSS

BUS. PH.

OTHER INS.   Y   N

INSURANCE AMOUNT

DESCRIPTION OF ITEM	UNIT COST OR MATERIAL PRICE ONLY			LABOR PRICE ONLY		
	UNIT	UNIT PRICE	TOTAL (COL A)	HOURS	RATE	TOTAL (COL B)
TOTALS						

TOTAL COLUMN A

THIS IS NOT AN ORDER TO REPAIR

The undersigned agrees to complete and guarantee repairs at a total of $

REPAIRER

STREET

CITY   PHONE

BY

ADJUSTER   DATE OF A/P

ADJ. LICENSE NO. (IF ANY)

SERVICE OFFICE NAME

GRAND TOTAL

NOTE: THIS FORM DOES NOT REPLACE THE NEED FOR FIELD NOTES, SKETCHES AND MEASUREMENTS

Wage Rates Used In This Book

Wage rates listed here and used in this book are representative of prevailing wages in the United States as of January 1991. Wage rates are in dollars per hour.

"Base Wage Per Hour" (Col. 1) includes items such as vacation pay and sick leave which are normally taxed as wages. Nationally, these benefits average 5.15% of the Base Wage Per Hour. This amount is paid by the Employer in addition to the Base Wage Per Hour.

"Liability Insurance and Employer Taxes" (Cols. 3 & 4) include national averages for state unemployment insurance (4.00%), federal unemployment insurance (0.80%), Social Security and Medicare tax (6.70%), liability insurance (2.29%), and Worker's Compensation Insurance which varies by trade. This total percentage (Col. 3) is applied to the sum of Base Wage Per Hour and Taxable Fringe Benefits (Col. 1 + Col. 2) and is listed in Dollars (Col. 4). This amount is paid by the Employer in addition to the Base Wage Per Hour and the Taxable Fringe Benefits.

"Non-Taxable Fringe Benefits" (Col. 5) include employer sponsored medical insurance and other benefits, which nationally average 4.55% of the Base Wage Per Hour.

"Total Hourly Cost Used In This Book" is the sum of Columns 1, 2, 4, & 5.

| | 1 | 2 | 3 | 4 | 5 | 6 |
| | Base Wage | Taxable Fringe Benefits | Liab'ty Insurance & Employer Taxes | | Non-taxable Fringe Benefits | Total Hourly Cost |
Trade	Per Hour	(5.15% of Base Wage)	%	$	(4.55% of Base Wage)	Used In This Book
Air Tool Operator	$15.30	$0.79	26.9%	$4.33	$0.70	$21.11
Bricklayer or Stone Mason	$15.85	$0.82	27.6%	$4.60	$0.72	$21.99
Bricktender	$11.70	$0.60	27.6%	$3.40	$0.53	$16.23
Carpenter	$14.85	$0.76	30.6%	$4.78	$0.68	$21.07
Cement Mason	$14.85	$0.76	24.9%	$3.89	$0.68	$20.18
Electrician, Journeyman Wireman	$18.25	$0.94	23.3%	$4.47	$0.83	$24.49
Fence Erector	$15.50	$0.80	26.9%	$4.38	$0.71	$21.39
Floor Layer: Carpet, Linoleum, Soft Tile	$14.95	$0.77	24.6%	$3.87	$0.68	$20.27
Floor Layer: Hardwood	$17.10	$0.88	24.6%	$4.42	$0.78	$23.18
Glazier	$15.00	$0.77	27.6%	$4.35	$0.68	$20.81
Laborer, General Construction	$11.50	$0.59	30.1%	$3.64	$0.52	$16.26
Lather	$15.45	$0.80	21.7%	$3.53	$0.70	$20.47
Marble and Terrazzo Setter	$15.65	$0.81	27.6%	$4.54	$0.71	$21.71
Painter, Brush	$13.90	$0.72	27.5%	$4.02	$0.63	$19.27
Painter, Spray-Gun	$14.35	$0.74	27.5%	$4.15	$0.65	$19.89
Paperhanger	$14.60	$0.75	27.5%	$4.22	$0.66	$20.24
Plasterer	$14.80	$0.76	27.6%	$4.30	$0.67	$20.53
Plumber	$18.40	$0.95	24.2%	$4.68	$0.84	$24.87
Reinforcing Ironworker	$15.90	$0.82	37.3%	$6.24	$0.72	$23.68
Roofer, Foreman	$13.90	$0.72	40.2%	$5.88	$0.63	$21.12
Roofer, Journeyman	$12.55	$0.65	40.2%	$5.30	$0.57	$19.07
Roofer, Hot Mop Pitch	$13.25	$0.68	40.2%	$5.60	$0.60	$20.14
Roofer, Wood Shingles	$13.85	$0.71	40.2%	$5.85	$0.63	$21.05
Sheet Metal Worker	$17.35	$0.89	26.0%	$4.74	$0.79	$23.78
Tile Setter	$15.20	$0.78	24.7%	$3.95	$0.69	$20.62
Tile Setter Helper	$12.00	$0.62	24.7%	$3.12	$0.55	$16.28
Tractor Operator	$17.05	$0.88	26.9%	$4.82	$0.78	$23.53
Truck Driver	$13.45	$0.69	28.5%	$4.03	$0.61	$18.79

Area Modification Factors

Construction costs are higher in some cities than in other cities. Use the factors on this page and the next to adapt the costs listed in this book to your job site. Multiply your estimated total project cost by the factor listed for the appropriate city to find your estimated building cost.

These factors were compiled by comparing the actual construction cost of residential, institutional and commercial buildings in 402 communities throughout the United States. Because these factors are based on completed project costs, they consider all construction cost variables, including labor, equipment and material cost, labor productivity, weather, job conditions, and markup.

Use the factor for the nearest or most comparable city. If the city you need is not listed in the table, use the factor for the appropriate state.

Note that these location factors are composites of many costs and will not necessarily be accurate when estimating the cost of a particular part of a building. But when used to modify all estimated costs on a job, they should improve the accuracy of your estimates.

Location	Factor	Location	Factor	Location	Factor
Alabama	**0.80**	El Cajon	1.20	**Connecticut**	**1.21**
Birmingham	0.82	Eureka	1.17	Bridgeport	1.24
Dothan	0.76	Fresno	1.09	Hartford	1.22
Florence	0.78	Inyokern	1.20	New Haven	1.20
Huntsville	0.80	Lancaster	1.28	New London	1.18
Mobile	0.83	Los Angeles	1.32	New Milford	1.21
Montgomery	0.80	Modesto	1.08	Norwich	1.17
Tuscaloosa	0.79	Mojave	1.23	Ridgefield	1.25
Alaska	**2.21**	Needles	1.23	Windam	1.17
Anchorage	1.47	Oakland	1.34	**Delaware**	**1.07**
Fairbanks	1.57	Ojai	1.22	Dover	1.06
Juneau	1.51	Oxnard	1.25	Wilmington	1.08
Kenai	1.35	Paso Robles	1.22	**District of Columbia**	**1.22**
Ketchikan	1.51	Piru	1.21	Washington D.C.	1.22
Sitka	1.53	Placerville	1.09	**Florida**	**0.90**
Arizona	**0.97**	Redding	1.05	Jacksonville	0.85
Casa Grande	1.01	Ridgecrest	1.17	Key West	0.94
Flagstaff	0.94	Sacramento	1.08	Miami	0.93
Kingman	0.94	San Bernardino	1.23	Orlando	0.99
Phoenix	1.03	San Diego	1.19	Pensacola	0.83
Sierra Vista	0.92	San Francisco	1.47	Tampa	0.91
Tucson	1.00	San Jose	1.32	**Georgia**	**0.82**
Yuma	0.94	Santa Ana	1.31	Albany	0.81
Arkansas	**0.82**	Santa Barbara	1.26	Atlanta	0.84
Fayetteville	0.81	Santa Cruz	1.32	Augusta	0.81
Fort Smith	0.81	Santa Maria	1.24	Brunswick	0.80
Jonesboro	0.80	Santa Rosa	1.19	Columbus	0.80
Little Rock	0.84	South Lake Tahoe	1.10	Macon	0.80
Texarkana	0.80	Tehachapi	1.22	Rome	0.79
California	**1.20**	Ventura	1.25	Savannah	0.80
Arrowhead	1.24	Victorville	1.23	Valdosta	0.80
Bakersfield	1.18	Yreka	1.05	**Guam**	**1.39**
Barstow	1.22	**Colorado**	**1.02**	Agana	1.39
Big Bear	1.25	Aspen-Vail	1.14		
Desert Center	1.31	Denver	0.99		
		Grand Junction	1.02		

Location	Factor	Location	Factor	Location	Factor
Hawaii	**1.49**	Mason City	0.91	**Maryland**	**1.00**
Hilo	1.60	Sioux City	0.90	Baltimore	1.02
Honolulu	1.52	Waterloo	0.92	Hagerstown	0.95
Kauai	1.63	**Kansas**	**0.86**	Salisbury	0.98
Kona	1.62	Garden City	0.81	Waldorf	1.00
Maui	1.59	Kansas City	0.96	**Massachusetts**	**1.23**
Idaho	**0.99**	Pittsburg	0.80	Boston	1.36
Boise	0.98	Salina	0.82	Fall River	1.25
Coeur D'Alene	1.04	Topeka	0.88	Worchester	1.22
Idaho Falls	0.98	Wichita	0.82	**Michigan**	**1.00**
Pocatello	1.02	**Kentucky**	**0.92**	Ann Arbor	1.04
Illinois	**1.05**	Ashland	0.96	Battle Creek	0.96
Bellville	1.04	Covington	0.98	Benton Harbor	0.99
Chicago	1.13	Louisville	0.89	Detroit	1.06
East St. Louis	1.06	Middlesboro	0.93	Flint	1.03
Moline	1.06	Owensboro	0.90	Grand Rapids	0.95
Springfield	1.03	Paducah	0.89	Jackson	1.01
Indiana	**0.96**	**Louisiana**	**0.85**	Lansing	1.04
Bloomington	0.94	Alexandria	0.80	Marquette	1.00
Evansville	0.92	Baton Rouge	0.89	Mt. Pleasant	0.99
Fort Wayne	0.92	Houma	0.87	Muskegon	0.94
Gary	1.02	Lafayette	0.85	Saginaw	0.98
Hammond	1.00	Lake Charles	0.88	Traverse City	0.98
Indianapolis	0.93	Marshall	0.83	Ypsilanti	1.07
Lafayette	0.94	Monroe	0.81	**Minnesota**	**1.02**
South Bend	0.95	New Orleans	0.88	Duluth	1.04
Terre Haute	0.96	Shreveport	0.83	Mankato	1.00
Iowa	**0.92**	**Maine**	**1.06**	Minneapolis	1.07
Bettendorf	0.97	Augusta	1.10	Rochester	1.02
Cedar Rapids	0.93	Bangor	1.04	St. Cloud	1.01
Council Bluffs	0.89	Brunswick	1.13	Worthington	0.95
Davenport	0.95	Lewiston	1.11	**Mississippi**	**0.80**
Des Moines	0.92	Portland	1.11	Corinth	0.78
Dubuque	0.92	Waterville	1.06	Greenville	0.79
				Greenwood	0.78

Gulfport	0.80
Hattiesburg	0.77
Jackson	0.78
Southaven	0.78
Missouri	**0.95**
Cape Girardeau	0.95
Columbia	0.94
Joplin	0.89
Kansas City	1.00
Kirksville	0.95
Rolla	0.89
St. Joseph	0.95
St. Louis	1.04
Sedalia	0.93
Springfield	0.92
Montana	**0.96**
Billings	0.96
Great Falls	0.97
Helena	0.96
Missoula	0.96
Nebraska	**0.89**
Grand Island	0.91
Lincoln	0.88
Omaha	0.91
Norfolk	0.91
North Platte	0.87
Scottsbluff	0.87
Nevada	**1.13**
Las Vegas	1.16
Reno	1.15
New Hampshire	**1.07**
Concord	1.07
Dover	1.07
Keene	1.05
Manchester	1.07
Nashua	1.09
Portsmouth	1.09
New Jersey	**1.21**
Asbury Park	1.23
Atlantic City	1.25
Burlington	1.22
Camden	1.23
Freehold	1.23
Gloucester	1.22
Newark	1.30
North Bergen	1.30
Trenton	1.23
Vineland	1.21
New Mexico	**0.89**
Albuquerque	0.84
Clovis	0.83
Santa Fe	0.87
Silver City	0.85
Taos	0.93
New York	**1.21**
Albany	1.07
Binghamton	1.05
Buffalo	1.06
Elmira	1.02
Jamestown	0.98
Nassau County	1.41
N.Y. City (Metro)	1.46
N.Y. City (Inner)	1.52
Orange County	1.35
Plattsburgh	1.01
Poughkeepsie	1.21
Rochester	1.04
Rockland County	1.36
Suffolk County	1.27
Syracuse	1.06
Westchester County	1.42
North Carolina	**0.81**
Asheville	0.81
Charlotte	0.81
Durham	0.82
Elizabeth City	0.82
Fayetteville	0.78
Greensboro	0.79
Greenville	0.78
Raleigh	0.83
Wilmington	0.79
Winston-Salem	0.79
North Dakota	**0.93**
Bismarck	0.96
Dickinson	0.95
Fargo	0.92
Ohio	**1.07**
Akron	1.12
Cincinnati	1.01
Cleveland	1.15
Columbus	1.00
Dayton	1.00
Findlay	1.03
Lorain	1.11
Mansfield	1.03
Toledo	1.12
Youngstown	1.06
Oklahoma	**0.84**
Ada	0.85
Ardmore	0.84
Bartlesville	0.84
Enid	0.83
Guymon	0.85
Lawton	0.83
McAlester	0.85
Muskogee	0.86
Oklahoma City	0.84
Shawnee	0.85
Stillwater	0.84
Tulsa	0.84
Woodward	0.84
Oregon	**1.02**
Bend	1.00
Coos Bay	1.03
Eugene	0.99
Portland	1.02
Pennsylvania	**1.06**
Allentown	1.17
Altoona	1.06
Bellefonte	1.08
Erie	1.09
Harrisburg	1.04
Johnstown	1.06
Lancaster	1.08
Philadelphia	1.21
Pittsburgh	1.10
Reading	1.06
Scranton	1.08
Wellsboro	1.07
York	1.03
Puerto Rico	**0.88**
Arecibo	0.88
Mayaguez	0.88
Old San Juan	0.89
Ponce	0.88
San Juan	0.87
Rhode Island	**1.16**
Providence	1.18
South Carolina	**0.81**
Aiken	0.80
Anderson	0.80
Beaufort	0.82
Charleston	0.84
Columbia	0.80
Florence	0.79
Greenville	0.81
Greenwood	0.80
Myrtle Beach	0.82
North Augusta	0.82
Orangeburg	0.80
Rockhill	0.80
Spartansburg	0.81
South Dakota	**0.95**
Pierre	0.98
Rapid City	0.96
Sioux Falls	0.98
Tennessee	**0.81**
Chattanooga	0.83
Clarksville	0.79
Columbia	0.81
Jackson	0.82
Johnson City	0.79
Kingsport	0.80
Knoxville	0.82
Memphis	0.82
Nashville	0.80
Oak Ridge	0.81
Texas	**0.83**
Abilene	0.85
Amarillo	0.87
Austin	0.81
Beaumont	0.85
Bryan	0.85
Corpus Christi	0.77
Dallas	0.77
Eagle Pass	0.78
El Campo	0.85
El Paso	0.82
Fort Worth	0.85
Harlingen	0.77
Houston	0.84
Junction	0.79
Laredo	0.77
Lubbock	0.88
Lufkin	0.84
Midland	0.90
Odessa	0.90
San Angelo	0.84
San Antonio	0.79
Sherman	0.84
Texas City	0.85
Tyler	0.82
Waco	0.81
Wichita Falls	0.85
Victoria	0.79
Utah	**0.94**
Cedar City	0.95
Salt Lake City	0.98
Vernal	0.97
Vermont	**1.07**
Bennington	1.09
Brattleboro	1.09
Burlington	1.06
Montpelier	1.06
Rutland	1.06
Virginia	**0.92**
Charlottesville	0.92
Harrisonburg	0.88
Newport News	0.91
Norfolk	0.93
Norton	0.92
Richmond	0.94
Virgin Islands	**1.16**
St. Thomas	1.17
St. Croix	1.15
Washington	**1.06**
Aberdeen	1.06
Bellingham	1.06
Cheney	1.05
Kennewick	1.07
Longview	1.05
Olympia	1.07
Port Angeles	1.07
Pullman	1.06
Seattle	1.07
Spokane	1.04
Yakima	1.06
West Virginia	**1.00**
Bluefield	0.96
Charleston	1.02
Fairmont	1.03
Huntington	0.99
Martinsburg	0.96
Parkersburg	1.01
Point Pleasant	1.03
Wheeling	1.08
Wisconsin	**1.00**
Eau Claire	1.01
Green Bay	0.97
Madison	1.01
Milwaukee	1.05
Reedsville	0.99
Superior	1.03
Wausau	0.98
Wyoming	**0.94**
Casper	0.95
Cheyenne	0.93
Cody	0.96

Crew Compositions & Wage Rates

Crew Number	Manhours Per Day	$/Hr.	$/Day	ACR Per Hour	Crew Number	Manhours Per Day	$/Hr.	$/Day	ACR Per Hour
A-1					**C-1**				
1 Air tool operator	8	$21.11	$168.89	$21.11	1 Carpenter	8	$21.07	$168.55	$21.07
A-2					**C-2**				
1 Air tool operator	8	$21.11	$168.89		1 Carpenter	8	$21.07	$168.55	
1 Laborer	8	$16.26	$130.04		1 Laborer	1	$16.26	$16.26	
Total	16		$298.94	$18.68	Total	9		$184.80	$20.53
A-3					**C-3**				
1 Air tool operator	8	$21.11	$168.89		1 Carpenter	8	$21.07	$168.55	
2 Laborers	16	$16.26	$260.08		1 Laborer	2	$16.26	$32.51	
Total	24		$428.98	$17.87	Total	10		$201.06	$20.11
A-4					**C-4**				
2 Air tool operators	16	$21.11	$337.79		1 Carpenter	8	$21.07	$168.55	
1 Laborer	8	$16.26	$130.04		1 Laborer	3	$16.26	$48.77	
Total	24		$467.83	$19.49	Total	11		$217.31	$19.76
B-1					**C-5**				
1 Bricklayer	8	$21.99	$175.90	$21.99	1 Carpenter	8	$21.07	$168.55	
B-2					1 Laborer	4	$16.26	$65.02	
1 Bricklayer	8	$21.99	$175.90		Total	12		$233.57	$19.46
1 Bricktender	8	$16.23	$129.84		**C-6**				
Total	16		$305.74	$19.11	1 Carpenter	8	$21.07	$168.55	
B-3					1 Laborer	5	$16.26	$81.28	
2 Bricklayers	16	$21.99	$351.80		Total	13		$249.82	$19.22
2 Bricktenders	16	$16.23	$259.69		**C-7**				
Total	32		$611.48	$19.11	1 Carpenter	8	$21.07	$168.55	
B-4					1 Laborer	6	$16.26	$97.53	
3 Bricklayers	24	$21.99	$527.70		Total	14		$266.08	$19.01
2 Bricktenders	16	$16.23	$259.69		**C-8**				
Total	40		$787.38	$19.68	1 Carpenter	8	$21.07	$168.55	
B-5					1 Laborer	7	$16.26	$113.79	
2 Bricklayers	16	$21.99	$351.80		Total	15		$282.34	$18.82
1 Bricktender	8	$16.23	$129.84		**C-9**				
Total	24		$481.64	$20.07	1 Carpenter	8	$21.07	$168.55	
B-6					1 Laborer	8	$16.26	$130.04	
3 Bricklayers	24	$21.99	$527.70		Total	16		$298.59	$18.66
3 Bricktenders	24	$16.23	$389.53						
Total	48		$917.23	$19.11					

15

Crew Compositions & Wage Rates

Crew Number	Manhours Per Day	Total Costs $/Hr.	Total Costs $/Day	Average Crew Rate (ACR) Per Hour	Crew Number	Manhours Per Day	Total Costs $/Hr.	Total Costs $/Day	Average Crew Rate (ACR) Per Hour
C-10					**C-19**				
2 Carpenters	16	$21.07	$337.10		4 Carpenters	32	$21.07	$674.19	
1 Laborer	1	$16.26	$16.26		1 Laborer	8	$16.26	$130.04	
Total	17		$353.35	**$20.79**	Total	40		$804.24	**$20.11**
C-11					**C-20**				
2 Carpenters	16	$21.07	$337.10		4 Carpenters	32	$21.07	$674.19	
1 Laborer	2	$16.26	$32.51		2 Laborers	16	$16.26	$260.08	
Total	18		$369.61	**$20.53**	Total	48		$934.28	**$19.46**
C-12					**C-21**				
2 Carpenters	16	$21.07	$337.10		2 Carpenters	16	$21.07	$337.10	
1 Laborer	3	$16.26	$48.77		2 Laborers	16	$16.26	$260.08	
Total	19		$385.86	**$20.31**	Total	32		$597.18	**$18.66**
C-13					**C-22**				
2 Carpenters	16	$21.07	$337.10		4 Carpenters	32	$21.07	$674.19	**$21.07**
1 Laborer	4	$16.26	$65.02		**D-1**				
Total	20		$402.12	**$20.11**	1 Cement mason	8	$20.18	$161.43	**$20.18**
C-14					**D-2**				
2 Carpenters	16	$21.07	$337.10		1 Cement mason	8	$20.18	$161.43	
1 Laborer	5	$16.26	$81.28		1 Laborer	8	$16.26	$130.04	
Total	21		$418.37	**$19.92**	Total	16		$291.47	**$18.22**
C-15					**D-3**				
2 Carpenters	16	$21.07	$337.10		1 Cement mason	8	$20.18	$161.43	
1 Laborer	6	$16.26	$97.53		2 Laborers	16	$16.26	$260.08	
Total	22		$434.63	**$19.76**	Total	24		$421.51	**$17.56**
C-16					**D-4**				
2 Carpenters	16	$21.07	$337.10		2 Cement masons	16	$20.18	$322.86	
1 Laborer	7	$16.26	$113.79		1 Laborer	8	$16.26	$130.04	
Total	23		$450.88	**$19.60**	Total	24		$452.90	**$18.87**
C-17					**D-5**				
2 Carpenters	16	$21.07	$337.10		3 Cement masons	24	$20.18	$484.28	
1 Laborer	8	$16.26	$130.04		4 Laborers	32	$16.26	$520.17	
Total	24		$467.14	**$19.46**	Total	56		$1,004	**$17.94**
C-18									
2 Carpenters	16	$21.07	$337.10	**$21.07**					

Crew Compositions & Wage Rates

Crew Number	Manhours Per Day	Total Costs $/Hr.	Total Costs $/Day	Average Crew Rate (ACR) Per Hour	Crew Number	Manhours Per Day	Total Costs $/Hr.	Total Costs $/Day	Average Crew Rate (ACR) Per Hour
D-6					**G-3**				
3 Cement masons	24	$20.18	$484.28		3 Glaziers	24	$20.81	$499.40	**$20.81**
5 Laborers	40	$16.26	$650.21		**H-1**				
Total	64		$1,134	**$17.73**	1 Fence erector	8	$21.39	$171.10	**$21.39**
D-7					**H-2**				
3 Cement masons	24	$20.18	$484.28		2 Fence erectors	16	$21.39	$342.20	
6 Laborers	48	$16.26	$780.25		1 Laborer	8	$16.26	$130.04	
Total	72		$1,265	**$17.56**	Total	24		$472.25	**$19.68**
E-1					**L-1**				
1 Electrician	8	$24.49	$195.93	**$24.49**	1 Laborer	8	$16.26	$130.04	**$16.26**
E-2					**L-2**				
2 Electricians	16	$24.49	$391.86	**$24.49**	2 Laborers	16	$16.26	$260.08	**$16.26**
E-3					**L-3**				
1 Electrician	8	$24.49	$195.93		2 Laborers	16	$16.26	$260.08	
1 Plumber	8	$24.87	$198.94		1 Carpenter	8	$21.07	$168.55	
Total	16		$394.87	**$24.68**	Total	24		$428.63	**$17.86**
E-4					**L-4**				
1 Electrician	8	$24.49	$195.93		2 Laborers	16	$16.26	$260.08	
1 Carpenter	8	$21.07	$168.55		2 Carpenters	16	$21.07	$337.10	
Total	16		$364.48	**$22.78**	Total	32		$597.18	**$18.66**
F-1					**L-5**				
1 Floorlayer	8	$20.27	$162.14	**$20.27**	4 Laborers	32	$16.26	$520.17	
F-2					1 Carpenter	8	$21.07	$168.55	
2 Floorlayers	16	$20.27	$324.28		Total	40		$688.72	**$17.22**
1 Laborer	2	$16.26	$32.51		**L-6**				
Total	18		$356.79	**$19.82**	4 Laborers	32	$16.26	$520.17	
F-3					2 Carpenters	16	$21.07	$337.10	
1 Floorlayer (hardwood)	8	$23.18	$185.46	**$23.18**	Total	48		$857.27	**$17.86**
F-4					**L-7**				
2 Floorlayers (hardwood)	16	$23.18	$370.91		5 Laborers	40	$16.26	$650.21	
1 Laborer	2	$16.26	$32.51		1 Carpenter	8	$21.07	$168.55	
Total	18		$403.42	**$22.41**	Total	48		$818.76	**$17.06**
G-1					**L-8**				
1 Glazier	8	$20.81	$166.47	**$20.81**	3 Laborers	24	$16.26	$390.13	**$16.26**
G-2					**L-9**				
2 Glaziers	16	$20.81	$332.93	**$20.81**	4 Laborers	32	$16.26	$520.17	**$16.26**

Crew Compositions & Wage Rates

Crew Number	Manhours Per Day	Total Costs $/Hr.	Total Costs $/Day	Average Crew Rate (ACR) Per Hour	Crew Number	Manhours Per Day	Total Costs $/Hr.	Total Costs $/Day	Average Crew Rate (ACR) Per Hour
N-1					**P-8**				
1 Painter (brush)	8	$19.27	$154.14	**$19.27**	1 Lather	8	$20.47	$163.79	
N-2					2 Plasterers	16	$20.53	$328.49	
2 Painters (brush)	16	$19.27	$308.28	**$19.27**	1 Laborer	8	$16.26	$130.04	
N-3					Total	32		$622.33	**$19.45**
1 Painter (spray)	8	$19.89	$159.13	**$19.89**	**P-9**				
N-4					1 Lather	10	$20.47	$204.74	
2 Painters (spray)	16	$19.89	$318.26	**$19.89**	2 Plasterers	16	$20.53	$328.49	
P-1					1 Laborer	8	$16.26	$130.04	
1 Lather	8	$20.47	$163.79	**$20.47**	Total	34		$663.27	**$19.51**
P-2					**Q-1**				
1 Lather	8	$20.47	$163.79		1 Paperhanger	8	$20.24	$161.90	**$20.24**
1 Laborer	4	$16.26	$65.02		**R-1**				
Total	12		$228.81	**$19.07**	1 Roofer (comp.)	8	$19.07	$152.58	**$19.07**
P-3					**R-2**				
1 Plasterer	8	$20.53	$164.25	**$20.53**	2 Roofers (comp.)	16	$19.07	$305.16	**$19.07**
P-4					**R-3**				
2 Plasterers	16	$20.53	$328.49		2 Roofers (comp.)	16	$19.07	$305.16	
1 Laborer	8	$16.26	$130.04		1 Laborer	1	$16.26	$16.26	
Total	24		$458.53	**$19.11**	Total	17		$321.41	**$18.91**
P-5					**R-4**				
3 Plasterers	24	$20.53	$492.74		2 Roofers (comp.)	16	$19.07	$305.16	
2 Laborers	16	$16.26	$260.08		1 Laborer	2	$16.26	$32.51	
Total	40		$752.82	**$18.82**	Total	18		$337.67	**$18.76**
P-6					**R-5**				
2 Plasterers	16	$20.53	$328.49		2 Roofers (comp.)	16	$19.07	$305.16	
2 Laborers	16	$16.26	$260.08		1 Laborer	3	$16.26	$48.77	
Total	32		$588.58	**$18.39**	Total	19		$353.92	**$18.63**
P-7					**R-6**				
1 Lather	6	$20.47	$122.84		2 Roofers (comp.)	16	$19.07	$305.16	
2 Plasterers	16	$20.53	$328.49		1 Laborer	4	$16.26	$65.02	
1 Laborer	8	$16.26	$130.04		Total	20		$370.18	**$18.51**
Total	30		$581.38	**$19.38**					

Crew Compositions & Wage Rates

Crew Number	Manhours Per Day	Total Costs $/Hr.	$/Day	Average Crew Rate (ACR) Per Hour	Crew Number	Manhours Per Day	Total Costs $/Hr.	$/Day	Average Crew Rate (ACR) Per Hour
R-7					**R-17**				
2 Roofers (comp.)	16	$19.07	$305.16		1 Roofer (foreman)	8	$21.12	$168.99	
1 Laborer	8	$16.26	$130.04		3 Roofers (pitch)	24	$20.14	$483.27	
Total	24		$435.20	**$18.13**	2 Laborers	16	$16.26	$260.08	
R-8					Total	48		$912.34	**$19.01**
1 Roofer (wood shingles)	8	$21.05	$168.38	**$21.05**	**R-18**				
R-9					2 Roofers (pitch)	16	$20.14	$322.18	
2 Roofers (wood shingles)	16	$21.05	$336.77	**$21.05**	1 Laborer	8	$16.26	$130.04	
R-10					Total	24		$452.22	**$18.84**
2 Roofers (wood shingles)	16	$21.05	$336.77		**S-1**				
1 Laborer	2	$16.26	$32.51		1 Plumber	8	$24.87	$198.94	**$24.87**
Total	18		$369.28	**$20.52**	**S-2**				
R-11					1 Plumber	8	$24.87	$198.94	
2 Roofers (wood shingles)	16	$21.05	$336.77		1 Laborer	8	$16.26	$130.04	
1 Laborer	3	$16.26	$48.77		Total	16		$328.98	**$20.56**
Total	19		$385.53	**$20.29**	**S-3**				
R-12					1 Plumber	8	$24.87	$198.94	
2 Roofers (wood shingles)	16	$21.05	$336.77		1 Electrician	8	$24.49	$195.93	
1 Laborer	4	$16.26	$65.02		Total	16		$394.87	**$24.68**
Total	20		$401.79	**$20.09**	**S-4**				
R-13					1 Plumber	8	$24.87	$198.94	
2 Roofers (wood shingles)	16	$21.05	$336.77		1 Laborer	8	$16.26	$130.04	
1 Laborer	5	$16.26	$81.28		1 Electrician	8	$24.49	$195.93	
Total	21		$418.04	**$19.91**	Total	24		$524.91	**$21.87**
R-14					**S-5**				
2 Roofers (wood shingles)	16	$21.05	$336.77		2 Plumbers	16	$24.87	$397.87	
1 Laborer	6	$16.26	$97.53		1 Laborer	8	$16.26	$130.04	
Total	22		$434.30	**$19.74**	1 Electrician	8	$24.49	$195.93	
R-15					Total	32		$723.84	**$22.62**
2 Roofers (wood shingles)	16	$21.05	$336.77		**S-6**				
1 Laborer	7	$16.26	$113.79		2 Plumbers	16	$24.87	$397.87	
Total	23		$450.55	**$19.59**	1 Laborer	8	$16.26	$130.04	
R-16					Total	24		$527.91	**$22.00**
1 Roofer (pitch)	8	$20.14	$161.09	**$20.14**					

Crew Compositions & Wage Rates

Crew Number	Manhours Per Day	$/Hr.	$/Day	Average Crew Rate (ACR) Per Hour	Crew Number	Manhours Per Day	$/Hr.	$/Day	Average Crew Rate (ACR) Per Hour
S-7					U-4				
3 Plumbers	24	$24.87	$596.81		1 Sheet metal worker	8	$23.78	$190.21	
1 Laborer	8	$16.26	$130.04		1 Laborer	8	$16.26	$130.04	
Total	32		$726.85	$22.71	Total	16		$320.25	$20.02
T-1					U-5				
1 Tile setter (ceramic)	8	$20.62	$164.98	$20.62	1 Sheet metal worker	8	$23.78	$190.21	
T-2					1 Laborer	8	$16.26	$130.04	
1 Tile setter (ceramic)	8	$20.62	$164.98		1 Electrician	4	$24.49	$97.97	
1 Tile setter (helper)	8	$16.28	$130.25		Total	20		$418.22	$20.91
Total	16		$295.22	$18.45	U-6				
T-3					2 Sheet metal workers	16	$23.78	$380.42	
2 Tile setter (ceramic)	16	$20.62	$329.95		1 Laborer	8	$16.26	$130.04	
1 Tile setter (helper)	8	$16.28	$130.25		Total	24		$510.46	$21.27
Total	24		$460.20	$19.17	V-1				
T-4					1 Tractor operator	8	$23.53	$188.21	
2 Tile setter (ceramic)	16	$20.62	$329.95		1 Laborer	4	$16.26	$65.02	
2 Tile setter (helper)	16	$16.28	$260.49		Total	12		$253.23	$21.10
Total	32		$590.44	$18.45	V-2				
U-1					1 Tractor operator	8	$23.53	$188.21	
1 Sheet metal worker	8	$23.78	$190.21	$23.78	1 Laborer	8	$16.26	$130.04	
U-2					Total	16		$318.25	$19.89
2 Sheet metal workers	16	$23.78	$380.42	$23.78					
U-3									
2 Sheet metal workers	16	$23.78	$380.42						
2 Laborers	16	$16.26	$260.08						
Total	32		$640.50	$20.02					

Abbreviations Used In This Book

ABS	acrylonitrile butadiene styrene	**D.S.A.**	double strength, A grade	**L**	length or long	**S.S.B.**	single strength, B quality
ACR	average crew rate	**dia.**	diameter	**LF**	linear feet	**std.**	standard
AGA	American Gas Association	**D.S.B.**	double strength, B grade	**LS**	lump sum	**SY**	square yard
AMP	ampere	**e.g.**	for example	**lb, lbs.**	pound(s)	**T**	thick
ASME	American Society of Mechanical Engineers	**EA, ea**	each	**MBF**	1000 board feet	**T&G**	tongue and groove
auto.	automatic	**etc.**	et cetera	**MSF**	1000 square feet	**"U"**	thermal conductivity
Avg.	average	**exp.**	exposure	**M**	1000	**U.I.**	united inches
Approx.	approximately	**ft.**	foot	**Max.**	maximum	**UL**	Underwriters Laboratories
Bdle.	bundle	**flt**	flight	**Mi**	miles	**U.S.G.**	United States Gypsum
BTU	British thermal unit	**FAS**	First and Select grade	**Min.**	minimum	**VLF**	vertical linear feet
BTUH	British thermal unit per hour	**F.H.A.**	Federal Housing Administration	**O.B.**	opposed blade	**W**	width or wide
C	100	**ga.**	gauge	**o.c.**	on center	**yr.**	year
cc	cubic centimeter	**gal**	gallon	**o.d.**	outside dimension		
Cu.	cubic	**galv.**	galvanized	**oz.**	ounce		
CF	cubic foot	**GFI**	ground fault interrupter	**pcs.**	pieces	**Symbols**	
CFM	cubic foot per minute	**GPM**	gallons per minute	**pkg.**	package	**/**	per
CLF	100 linear feet	**GPH**	gallons per hour	**PR**	pair	**%**	percent
CSF	100 square feet	**H**	height or high	**PSI**	per square inch	**"**	inches
CSY	100 square yards	**Hr.**	hour	**PVC**	polyvinyl chloride	**'**	foot or feet
CWT	100 pounds	**HP, hp**	horsepower	**Qt.**	quart	**x**	by
CY	cubic yard	**HVAC**	heating, ventilating, air conditioning	**R/L**	random length	**°**	degree
Const.	construction	**i.d.**	inside diameter	**R/W/L**	random width and length	**#**	number or pounds
Corr.	corrugated	**i.e.**	that is	**R.E.**	rounded edge	**$**	dollar
Ctn	carton	**I.P.S.**	iron pipe size	**R&R**	remove and reset	**±**	plus or minus
d	penny	**KD**	knocked down	**S4S**	surfaced-four-sides		
D	deep or depth	**KW, kw**	kilowatts	**SA**	sack		
				SF	square foot		
				SQ, Sq.	1 square or 100 square feet		

For abbreviations of crew members please see Wage Rate Chart.

			Costs Based On Large Volume							Costs Based On Small Volume						
Description	Oper	Unit	Crew Size	Avg Day Prod	Avg Mat'l Unit Cost	Avg Equip Unit Cost	Avg Labor Unit Cost	Avg Total Unit Cost	Avg Price Incl O&P	Crew Size	Avg Day Prod	Avg Mat'l Unit Cost	Avg Equip Unit Cost	Avg Labor Unit Cost	Avg Total Unit Cost	Avg Price Incl O&P

Acoustical treatment

See also Suspended Ceiling Systems

Ceiling and wall tile

Adhesive set

Description	Oper	Unit	Crew Size	Avg Day Prod	Avg Mat'l Unit Cost	Avg Equip Unit Cost	Avg Labor Unit Cost	Avg Total Unit Cost	Avg Price Incl O&P	Crew Size	Avg Day Prod	Avg Mat'l Unit Cost	Avg Equip Unit Cost	Avg Labor Unit Cost	Avg Total Unit Cost	Avg Price Incl O&P
Tile only, no grid system	Demo	SF	L-2	1300	---	---	0.25	0.25	0.37	L-2	1040	---	---	0.31	0.31	0.47
Adhesive set tile with furring strip	Demo	SF	L-2	1710	---	---	0.19	0.19	0.28	L-2	1368	---	---	0.24	0.24	0.35

Mineral fiber, vinyl coated, tile only
Applied in square pattern by adhesive to solid backing; 5% tile waste
1/2" thick, 12" x 12" or 12" x 24"

Description	Oper	Unit	Crew Size	Avg Day Prod	Avg Mat'l Unit Cost	Avg Equip Unit Cost	Avg Labor Unit Cost	Avg Total Unit Cost	Avg Price Incl O&P	Crew Size	Avg Day Prod	Avg Mat'l Unit Cost	Avg Equip Unit Cost	Avg Labor Unit Cost	Avg Total Unit Cost	Avg Price Incl O&P
Economy, mini perforated	Inst	SF	C-10	880	0.57	---	0.41	0.98	1.18	C-10	704	0.65	---	0.51	1.16	1.41
Standard, random perforated	Inst	SF	C-10	880	0.72	---	0.41	1.12	1.32	C-10	704	0.81	---	0.51	1.32	1.57
Designer, swirl perforation	Inst	SF	C-10	880	0.75	---	0.41	1.15	1.35	C-10	704	0.85	---	0.51	1.36	1.61
Deluxe, sculptured face	Inst	SF	C-10	880	0.88	---	0.41	1.29	1.49	C-10	704	1.00	---	0.51	1.51	1.76

5/8" thick, 12" x 12" or 12" x 24"

Description	Oper	Unit	Crew Size	Avg Day Prod	Avg Mat'l Unit Cost	Avg Equip Unit Cost	Avg Labor Unit Cost	Avg Total Unit Cost	Avg Price Incl O&P	Crew Size	Avg Day Prod	Avg Mat'l Unit Cost	Avg Equip Unit Cost	Avg Labor Unit Cost	Avg Total Unit Cost	Avg Price Incl O&P
Economy, mini perforated	Inst	SF	C-10	880	0.68	---	0.41	1.09	1.29	C-10	704	0.78	---	0.51	1.28	1.53
Standard, random perforated	Inst	SF	C-10	880	0.83	---	0.41	1.23	1.43	C-10	704	0.94	---	0.51	1.45	1.69
Designer, swirl perforation	Inst	SF	C-10	880	0.86	---	0.41	1.26	1.46	C-10	704	0.98	---	0.51	1.48	1.73
Deluxe, sculptured face	Inst	SF	C-10	880	0.99	---	0.41	1.40	1.60	C-10	704	1.13	---	0.51	1.63	1.88

3/4" thick, 12" x 12" or 12" x 24"

Description	Oper	Unit	Crew Size	Avg Day Prod	Avg Mat'l Unit Cost	Avg Equip Unit Cost	Avg Labor Unit Cost	Avg Total Unit Cost	Avg Price Incl O&P	Crew Size	Avg Day Prod	Avg Mat'l Unit Cost	Avg Equip Unit Cost	Avg Labor Unit Cost	Avg Total Unit Cost	Avg Price Incl O&P
Economy, mini perforated	Inst	SF	C-10	880	0.74	---	0.41	1.14	1.34	C-10	704	0.84	---	0.51	1.35	1.59
Standard, random perforated	Inst	SF	C-10	880	0.88	---	0.41	1.29	1.49	C-10	704	1.00	---	0.51	1.51	1.76
Designer, swirl perforation	Inst	SF	C-10	880	0.91	---	0.41	1.32	1.52	C-10	704	1.04	---	0.51	1.55	1.79
Deluxe, sculptured face	Inst	SF	C-10	880	1.05	---	0.41	1.45	1.65	C-10	704	1.19	---	0.51	1.70	1.94

Applied by adhesive to furring strips

Description	Oper	Unit	Crew Size	Avg Day Prod	Avg Mat'l Unit Cost	Avg Equip Unit Cost	Avg Labor Unit Cost	Avg Total Unit Cost	Avg Price Incl O&P	Crew Size	Avg Day Prod	Avg Mat'l Unit Cost	Avg Equip Unit Cost	Avg Labor Unit Cost	Avg Total Unit Cost	Avg Price Incl O&P
reduce output & increase cost	Inst	SF	C-10	80	---	---	0.04	0.04	0.06	C-10	64	---	---	0.05	0.05	0.08

Stapled

Description	Oper	Unit	Crew Size	Avg Day Prod	Avg Mat'l Unit Cost	Avg Equip Unit Cost	Avg Labor Unit Cost	Avg Total Unit Cost	Avg Price Incl O&P	Crew Size	Avg Day Prod	Avg Mat'l Unit Cost	Avg Equip Unit Cost	Avg Labor Unit Cost	Avg Total Unit Cost	Avg Price Incl O&P
Tile only, no grid system	Demo	SF	L-2	1170	---	---	0.28	0.28	0.41	L-2	936	---	---	0.35	0.35	0.52
Stapled tile with furring strip	Demo	SF	L-2	1540	---	---	0.21	0.21	0.31	L-2	1232	---	---	0.27	0.27	0.39

Mineral fiber, vinyl coated, tile only.
Applied in square pattern by staples, nails, or clips; 5% tile waste
1/2" thick, 12" x 12" or 12" x 24"

Description	Oper	Unit	Crew Size	Avg Day Prod	Avg Mat'l Unit Cost	Avg Equip Unit Cost	Avg Labor Unit Cost	Avg Total Unit Cost	Avg Price Incl O&P	Crew Size	Avg Day Prod	Avg Mat'l Unit Cost	Avg Equip Unit Cost	Avg Labor Unit Cost	Avg Total Unit Cost	Avg Price Incl O&P
Economy, mini perforated	Inst	SF	C-10	960	0.61	---	0.37	0.98	1.16	C-10	768	0.69	---	0.47	1.15	1.38
Standard, random perforated	Inst	SF	C-10	960	0.46	---	0.37	0.83	1.02	C-10	768	0.53	---	0.47	0.99	1.22

Acoustical and insulating tile

1. **Dimensions**.

   a. Acoustical tile. 1/2" thick x 12" x 12", 24".

   b. Insulating tile, decorative, 1/2" thick x 12" x 12", 24"; 1/2" thick x 16" x 16", 32".

2. **Installation.** Tile may be applied to existing plaster (if joist spacing is suitable) or to wood furring strips. Tile may have a square edge or flange. Depending on type and shape of tile, adhesive, staples, nails or clips are available for attaching tile.

3. **Estimating technique.** Determine area and add 5-10% for waste.

Notes on Material Pricing. The material prices listed below were used to compile the Average Material Cost/Unit on the following pages:

1) Adhesive ($12.00/Gal)

12" x 12"	1.25 Gal/CSF
12" x 24"	0.95 Gal/CSF
16" x 16"	0.75 Gal/CSF
16" x 32"	0.55 Gal/CSF

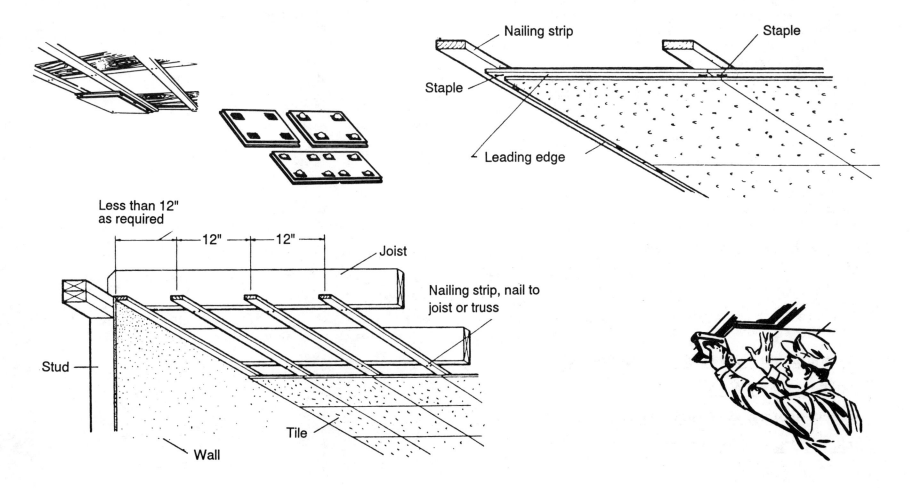

Description	Oper	Unit	Costs Based On Large Volume							Costs Based On Small Volume						
			Crew Size	Avg Day Prod	Avg Mat'l Unit Cost	Avg Equip Unit Cost	Avg Labor Unit Cost	Avg Total Unit Cost	Avg Price Incl O&P	Crew Size	Avg Day Prod	Avg Mat'l Unit Cost	Avg Equip Unit Cost	Avg Labor Unit Cost	Avg Total Unit Cost	Avg Price Incl O&P

Acoustical treatment (continued)

Description	Oper	Unit	Crew Size	Avg Day Prod	Avg Mat'l Unit Cost	Avg Equip Unit Cost	Avg Labor Unit Cost	Avg Total Unit Cost	Avg Price Incl O&P	Crew Size	Avg Day Prod	Avg Mat'l Unit Cost	Avg Equip Unit Cost	Avg Labor Unit Cost	Avg Total Unit Cost	Avg Price Incl O&P
Designer, swirl perforation	Inst	SF	C-10	960	0.65	---	0.37	1.02	1.20	C-10	768	0.74	---	0.47	1.20	1.43
Deluxe, sculptured face	Inst	SF	C-10	960	0.78	---	0.37	1.15	1.34	C-10	768	0.89	---	0.47	1.35	1.58
5/8" thick, 12" x 12" or 12" x 24"																
Economy, mini perforated	Inst	SF	C-10	960	0.72	---	0.37	1.09	1.27	C-10	768	0.81	---	0.47	1.28	1.51
Standard, random perforated	Inst	SF	C-10	960	0.57	---	0.37	0.94	1.13	C-10	768	0.65	---	0.47	1.12	1.34
Designer, swirl perforation	Inst	SF	C-10	960	0.76	---	0.37	1.13	1.31	C-10	768	0.86	---	0.47	1.33	1.56
Deluxe, sculptured face	Inst	SF	C-10	960	0.89	---	0.37	1.26	1.45	C-10	768	1.01	---	0.47	1.48	1.71
3/4" thick, 12" x 12" or 12" x 24"																
Economy, mini perforated	Inst	SF	C-10	960	0.77	---	0.37	1.14	1.32	C-10	768	0.88	---	0.47	1.34	1.57
Standard, random perforated	Inst	SF	C-10	960	0.63	---	0.37	1.00	1.18	C-10	768	0.71	---	0.47	1.18	1.41
Designer, swirl perforation	Inst	SF	C-10	960	0.81	---	0.37	1.19	1.37	C-10	768	0.93	---	0.47	1.39	1.62
Deluxe, sculptured face	Inst	SF	C-10	960	0.95	---	0.37	1.32	1.50	C-10	768	1.08	---	0.47	1.54	1.77
Applied by staples, nails or clips to furring strips,																
reduce output & increase cost	Inst	SF	C-10	80	---	---	0.03	0.03	0.05	C-10	64	---	---	0.04	0.04	0.06

Tile patterns, effect on labor

Description	Oper	Unit	Crew Size	Avg Day Prod	Avg Mat'l Unit Cost	Avg Equip Unit Cost	Avg Labor Unit Cost	Avg Total Unit Cost	Avg Price Incl O&P	Crew Size	Avg Day Prod	Avg Mat'l Unit Cost	Avg Equip Unit Cost	Avg Labor Unit Cost	Avg Total Unit Cost	Avg Price Incl O&P
Herringbone, Reduce output	Inst	%	C-10	25	---	---	---	---	—	C-10	20	---	---	---	---	—
Diagonal, Reduce output	Inst	%	C-10	20	---	---	---	---	—	C-10	16	---	---	---	---	—
Ashlar, Reduce output	Inst	%	C-10	30	---	---	---	---	—	C-10	24	---	---	---	---	—

Furring strips, 8% waste included

Over wood

Description	Oper	Unit	Crew Size	Avg Day Prod	Avg Mat'l Unit Cost	Avg Equip Unit Cost	Avg Labor Unit Cost	Avg Total Unit Cost	Avg Price Incl O&P	Crew Size	Avg Day Prod	Avg Mat'l Unit Cost	Avg Equip Unit Cost	Avg Labor Unit Cost	Avg Total Unit Cost	Avg Price Incl O&P
1" x 4", 12" o.c.	Inst	SF	C-10	1600	0.14	---	0.22	0.37	0.48	C-10	1280	0.16	---	0.28	0.44	0.58
1" x 4", 16" o.c.	Inst	SF	C-10	1920	0.11	---	0.19	0.30	0.39	C-10	1536	0.13	---	0.23	0.36	0.47
Over plaster																
1" x 4", 12" o.c.	Inst	SF	C-10	1280	0.14	---	0.28	0.42	0.56	C-10	1024	0.16	---	0.35	0.51	0.68
1" x 4", 16" o.c.	Inst	SF	C-10	1600	0.11	---	0.22	0.33	0.44	C-10	1280	0.13	---	0.28	0.40	0.54

Air conditioning. See HVAC, Air Conditioning.

Backfill. See Site Work.

Bath accessories. See Plumbing.

Block, concrete. See Masonry.

Brick. See Masonry.

| | | | Costs Based On Large Volume | | | | | | | Costs Based On Small Volume | | | | | | |
Description	Oper	Unit	Crew Size	Avg Day Prod	Avg Mat'l Unit Cost	Avg Equip Unit Cost	Avg Labor Unit Cost	Avg Total Unit Cost	Avg Price Incl O&P	Crew Size	Avg Day Prod	Avg Mat'l Unit Cost	Avg Equip Unit Cost	Avg Labor Unit Cost	Avg Total Unit Cost	Avg Price Incl O&P

Cabinets
Labor to hang and fit.

Kitchen
3' x 4', wood; base,

| wall, or peninsula | Demo | EA | L-2 | 25.0 | --- | --- | 10.40 | 10.40 | **15.40** | L-2 | 17.5 | --- | --- | 14.86 | 14.86 | **22.00** |

Kitchen; all hardware included

Base; 35" H, 24" D; no tops
12" W, 1 door, 1 drawer

High quality workmanship	Inst	EA	C-9	16.0	136.25	---	18.66	154.91	**164.06**	C-9	11.2	163.50	---	26.66	190.16	**203.22**
Good quality workmanship	Inst	EA	C-9	24.0	111.25	---	12.44	123.69	**129.79**	C-9	16.8	133.50	---	17.77	151.27	**159.98**
Average quality workmanship	Inst	EA	C-9	24.0	85.00	---	12.44	97.44	**103.54**	C-9	16.8	102.00	---	17.77	119.77	**128.48**

15" W, 1 door, 1 drawer

High quality workmanship	Inst	EA	C-9	16.0	143.75	---	18.66	162.41	**171.56**	C-9	11.2	172.50	---	26.66	199.16	**212.22**
Good quality workmanship	Inst	EA	C-9	24.0	118.75	---	12.44	131.19	**137.29**	C-9	16.8	142.50	---	17.77	160.27	**168.98**
Average quality workmanship	Inst	EA	C-9	24.0	90.00	---	12.44	102.44	**108.54**	C-9	16.8	108.00	---	17.77	125.77	**134.48**

18" W, 1 door, 1 drawer

High quality workmanship	Inst	EA	C-9	16.0	148.75	---	18.66	167.41	**176.56**	C-9	11.2	178.50	---	26.66	205.16	**218.22**
Good quality workmanship	Inst	EA	C-9	24.0	122.50	---	12.44	134.94	**141.04**	C-9	16.8	147.00	---	17.77	164.77	**173.48**
Average quality workmanship	Inst	EA	C-9	24.0	93.75	---	12.44	106.19	**112.29**	C-9	16.8	112.50	---	17.77	130.27	**138.98**

21" W, 1 door, 1 drawer

High quality workmanship	Inst	EA	C-9	13.8	165.00	---	21.64	186.64	**197.24**	C-9	9.7	198.00	---	30.91	228.91	**244.06**
Good quality workmanship	Inst	EA	C-9	21.6	136.25	---	13.82	150.07	**156.85**	C-9	15.1	163.50	---	19.75	183.25	**192.92**
Average quality workmanship	Inst	EA	C-9	21.6	105.00	---	13.82	118.82	**125.60**	C-9	15.1	126.00	---	19.75	145.75	**155.42**

24" W, 1 door, 1 drawer

High quality workmanship	Inst	EA	C-9	13.8	181.25	---	21.64	202.89	**213.49**	C-9	9.7	217.50	---	30.91	248.41	**263.56**
Good quality workmanship	Inst	EA	C-9	21.6	131.25	---	13.82	145.07	**151.85**	C-9	15.1	157.50	---	19.75	177.25	**186.92**
Average quality workmanship	Inst	EA	C-9	21.6	148.75	---	13.82	162.57	**169.35**	C-9	15.1	178.50	---	19.75	198.25	**207.92**

30" W, 2 doors, 2 drawers

High quality workmanship	Inst	EA	C-9	10.4	196.25	---	28.71	224.96	**239.03**	C-9	7.3	235.50	---	41.02	276.52	**296.61**
Good quality workmanship	Inst	EA	C-9	15.4	170.00	---	19.39	189.39	**198.89**	C-9	10.8	204.00	---	27.70	231.70	**245.27**
Average quality workmanship	Inst	EA	C-9	15.4	128.75	---	19.39	148.14	**157.64**	C-9	10.8	154.50	---	27.70	182.20	**195.77**

Cabinets

Top quality cabinets are built with the structural stability of fine furniture. Framing stock is kiln dried and a full 1" thick. Cabinets will have backs, usually 5-ply 3/16"-thick plywood, with all backs and interiors finished. Frames should be constructed of hardwood with mortise and tenon joints; corner blocks should be used on all four corners of all base cabinets. Doors are usually of select 7/16" thick solid core construction using semi-concealed hinges. End panels are 1/2" thick and attached to frames with mortise and tenon joints, glued and pinned under pressure. Panels should also be dadoed to receive tops and bottoms of wall cabinets. Shelves are adjustable with veneer faces and front edges. Hardware includes magnetic catches, heavy duty die cast pulls and hinges, and ball-bearing suspension system. Finish is scratch and stain resistant. First coat is hand-wiped stain, followed by sealer coat. Lastly, a synthetic varnish having plastic laminate characteristics is applied.

Average quality cabinets feature hardwood frame construction with plywood backs and veneered plywood end panels. Joints are mortise and tenon, and glued. Doors are solid core attached with exposed hinges that are self-closing. Shelves are adjustable, and drawers ride on a ball-bearing side suspension glide system. Finish is usually three coats including stain, sealer, and a mar-resistant top coat for easy cleaning.

Economy quality cabinets feature pine construction with joints glued under pressure. Doors, drawers fronts, and side or end panels are constructed of either 1/2"-thick wood composition board or 1/2"-thick veneered pine. Face frames are 3/4"-thick wood composition board or 3/4"-thick pine. Features include adjustable shelves, hinge straps, and a three-point suspension system on drawers (using nylon rollers). Finish consists of a filler coat, base coat, and final polyester top coat.

Kitchen Cabinet Installation Procedure:

Develop a layout plan — measure and write down the following:

1. Floor space.
2. Height and width of all walls.
3. Location of electrical outlets.
4. Size and position of doors, windows, and vents.
5. Location of any posts or pillars. Walls must be prepared if chair rails or baseboards are located where cabinets will be installed.
6. Common height and depth of base cabinets (including 1" for countertops) and wall cabinets.

Then ask yourself:

1. Will what you want fit into the available space?
2. Is there enough counter space on both sides of all appliances and sinks? The kitchen is a series of work centers, each with a major appliance as its hub, and each needing adequate counter space.

   a. Fresh and frozen food center — Refrigerator-Freezer

   b. Clean-up center — Sink with disposal-dishwasher

   c. Cooking center — Range-oven

Rule of Thumb: Not less than 4-1/2" nor more than 5-1/2" of counter surface between refrigerator and sink. Not less than 3' nor more than 4' between sink and range.

3. Will the sink workspace fit neatly in front of a window?
4. Is the distance in the work triangle — distance from sink, to refrigerator, to range area — less than 25 feet total? The kitchen triangle is the most efficient step-saver; it means placing each major center at approximately equidistant triangle points. The ideal triangle is 22 feet total. It should never be less than 13 feet total.
5. Where are the centers located? A logical working and walking pattern is from refrigerator to sink to range. The refrigerator should be at a triangle point near a door — minimizing steps to bring in groceries and also reducing traffic that could interfere with food preparation. The range should be at a triangle point near the serving and dining area. The sink is located between the two. The refrigerator should be located far enough from the range so that the heat will not affect the refrigerator's cooling efficiency.
6. Does the plan allow for lighting the range and sink work centers and for ventilating the range center?
7. Does the plan allow for opened doors (such as cabinet doors during food preparation or open entrance/exit doors) that won't interfere with access to an appliance? To clear appliances and cabinets, a door opening should not be less than 30" from the corner, since such equipment is 24" to 28" in depth. A clearance of 48" is necessary when a range is next to a door.

Next locate the wall studs with a stud finder or hammer, since all cabinets attached to walls are done so with screws, never nails. Also remove chair rails or baseboards where they conflict with cabinets.

Wall Cabinets: From the highest point on the floor measure up approximately 84" to determine the top of wall cabinets. Using two #10 x 2-1/2" wood screws, drill through hanging strips built into cabinet backs at top and bottom. Use a level to assure that cabinets and doors are aligned, then tighten screws.

Base Cabinets: Start with a corner unit and a base unit on each side of the corner unit. Place this combination in the corner and work out from both sides. Use "C" clamps when connecting cabinets to draw adjoining units into alignment. With the front face plumb and the unit level from front to back and across the front edge, attach unit to wall studs by screwing through the hanging strips. To attach adjoining cabinets, drill two holes in the vertical side of one cabinet, inside the door (near top and bottom) and just into the stile of the adjoining cabinet.

Base

Base

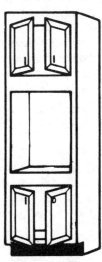

Oven cabinet

Sink front

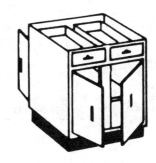

			Costs Based On Large Volume							Costs Based On Small Volume						
Description	Oper	Unit	Crew Size	Avg Day Prod	Avg Mat'l Unit Cost	Avg Equip Unit Cost	Avg Labor Unit Cost	Avg Total Unit Cost	Avg Price Incl O&P	Crew Size	Avg Day Prod	Avg Mat'l Unit Cost	Avg Equip Unit Cost	Avg Labor Unit Cost	Avg Total Unit Cost	Avg Price Incl O&P

Cabinets (continued)

36" W, 2 doors, 2 drawers

Description	Oper	Unit	Crew Size	Avg Day Prod	Avg Mat'l Unit Cost	Avg Equip Unit Cost	Avg Labor Unit Cost	Avg Total Unit Cost	Avg Price Incl O&P	Crew Size	Avg Day Prod	Avg Mat'l Unit Cost	Avg Equip Unit Cost	Avg Labor Unit Cost	Avg Total Unit Cost	Avg Price Incl O&P
High quality workmanship	Inst	EA	C-9	10.4	226.25	---	28.71	254.96	269.03	C-9	7.3	271.50	---	41.02	312.52	332.61
Good quality workmanship	Inst	EA	C-9	15.4	186.25	---	19.39	205.64	215.14	C-9	10.8	223.50	---	27.70	251.20	264.77
Average quality workmanship	Inst	EA	C-9	15.4	142.50	---	19.39	161.89	171.39	C-9	10.8	171.00	---	27.70	198.70	212.27
42" W, 2 doors, 2 drawers																
High quality workmanship	Inst	EA	C-9	8.0	261.25	---	37.32	298.57	316.86	C-9	5.6	313.50	---	53.32	366.82	392.95
Good quality workmanship	Inst	EA	C-9	12.0	216.25	---	24.88	241.13	253.33	C-9	8.4	259.50	---	35.55	295.05	312.46
Average quality workmanship	Inst	EA	C-9	12.0	163.75	---	24.88	188.63	200.83	C-9	8.4	196.50	---	35.55	232.05	249.46
48" W, 4 doors, 2 drawers																
High quality workmanship	Inst	EA	C-9	8.0	296.25	---	37.32	333.57	351.86	C-9	5.6	355.50	---	53.32	408.82	434.95
Good quality workmanship	Inst	EA	C-9	12.0	245.00	---	24.88	269.88	282.08	C-9	8.4	294.00	---	35.55	329.55	346.96
Average quality workmanship	Inst	EA	C-9	12.0	185.00	---	24.88	209.88	222.08	C-9	8.4	222.00	---	35.55	257.55	274.96
Base corner, blind, 35" H, 24" D, no tops																
36" W, 1 door, 1 drawer																
High quality workmanship	Inst	EA	C-9	11.2	165.00	---	26.66	191.66	204.72	C-9	7.8	198.00	---	38.09	236.09	254.75
Good quality workmanship	Inst	EA	C-9	16.2	131.25	---	18.43	149.68	158.71	C-9	11.3	157.50	---	26.33	183.83	196.73
Average quality workmanship	Inst	EA	C-9	16.2	105.00	---	18.43	123.43	132.46	C-9	11.3	126.00	---	26.33	152.33	165.23
42" W, 1 door, 1 drawer																
High quality workmanship	Inst	EA	C-9	8.8	181.25	---	33.93	215.18	231.81	C-9	6.2	217.50	---	48.47	265.97	289.72
Good quality workmanship	Inst	EA	C-9	13.8	150.00	---	21.64	171.64	182.24	C-9	9.7	180.00	---	30.91	210.91	226.06
Average quality workmanship	Inst	EA	C-9	13.8	113.75	---	21.64	135.39	145.99	C-9	9.7	136.50	---	30.91	167.41	182.56
36" x 36" (Lazy susan)																
High quality workmanship	Inst	EA	C-9	11.2	245.00	---	26.66	271.66	284.72	C-9	7.8	294.00	---	38.09	332.09	350.75
Good quality workmanship	Inst	EA	C-9	16.2	202.50	---	18.43	220.93	229.96	C-9	11.3	243.00	---	26.33	269.33	282.23
Average quality workmanship	Inst	EA	C-9	16.2	155.00	---	18.43	173.43	182.46	C-9	11.3	186.00	---	26.33	212.33	225.23
Utility closets, 81" to 85" H, 24" W, 24" D																
High quality workmanship	Inst	EA	C-9	8.0	442.50	---	37.32	479.82	498.11	C-9	5.6	531.00	---	53.32	584.32	610.45
Good quality workmanship	Inst	EA	C-9	12.0	345.00	---	24.88	369.88	382.08	C-9	8.4	414.00	---	35.55	449.55	466.96
Average quality workmanship	Inst	EA	C-9	12.0	262.50	---	24.88	287.38	299.58	C-9	8.4	315.00	---	35.55	350.55	367.96

Drawer base

35" H, 24" D 4 drawers, no tops

18" W

Description	Oper	Unit	Crew Size	Avg Day Prod	Avg Mat'l Unit Cost	Avg Equip Unit Cost	Avg Labor Unit Cost	Avg Total Unit Cost	Avg Price Incl O&P	Crew Size	Avg Day Prod	Avg Mat'l Unit Cost	Avg Equip Unit Cost	Avg Labor Unit Cost	Avg Total Unit Cost	Avg Price Incl O&P
High quality workmanship	Inst	EA	C-9	16.0	217.50	---	18.66	236.16	245.31	C-9	11.2	261.00	---	26.66	287.66	300.72

Description	Oper	Unit	Costs Based On Large Volume							Costs Based On Small Volume						
			Crew Size	Avg Day Prod	Avg Mat'l Unit Cost	Avg Equip Unit Cost	Avg Labor Unit Cost	Avg Total Unit Cost	Avg Price Incl O&P	Crew Size	Avg Day Prod	Avg Mat'l Unit Cost	Avg Equip Unit Cost	Avg Labor Unit Cost	Avg Total Unit Cost	Avg Price Incl O&P

Cabinets (continued)

Description	Oper	Unit	Crew Size	Day Prod	Mat'l	Equip	Labor	Total	Price O&P	Crew Size	Day Prod	Mat'l	Equip	Labor	Total	Price O&P
Good quality workmanship	Inst	EA	C-9	24.0	186.25	---	12.44	198.69	204.79	C-9	16.8	223.50	---	17.77	241.27	249.98
Average quality workmanship	Inst	EA	C-9	24.0	142.50	---	12.44	154.94	161.04	C-9	16.8	171.00	---	17.77	188.77	197.48
24" W																
High quality workmanship	Inst	EA	C-9	13.8	243.75	---	21.64	265.39	275.99	C-9	9.7	292.50	---	30.91	323.41	338.56
Good quality workmanship	Inst	EA	C-9	21.6	201.25	---	13.82	215.07	221.85	C-9	15.1	241.50	---	19.75	261.25	270.92
Average quality workmanship	Inst	EA	C-9	21.6	153.75	---	13.82	167.57	174.35	C-9	15.1	184.50	---	19.75	204.25	213.92

Island cabinets
Island base cabinets, 35" H, 24" D
24" W, 1 door both sides

Description	Oper	Unit	Crew Size	Day Prod	Mat'l	Equip	Labor	Total	Price O&P	Crew Size	Day Prod	Mat'l	Equip	Labor	Total	Price O&P
High quality workmanship	Inst	EA	C-9	13.0	322.50	---	22.97	345.47	356.72	C-9	9.1	387.00	---	32.81	419.81	435.89
Good quality workmanship	Inst	EA	C-9	19.5	265.00	---	15.31	280.31	287.82	C-9	13.7	318.00	---	21.87	339.87	350.59
Average quality workmanship	Inst	EA	C-9	19.5	203.75	---	15.31	219.06	226.57	C-9	13.7	244.50	---	21.87	266.37	277.09
30" W, 2 doors both sides																
High quality workmanship	Inst	EA	C-9	10.0	341.25	---	29.86	371.11	385.74	C-9	7.0	409.50	---	42.66	452.16	473.06
Good quality workmanship	Inst	EA	C-9	15.0	281.25	---	19.91	301.16	310.91	C-9	10.5	337.50	---	28.44	365.94	379.87
Average quality workmanship	Inst	EA	C-9	15.0	213.75	---	19.91	233.66	243.41	C-9	10.5	256.50	---	28.44	284.94	298.87
36" W, 2 doors both sides																
High quality workmanship	Inst	EA	C-9	10.0	358.75	---	29.86	388.61	403.24	C-9	7.0	430.50	---	42.66	473.16	494.06
Good quality workmanship	Inst	EA	C-9	15.0	296.25	---	19.91	316.16	325.91	C-9	10.5	355.50	---	28.44	383.94	397.87
Average quality workmanship	Inst	EA	C-9	15.0	225.00	---	19.91	244.91	254.66	C-9	10.5	270.00	---	28.44	298.44	312.37
48" W, 4 doors both sides																
High quality workmanship	Inst	EA	C-9	7.5	385.00	---	39.81	424.81	444.32	C-9	5.3	462.00	---	56.87	518.87	546.74
Good quality workmanship	Inst	EA	C-9	11.5	317.50	---	25.96	343.46	356.19	C-9	8.1	381.00	---	37.09	418.09	436.27
Average quality workmanship	Inst	EA	C-9	11.5	241.25	---	25.96	267.21	279.94	C-9	8.1	289.50	---	37.09	326.59	344.77

Corner island base cabinets, 35" H, 24" D
42" W, 4 doors, 2 drawers

Description	Oper	Unit	Crew Size	Day Prod	Mat'l	Equip	Labor	Total	Price O&P	Crew Size	Day Prod	Mat'l	Equip	Labor	Total	Price O&P
High quality workmanship	Inst	EA	C-9	7.5	341.25	---	39.81	381.06	400.57	C-9	5.3	409.50	---	56.87	466.37	494.24
Good quality workmanship	Inst	EA	C-9	11.5	278.75	---	25.96	304.71	317.44	C-9	8.1	334.50	---	37.09	371.59	389.77
Average quality workmanship	Inst	EA	C-9	11.5	210.00	---	25.96	235.96	248.69	C-9	8.1	252.00	---	37.09	289.09	307.27
48" W, 6 doors, 2 drawers																
High quality workmanship	Inst	EA	C-9	6.0	357.50	---	49.77	407.27	431.65	C-9	4.2	429.00	---	71.09	500.09	534.93
Good quality workmanship	Inst	EA	C-9	9.0	295.00	---	33.18	328.18	344.43	C-9	6.3	354.00	---	47.40	401.40	424.62
Average quality workmanship	Inst	EA	C-9	9.0	223.75	---	33.18	256.93	273.18	C-9	6.3	268.50	---	47.40	315.90	339.12

			Costs Based On Large Volume							Costs Based On Small Volume					
Description	Oper	Unit	Crew Size	Avg Day Prod	Avg Mat'l Unit Cost	Avg Equip Unit Cost	Avg Labor Unit Cost	Avg Total Unit Cost	Avg Price Incl O&P	Crew Size	Avg Day Prod	Avg Mat'l Unit Cost	Avg Equip Unit Cost	Avg Labor Unit Cost	Avg Total Unit Cost	Avg Price Incl O&P

Cabinets (continued)

Hanging corner island wall, 18" H, 24" D
18" W, 3 doors

Description	Oper	Unit	Crew Size	Avg Day Prod	Avg Mat'l Unit Cost	Avg Equip Unit Cost	Avg Labor Unit Cost	Avg Total Unit Cost	Avg Price Incl O&P	Crew Size	Avg Day Prod	Avg Mat'l Unit Cost	Avg Equip Unit Cost	Avg Labor Unit Cost	Avg Total Unit Cost	Avg Price Incl O&P
High quality workmanship	Inst	EA	C-9	11.0	247.50	---	27.14	274.64	287.95	C-9	7.7	297.00	---	38.78	335.78	354.78
Good quality workmanship	Inst	EA	C-9	16.5	181.25	---	18.10	199.35	208.21	C-9	11.6	217.50	---	25.85	243.35	256.02
Average quality workmanship	Inst	EA	C-9	16.5	138.75	---	18.10	156.85	165.71	C-9	11.6	166.50	---	25.85	192.35	205.02
24" W, 3 doors																
High quality workmanship	Inst	EA	C-9	11.0	255.00	---	27.14	282.14	295.45	C-9	7.7	306.00	---	38.78	344.78	363.78
Good quality workmanship	Inst	EA	C-9	16.5	188.75	---	18.10	206.85	215.71	C-9	11.6	226.50	---	25.85	252.35	265.02
Average quality workmanship	Inst	EA	C-9	16.5	142.50	---	18.10	160.60	169.46	C-9	11.6	171.00	---	25.85	196.85	209.52
30" W, 3 doors																
High quality workmanship	Inst	EA	C-9	8.5	262.50	---	35.13	297.63	314.84	C-9	6.0	315.00	---	50.18	365.18	389.77
Good quality workmanship	Inst	EA	C-9	12.7	196.25	---	23.51	219.76	231.28	C-9	8.9	235.50	---	33.59	269.09	285.55
Average quality workmanship	Inst	EA	C-9	12.7	150.00	---	23.51	173.51	185.03	C-9	8.9	180.00	---	33.59	213.59	230.05
Hanging island cabinets																
18" H, 12" D																
30" W, 2 doors both sides																
High quality workmanship	Inst	EA	C-9	9.4	262.50	---	31.76	294.26	309.83	C-9	6.6	315.00	---	45.38	360.38	382.61
Good quality workmanship	Inst	EA	C-9	14.0	192.50	---	21.33	213.83	224.28	C-9	9.8	231.00	---	30.47	261.47	276.40
Average quality workmanship	Inst	EA	C-9	14.0	146.25	---	21.33	167.58	178.03	C-9	9.8	175.50	---	30.47	205.97	220.90
36" W, 2 doors both sides																
High quality workmanship	Inst	EA	C-9	9.4	270.00	---	31.76	301.76	317.33	C-9	6.6	324.00	---	45.38	369.38	391.61
Good quality workmanship	Inst	EA	C-9	14.0	198.75	---	21.33	220.08	230.53	C-9	9.8	238.50	---	30.47	268.97	283.90
Average quality workmanship	Inst	EA	C-9	14.0	151.25	---	21.33	172.58	183.03	C-9	9.8	181.50	---	30.47	211.97	226.90
24" H, 12" D																
24" W, 2 doors both sides																
High quality workmanship	Inst	EA	C-9	11.0	247.50	---	27.14	274.64	287.95	C-9	7.7	297.00	---	38.78	335.78	354.78
Good quality workmanship	Inst	EA	C-9	16.5	172.50	---	18.10	190.60	199.46	C-9	11.6	207.00	---	25.85	232.85	245.52
Average quality workmanship	Inst	EA	C-9	16.5	127.50	---	18.10	145.60	154.46	C-9	11.6	153.00	---	25.85	178.85	191.52
30" W, 2 doors both sides																
High quality workmanship	Inst	EA	C-9	8.5	257.50	---	35.13	292.63	309.84	C-9	6.0	309.00	---	50.18	359.18	383.77
Good quality workmanship	Inst	EA	C-9	12.7	186.25	---	23.51	209.76	221.28	C-9	8.9	223.50	---	33.59	257.09	273.55
Average quality workmanship	Inst	EA	C-9	12.7	137.50	---	23.51	161.01	172.53	C-9	8.9	165.00	---	33.59	198.59	215.05

			Costs Based On Large Volume							Costs Based On Small Volume						
Description	Oper	Unit	Crew Size	Avg Day Prod	Avg Mat'l Unit Cost	Avg Equip Unit Cost	Avg Labor Unit Cost	Avg Total Unit Cost	Avg Price Incl O&P	Crew Size	Avg Day Prod	Avg Mat'l Unit Cost	Avg Equip Unit Cost	Avg Labor Unit Cost	Avg Total Unit Cost	Avg Price Incl O&P

Cabinets (continued)

36" W, 2 doors both sides

Description	Oper	Unit	Crew Size	Avg Day Prod	Avg Mat'l Unit Cost	Avg Equip Unit Cost	Avg Labor Unit Cost	Avg Total Unit Cost	Avg Price Incl O&P	Crew Size	Avg Day Prod	Avg Mat'l Unit Cost	Avg Equip Unit Cost	Avg Labor Unit Cost	Avg Total Unit Cost	Avg Price Incl O&P
High quality workmanship	Inst	EA	C-9	8.5	270.00	---	35.13	305.13	322.34	C-9	6.0	324.00	---	50.18	374.18	398.77
Good quality workmanship	Inst	EA	C-9	12.7	198.75	---	23.51	222.26	233.78	C-9	8.9	238.50	---	33.59	272.09	288.55
Average quality workmanship	Inst	EA	C-9	12.7	151.25	---	23.51	174.76	186.28	C-9	8.9	181.50	---	33.59	215.09	231.55

42" W, 2 doors both sides

High quality workmanship	Inst	EA	C-9	6.5	326.25	---	45.94	372.19	394.70	C-9	4.6	391.50	---	65.62	457.12	489.28
Good quality workmanship	Inst	EA	C-9	9.8	240.00	---	30.47	270.47	285.40	C-9	6.9	288.00	---	43.53	331.53	352.85
Average quality workmanship	Inst	EA	C-9	9.8	183.75	---	30.47	214.22	229.15	C-9	6.9	220.50	---	43.53	264.03	285.35

48" W, 4 doors both sides

High quality workmanship	Inst	EA	C-9	6.5	391.25	---	45.94	437.19	459.70	C-9	4.6	469.50	---	65.62	535.12	567.28
Good quality workmanship	Inst	EA	C-9	9.8	288.75	---	30.47	319.22	334.15	C-9	6.9	346.50	---	43.53	390.03	411.35
Average quality workmanship	Inst	EA	C-9	9.8	220.00	---	30.47	250.47	265.40	C-9	6.9	264.00	---	43.53	307.53	328.85

Oven cabinets

81" to 85" H, 24" D, 27" W

High quality workmanship	Inst	EA	C-9	8.0	472.50	---	37.32	509.82	528.11	C-9	5.6	567.00	---	53.32	620.32	646.45
Good quality workmanship	Inst	EA	C-9	12.0	367.50	---	24.88	392.38	404.58	C-9	8.4	441.00	---	35.55	476.55	493.96
Average quality workmanship	Inst	EA	C-9	12.0	281.25	---	24.88	306.13	318.33	C-9	8.4	337.50	---	35.55	373.05	390.46

Sink/range

Base cabinets, 35" H 24" D, 2 doors, no tops included

30" W

High quality workmanship	Inst	EA	C-9	12.0	153.75	---	24.88	178.63	190.83	C-9	8.4	184.50	---	35.55	220.05	237.46
Good quality workmanship	Inst	EA	C-9	18.0	126.25	---	16.59	142.84	150.97	C-9	12.6	151.50	---	23.70	175.20	186.81
Average quality workmanship	Inst	EA	C-9	18.0	96.25	---	16.59	112.84	120.97	C-9	12.6	115.50	---	23.70	139.20	150.81

36" W

High quality workmanship	Inst	EA	C-9	12.0	163.75	---	24.88	188.63	200.83	C-9	8.4	196.50	---	35.55	232.05	249.46
Good quality workmanship	Inst	EA	C-9	18.0	136.25	---	16.59	152.84	160.97	C-9	12.6	163.50	---	23.70	187.20	198.81
Average quality workmanship	Inst	EA	C-9	18.0	103.75	---	16.59	120.34	128.47	C-9	12.6	124.50	---	23.70	148.20	159.81

42" W

High quality workmanship	Inst	EA	C-9	9.2	181.25	---	32.46	213.71	229.61	C-9	6.4	217.50	---	46.37	263.87	286.58
Good quality workmanship	Inst	EA	C-9	13.8	150.00	---	21.64	171.64	182.24	C-9	9.7	180.00	---	30.91	210.91	226.06
Average quality workmanship	Inst	EA	C-9	13.8	113.75	---	21.64	135.39	145.99	C-9	9.7	136.50	---	30.91	167.41	182.56

			Costs Based On Large Volume							Costs Based On Small Volume						
Description	Oper	Unit	Crew Size	Avg Day Prod	Avg Mat'l Unit Cost	Avg Equip Unit Cost	Avg Labor Unit Cost	Avg Total Unit Cost	Avg Price Incl O&P	Crew Size	Avg Day Prod	Avg Mat'l Unit Cost	Avg Equip Unit Cost	Avg Labor Unit Cost	Avg Total Unit Cost	Avg Price Incl O&P

Cabinets (continued)

48" W

Description	Oper	Unit	Crew Size	Avg Day Prod	Avg Mat'l Unit Cost	Avg Equip Unit Cost	Avg Labor Unit Cost	Avg Total Unit Cost	Avg Price Incl O&P	Crew Size	Avg Day Prod	Avg Mat'l Unit Cost	Avg Equip Unit Cost	Avg Labor Unit Cost	Avg Total Unit Cost	Avg Price Incl O&P
High quality workmanship	Inst	EA	C-9	9.2	205.00	---	32.46	237.46	253.36	C-9	6.4	246.00	---	46.37	292.37	315.08
Good quality workmanship	Inst	EA	C-9	13.8	168.75	---	21.64	190.39	200.99	C-9	9.7	202.50	---	30.91	233.41	248.56
Average quality workmanship	Inst	EA	C-9	13.8	128.75	---	21.64	150.39	160.99	C-9	9.7	154.50	---	30.91	185.41	200.56

Sink/range fronts, 35" H, 2 doors
30" W

Description	Oper	Unit	Crew Size	Avg Day Prod	Avg Mat'l Unit Cost	Avg Equip Unit Cost	Avg Labor Unit Cost	Avg Total Unit Cost	Avg Price Incl O&P	Crew Size	Avg Day Prod	Avg Mat'l Unit Cost	Avg Equip Unit Cost	Avg Labor Unit Cost	Avg Total Unit Cost	Avg Price Incl O&P
High quality workmanship	Inst	EA	C-9	13.8	96.25	---	21.64	117.89	128.49	C-9	9.7	115.50	---	30.91	146.41	161.56
Good quality workmanship	Inst	EA	C-9	20.0	80.00	---	14.93	94.93	102.25	C-9	14.0	96.00	---	21.33	117.33	127.78
Average quality workmanship	Inst	EA	C-9	20.0	61.25	---	14.93	76.18	83.50	C-9	14.0	73.50	---	21.33	94.83	105.28

36" W

Description	Oper	Unit	Crew Size	Avg Day Prod	Avg Mat'l Unit Cost	Avg Equip Unit Cost	Avg Labor Unit Cost	Avg Total Unit Cost	Avg Price Incl O&P	Crew Size	Avg Day Prod	Avg Mat'l Unit Cost	Avg Equip Unit Cost	Avg Labor Unit Cost	Avg Total Unit Cost	Avg Price Incl O&P
High quality workmanship	Inst	EA	C-9	13.8	108.75	---	21.64	130.39	140.99	C-9	9.7	130.50	---	30.91	161.41	176.56
Good quality workmanship	Inst	EA	C-9	20.0	90.00	---	14.93	104.93	112.25	C-9	14.0	108.00	---	21.33	129.33	139.78
Average quality workmanship	Inst	EA	C-9	20.0	70.00	---	14.93	84.93	92.25	C-9	14.0	84.00	---	21.33	105.33	115.78

42" W

Description	Oper	Unit	Crew Size	Avg Day Prod	Avg Mat'l Unit Cost	Avg Equip Unit Cost	Avg Labor Unit Cost	Avg Total Unit Cost	Avg Price Incl O&P	Crew Size	Avg Day Prod	Avg Mat'l Unit Cost	Avg Equip Unit Cost	Avg Labor Unit Cost	Avg Total Unit Cost	Avg Price Incl O&P
High quality workmanship	Inst	EA	C-9	10.6	118.75	---	28.17	146.92	160.72	C-9	7.4	142.50	---	40.24	182.74	202.46
Good quality workmanship	Inst	EA	C-9	16.0	98.75	---	18.66	117.41	126.56	C-9	11.2	118.50	---	26.66	145.16	158.22
Average quality workmanship	Inst	EA	C-9	16.0	75.00	---	18.66	93.66	102.81	C-9	11.2	90.00	---	26.66	116.66	129.72

48" W

Description	Oper	Unit	Crew Size	Avg Day Prod	Avg Mat'l Unit Cost	Avg Equip Unit Cost	Avg Labor Unit Cost	Avg Total Unit Cost	Avg Price Incl O&P	Crew Size	Avg Day Prod	Avg Mat'l Unit Cost	Avg Equip Unit Cost	Avg Labor Unit Cost	Avg Total Unit Cost	Avg Price Incl O&P
High quality workmanship	Inst	EA	C-9	10.6	125.00	---	28.17	153.17	166.97	C-9	7.4	150.00	---	40.24	190.24	209.96
Good quality workmanship	Inst	EA	C-9	16.0	107.50	---	18.66	126.16	135.31	C-9	11.2	129.00	---	26.66	155.66	168.72
Average quality workmanship	Inst	EA	C-9	16.0	81.25	---	18.66	99.91	109.06	C-9	11.2	97.50	---	26.66	124.16	137.22

Wall cabinets
15" H, 12" D, 2 doors, Refrigerator Cabinets
30" W

Description	Oper	Unit	Crew Size	Avg Day Prod	Avg Mat'l Unit Cost	Avg Equip Unit Cost	Avg Labor Unit Cost	Avg Total Unit Cost	Avg Price Incl O&P	Crew Size	Avg Day Prod	Avg Mat'l Unit Cost	Avg Equip Unit Cost	Avg Labor Unit Cost	Avg Total Unit Cost	Avg Price Incl O&P
High quality workmanship	Inst	EA	C-9	10.4	108.75	---	28.71	137.46	151.53	C-9	7.3	130.50	---	41.02	171.52	191.61
Good quality workmanship	Inst	EA	C-9	15.6	80.00	---	19.14	99.14	108.52	C-9	10.9	96.00	---	27.34	123.34	136.74
Average quality workmanship	Inst	EA	C-9	15.6	61.25	---	19.14	80.39	89.77	C-9	10.9	73.50	---	27.34	100.84	114.24

33" W

Description	Oper	Unit	Crew Size	Avg Day Prod	Avg Mat'l Unit Cost	Avg Equip Unit Cost	Avg Labor Unit Cost	Avg Total Unit Cost	Avg Price Incl O&P	Crew Size	Avg Day Prod	Avg Mat'l Unit Cost	Avg Equip Unit Cost	Avg Labor Unit Cost	Avg Total Unit Cost	Avg Price Incl O&P
High quality workmanship	Inst	EA	C-9	10.4	112.50	---	28.71	141.21	155.28	C-9	7.3	135.00	---	41.02	176.02	196.11
Good quality workmanship	Inst	EA	C-9	15.6	82.50	---	19.14	101.64	111.02	C-9	10.9	99.00	---	27.34	126.34	139.74
Average quality workmanship	Inst	EA	C-9	15.6	62.50	---	19.14	81.64	91.02	C-9	10.9	75.00	---	27.34	102.34	115.74

			Costs Based On Large Volume							Costs Based On Small Volume					
Description	Oper	Unit	Crew Size	Avg Day Prod	Avg Mat'l Unit Cost	Avg Equip Unit Cost	Avg Labor Unit Cost	Avg Total Unit Cost	Avg Price Incl O&P	Crew Size	Avg Day Prod	Avg Mat'l Unit Cost	Avg Equip Unit Cost	Avg Labor Unit Cost	Avg Total Unit Cost	Avg Price Incl O&P

Cabinets (continued)

36" W

Description	Oper	Unit	Crew Size	Avg Day Prod	Avg Mat'l Unit Cost	Avg Equip Unit Cost	Avg Labor Unit Cost	Avg Total Unit Cost	Avg Price Incl O&P	Crew Size	Avg Day Prod	Avg Mat'l Unit Cost	Avg Equip Unit Cost	Avg Labor Unit Cost	Avg Total Unit Cost	Avg Price Incl O&P
High quality workmanship	Inst	EA	C-9	10.4	118.75	---	28.71	147.46	161.53	C-9	7.3	142.50	---	41.02	183.52	203.61
Good quality workmanship	Inst	EA	C-9	15.6	87.50	---	19.14	106.64	116.02	C-9	10.9	105.00	---	27.34	132.34	145.74
Average quality workmanship	Inst	EA	C-9	15.6	66.25	---	19.14	85.39	94.77	C-9	10.9	79.50	---	27.34	106.84	120.24

21" H, 12" D, Range Cabinets
24" W, 1 door

Description	Oper	Unit	Crew Size	Avg Day Prod	Avg Mat'l Unit Cost	Avg Equip Unit Cost	Avg Labor Unit Cost	Avg Total Unit Cost	Avg Price Incl O&P	Crew Size	Avg Day Prod	Avg Mat'l Unit Cost	Avg Equip Unit Cost	Avg Labor Unit Cost	Avg Total Unit Cost	Avg Price Incl O&P
High quality workmanship	Inst	EA	C-9	10.6	111.25	---	28.17	139.42	153.22	C-9	7.4	133.50	---	40.24	173.74	193.46
Good quality workmanship	Inst	EA	C-9	15.8	82.50	---	18.90	101.40	110.66	C-9	11.1	99.00	---	27.00	126.00	139.23
Average quality workmanship	Inst	EA	C-9	15.8	62.50	---	18.90	81.40	90.66	C-9	11.1	75.00	---	27.00	102.00	115.23

30" W, 2 doors

Description	Oper	Unit	Crew Size	Avg Day Prod	Avg Mat'l Unit Cost	Avg Equip Unit Cost	Avg Labor Unit Cost	Avg Total Unit Cost	Avg Price Incl O&P	Crew Size	Avg Day Prod	Avg Mat'l Unit Cost	Avg Equip Unit Cost	Avg Labor Unit Cost	Avg Total Unit Cost	Avg Price Incl O&P
High quality workmanship	Inst	EA	C-9	10.6	123.75	---	28.17	151.92	165.72	C-9	7.4	148.50	---	40.24	188.74	208.46
Good quality workmanship	Inst	EA	C-9	15.8	91.25	---	18.90	110.15	119.41	C-9	11.1	109.50	---	27.00	136.50	149.73
Average quality workmanship	Inst	EA	C-9	15.8	70.00	---	18.90	88.90	98.16	C-9	11.1	84.00	---	27.00	111.00	124.23

33" W, 2 doors

Description	Oper	Unit	Crew Size	Avg Day Prod	Avg Mat'l Unit Cost	Avg Equip Unit Cost	Avg Labor Unit Cost	Avg Total Unit Cost	Avg Price Incl O&P	Crew Size	Avg Day Prod	Avg Mat'l Unit Cost	Avg Equip Unit Cost	Avg Labor Unit Cost	Avg Total Unit Cost	Avg Price Incl O&P
High quality workmanship	Inst	EA	C-9	10.0	128.75	---	29.86	158.61	173.24	C-9	7.0	154.50	---	42.66	197.16	218.06
Good quality workmanship	Inst	EA	C-9	15.0	95.00	---	19.91	114.91	124.66	C-9	10.5	114.00	---	28.44	142.44	156.37
Average quality workmanship	Inst	EA	C-9	15.0	72.50	---	19.91	92.41	102.16	C-9	10.5	87.00	---	28.44	115.44	129.37

36" W, 2 doors

Description	Oper	Unit	Crew Size	Avg Day Prod	Avg Mat'l Unit Cost	Avg Equip Unit Cost	Avg Labor Unit Cost	Avg Total Unit Cost	Avg Price Incl O&P	Crew Size	Avg Day Prod	Avg Mat'l Unit Cost	Avg Equip Unit Cost	Avg Labor Unit Cost	Avg Total Unit Cost	Avg Price Incl O&P
High quality workmanship	Inst	EA	C-9	10.0	137.50	---	29.86	167.36	181.99	C-9	7.0	165.00	---	42.66	207.66	228.56
Good quality workmanship	Inst	EA	C-9	15.0	100.00	---	19.91	119.91	129.66	C-9	10.5	120.00	---	28.44	148.44	162.37
Average quality workmanship	Inst	EA	C-9	15.0	76.25	---	19.91	96.16	105.91	C-9	10.5	91.50	---	28.44	119.94	133.87

42" W, 2 doors

Description	Oper	Unit	Crew Size	Avg Day Prod	Avg Mat'l Unit Cost	Avg Equip Unit Cost	Avg Labor Unit Cost	Avg Total Unit Cost	Avg Price Incl O&P	Crew Size	Avg Day Prod	Avg Mat'l Unit Cost	Avg Equip Unit Cost	Avg Labor Unit Cost	Avg Total Unit Cost	Avg Price Incl O&P
High quality workmanship	Inst	EA	C-9	10.0	150.00	---	29.86	179.86	194.49	C-9	7.0	180.00	---	42.66	222.66	243.56
Good quality workmanship	Inst	EA	C-9	15.0	111.25	---	19.91	131.16	140.91	C-9	10.5	133.50	---	28.44	161.94	175.87
Average quality workmanship	Inst	EA	C-9	15.0	85.00	---	19.91	104.91	114.66	C-9	10.5	102.00	---	28.44	130.44	144.37

48" W, 2 doors

Description	Oper	Unit	Crew Size	Avg Day Prod	Avg Mat'l Unit Cost	Avg Equip Unit Cost	Avg Labor Unit Cost	Avg Total Unit Cost	Avg Price Incl O&P	Crew Size	Avg Day Prod	Avg Mat'l Unit Cost	Avg Equip Unit Cost	Avg Labor Unit Cost	Avg Total Unit Cost	Avg Price Incl O&P
High quality workmanship	Inst	EA	C-9	8.0	170.00	---	37.32	207.32	225.61	C-9	5.6	204.00	---	53.32	257.32	283.45
Good quality workmanship	Inst	EA	C-9	12.0	126.25	---	24.88	151.13	163.33	C-9	8.4	151.50	---	35.55	187.05	204.46
Average quality workmanship	Inst	EA	C-9	12.0	96.25	---	24.88	121.13	133.33	C-9	8.4	115.50	---	35.55	151.05	168.46

Description	Oper	Unit	Costs Based On Large Volume							Costs Based On Small Volume						
			Crew Size	Avg Day Prod	Avg Mat'l Unit Cost	Avg Equip Unit Cost	Avg Labor Unit Cost	Avg Total Unit Cost	Avg Price Incl O&P	Crew Size	Avg Day Prod	Avg Mat'l Unit Cost	Avg Equip Unit Cost	Avg Labor Unit Cost	Avg Total Unit Cost	Avg Price Incl O&P

Cabinets (continued)

30" H, 12" D
12" W, 1 door

Description	Oper	Unit	Crew Size	Avg Day Prod	Avg Mat'l Unit Cost	Avg Equip Unit Cost	Avg Labor Unit Cost	Avg Total Unit Cost	Avg Price Incl O&P	Crew Size	Avg Day Prod	Avg Mat'l Unit Cost	Avg Equip Unit Cost	Avg Labor Unit Cost	Avg Total Unit Cost	Avg Price Incl O&P
High quality workmanship	Inst	EA	C-9	13.6	97.50	---	21.96	119.46	130.21	C-9	9.5	117.00	---	31.36	148.36	163.73
Good quality workmanship	Inst	EA	C-9	20.5	71.25	---	14.57	85.82	92.95	C-9	14.4	85.50	---	20.81	106.31	116.50
Average quality workmanship	Inst	EA	C-9	20.5	53.75	---	14.57	68.32	75.45	C-9	14.4	64.50	---	20.81	85.31	95.50

15" W, 1 door

Description	Oper	Unit	Crew Size	Avg Day Prod	Avg Mat'l Unit Cost	Avg Equip Unit Cost	Avg Labor Unit Cost	Avg Total Unit Cost	Avg Price Incl O&P	Crew Size	Avg Day Prod	Avg Mat'l Unit Cost	Avg Equip Unit Cost	Avg Labor Unit Cost	Avg Total Unit Cost	Avg Price Incl O&P
High quality workmanship	Inst	EA	C-9	13.6	107.50	---	21.96	129.46	140.21	C-9	9.5	129.00	---	31.36	160.36	175.73
Good quality workmanship	Inst	EA	C-9	20.5	78.75	---	14.57	93.32	100.45	C-9	14.4	94.50	---	20.81	115.31	125.50
Average quality workmanship	Inst	EA	C-9	20.5	60.00	---	14.57	74.57	81.70	C-9	14.4	72.00	---	20.81	92.81	103.00

18" W, 1 door

Description	Oper	Unit	Crew Size	Avg Day Prod	Avg Mat'l Unit Cost	Avg Equip Unit Cost	Avg Labor Unit Cost	Avg Total Unit Cost	Avg Price Incl O&P	Crew Size	Avg Day Prod	Avg Mat'l Unit Cost	Avg Equip Unit Cost	Avg Labor Unit Cost	Avg Total Unit Cost	Avg Price Incl O&P
High quality workmanship	Inst	EA	C-9	13.6	112.50	---	21.96	134.46	145.21	C-9	9.5	135.00	---	31.36	166.36	181.73
Good quality workmanship	Inst	EA	C-9	20.5	82.50	---	14.57	97.07	104.20	C-9	14.4	99.00	---	20.81	119.81	130.00
Average quality workmanship	Inst	EA	C-9	20.5	62.50	---	14.57	77.07	84.20	C-9	14.4	75.00	---	20.81	95.81	106.00

21" W, 1 door

Description	Oper	Unit	Crew Size	Avg Day Prod	Avg Mat'l Unit Cost	Avg Equip Unit Cost	Avg Labor Unit Cost	Avg Total Unit Cost	Avg Price Incl O&P	Crew Size	Avg Day Prod	Avg Mat'l Unit Cost	Avg Equip Unit Cost	Avg Labor Unit Cost	Avg Total Unit Cost	Avg Price Incl O&P
High quality workmanship	Inst	EA	C-9	12.0	118.75	---	24.88	143.63	155.83	C-9	8.4	142.50	---	35.55	178.05	195.46
Good quality workmanship	Inst	EA	C-9	18.0	87.50	---	16.59	104.09	112.22	C-9	12.6	105.00	---	23.70	128.70	140.31
Average quality workmanship	Inst	EA	C-9	18.0	66.25	---	16.59	82.84	90.97	C-9	12.6	79.50	---	23.70	103.20	114.81

24" W, 1 door

Description	Oper	Unit	Crew Size	Avg Day Prod	Avg Mat'l Unit Cost	Avg Equip Unit Cost	Avg Labor Unit Cost	Avg Total Unit Cost	Avg Price Incl O&P	Crew Size	Avg Day Prod	Avg Mat'l Unit Cost	Avg Equip Unit Cost	Avg Labor Unit Cost	Avg Total Unit Cost	Avg Price Incl O&P
High quality workmanship	Inst	EA	C-9	12.0	132.50	---	24.88	157.38	169.58	C-9	8.4	159.00	---	35.55	194.55	211.96
Good quality workmanship	Inst	EA	C-9	18.0	97.50	---	16.59	114.09	122.22	C-9	12.6	117.00	---	23.70	140.70	152.31
Average quality workmanship	Inst	EA	C-9	18.0	73.75	---	16.59	90.34	98.47	C-9	12.6	88.50	---	23.70	112.20	123.81

27" W, 2 doors

Description	Oper	Unit	Crew Size	Avg Day Prod	Avg Mat'l Unit Cost	Avg Equip Unit Cost	Avg Labor Unit Cost	Avg Total Unit Cost	Avg Price Incl O&P	Crew Size	Avg Day Prod	Avg Mat'l Unit Cost	Avg Equip Unit Cost	Avg Labor Unit Cost	Avg Total Unit Cost	Avg Price Incl O&P
High quality workmanship	Inst	EA	C-9	12.0	146.25	---	24.88	171.13	183.33	C-9	8.4	175.50	---	35.55	211.05	228.46
Good quality workmanship	Inst	EA	C-9	18.0	107.50	---	16.59	124.09	132.22	C-9	12.6	129.00	---	23.70	152.70	164.31
Average quality workmanship	Inst	EA	C-9	18.0	82.50	---	16.59	99.09	107.22	C-9	12.6	99.00	---	23.70	122.70	134.31

30" W, 2 doors

Description	Oper	Unit	Crew Size	Avg Day Prod	Avg Mat'l Unit Cost	Avg Equip Unit Cost	Avg Labor Unit Cost	Avg Total Unit Cost	Avg Price Incl O&P	Crew Size	Avg Day Prod	Avg Mat'l Unit Cost	Avg Equip Unit Cost	Avg Labor Unit Cost	Avg Total Unit Cost	Avg Price Incl O&P
High quality workmanship	Inst	EA	C-9	9.0	155.00	---	33.18	188.18	204.43	C-9	6.3	186.00	---	47.40	233.40	256.62
Good quality workmanship	Inst	EA	C-9	13.5	115.00	---	22.12	137.12	147.96	C-9	9.5	138.00	---	31.60	169.60	185.08
Average quality workmanship	Inst	EA	C-9	13.5	86.25	---	22.12	108.37	119.21	C-9	9.5	103.50	---	31.60	135.10	150.58

33" W, 2 doors

Description	Oper	Unit	Crew Size	Avg Day Prod	Avg Mat'l Unit Cost	Avg Equip Unit Cost	Avg Labor Unit Cost	Avg Total Unit Cost	Avg Price Incl O&P	Crew Size	Avg Day Prod	Avg Mat'l Unit Cost	Avg Equip Unit Cost	Avg Labor Unit Cost	Avg Total Unit Cost	Avg Price Incl O&P
High quality workmanship	Inst	EA	C-9	9.0	163.75	---	33.18	196.93	213.18	C-9	6.3	196.50	---	47.40	243.90	267.12
Good quality workmanship	Inst	EA	C-9	13.5	121.25	---	22.12	143.37	154.21	C-9	9.5	145.50	---	31.60	177.10	192.58
Average quality workmanship	Inst	EA	C-9	13.5	92.50	---	22.12	114.62	125.46	C-9	9.5	111.00	---	31.60	142.60	158.08

			Costs Based On Large Volume							Costs Based On Small Volume						
Description	Oper	Unit	Crew Size	Avg Day Prod	Avg Mat'l Unit Cost	Avg Equip Unit Cost	Avg Labor Unit Cost	Avg Total Unit Cost	Avg Price Incl O&P	Crew Size	Avg Day Prod	Avg Mat'l Unit Cost	Avg Equip Unit Cost	Avg Labor Unit Cost	Avg Total Unit Cost	Avg Price Incl O&P

Cabinets (continued)

36" W, 2 doors

Description	Oper	Unit	Crew Size	Avg Day Prod	Avg Mat'l Unit Cost	Avg Equip Unit Cost	Avg Labor Unit Cost	Avg Total Unit Cost	Avg Price Incl O&P	Crew Size	Avg Day Prod	Avg Mat'l Unit Cost	Avg Equip Unit Cost	Avg Labor Unit Cost	Avg Total Unit Cost	Avg Price Incl O&P
High quality workmanship	Inst	EA	C-9	9.0	170.00	---	33.18	203.18	219.43	C-9	6.3	204.00	---	47.40	251.40	274.62
Good quality workmanship	Inst	EA	C-9	13.5	126.25	---	22.12	148.37	159.21	C-9	9.5	151.50	---	31.60	183.10	198.58
Average quality workmanship	Inst	EA	C-9	13.5	96.25	---	22.12	118.37	129.21	C-9	9.5	115.50	---	31.60	147.10	162.58
42" W, 2 doors																
High quality workmanship	Inst	EA	C-9	7.0	185.00	---	42.66	227.66	248.56	C-9	4.9	222.00	---	60.94	282.94	312.80
Good quality workmanship	Inst	EA	C-9	10.5	136.25	---	28.44	164.69	178.62	C-9	7.4	163.50	---	40.62	204.12	224.03
Average quality workmanship	Inst	EA	C-9	10.5	103.75	---	28.44	132.19	146.12	C-9	7.4	124.50	---	40.62	165.12	185.03
48" W, 2 doors																
High quality workmanship	Inst	EA	C-9	7.0	218.75	---	42.66	261.41	282.31	C-9	4.9	262.50	---	60.94	323.44	353.30
Good quality workmanship	Inst	EA	C-9	10.5	161.25	---	28.44	189.69	203.62	C-9	7.4	193.50	---	40.62	234.12	254.03
Average quality workmanship	Inst	EA	C-9	10.5	122.50	---	28.44	150.94	164.87	C-9	7.4	147.00	---	40.62	187.62	207.53
24" W, blind corner unit, 1 door																
High quality workmanship	Inst	EA	C-9	9.5	118.75	---	31.43	150.18	165.58	C-9	6.7	142.50	---	44.90	187.40	209.40
Good quality workmanship	Inst	EA	C-9	14.0	93.75	---	21.33	115.08	125.53	C-9	9.8	112.50	---	30.47	142.97	157.90
Average quality workmanship	Inst	EA	C-9	14.0	91.25	---	21.33	112.58	123.03	C-9	9.8	109.50	---	30.47	139.97	154.90
24" x 24" angle corner unit, stationary																
High quality workmanship	Inst	EA	C-9	9.5	170.00	---	31.43	201.43	216.83	C-9	6.7	204.00	---	44.90	248.90	270.90
Good quality workmanship	Inst	EA	C-9	14.0	125.00	---	21.33	146.33	156.78	C-9	9.8	150.00	---	30.47	180.47	195.40
Average quality workmanship	Inst	EA	C-9	14.0	97.50	---	21.33	118.83	129.28	C-9	9.8	117.00	---	30.47	147.47	162.40
24" x 24" angle corner unit, lazy susan																
High quality workmanship	Inst	EA	C-9	9.5	252.50	---	31.43	283.93	299.33	C-9	6.7	303.00	---	44.90	347.90	369.90
Good quality workmanship	Inst	EA	C-9	14.0	205.00	---	21.33	226.33	236.78	C-9	9.8	246.00	---	30.47	276.47	291.40
Average quality workmanship	Inst	EA	C-9	14.0	140.00	---	21.33	161.33	171.78	C-9	9.8	168.00	---	30.47	198.47	213.40

Vanity cabinet and sink top

Description	Oper	Unit	Crew Size	Avg Day Prod	Avg Mat'l Unit Cost	Avg Equip Unit Cost	Avg Labor Unit Cost	Avg Total Unit Cost	Avg Price Incl O&P	Crew Size	Avg Day Prod	Avg Mat'l Unit Cost	Avg Equip Unit Cost	Avg Labor Unit Cost	Avg Total Unit Cost	Avg Price Incl O&P
Disconnect plumbing and remove to dumpster	Demo	EA	L-2	16.0	---	---	16.26	16.26	24.06	L-2	11.2		---	23.22	23.22	34.37
Remove old unit, replace with new unit, reconnect plumbing	Demo	EA	S-2	7.0	---	---	47.00	47.00	68.62	S-2	4.9		---	67.14	67.14	98.02

			Costs Based On Large Volume						Costs Based On Small Volume					
Description	Oper	Unit	Crew Size	Avg Day Prod	Avg Mat'l Unit Cost	Avg Equip Unit Cost	Avg Labor Unit Cost	Avg Total Unit Cost	Avg Price Incl O&P	Crew Size	Avg Day Prod	Avg Mat'l Unit Cost	Avg Equip Unit Cost	Avg Labor Unit Cost	Avg Total Unit Cost	Avg Price Incl O&P

Cabinets (continued)

Vanity units, with marble tops, good quality fittings and faucets; hardware, deluxe, finished models

Stained ash and birch. Composition construction is primed. 2 door units

Description	Oper	Unit	Crew Size	Avg Day Prod	Avg Mat'l Unit Cost	Avg Equip Unit Cost	Avg Labor Unit Cost	Avg Total Unit Cost	Avg Price Incl O&P	Crew Size	Avg Day Prod	Avg Mat'l Unit Cost	Avg Equip Unit Cost	Avg Labor Unit Cost	Avg Total Unit Cost	Avg Price Incl O&P
20" x 16"																
Ash	Inst	EA	S-2	3.2	282.50	---	102.81	385.31	432.60	S-2	2.2	339.00	---	146.86	485.86	553.42
Birch	Inst	EA	S-2	3.2	257.50	---	102.81	360.31	407.60	S-2	2.2	309.00	---	146.86	455.86	523.42
Composition construction	Inst	EA	S-2	3.2	195.00	---	102.81	297.81	345.10	S-2	2.2	234.00	---	146.86	380.86	448.42
25" x 19"																
Ash	Inst	EA	S-2	3.2	328.75	---	102.81	431.56	478.85	S-2	2.2	394.50	---	146.86	541.36	608.92
Birch	Inst	EA	S-2	3.2	300.00	---	102.81	402.81	450.10	S-2	2.2	360.00	---	146.86	506.86	574.42
Composition construction	Inst	EA	S-2	3.2	213.75	---	102.81	316.56	363.85	S-2	2.2	256.50	---	146.86	403.36	470.92
31" x 19"																
Ash	Inst	EA	S-2	3.2	391.25	---	102.81	494.06	541.35	S-2	2.2	469.50	---	146.86	616.36	683.92
Birch	Inst	EA	S-2	3.2	355.00	---	102.81	457.81	505.10	S-2	2.2	426.00	---	146.86	572.86	640.42
Composition construction	Inst	EA	S-2	3.2	253.75	---	102.81	356.56	403.85	S-2	2.2	304.50	---	146.86	451.36	518.92
35" x 19"																
Ash	Inst	EA	S-2	3.0	422.50	---	109.66	532.16	582.60	S-2	2.1	507.00	---	156.66	663.66	735.72
Birch	Inst	EA	S-2	3.0	382.50	---	109.66	492.16	542.60	S-2	2.1	459.00	---	156.66	615.66	687.72
Composition construction	Inst	EA	S-2	3.0	276.25	---	109.66	385.91	436.35	S-2	2.1	331.50	---	156.66	488.16	560.22
For drawers in any above unit, ADD per drawer	Inst	EA	S-2	---	37.50	---	---	37.50	37.50	S-2	---	45.00	---	---	45.00	45.00
2 door cutback units with 3 drawers																
36" x 19"																
Ash	Inst	EA	S-2	3.0	490.00	---	109.66	599.66	650.10	S-2	2.1	588.00	---	156.66	744.66	816.72
Birch	Inst	EA	S-2	3.0	440.00	---	109.66	549.66	600.10	S-2	2.1	528.00	---	156.66	684.66	756.72
Composition construction	Inst	EA	S-2	3.0	328.75	---	109.66	438.41	488.85	S-2	2.1	394.50	---	156.66	551.16	623.22
49" x 19"																
Ash	Inst	EA	S-2	3.0	612.50	---	109.66	722.16	772.60	S-2	2.1	735.00	---	156.66	891.66	963.72
Birch	Inst	EA	S-2	3.0	530.00	---	109.66	639.66	690.10	S-2	2.1	636.00	---	156.66	792.66	864.72
Composition construction	Inst	EA	S-2	3.0	396.25	---	109.66	505.91	556.35	S-2	2.1	475.50	---	156.66	632.16	704.22
60" x 19"																
Ash	Inst	EA	S-2	2.4	772.50	---	137.07	909.57	972.63	S-2	1.7	927.00	---	195.82	1122.82	1212.90
Birch	Inst	EA	S-2	2.4	641.25	---	137.07	778.32	841.38	S-2	1.7	769.50	---	195.82	965.32	1055.40
Composition construction	Inst	EA	S-2	2.4	498.75	---	137.07	635.82	698.88	S-2	1.7	598.50	---	195.82	794.32	884.40

			Costs Based On Large Volume							Costs Based On Small Volume						
Description	Oper	Unit	Crew Size	Avg Day Prod	Avg Mat'l Unit Cost	Avg Equip Unit Cost	Avg Labor Unit Cost	Avg Total Unit Cost	Avg Price Incl O&P	Crew Size	Avg Day Prod	Avg Mat'l Unit Cost	Avg Equip Unit Cost	Avg Labor Unit Cost	Avg Total Unit Cost	Avg Price Incl O&P

Cabinets (continued)

Corner unit, 1 door
22" x 22"

Description	Oper	Unit	Crew Size	Avg Day Prod	Avg Mat'l Unit Cost	Avg Equip Unit Cost	Avg Labor Unit Cost	Avg Total Unit Cost	Avg Price Incl O&P	Crew Size	Avg Day Prod	Avg Mat'l Unit Cost	Avg Equip Unit Cost	Avg Labor Unit Cost	Avg Total Unit Cost	Avg Price Incl O&P
Ash	Inst	EA	S-2	3.2	363.75	---	102.81	466.56	513.85	S-2	2.2	436.50	---	146.86	583.36	650.92
Birch	Inst	EA	S-2	3.2	343.75	---	102.81	446.56	493.85	S-2	2.2	412.50	---	146.86	559.36	626.92
Composition construction	Inst	EA	S-2	3.2	280.00	---	102.81	382.81	430.10	S-2	2.2	336.00	---	146.86	482.86	550.42
Adjustments																
To remove and reset																
Vanity units with tops	Inst	EA	S-1	2.4	---	---	82.89	82.89	121.02	S-1	1.7	---	---	118.41	118.41	172.88
Top only	Inst	EA	S-1	4.0	---	---	49.73	49.73	72.61	S-1	2.8	---	---	71.05	71.05	103.73
To install rough-in	Inst	EA	S-2	4.0	---	---	82.24	82.24	120.08	S-2	2.8	---	---	117.49	117.49	171.54

Ceramic tile

Countertop / backsplash, ceramic
Adhesive set with backmounted tile

Description	Oper	Unit	Crew Size	Avg Day Prod	Avg Mat'l Unit Cost	Avg Equip Unit Cost	Avg Labor Unit Cost	Avg Total Unit Cost	Avg Price Incl O&P	Crew Size	Avg Day Prod	Avg Mat'l Unit Cost	Avg Equip Unit Cost	Avg Labor Unit Cost	Avg Total Unit Cost	Avg Price Incl O&P
1" x 1"	Inst	SF	T-2	70	2.79	---	4.22	7.01	8.95	T-2	46	3.28	---	6.49	9.77	12.75
4-1/4" x 4-1/4"	Inst	SF	T-2	80	2.59	---	3.69	6.28	7.98	T-2	52	3.04	---	5.68	8.71	11.33

Cove / base, ceramic
Adhesive set with unmounted tile

Description	Oper	Unit	Crew Size	Avg Day Prod	Avg Mat'l Unit Cost	Avg Equip Unit Cost	Avg Labor Unit Cost	Avg Total Unit Cost	Avg Price Incl O&P	Crew Size	Avg Day Prod	Avg Mat'l Unit Cost	Avg Equip Unit Cost	Avg Labor Unit Cost	Avg Total Unit Cost	Avg Price Incl O&P
4-1/4" x 4-1/4"	Inst	SF	T-2	80	3.14	---	3.69	6.83	8.53	T-2	52	3.69	---	5.68	9.36	11.97
6" x 4-1/4"	Inst	SF	T-2	95	2.65	---	3.11	5.75	7.18	T-2	62	3.11	---	4.78	7.89	10.09
Conventional mortar with unmounted tile																
4-1/4" x 4-1/4"	Inst	SF	T-2	45	2.97	---	6.56	9.53	12.55	T-2	29	3.48	---	10.09	13.58	18.22
6" x 4-1/4"	Inst	SF	T-2	50	2.59	---	5.90	8.49	11.21	T-2	33	3.04	---	9.08	12.12	16.30
Dry-set mortar with unmounted tile																
4-1/4" x 4-1/4"	Inst	SF	T-2	65	2.99	---	4.54	7.53	9.62	T-2	42	3.51	---	6.99	10.50	13.71
6" x 4-1/4"	Inst	SF	T-2	75	2.60	---	3.94	6.54	8.35	T-2	49	3.05	---	6.06	9.11	11.89

Floors, ceramic
1" x 1"

Description	Oper	Unit	Crew Size	Avg Day Prod	Avg Mat'l Unit Cost	Avg Equip Unit Cost	Avg Labor Unit Cost	Avg Total Unit Cost	Avg Price Incl O&P	Crew Size	Avg Day Prod	Avg Mat'l Unit Cost	Avg Equip Unit Cost	Avg Labor Unit Cost	Avg Total Unit Cost	Avg Price Incl O&P
Adhesive or dry-set base	Demo	SF	L-2	550	---	---	0.47	0.47	0.70	L-2	358	---	---	0.73	0.73	1.08
Conventional mortar base	Demo	SF	L-2	475	---	---	0.55	0.55	0.81	L-2	309	---	---	0.84	0.84	1.25

Ceramic tile

1. **Dimensions.** There are many sizes of ceramic tile. Only 4-1/4" x 4-1/4" and 1" x 1" will be discussed here.

   a. 4-1/4" x 4-1/4" tile is furnished both unmounted and back-mounted. Back-mounted tile are usually furnished in sheets of 12 tile.

   b. 1" x 1" mosaic tile is furnished face-mounted and back-mounted in sheets; normally, 2'-0" x 1'-0".

2. **Installation.** Three methods:

   a. Conventional. Utilizes portland cement, sand and wet tile grout.

   b. Dry-set. Utilizes dry-set mix and dry tile grout mix.

   c. Organic adhesive. Utilizes adhesive and dry tile grout mix.

   The conventional method is the most expensive and is used less frequently than the other methods.

3. **Estimating technique.** For tile, determine area and add 5-10% for waste. For cove, base or trim, determine LF and add 5-10% for waste.

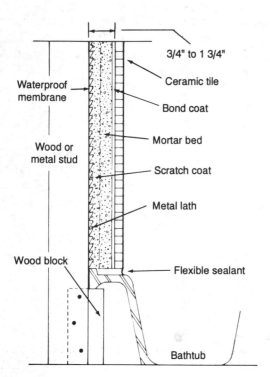

Cross section of bathtub wall using cement mortar

Wood or steel stud construction with plaster above tile wainscot

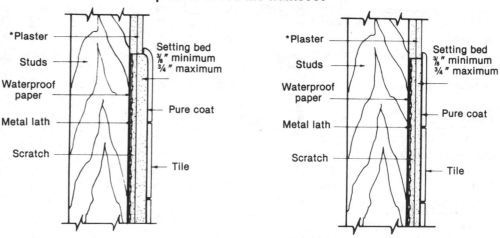

*This installation would be similar if wall finish above tile wainscot were of other material such as wallboard, plywood, etc.

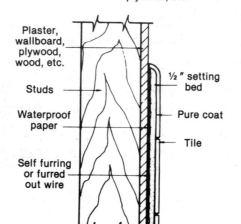

Wood or steel stud construction with solid covered backing

Courtesy: *Ceramic Tile Institute of America*
700 N. Virgil Ave., Ste 300
Los Angeles, CA 90029

			Costs Based On Large Volume							Costs Based On Small Volume						
Description	Oper	Unit	Crew Size	Avg Day Prod	Avg Mat'l Unit Cost	Avg Equip Unit Cost	Avg Labor Unit Cost	Avg Total Unit Cost	Avg Price Incl O&P	Crew Size	Avg Day Prod	Avg Mat'l Unit Cost	Avg Equip Unit Cost	Avg Labor Unit Cost	Avg Total Unit Cost	Avg Price Incl O&P

Ceramic tile (continued)

Adhesive set with backmounted tile

1" x 1"	Inst	SF	T-2	120	2.79	---	2.46	5.25	**6.39**	T-2	78	3.28	---	3.78	7.07	**8.81**
4-1/4" x 4-1/4"	Inst	SF	T-2	130	2.59	---	2.27	4.86	**5.90**	T-2	85	3.04	---	3.49	6.53	**8.14**

Conventional mortar set with backmounted tile

1" x 1"	Inst	SF	T-2	60	2.62	---	4.92	7.54	**9.81**	T-2	39	3.08	---	7.57	10.65	**14.13**
4-1/4" x 4-1/4"	Inst	SF	T-2	70	2.31	---	4.22	6.53	**8.47**	T-2	46	2.71	---	6.49	9.20	**12.19**

Dry-set mortar with backmounted tile

1" x 1"	Inst	SF	T-2	95	2.66	---	3.11	5.76	**7.19**	T-2	62	3.12	---	4.78	7.90	**10.10**
4-1/4" x 4-1/4"	Inst	SF	T-2	105	2.32	---	2.81	5.13	**6.43**	T-2	68	2.73	---	4.33	7.05	**9.04**

Wainscot cap, ceramic

Adhesive set with unmounted tile

2" x 6"	Inst	SF	T-2	120	1.10	---	2.46	3.56	**4.70**	T-2	78	1.30	---	3.78	5.08	**6.82**

Conventional mortar set with unmounted tile

2" x 6"	Inst	SF	T-2	75	1.08	---	3.94	5.02	**6.83**	T-2	49	1.27	---	6.06	7.32	**10.11**

Dry-set mortar with unmounted tile

2" x 6"	Inst	SF	T-2	100	1.08	---	2.95	4.03	**5.39**	T-2	65	1.27	---	4.54	5.81	**7.90**

Walls, 1" x 1" or 4-1/4" x 4-1/4"

Adhesive or dry-set base	Demo	SF	L-2	480	---	---	0.54	0.54	**0.80**	L-2	312	---	---	0.83	0.83	**1.23**
Conventional mortar base	Demo	SF	L-2	400	---	---	0.65	0.65	**0.96**	L-2	260	---	---	1.00	1.00	**1.48**

Adhesive set with backmounted tile

1" x 1"	Inst	SF	T-2	100	2.79	---	2.95	5.75	**7.10**	T-2	65	3.28	---	4.54	7.82	**9.91**
4-1/4" x 4-1/4"	Inst	SF	T-2	110	2.59	---	2.68	5.27	**6.51**	T-2	72	3.04	---	4.13	7.17	**9.07**

Conventional mortar set with backmounted tile

1" x 1"	Inst	SF	T-2	50	2.62	---	5.90	8.53	**11.24**	T-2	33	3.08	---	9.08	12.16	**16.34**
4-1/4" x 4-1/4"	Inst	SF	T-2	60	2.31	---	4.92	7.23	**9.50**	T-2	39	2.71	---	7.57	10.28	**13.77**

Dry-set mortar with backmounted tile

1" x 1"	Inst	SF	T-2	80	2.66	---	3.69	6.35	**8.04**	T-2	52	3.12	---	5.68	8.80	**11.41**
4-1/4" x 4-1/4"	Inst	SF	T-2	90	2.32	---	3.28	5.60	**7.11**	T-2	59	2.73	---	5.05	7.77	**10.09**

			Costs Based On Large Volume							Costs Based On Small Volume						
Description	Oper	Unit	Crew Size	Avg Day Prod	Avg Mat'l Unit Cost	Avg Equip Unit Cost	Avg Labor Unit Cost	Avg Total Unit Cost	Avg Price Incl O&P	Crew Size	Avg Day Prod	Avg Mat'l Unit Cost	Avg Equip Unit Cost	Avg Labor Unit Cost	Avg Total Unit Cost	Avg Price Incl O&P

Chimneys. See Masonry.

Closet doors. See Doors.

Columns

See also Framing.

Aluminum, extruded; self supporting. Designed as decorative, loadbearing elements for porches, entrances, colonnades, etc.; primed, knocked-down, and carton packed complete with cap and base.

Standard cap and base

Description	Oper	Unit	Crew Size	Avg Day Prod	Avg Mat'l Unit Cost	Avg Equip Unit Cost	Avg Labor Unit Cost	Avg Total Unit Cost	Avg Price Incl O&P	Crew Size	Avg Day Prod	Avg Mat'l Unit Cost	Avg Equip Unit Cost	Avg Labor Unit Cost	Avg Total Unit Cost	Avg Price Incl O&P
8" dia. x 8' to 12' H	Inst	EA	C-17	6.0	106.25	49.94	77.86	234.05	272.20	C-17	4.50	123.25	66.59	103.81	293.65	344.52
10" dia. x 8' to 12' H	Inst	EA	C-17	6.0	156.25	49.94	77.86	284.05	322.20	C-17	4.50	181.25	66.59	103.81	351.65	402.52
10" dia. x 16' to 20' H	Inst	EA	C-17	4.4	262.50	68.11	106.17	436.77	488.80	C-17	3.30	304.50	90.81	141.56	536.87	606.23
12" dia. x 9' to 12' H	Inst	EA	C-17	6.0	217.50	49.94	77.86	345.30	383.45	C-17	4.50	252.30	66.59	103.81	422.70	473.57
12" dia. x 16' to 24' H	Inst	EA	C-17	4.4	387.50	68.11	106.17	561.77	613.80	C-17	3.30	449.50	90.81	141.56	681.87	751.23

Corinthian cap and decorative base

Description	Oper	Unit	Crew Size	Avg Day Prod	Avg Mat'l Unit Cost	Avg Equip Unit Cost	Avg Labor Unit Cost	Avg Total Unit Cost	Avg Price Incl O&P	Crew Size	Avg Day Prod	Avg Mat'l Unit Cost	Avg Equip Unit Cost	Avg Labor Unit Cost	Avg Total Unit Cost	Avg Price Incl O&P
8" dia. x 8' to 12' H	Inst	EA	C-17	6.0	281.25	49.94	77.86	409.05	447.20	C-17	4.50	326.25	66.59	103.81	496.65	547.52
10" dia. x 8' to 12' H	Inst	EA	C-17	6.0	406.25	49.94	77.86	534.05	572.20	C-17	4.50	471.25	66.59	103.81	641.65	692.52
10" dia. x 16' to 20' H	Inst	EA	C-17	4.4	512.50	68.11	106.17	686.77	738.80	C-17	3.30	594.50	90.81	141.56	826.87	896.23
12" dia. x 9' to 12' H	Inst	EA	C-17	6.0	662.50	49.94	77.86	790.30	828.45	C-17	4.50	768.50	66.59	103.81	938.90	989.77
12" dia. x 16' to 24' H	Inst	EA	C-17	4.4	843.75	68.11	106.17	1018.02	1070.05	C-17	3.30	978.75	90.81	141.56	1211.12	1280.48

Brick. See Masonry.

Wood

Treated, No. 1 common and better white pine, T&G construction. Designed as decorative, loadbearing elements for porches, entrances, colonnades, etc.; primed, knocked-down, and carton packed complete with cap and base.

Plain column with standard cap

Description	Oper	Unit	Crew Size	Avg Day Prod	Avg Mat'l Unit Cost	Avg Equip Unit Cost	Avg Labor Unit Cost	Avg Total Unit Cost	Avg Price Incl O&P	Crew Size	Avg Day Prod	Avg Mat'l Unit Cost	Avg Equip Unit Cost	Avg Labor Unit Cost	Avg Total Unit Cost	Avg Price Incl O&P
8" dia. x 8' to 12' H	Inst	EA	C-17	5.0	162.50	59.93	93.43	315.86	361.64	C-17	3.75	188.50	79.91	124.57	392.98	454.02
10" dia. x 8' to 12' H	Inst	EA	C-17	5.0	193.75	59.93	93.43	347.11	392.89	C-17	3.75	224.75	79.91	124.57	429.23	490.27
12" dia. x 8' to 12' H	Inst	EA	C-17	4.5	262.50	66.59	103.81	432.90	483.77	C-17	3.38	304.50	88.79	138.41	531.70	599.52
14" dia. x 12' to 16' H	Inst	EA	C-17	3.5	687.50	85.62	133.47	906.59	971.99	C-17	2.63	797.50	114.2	177.96	1089.62	1176.8
16" dia. x 18' to 20' H	Inst	EA	C-17	3.2	1125.00	93.65	145.98	1364.63	1436.2	C-17	2.40	1305.00	124.9	194.64	1624.50	1719.9

			Costs Based On Large Volume							Costs Based On Small Volume						
Description	Oper	Unit	Crew Size	Avg Day Prod	Avg Mat'l Unit Cost	Avg Equip Unit Cost	Avg Labor Unit Cost	Avg Total Unit Cost	Avg Price Incl O&P	Crew Size	Avg Day Prod	Avg Mat'l Unit Cost	Avg Equip Unit Cost	Avg Labor Unit Cost	Avg Total Unit Cost	Avg Price Incl O&P

Columns (continued)

Plain column with Corinthian cap

Description	Oper	Unit	Crew Size	Avg Day Prod	Avg Mat'l Unit Cost	Avg Equip Unit Cost	Avg Labor Unit Cost	Avg Total Unit Cost	Avg Price Incl O&P	Crew Size	Avg Day Prod	Avg Mat'l Unit Cost	Avg Equip Unit Cost	Avg Labor Unit Cost	Avg Total Unit Cost	Avg Price Incl O&P
8" dia. x 8' to 12' H	Inst	EA	C-17	5.0	475.00	59.93	93.43	628.36	674.14	C-17	3.75	551.00	79.91	124.57	755.48	816.52
10" dia. x 8' to 12' H	Inst	EA	C-17	5.0	550.00	59.93	93.43	703.36	749.14	C-17	3.75	638.00	79.91	124.57	842.48	903.52
12" dia. x 8' to 12' H	Inst	EA	C-17	4.5	762.50	66.59	103.81	932.90	983.77	C-17	3.38	884.50	88.79	138.41	1111.70	1179.5
14" dia. x 12' to 16' H	Inst	EA	C-17	3.5	1218.75	85.62	133.47	1437.84	1503.2	C-17	2.63	1413.75	114.2	177.96	1705.87	1793.1
16" dia. x 18' to 20' H	Inst	EA	C-17	3.2	1875.00	93.65	145.98	2114.63	2186.2	C-17	2.40	2175.00	124.9	194.64	2494.50	2589.9

Concrete, Cast-In-Place

Footings, with air tools; reinforced

Description	Oper	Unit	Crew Size	Avg Day Prod	Avg Mat'l Unit Cost	Avg Equip Unit Cost	Avg Labor Unit Cost	Avg Total Unit Cost	Avg Price Incl O&P	Crew Size	Avg Day Prod	Avg Mat'l Unit Cost	Avg Equip Unit Cost	Avg Labor Unit Cost	Avg Total Unit Cost	Avg Price Incl O&P
8" T x 12" W	Demo	LF	A-4	140	---	0.63	3.34	3.97	5.54	A-4	119	---	0.74	3.93	4.67	6.51
8" T x 16" W	Demo	LF	A-4	132	---	0.66	3.54	4.21	5.87	A-4	112	---	0.78	4.17	4.95	6.91
8" T x 20" W	Demo	LF	A-4	124	---	0.71	3.77	4.48	6.25	A-4	105	---	0.83	4.44	5.27	7.36
12" T x 12" W	Demo	LF	A-4	126	---	0.69	3.71	4.41	6.15	A-4	107	---	0.82	4.37	5.19	7.24
12" T x 16" W	Demo	LF	A-4	120	---	0.73	3.90	4.63	6.46	A-4	102	---	0.86	4.59	5.44	7.60
12" T x 20" W	Demo	LF	A-4	114	---	0.77	4.10	4.87	6.80	A-4	97	---	0.90	4.83	5.73	8.00
12" T x 24" W	Demo	LF	A-4	108	---	0.81	4.33	5.14	7.18	A-4	92	---	0.95	5.10	6.05	8.44

Footings, per LF poured footing

Forming, 4 uses

Description	Oper	Unit	Crew Size	Avg Day Prod	Avg Mat'l Unit Cost	Avg Equip Unit Cost	Avg Labor Unit Cost	Avg Total Unit Cost	Avg Price Incl O&P	Crew Size	Avg Day Prod	Avg Mat'l Unit Cost	Avg Equip Unit Cost	Avg Labor Unit Cost	Avg Total Unit Cost	Avg Price Incl O&P
2" x 6"	Inst	LF	C-17	225	0.38	---	2.08	2.45	3.47	C-17	207	0.43	---	2.26	2.69	3.80
2" x 8"	Inst	LF	C-17	200	0.49	---	2.34	2.82	3.97	C-17	184	0.56	---	2.54	3.10	4.34
2" x 12"	Inst	LF	C-17	180	0.71	---	2.60	3.31	4.58	C-17	166	0.82	---	2.82	3.64	5.02

Grading, finish by hand

Description	Oper	Unit	Crew Size	Avg Day Prod	Avg Mat'l Unit Cost	Avg Equip Unit Cost	Avg Labor Unit Cost	Avg Total Unit Cost	Avg Price Incl O&P	Crew Size	Avg Day Prod	Avg Mat'l Unit Cost	Avg Equip Unit Cost	Avg Labor Unit Cost	Avg Total Unit Cost	Avg Price Incl O&P
6", 8", 12" T x 12" W	Inst	LF	L-1	345	---	---	0.38	0.38	0.56	L-1	311	---	---	0.42	0.42	0.62
6", 8", 12" T x 16" W	Inst	LF	L-1	295	---	---	0.44	0.44	0.65	L-1	266	---	---	0.49	0.49	0.72
6", 8", 12" T x 20" W	Inst	LF	L-1	260	---	---	0.50	0.50	0.74	L-1	234	---	---	0.56	0.56	0.82
12" T x 24" W	Inst	LF	L-1	230	---	---	0.57	0.57	0.84	L-1	207	---	---	0.63	0.63	0.93

Reinforcing steel in place. Lap, waste, and tie wire included

Description	Oper	Unit	Crew Size	Avg Day Prod	Avg Mat'l Unit Cost	Avg Equip Unit Cost	Avg Labor Unit Cost	Avg Total Unit Cost	Avg Price Incl O&P	Crew Size	Avg Day Prod	Avg Mat'l Unit Cost	Avg Equip Unit Cost	Avg Labor Unit Cost	Avg Total Unit Cost	Avg Price Incl O&P
Two (No. 3) 3/8" rods	Inst	LF	L-1	590	0.29	---	0.22	0.51	0.62	L-1	531	0.33	---	0.24	0.58	0.70
Two (No. 4) 1/2" rods	Inst	LF	L-1	560	0.44	---	0.23	0.67	0.79	L-1	504	0.51	---	0.26	0.77	0.89
Two (No. 5) 5/8" rods	Inst	LF	L-1	540	0.64	---	0.24	0.88	1.00	L-1	486	0.74	---	0.27	1.01	1.14
Three (No. 3) 3/8" rods	Inst	LF	L-1	460	0.44	---	0.28	0.72	0.85	L-1	414	0.50	---	0.31	0.82	0.97
Three (No. 4) 1/2" rods	Inst	LF	L-1	440	0.66	---	0.30	0.96	1.10	L-1	396	0.76	---	0.33	1.09	1.25
Three (No. 5) 5/8" rods	Inst	LF	L-1	420	0.96	---	0.31	1.27	1.42	L-1	378	1.11	---	0.34	1.45	1.62

Concrete

Concrete Footings

Dimensions. 6" T, 8" T, 12" T x 12" W, 16" W, 20" W; 12" T x 24" W.

Installation.

1. Forms. 2" side forms equal in height to the thickness of the footing. 2" x 4" stakes 4'-0" o.c. — no less than the thickness of the footing. 2" x 4" bracing for stakes 8'-0" o.c. for 6" and 8" thick footings and 4'-0" o.c. for 12" thick footings. 1" x 2" or 1" x 3" spreaders 4'-0" o.c.

2. Concrete. 1-2-4 mix utilized in this section.

3. Reinforcing Steel. Various sizes — usually #3, #4, or #5 straight rods with end ties only are used.

Notes — Labor.

1. Forming. Output based on a crew of two carpenters and one laborer.

2. Grading, Finish. Output based on what one laborer can do in one day.

3. Reinforcing Steel. Output based on what one laborer or one ironworker can do in one day.

4. Concrete. Output based on a crew of two laborers and one carpenter.

5. Forms, Wrecking and Cleaning. Output based on what one laborer can do in one day.

Estimating Technique. Determine linear feet of footing.

Concrete Foundations

Dimensions. 8" T, 12" T x 4' H, 8' H, or 12' H.

Installation.

1. Forms, 4' x 8' panels comprised of 3/4" form grade plywood backed with 2" x 4" studs and sills (studs approximately 16" o.c.), three sets and six sets of 2" x 4" wales for 4', 8' and 12' high walls. 2" x 4" wales for 4', 8' and 12' high walls. 2" x 4" diagonal braces (with stakes) 12'-0" o.c. one side. Snap ties spaced 22" o.c., 20" o.c. and 17" o.c. along each wale for 4', 8' and 12' high walls. Paraffin oil coating for forms. Twelve uses are estimated for panels; twenty uses for wales and braces; snap ties are used only once.

2. Concrete. 1-2-4 mix.

3. Reinforcing Steel. Sizes #3 to #7. Bars are straight except dowels which may on occasion be bent rods.

Notes — Labor.

1. Concrete, Placing. Output based on a crew of one carpenter and five laborers.

2. Forming. Output based on a crew of four carpenters and one laborer.

3. Reinforcing Steel Rods. Output based on what two laborers or two ironworkers can do in one day.

4. Wrecking and Cleaning Forms. Output based on what two laborers can do in one day.

Estimating Technique. Determine linear feet of wall if wall is 8" or 12" x 4', 8' or 12' or determine square feet of wall. The linear feet of rods must be determined and the cost added.

Concrete Interior Floor Finishes

Dimensions. 3-1/2", 4", 5", 6" thick x various areas.

Installation.

1. Forms. A wood form may or may not be required. A foundation wall may serve as a form for both basement and first floor slabs. In this section, only 2" x 4" and 2" x 6" side forms with stakes 4'-0" o.c. are considered.

2. Finish Grading. Dirt or gravel.

3. Screeds. Wood strips placed in area where concrete is to be placed. The concrete when placed will be finished even with top of the screeds. Screeds must be pulled before concrete sets up and the void filled with concrete. 2" x 2" and 2" x 4" screeds with 2" x 2" stakes 6'-0" o.c. will be covered in this section.

4. Steel Reinforcing. Items to be covered are: #3, #4, #5 rods; 6 x 6/10-10 and 6 x 6/6-6 welded wire mesh.

5. Concrete. 1-2-4 mix.

Notes — Labor.

1. Forms and Screeds. Output based on a crew of two masons and one laborer.

2. Finish Grading. Output based on what one laborer can do in one day.

3. Reinforcing. Output based on what two laborers or two ironworkers can do in one day.

4. Concrete, Place and Finish. Output based on three cement masons and five laborers as a crew.

5. Wrecking and Cleaning Forms. Output based on what one laborer can do in one day.

Estimating Technique.

1. Finish grading, mesh, and concrete. Determine area and add waste.

2. Forms, screeds and rods. Determine LF.

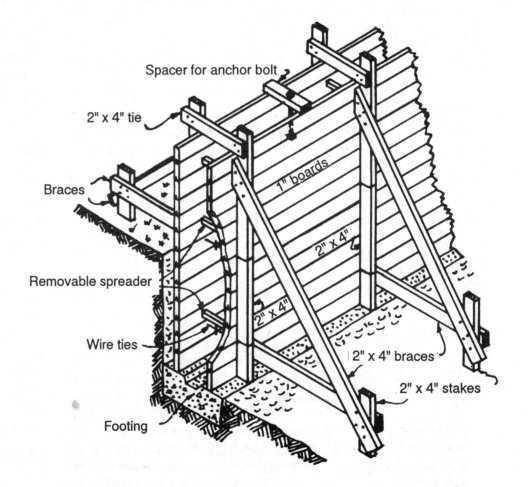

Spacer for anchor bolt

2" x 4" tie

1" boards

Braces

2" x 4"

Removable spreader

2" x 4"

Wire ties

2" x 4" braces

2" x 4" stakes

Footing

			Costs Based On Large Volume							Costs Based On Small Volume						
Description	Oper	Unit	Crew Size	Avg Day Prod	Avg Mat'l Unit Cost	Avg Equip Unit Cost	Avg Labor Unit Cost	Avg Total Unit Cost	Avg Price Incl O&P	Crew Size	Avg Day Prod	Avg Mat'l Unit Cost	Avg Equip Unit Cost	Avg Labor Unit Cost	Avg Total Unit Cost	Avg Price Incl O&P

Concrete, Cast-In-Place (continued)

Concrete, pour from truck into forms; using 3000 PSI, 1-1/2" aggregate, 7.7 sack mix

Description	Oper	Unit	Crew	Prod	Mat'l	Equip	Labor	Total	O&P	Crew	Prod	Mat'l	Equip	Labor	Total	O&P
6" T x 12" W (54.0 LF/CY)	Inst	LF	L-3	615	0.99	---	0.70	1.69	2.02	L-3	554	1.14	---	0.77	1.91	2.28
6" T x 16" W (40.5 LF/CY)	Inst	LF	L-3	560	1.32	---	0.77	2.09	2.45	L-3	504	1.52	---	0.85	2.37	2.78
6" T x 20" W (32.3 LF/CY)	Inst	LF	L-3	545	1.65	---	0.79	2.44	2.82	L-3	491	1.90	---	0.87	2.78	3.20
8" T x 12" W (40.5 LF/CY)	Inst	LF	L-3	560	1.32	---	0.77	2.09	2.45	L-3	504	1.52	---	0.85	2.37	2.78
8" T x 16" W (30.4 LF/CY)	Inst	LF	L-3	510	1.76	---	0.84	2.60	3.00	L-3	459	2.02	---	0.93	2.96	3.40
8" T x 20" W (24.3 LF/CY)	Inst	LF	L-3	485	2.20	---	0.88	3.08	3.51	L-3	437	2.53	---	0.98	3.51	3.98
12" T x 12" W (27.0 LF/CY)	Inst	LF	L-3	535	1.98	---	0.80	2.78	3.17	L-3	482	2.28	---	0.89	3.17	3.59
12" T x 16" W (20.3 LF/CY)	Inst	LF	L-3	460	2.63	---	0.93	3.56	4.01	L-3	414	3.03	---	1.04	4.06	4.56
12" T x 20" W (16.2 LF/CY)	Inst	LF	L-3	420	3.30	---	1.02	4.32	4.81	L-3	378	3.79	---	1.13	4.93	5.47
12" T x 24" W (13.5 LF/CY)	Inst	LF	L-3	485	3.96	---	0.88	4.84	5.27	L-3	437	4.55	---	0.98	5.53	6.01
Forms, wreck, remove, and clean																
2" x 6"	Inst	LF	L-1	175	---	---	0.74	0.74	1.10	L-1	158	---	---	0.83	0.83	1.22
2" x 8"	Inst	LF	L-1	160	---	---	0.81	0.81	1.20	L-1	144	---	---	0.90	0.81	1.34
2" x 12"	Inst	LF	L-1	145	---	---	0.90	0.90	1.33	L-1	131	---	---	1.00	0.90	1.47

Foundations and retaining walls
with air tools, per LF wall

With reinforcing

4'-0" H

Description	Oper	Unit	Crew	Prod	Mat'l	Equip	Labor	Total	O&P	Crew	Prod	Mat'l	Equip	Labor	Total	O&P
8" T	Demo	SF	A-4	80	---	1.09	5.85	6.94	9.69	A-4	68	---	1.29	6.88	8.17	11.40
12" T	Demo	SF	A-4	58	---	1.51	8.07	9.58	13.37	A-4	49	---	1.78	9.49	11.26	15.73

8'-0" H

8" T	Demo	SF	A-4	74	---	1.18	6.32	7.50	10.48	A-4	63	---	1.39	7.44	8.83	12.33
12" T	Demo	SF	A-4	52	---	1.68	9.00	10.68	14.91	A-4	44	---	1.98	10.58	12.56	17.54

12"-0" H

8" T	Demo	SF	A-4	68	---	1.29	6.88	8.17	11.40	A-4	58	---	1.51	8.09	9.61	13.41
12" T	Demo	SF	A-4	44	---	1.99	10.63	12.62	17.62	A-4	37	---	2.34	12.51	14.85	20.73

Without reinforcing

4'-0" H

8" T	Demo	SF	A-4	92	---	0.95	5.09	6.04	8.43	A-4	78	---	1.12	5.98	7.10	9.91
12" T	Demo	SF	A-4	68	---	1.29	6.88	8.17	11.40	A-4	58	---	1.51	8.09	9.61	13.41

8'-0" H

8" T	Demo	SF	A-4	84	---	1.04	5.57	6.61	9.23	A-4	71	---	1.23	6.55	7.78	10.86
12" T	Demo	SF	A-4	60	---	1.46	7.80	9.26	12.92	A-4	51	---	1.72	9.17	10.89	15.20

			Costs Based On Large Volume							Costs Based On Small Volume					

Description	Oper	Unit	Crew Size	Avg Day Prod	Avg Mat'l Unit Cost	Avg Equip Unit Cost	Avg Labor Unit Cost	Avg Total Unit Cost	Avg Price Incl O&P	Crew Size	Avg Day Prod	Avg Mat'l Unit Cost	Avg Equip Unit Cost	Avg Labor Unit Cost	Avg Total Unit Cost	Avg Price Incl O&P

Concrete, Cast-In-Place (continued)

12'-0" H

Description	Oper	Unit	Crew Size	Avg Day Prod	Avg Mat'l Unit Cost	Avg Equip Unit Cost	Avg Labor Unit Cost	Avg Total Unit Cost	Avg Price Incl O&P	Crew Size	Avg Day Prod	Avg Mat'l Unit Cost	Avg Equip Unit Cost	Avg Labor Unit Cost	Avg Total Unit Cost	Avg Price Incl O&P
8" T	Demo	SF	A-4	76	---		1.15	6.16	7.31	**10.20**	A-4	65	---		1.36	7.24	8.60	**12.00**
12" T	Demo	SF	A-4	52	---		1.68	9.00	10.68	**14.91**	A-4	44	---		1.98	10.58	12.56	**17.54**

Forming only. Material price includes panel forms, wales, braces, snap ties, paraffin oil, and nails

8" or 12" T x 4'-0" H

Description	Oper	Unit	Crew Size	Avg Day Prod	Avg Mat'l Unit Cost	Avg Equip Unit Cost	Avg Labor Unit Cost	Avg Total Unit Cost	Avg Price Incl O&P	Crew Size	Avg Day Prod	Avg Mat'l Unit Cost	Avg Equip Unit Cost	Avg Labor Unit Cost	Avg Total Unit Cost	Avg Price Incl O&P
Make (@ 12 uses)	Inst	SF	C-19	685	0.68	---	0.10	0.77	**0.82**	C-19	630	0.78	---	0.11	0.88	**0.94**
Erect and coat	Inst	SF	C-19	410	0.03	---	1.96	1.99	**2.96**	C-19	377	0.04	---	2.13	2.17	**3.21**
Wreck and clean	Inst	SF	L-2	370	---		0.70	0.70	**1.04**	L-2	333	---		0.78	0.78	**1.16**

8" or 12" T x 8'-0" H

Description	Oper	Unit	Crew Size	Avg Day Prod	Avg Mat'l Unit Cost	Avg Equip Unit Cost	Avg Labor Unit Cost	Avg Total Unit Cost	Avg Price Incl O&P	Crew Size	Avg Day Prod	Avg Mat'l Unit Cost	Avg Equip Unit Cost	Avg Labor Unit Cost	Avg Total Unit Cost	Avg Price Incl O&P
Make (@ 12 uses)	Inst	SF	C-19	655	0.62	---	0.10	0.72	**0.77**	C-19	603	0.72	---	0.11	0.83	**0.88**
Erect and coat	Inst	SF	C-19	340	0.03	---	2.37	2.40	**3.56**	C-19	313	0.04	---	2.57	2.61	**3.87**
Wreck and clean	Inst	SF	L-2	305	---		0.85	0.85	**1.26**	L-2	275	---		0.95	0.95	**1.40**

8" or 12" T x 12'-0" H

Description	Oper	Unit	Crew Size	Avg Day Prod	Avg Mat'l Unit Cost	Avg Equip Unit Cost	Avg Labor Unit Cost	Avg Total Unit Cost	Avg Price Incl O&P	Crew Size	Avg Day Prod	Avg Mat'l Unit Cost	Avg Equip Unit Cost	Avg Labor Unit Cost	Avg Total Unit Cost	Avg Price Incl O&P
Make (@ 12 uses)	Inst	SF	C-19	670	0.65	---	0.10	0.76	**0.80**	C-19	616	0.75	---	0.11	0.86	**0.92**
Erect and coat	Inst	SF	C-19	290	0.03	---	2.77	2.81	**4.17**	C-19	267	0.04	---	3.01	3.05	**4.53**
Wreck and clean	Inst	SF	L-2	260	---		1.00	1.00	**1.48**	L-2	234	---		1.11	1.11	**1.64**

Reinforcing steel rods

5% waste included; pricing based on LF of rod

Description	Oper	Unit	Crew Size	Avg Day Prod	Avg Mat'l Unit Cost	Avg Equip Unit Cost	Avg Labor Unit Cost	Avg Total Unit Cost	Avg Price Incl O&P	Crew Size	Avg Day Prod	Avg Mat'l Unit Cost	Avg Equip Unit Cost	Avg Labor Unit Cost	Avg Total Unit Cost	Avg Price Incl O&P
No. 3 (3/8" rod)	Inst	LF	L-2	1390	0.14	---	0.19	0.33	**0.42**	L-2	1251	0.16	---	0.21	0.37	**0.47**
No. 4 (1/2" rod)	Inst	LF	L-2	1330	0.21	---	0.20	0.41	**0.50**	L-2	1197	0.24	---	0.22	0.46	**0.56**
No. 5 (5/8" rod)	Inst	LF	L-2	1260	0.31	---	0.21	0.51	**0.61**	L-2	1134	0.35	---	0.23	0.58	**0.69**
No. 6 (3/4" rod)	Inst	LF	L-2	1130	0.39	---	0.23	0.63	**0.74**	L-2	1017	0.45	---	0.26	0.71	**0.83**
No. 7 (7/8" rod)	Inst	LF	L-2	1000	0.49	---	0.26	0.75	**0.88**	L-2	900	0.57	---	0.29	0.86	**1.00**

Concrete, placed from truck into forms. 5% waste included using 3000 PSI, 1-1/2" aggregate, 5.7 sack mix

Description	Oper	Unit	Crew Size	Avg Day Prod	Avg Mat'l Unit Cost	Avg Equip Unit Cost	Avg Labor Unit Cost	Avg Total Unit Cost	Avg Price Incl O&P	Crew Size	Avg Day Prod	Avg Mat'l Unit Cost	Avg Equip Unit Cost	Avg Labor Unit Cost	Avg Total Unit Cost	Avg Price Incl O&P
8" T x 4'-0" H (10.12 LF/CY)	Inst	LF	L-7	690	5.28	---	1.19	6.47	**7.04**	L-7	621	6.07	---	1.32	7.39	**8.02**
8" T x 4'-0" H (40.5 SF/CY)	Inst	SF	L-7	2760	1.32	---	0.30	1.62	**1.76**	L-7	2484	1.52	---	0.33	1.85	**2.01**
8" T x 8'-0" H (5.06 LF/CY)	Inst	LF	L-7	345	10.56	---	2.37	12.94	**14.07**	L-7	311	12.15	---	2.64	14.78	**16.05**
8" T x 8'-0" H (40.5 SF/CY)	Inst	SF	L-7	2760	1.32	---	0.30	1.62	**1.76**	L-7	2484	1.52	---	0.33	1.85	**2.01**
8" T x 12'-0" H (3.37 LF/CY)	Inst	LF	L-7	230	15.86	---	3.56	19.42	**21.13**	L-7	207	18.24	---	3.96	22.19	**24.09**
8" T x 12'-0" H (40.5 SF/CY)	Inst	SF	L-7	2760	1.32	---	0.30	1.62	**1.76**	L-7	2484	1.52	---	0.33	1.85	**2.01**
12" T x 4'-0" H (6.75 LF/CY)	Inst	LF	L-7	463	7.92	---	1.77	9.69	**10.53**	L-7	417	9.11	---	1.96	11.07	**12.01**
12" T x 4'-0" H (27.0 LF/CY)	Inst	SF	L-7	1850	1.98	---	0.44	2.42	**2.63**	L-7	1665	2.28	---	0.49	2.77	**3.00**
12" T x 8'-0" H (3.38 LF/CY)	Inst	LF	L-7	231	15.81	---	3.54	19.36	**21.06**	L-7	208	18.18	---	3.94	22.12	**24.01**

			Costs Based On Large Volume							Costs Based On Small Volume					
Description	Oper	Unit	Crew Size	Avg Day Prod	Avg Mat'l Unit Cost	Avg Equip Unit Cost	Avg Labor Unit Cost	Avg Total Unit Cost	Avg Price Incl O&P	Crew Size	Avg Day Prod	Avg Mat'l Unit Cost	Avg Equip Unit Cost	Avg Labor Unit Cost	Avg Total Unit Cost	Avg Price Incl O&P

Concrete, Cast-In-Place (continued)

12" T x 8'-0" H (27.0 SF/CY)	Inst	SF	L-7	1850	1.98	---	0.44	2.42	2.63	L-7	1665	2.28	---	0.49	2.77	3.00
12" T x 12'-0" H (2.25 LF/CY)	Inst	LF	L-7	154	23.75	---	5.32	29.07	31.62	L-7	139	27.32	---	5.91	33.22	36.06
12" T x 12'-0" H (27.0 SF/CY)	Inst	SF	L-7	1850	1.98	---	0.44	2.42	2.63	L-7	1665	2.28	---	0.49	2.77	3.00

Slabs and interior floor finishes
Slabs
with air tools
With reinforcing (6 x 6/10 x 10)

4" T	Demo	SF	A-4	380	---	0.17	1.23	1.40	1.98	A-4	323	---	0.19	1.45	1.64	2.32
5" T	Demo	SF	A-4	330	---	0.19	1.42	1.61	2.27	A-4	281	---	0.22	1.67	1.89	2.68
6" T	Demo	SF	A-4	290	---	0.22	1.61	1.83	2.59	A-4	247	---	0.25	1.90	2.15	3.04

Without reinforcing

4" T	Demo	SF	A-4	550	---	0.11	0.85	0.96	1.36	A-4	468	---	0.13	1.00	1.14	1.61
5" T	Demo	SF	A-4	475	---	0.13	0.98	1.12	1.58	A-4	404	---	0.16	1.16	1.31	1.86
6" T	Demo	SF	A-4	420	---	0.15	1.11	1.26	1.79	A-4	357	---	0.18	1.31	1.49	2.10

Forming; 4 uses

2" x 4" (4" T slab)	Inst	LF	D-4	580	0.14	---	0.78	0.93	1.29	D-4	534	0.17	---	0.85	1.01	1.41
2" x 6" (5" T and 6" T slab)	Inst	LF	D-4	580	0.18	---	0.78	0.96	1.33	D-4	534	0.21	---	0.85	1.06	1.46

Grading

Dirt, cut and fill, +/- 1/10 ft.	Inst	SF	L-1	800	---	---	0.16	0.16	0.24	L-1	720	---	---	0.18	0.18	0.27
Gravel, 3/4" to 1-1/2" stone	Inst	SF	L-1	750	0.26	---	0.17	0.43	0.51	L-1	675	0.29	---	0.19	0.49	0.58

Screeds; 3 uses

2" x 2"	Inst	LF	D-4	835	0.05	---	0.54	0.59	0.85	D-4	768	0.06	---	0.59	0.65	0.93
2" x 4"	Inst	LF	D-4	760	0.10	---	0.60	0.69	0.97	D-4	699	0.11	---	0.65	0.76	1.06

Reinforcing steel rods. 5% waste included. Pricing based on LF of rod or SF of mesh

No. 3 (3/8" rod)	Inst	LF	L-2	1480	0.14	---	0.18	0.31	0.40	L-2	1332	0.16	---	0.20	0.35	0.45
No. 4 (1/2" rod)	Inst	LF	L-2	1410	0.21	---	0.18	0.40	0.48	L-2	1269	0.24	---	0.20	0.45	0.55
No. 5 (5/8" rod)	Inst	LF	L-2	1340	0.31	---	0.19	0.50	0.59	L-2	1206	0.35	---	0.22	0.57	0.67
6 x 6/10-10 @21 lbs/CSF	Inst	SF	L-2	5250	0.11	---	0.05	0.16	0.18	L-2	4725	0.13	---	0.06	0.18	0.21
6 x 6/6-6 @42 lbs/CSF	Inst	SF	L-2	4500	0.17	---	0.06	0.22	0.25	L-2	4050	0.19	---	0.06	0.25	0.28

Concrete, pour and finish (steel trowel), 5% waste included. Using 2500 PSI, 1" aggregate, 5.5 sack mix

3-1/2" T (92.57 SF/CY)	Inst	SF	D-6	2805	0.58	---	0.40	0.98	1.17	D-6	2581	0.66	---	0.44	1.10	1.31
4" T (81.00 SF/CY)	Inst	SF	D-6	2720	0.66	---	0.42	1.08	1.27	D-6	2502	0.76	---	0.45	1.21	1.43
5" T (64.80 SF/CY)	Inst	SF	D-6	2590	0.82	---	0.44	1.26	1.47	D-6	2383	0.95	---	0.48	1.42	1.65
6" T (54.00 SF/CY)	Inst	SF	D-6	2460	0.99	---	0.46	1.45	1.67	D-6	2263	1.14	---	0.50	1.64	1.88

| | | | Costs Based On Large Volume | | | | | | | Costs Based On Small Volume | | | | | | |
Description	Oper	Unit	Crew Size	Avg Day Prod	Avg Mat'l Unit Cost	Avg Equip Unit Cost	Avg Labor Unit Cost	Avg Total Unit Cost	Avg Price Incl O&P	Crew Size	Avg Day Prod	Avg Mat'l Unit Cost	Avg Equip Unit Cost	Avg Labor Unit Cost	Avg Total Unit Cost	Avg Price Incl O&P

Concrete, Cast-In-Place (continued)

Forms, wreck and clean

Description	Oper	Unit	Crew Size	Avg Day Prod	Avg Mat'l Unit Cost	Avg Equip Unit Cost	Avg Labor Unit Cost	Avg Total Unit Cost	Avg Price Incl O&P	Crew Size	Avg Day Prod	Avg Mat'l Unit Cost	Avg Equip Unit Cost	Avg Labor Unit Cost	Avg Total Unit Cost	Avg Price Incl O&P
2" x 4"	Inst	LF	L-1	360	---	---	0.36	0.36	0.53	L-1	324	---	---	0.40	0.40	0.59
2" x 6"	Inst	LF	L-1	340	---	---	0.38	0.38	0.57	L-1	306	---	---	0.42	0.42	0.63

Ready mix, prices typical to most cities. Material cost only. Ten mile haul with more than 7.25 CY

Using 1-1/2" aggregate

Description	Oper	Unit	Crew Size	Avg Day Prod	Avg Mat'l Unit Cost	Avg Equip Unit Cost	Avg Labor Unit Cost	Avg Total Unit Cost	Avg Price Incl O&P	Crew Size	Avg Day Prod	Avg Mat'l Unit Cost	Avg Equip Unit Cost	Avg Labor Unit Cost	Avg Total Unit Cost	Avg Price Incl O&P
2000 PSI, 4.8 sack mix	Inst	CY	---	---	47.54	---	---	---	—	---	---	54.67	---	---	---	—
2500 PSI, 5.2 sack mix	Inst	CY	---	---	49.03	---	---	---	—	---	---	56.39	---	---	---	—
3000 PSI, 5.7 sack mix	Inst	CY	---	---	50.90	---	---	---	—	---	---	58.54	---	---	---	—
3500 PSI, 6.3 sack mix	Inst	CY	---	---	52.12	---	---	---	—	---	---	59.93	---	---	---	—
4000 PSI, 6.9 sack mix	Inst	CY	---	---	55.33	---	---	---	—	---	---	63.63	---	---	---	—
Using 1" aggregate																
2000 PSI, 5.0 sack mix	Inst	CY	---	---	48.28	---	---	---	—	---	---	55.53	---	---	---	—
2500 PSI, 5.5 sack mix	Inst	CY	---	---	50.15	---	---	---	—	---	---	57.67	---	---	---	—
3000 PSI, 6.0 sack mix	Inst	CY	---	---	52.01	---	---	---	—	---	---	59.82	---	---	---	—
3500 PSI, 6.6 sack mix	Inst	CY	---	---	54.25	---	---	---	—	---	---	62.39	---	---	---	—
4000 PSI, 7.1 sack mix	Inst	CY	---	---	56.12	---	---	---	—	---	---	64.54	---	---	---	—
Using 3/8" aggregate																
2000 PSI, 6.0 sack mix	Inst	CY	---	---	52.01	---	---	---	—	---	---	59.82	---	---	---	—
2500 PSI, 6.5 sack mix	Inst	CY	---	---	53.88	---	---	---	—	---	---	61.97	---	---	---	—
3000 PSI, 7.2 sack mix	Inst	CY	---	---	56.49	---	---	---	—	---	---	64.96	---	---	---	—
3500 PSI, 7.9 sack mix	Inst	CY	---	---	59.11	---	---	---	—	---	---	67.97	---	---	---	—
4000 PSI, 8.5 sack mix	Inst	CY	---	---	62.37	---	---	---	—	---	---	71.72	---	---	---	—
Adjustments																
Delivery over 10 miles, ADD	Inst	Mi	---	---	0.59	---	---	---	—	---	---	0.68	---	---	---	—
High early strength, ADD																
5.0 sack mix	Inst	CY	---	---	7.63	---	---	---	—	---	---	8.78	---	---	---	—
6.0 sack mix	Inst	CY	---	---	9.55	---	---	---	—	---	---	10.98	---	---	---	—
Lightweight aggregate, ADD																
Mix from truck to forms	Inst	CY	---	---	39.60	---	---	---	—	---	---	45.54	---	---	---	—
Pump mix	Inst	CY	---	---	41.80	---	---	---	—	---	---	48.07	---	---	---	—
For 1% Calcium chloride	Inst	SA	---	---	0.28	---	---	---	—	---	---	0.32	---	---	---	—
For 2% Calcium chloride	Inst	SA	---	---	0.39	---	---	---	—	---	---	0.44	---	---	---	—
Chemical compensated shrinkage (WRDA Admix), ADD	Inst	SA	---	---	0.22	---	---	---	—	---	---	0.25	---	---	---	—

			Costs Based On Large Volume							Costs Based On Small Volume					
Description	Oper	Unit	Crew Size	Avg Day Prod	Avg Mat'l Unit Cost	Avg Equip Unit Cost	Avg Labor Unit Cost	Avg Total Unit Cost	Avg Price Incl O&P	Crew Size	Avg Day Prod	Avg Mat'l Unit Cost	Avg Equip Unit Cost	Avg Labor Unit Cost	Avg Total Unit Cost	Avg Price Incl O&P

Concrete, Cast-In-Place (continued)

Loads 7.25 CY or less, ADD

Description	Oper	Unit	Crew Size	Avg Day Prod	Avg Mat'l Unit Cost	Avg Equip Unit Cost	Avg Labor Unit Cost	Avg Total Unit Cost	Avg Price Incl O&P	Crew Size	Avg Day Prod	Avg Mat'l Unit Cost	Avg Equip Unit Cost	Avg Labor Unit Cost	Avg Total Unit Cost	Avg Price Incl O&P
7.24 CY	Inst	LS	---	---	4.40	---	---	---	—	---	---	5.06	---	---	---	—
6.0 CY	Inst	LS	---	---	15.40	---	---	---	—	---	---	17.71	---	---	---	—
5.0 CY	Inst	LS	---	---	24.20	---	---	---	—	---	---	27.83	---	---	---	—
4.0 CY	Inst	LS	---	---	33.00	---	---	---	—	---	---	37.95	---	---	---	—
3.0 CY	Inst	LS	---	---	41.80	---	---	---	—	---	---	48.07	---	---	---	—
2.0 CY	Inst	LS	---	---	50.60	---	---	---	—	---	---	58.19	---	---	---	—
1.0 CY	Inst	LS	---	---	59.40	---	---	---	—	---	---	68.31	---	---	---	—
Coloring of concrete, ADD																
Light or sand colors	Inst	CY	---	---	17.33	---	---	---	—	---	---	19.92	---	---	---	—
Medium or buff colors	Inst	CY	---	---	25.58	---	---	---	—	---	---	29.41	---	---	---	—
Dark colors	Inst	CY	---	---	38.50	---	---	---	—	---	---	44.28	---	---	---	—
Green	Inst	CY	---	---	42.90	---	---	---	—	---	---	49.34	---	---	---	—
Standby charge for time in excess of																
5 Min/CY, ADD	Inst	Min	---	---	1.10	---	---	---	—	---	---	1.27	---	---	---	—

Concrete block. See Masonry.

Countertops

Formica, one piece tops; straight, "L", or "U" shape; surfaced with laminated plastic cemented to particleboard base

Post formed countertop with raised front drip edge
25" W, 1-1/2" H front edge, 4" H
coved backsplash

Description	Oper	Unit	Crew Size	Avg Day Prod	Avg Mat'l Unit Cost	Avg Equip Unit Cost	Avg Labor Unit Cost	Avg Total Unit Cost	Avg Price Incl O&P	Crew Size	Avg Day Prod	Avg Mat'l Unit Cost	Avg Equip Unit Cost	Avg Labor Unit Cost	Avg Total Unit Cost	Avg Price Incl O&P
Satin/suede patterns	Inst	LF	C-18	33	7.31	---	10.22	17.53	**22.53**	C-18	21	8.48	---	15.72	24.20	**31.90**
Solid patterns	Inst	LF	C-18	33	8.00	---	10.22	18.22	**23.22**	C-18	21	9.28	---	15.72	25.00	**32.70**
Specialty finish patterns	Inst	LF	C-18	33	9.50	---	10.22	19.72	**24.72**	C-18	21	11.02	---	15.72	26.74	**34.44**
Wood tone patterns	Inst	LF	C-18	33	10.00	---	10.22	20.22	**25.22**	C-18	21	11.60	---	15.72	27.32	**35.02**
Double roll top, 1-1/2" H front and back edges, no backsplash																
26" W																
Satin/suede patterns	Inst	LF	C-18	35	7.31	---	9.63	16.94	**21.66**	C-18	23	8.48	---	14.82	23.30	**30.56**
Solid patterns	Inst	LF	C-18	35	8.00	---	9.63	17.63	**22.35**	C-18	23	9.28	---	14.82	24.10	**31.36**
Specialty finish patterns	Inst	LF	C-18	35	9.50	---	9.63	19.13	**23.85**	C-18	23	11.02	---	14.82	25.84	**33.10**
Wood tone patterns	Inst	LF	C-18	35	10.00	---	9.63	19.63	**24.35**	C-18	23	11.60	---	14.82	26.42	**33.68**

			Costs Based On Large Volume							Costs Based On Small Volume						

Description	Oper	Unit	Crew Size	Avg Day Prod	Avg Mat'l Unit Cost	Avg Equip Unit Cost	Avg Labor Unit Cost	Avg Total Unit Cost	Avg Price Incl O&P	Crew Size	Avg Day Prod	Avg Mat'l Unit Cost	Avg Equip Unit Cost	Avg Labor Unit Cost	Avg Total Unit Cost	Avg Price Incl O&P

Countertops (continued)

36" W

Description	Oper	Unit	Crew Size	Avg Day Prod	Avg Mat'l Unit Cost	Avg Equip Unit Cost	Avg Labor Unit Cost	Avg Total Unit Cost	Avg Price Incl O&P	Crew Size	Avg Day Prod	Avg Mat'l Unit Cost	Avg Equip Unit Cost	Avg Labor Unit Cost	Avg Total Unit Cost	Avg Price Incl O&P
Satin/suede patterns	Inst	LF	C-18	32	11.88	---	10.53	22.41	27.57	C-18	21	13.78	---	16.21	29.98	37.92
Solid patterns	Inst	LF	C-18	32	12.56	---	10.53	23.10	28.26	C-18	21	14.57	---	16.21	30.78	38.72
Specialty finish patterns	Inst	LF	C-18	32	14.06	---	10.53	24.60	29.76	C-18	21	16.31	---	16.21	32.52	40.46
Wood tone patterns	Inst	LF	C-18	32	14.75	---	10.53	25.28	30.45	C-18	21	17.11	---	16.21	33.32	41.26

Post formed countertop with square edge veneer front

25" W, 1-1/2" H front edge, 4" H coved backsplash

Description	Oper	Unit	Crew Size	Avg Day Prod	Avg Mat'l Unit Cost	Avg Equip Unit Cost	Avg Labor Unit Cost	Avg Total Unit Cost	Avg Price Incl O&P	Crew Size	Avg Day Prod	Avg Mat'l Unit Cost	Avg Equip Unit Cost	Avg Labor Unit Cost	Avg Total Unit Cost	Avg Price Incl O&P
Satin/suede patterns	Inst	LF	C-18	33	10.69	---	10.22	20.90	25.91	C-18	21	12.40	---	15.72	28.11	35.81
Solid patterns	Inst	LF	C-18	33	11.38	---	10.22	21.59	26.60	C-18	21	13.20	---	15.72	28.91	36.61
Specialty finish patterns	Inst	LF	C-18	33	12.88	---	10.22	23.09	28.10	C-18	21	14.94	---	15.72	30.65	38.35
Wood tone patterns	Inst	LF	C-18	33	13.50	---	10.22	23.72	28.72	C-18	21	15.66	---	15.72	31.38	39.08

Self edge countertop with square edge veneer front

25" W, 1-1/2" H front edge, 4" H coved backsplash (@90 degree angle to deck)

Description	Oper	Unit	Crew Size	Avg Day Prod	Avg Mat'l Unit Cost	Avg Equip Unit Cost	Avg Labor Unit Cost	Avg Total Unit Cost	Avg Price Incl O&P	Crew Size	Avg Day Prod	Avg Mat'l Unit Cost	Avg Equip Unit Cost	Avg Labor Unit Cost	Avg Total Unit Cost	Avg Price Incl O&P
Satin/suede patterns	Inst	LF	C-18	33	34.44	---	10.22	44.65	49.66	C-18	21	39.95	---	15.72	55.66	63.36
Solid patterns	Inst	LF	C-18	33	35.00	---	10.22	45.22	50.22	C-18	21	40.60	---	15.72	56.32	64.02
Specialty finish patterns	Inst	LF	C-18	33	36.13	---	10.22	46.34	51.35	C-18	21	41.91	---	15.72	57.62	65.32
Wood tone patterns	Inst	LF	C-18	33	37.94	---	10.22	48.15	53.16	C-18	21	44.01	---	15.72	59.72	67.42

Material cost adjustments, ADD

Diagonal corner cut

Description	Oper	Unit	Crew Size	Avg Day Prod	Avg Mat'l Unit Cost	Avg Equip Unit Cost	Avg Labor Unit Cost	Avg Total Unit Cost	Avg Price Incl O&P	Crew Size	Avg Day Prod	Avg Mat'l Unit Cost	Avg Equip Unit Cost	Avg Labor Unit Cost	Avg Total Unit Cost	Avg Price Incl O&P
Standard	Inst	EA	---	---	29.56	---	---	29.56	29.56	---	---	34.29	---	---	34.29	34.29
With plateau shelf	Inst	EA	---	---	130.56	---	---	130.56	130.56	---	---	151.45	---	---	151.45	151.45
Radius corner cut																
3" or 6"	Inst	EA	---	---	17.25	---	---	17.25	17.25	---	---	20.01	---	---	20.01	20.01
12"	Inst	EA	---	---	17.25	---	---	17.25	17.25	---	---	20.01	---	---	20.01	20.01
Quarter radius end	Inst	EA	---	---	17.25	---	---	17.25	17.25	---	---	20.01	---	---	20.01	20.01
Half radius end	Inst	EA	---	---	33.50	---	---	33.50	33.50	---	---	38.86	---	---	38.86	38.86
Miter corner, shop assembled,	Inst	EA	---	---	27.56	---	---	27.56	27.56	---	---	31.97	---	---	31.97	31.97
Splicing any top or leg 12' or longer, shop assembled	Inst	EA	---	---	31.50	---	---	31.50	31.50	---	---	36.54	---	---	36.54	36.54

			Costs Based On Large Volume							Costs Based On Small Volume					
Description	Oper	Unit	Crew Size	Avg Day Prod	Avg Mat'l Unit Cost	Avg Equip Unit Cost	Avg Labor Unit Cost	Avg Total Unit Cost	Avg Price Incl O&P	Crew Size	Avg Day Prod	Avg Mat'l Unit Cost	Avg Equip Unit Cost	Avg Labor Unit Cost	Avg Total Unit Cost	Avg Price Incl O&P

Countertops (continued)

Endsplash with finished sides

and edges	Inst	EA	---	---	24.63	---	---	24.63	24.63	---	---	28.57	---	---	28.57	28.57
Sink or range cutout	Inst	EA	---	---	6.88	---	---	6.88	6.88	---	---	7.98	---	---	7.98	7.98

Ceramic tile

Countertop/backsplash, ceramic
Adhesive set with backmounted tile

1" x 1"	Inst	SF	T-2	70	2.79	---	4.22	7.01	8.95	T-2	46	3.28	---	6.49	9.77	12.75
4-1/4" x 4-1/4"	Inst	SF	T-2	80	2.59	---	3.69	6.28	7.98	T-2	52	3.04	---	5.68	8.71	11.33

Cove/base, ceramic
Adhesive set with unmounted tile

4-1/4" x 4-1/4"	Inst	LF	T-2	80	3.14	---	3.69	6.83	8.53	T-2	52	3.69	---	5.68	9.36	11.97
6" x 4-1/4"	Inst	LF	T-2	95	2.65	---	3.11	5.75	7.18	T-2	62	3.11	---	4.78	7.89	10.09

Conventional/mortar set with unmounted tile

4-1/4" x 4-1/4"	Inst	LF	T-2	45	2.97	---	6.56	9.53	12.55	T-2	29	3.48	---	10.09	13.58	18.22
6" x 4-1/4"	Inst	LF	T-2	50	2.59	---	5.90	8.49	11.21	T-2	33	3.04	---	9.08	12.12	16.30

Dry-set mortar with unmounted tile

4-1/4" x 4-1/4"	Inst	LF	T-2	65	2.99	---	4.54	7.53	9.62	T-2	42	3.51	---	6.99	10.50	13.71
6" x 4-1/4"	Inst	LF	T-2	75	2.60	---	3.94	6.54	8.35	T-2	49	3.05	---	6.06	9.11	11.89

Wood, butcher block construction throughout top; custom; straight, "L", or "U" shapes

Self-edge top; 26" W, with 4" backsplash,

1-1/2" front and back edges	Inst	LF	C-18	33	50.00	---	10.22	60.22	65.22	C-18	21	58.00	---	15.72	73.72	81.42
Miter corner	Inst	EA	---	---	15.00	---	---	15.00	15.00	---	---	17.00	---	---	17.00	17.00
Sink or surface saver cutout	Inst	EA	---	---	45.00	---	---	45.00	45.00	---	---	51.00	---	---	51.00	51.00
45 degree plateau corner	Inst	EA	---	---	225.00	---	---	225.00	225.00	---	---	255.00	---	---	255.00	255.00
End splash	Inst	EA	---	---	52.50	---	---	52.50	52.50	---	---	59.50	---	---	59.50	59.50

Cupolas

Natural finish redwood, aluminum roof

24" x 24", 25" H	Inst	EA	C-1	8.0	162.50	---	21.07	183.57	193.89	C-1	6.0	195.00	---	28.09	223.09	236.86
30" x 30", 30" H	Inst	EA	C-1	8.0	212.50	---	21.07	233.57	243.89	C-1	6.0	255.00	---	28.09	283.09	296.86
35" x 35", 33" H	Inst	EA	C-1	5.4	306.25	---	31.21	337.46	352.76	C-1	4.1	367.50	---	41.62	409.12	429.51

Description	Oper	Unit	Costs Based On Large Volume							Costs Based On Small Volume						
			Crew Size	Avg Day Prod	Avg Mat'l Unit Cost	Avg Equip Unit Cost	Avg Labor Unit Cost	Avg Total Unit Cost	Avg Price Incl O&P	Crew Size	Avg Day Prod	Avg Mat'l Unit Cost	Avg Equip Unit Cost	Avg Labor Unit Cost	Avg Total Unit Cost	Avg Price Incl O&P

Cupolas (continued)

Natural finish redwood, copper roof

Description	Oper	Unit	Crew Size	Avg Day Prod	Avg Mat'l Unit Cost	Avg Equip Unit Cost	Avg Labor Unit Cost	Avg Total Unit Cost	Avg Price Incl O&P	Crew Size	Avg Day Prod	Avg Mat'l Unit Cost	Avg Equip Unit Cost	Avg Labor Unit Cost	Avg Total Unit Cost	Avg Price Incl O&P
22" x 22", 33" H	Inst	EA	C-1	8.0	200.00	---	21.07	221.07	231.39	C-1	6.0	240.00	---	28.09	268.09	281.86
25" x 25", 39" H	Inst	EA	C-1	8.0	256.25	---	21.07	277.32	287.64	C-1	6.0	307.50	---	28.09	335.59	349.36
35" x 35", 33" H	Inst	EA	C-1	5.4	375.00	---	31.21	406.21	421.51	C-1	4.1	450.00	---	41.62	491.62	512.01

Natural finish redwood, octagonal shape copper roof

Description	Oper	Unit	Crew Size	Avg Day Prod	Avg Mat'l Unit Cost	Avg Equip Unit Cost	Avg Labor Unit Cost	Avg Total Unit Cost	Avg Price Incl O&P	Crew Size	Avg Day Prod	Avg Mat'l Unit Cost	Avg Equip Unit Cost	Avg Labor Unit Cost	Avg Total Unit Cost	Avg Price Incl O&P
31" W, 37" H	Inst	EA	C-1	5.4	443.75	---	31.21	474.96	490.25	C-1	4.1	532.50	---	41.62	574.12	594.51
35" W, 43" H	Inst	EA	C-1	4.0	600.00	---	42.14	642.14	662.78	C-1	3.0	720.00	---	56.18	776.18	803.71

Weathervanes for above, aluminum

Description	Oper	Unit	Crew Size	Avg Day Prod	Avg Mat'l Unit Cost	Avg Equip Unit Cost	Avg Labor Unit Cost	Avg Total Unit Cost	Avg Price Incl O&P	Crew Size	Avg Day Prod	Avg Mat'l Unit Cost	Avg Equip Unit Cost	Avg Labor Unit Cost	Avg Total Unit Cost	Avg Price Incl O&P
18" H, black finish	Inst	EA	---	---	43.75	---	---	43.75	43.75	---	---	52.50	---	---	52.50	52.50
24" H, black finish	Inst	EA	---	---	56.25	---	---	56.25	56.25	---	---	67.50	---	---	67.50	67.50
36" H, black and gold finish	Inst	EA	---	---	87.50	---	---	87.50	87.50	---	---	105.00	---	---	105.00	105.00

Demolition. See Site Work.

Doors

Closet door systems

Labor cost is for hanging and fitting doors and hardware

Bi-folding unit, hardware and pine fascia trim included

Unfinished

Birch, flush face, 1-3/8" T hollow core

Description	Oper	Unit	Crew Size	Avg Day Prod	Avg Mat'l Unit Cost	Avg Equip Unit Cost	Avg Labor Unit Cost	Avg Total Unit Cost	Avg Price Incl O&P	Crew Size	Avg Day Prod	Avg Mat'l Unit Cost	Avg Equip Unit Cost	Avg Labor Unit Cost	Avg Total Unit Cost	Avg Price Incl O&P
2'-0" x 6'-8", 2 doors	Inst	Set	C-18	10.0	68.90	---	33.71	102.61	119.13	C-18	7.5	79.50	---	44.95	124.45	146.47
4'-0" x 6'-8", 4 doors	Inst	Set	C-18	8.0	126.10	---	42.14	168.24	188.88	C-18	6.0	145.50	---	56.18	201.68	229.21
6'-0" x 6'-8", 4 doors	Inst	Set	C-18	8.0	157.30	---	42.14	199.44	220.08	C-18	6.0	181.50	---	56.18	237.68	265.21
8'-0" x 6'-8", 4 doors	Inst	Set	C-18	8.0	210.60	---	42.14	252.74	273.38	C-18	6.0	243.00	---	56.18	299.18	326.71
4'-0" x 8'-0", 4 doors	Inst	Set	C-18	8.0	188.50	---	42.14	230.64	251.28	C-18	6.0	217.50	---	56.18	273.68	301.21
6'-0" x 8'-0", 4 doors	Inst	Set	C-18	8.0	235.30	---	42.14	277.44	298.08	C-18	6.0	271.50	---	56.18	327.68	355.21
8'-0" x 8'-0", 4 doors	Inst	Set	C-18	8.0	300.30	---	42.14	342.44	363.08	C-18	6.0	346.50	---	56.18	402.68	430.21
Lauan, flush face, 1-3/8" T thick hollow core																
2'-0" x 6'-8", 2 doors	Inst	Set	C-18	10.0	59.80	---	33.71	93.51	110.03	C-18	7.5	69.00	---	44.95	113.95	135.97
4'-0" x 6'-8", 4 doors	Inst	Set	C-18	8.0	107.90	---	42.14	150.04	170.68	C-18	6.0	124.50	---	56.18	180.68	208.21
6'-0" x 6'-8", 4 doors	Inst	Set	C-18	8.0	132.60	---	42.14	174.74	195.38	C-18	6.0	153.00	---	56.18	209.18	236.71

Doors

Closet Door Systems

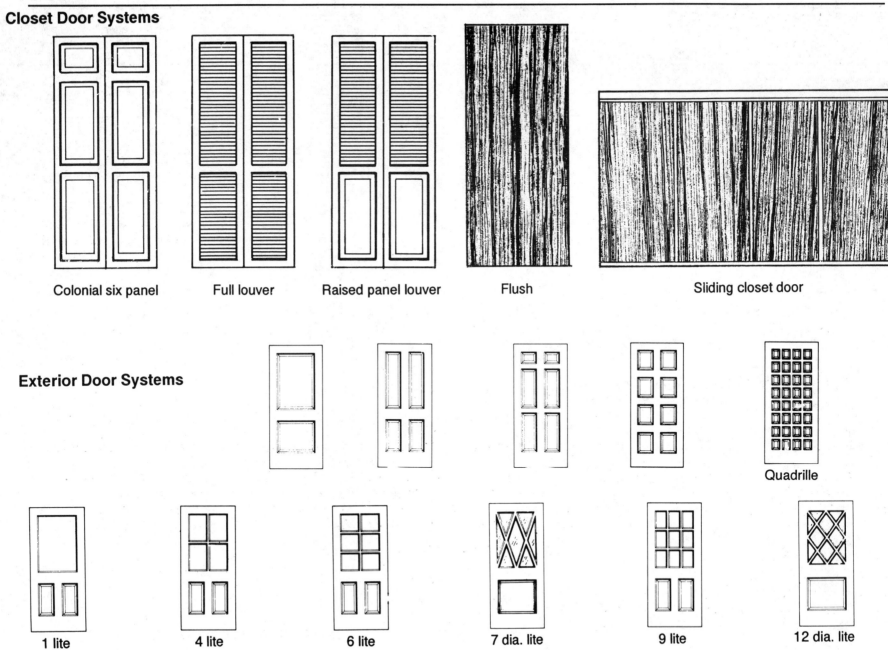

Colonial six panel   Full louver   Raised panel louver   Flush   Sliding closet door

Exterior Door Systems

Quadrille

1 lite   4 lite   6 lite   7 dia. lite   9 lite   12 dia. lite

Interior Door Systems

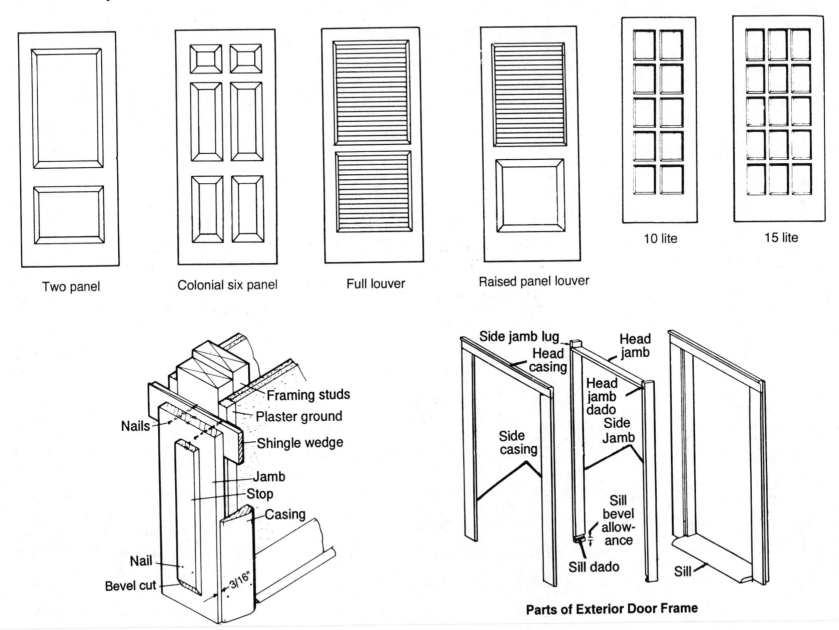

Two panel

Colonial six panel

Full louver

Raised panel louver

10 lite

15 lite

Framing studs

Plaster ground

Shingle wedge

Nails

Jamb

Stop

Casing

Nail

Bevel cut

3/16"

Side jamb lug

Head casing

Head jamb

Head jamb dado

Side Jamb

Side casing

Sill bevel allowance

Sill dado

Sill

Parts of Exterior Door Frame

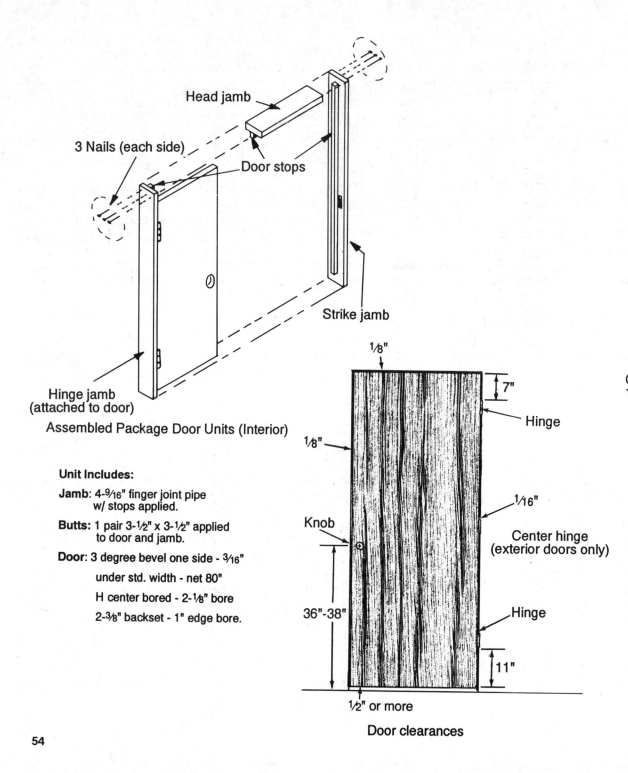

3 Nails (each side)

Head jamb

Door stops

Strike jamb

Hinge jamb
(attached to door)

Assembled Package Door Units (Interior)

Unit Includes:

Jamb: 4-9/16" finger joint pipe
w/ stops applied.

Butts: 1 pair 3-1/2" x 3-1/2" applied
to door and jamb.

Door: 3 degree bevel one side - 3/16"

under std. width - net 80"

H center bored - 2-1/8" bore

2-3/8" backset - 1" edge bore.

1/8"

1/8"

Knob

36"-38"

1/2" or more

7"

Hinge

1/16"

Center hinge
(exterior doors only)

Hinge

11"

Door clearances

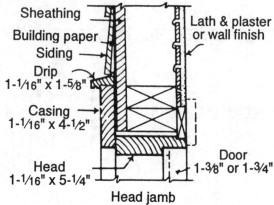

Sheathing

Building paper

Siding

Drip
1-1/16" x 1-5/8"

Casing
1-1/16" x 4-1/2"

Head
1-1/16" x 5-1/4"

Lath & plaster
or wall finish

Door
1-3/8" or 1-3/4"

Head jamb

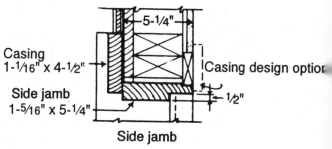

Casing
1-1/16" x 4-1/2"

Side jamb
1-5/16" x 5-1/4"

5-1/4"

Casing design option

1/2"

Side jamb

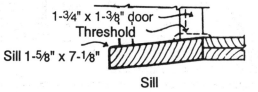

1-3/4" x 1-3/8" door

Threshold

Sill 1-5/8" x 7-1/8"

Sill

			Costs Based On Large Volume							Costs Based On Small Volume						
Description	Oper	Unit	Crew Size	Avg Day Prod	Avg Mat'l Unit Cost	Avg Equip Unit Cost	Avg Labor Unit Cost	Avg Total Unit Cost	Avg Price Incl O&P	Crew Size	Avg Day Prod	Avg Mat'l Unit Cost	Avg Equip Unit Cost	Avg Labor Unit Cost	Avg Total Unit Cost	Avg Price Incl O&P

Doors (continued)

Description	Oper	Unit	Crew Size	Avg Day Prod	Avg Mat'l Unit Cost	Avg Equip Unit Cost	Avg Labor Unit Cost	Avg Total Unit Cost	Avg Price Incl O&P	Crew Size	Avg Day Prod	Avg Mat'l Unit Cost	Avg Equip Unit Cost	Avg Labor Unit Cost	Avg Total Unit Cost	Avg Price Incl O&P
8'-0" x 6'-8", 4 doors	Inst	Set	C-18	8.0	178.10	---	42.14	220.24	240.88	C-18	6.0	205.50	---	56.18	261.68	289.21
4'-0" x 8'-0", 4 doors	Inst	Set	C-18	8.0	159.90	---	42.14	202.04	222.68	C-18	6.0	184.50	---	56.18	240.68	268.21
6'-0" x 8'-0", 4 doors	Inst	Set	C-18	8.0	197.60	---	42.14	239.74	260.38	C-18	6.0	228.00	---	56.18	284.18	311.71
8'-0" x 8'-0", 4 doors	Inst	Set	C-18	8.0	253.50	---	42.14	295.64	316.28	C-18	6.0	292.50	---	56.18	348.68	376.21
Ponderosa pine, colonial raised panel (2/door), 1-3/8" T solid core																
2'-0" x 6'-8", 2 doors	Inst	Set	C-18	10.0	154.70	---	33.71	188.41	204.93	C-18	7.5	178.50	---	44.95	223.45	245.47
4'-0" x 6'-8", 4 doors	Inst	Set	C-18	8.0	306.80	---	42.14	348.94	369.58	C-18	6.0	354.00	---	56.18	410.18	437.71
6'-0" x 6'-8", 4 doors	Inst	Set	C-18	8.0	378.30	---	42.14	420.44	441.08	C-18	6.0	436.50	---	56.18	492.68	520.21
8'-0" x 6'-8", 4 doors	Inst	Set	C-18	8.0	417.30	---	42.14	459.44	480.08	C-18	6.0	481.50	---	56.18	537.68	565.21
Ponderosa pine, raised panel louver, 1-3/8" T solid core																
2'-0" x 6'-8", 2 doors	Inst	Set	C-18	10.0	137.80	---	33.71	171.51	188.03	C-18	7.5	159.00	---	44.95	203.95	225.97
4'-0" x 6'-8", 4 doors	Inst	Set	C-18	8.0	253.50	---	42.14	295.64	316.28	C-18	6.0	292.50	---	56.18	348.68	376.21
6'-0" x 6'-8", 4 doors	Inst	Set	C-18	8.0	295.10	---	42.14	337.24	357.88	C-18	6.0	340.50	---	56.18	396.68	424.21
Hardboard, flush face, 1-3/8" T hollow core																
2'-0" x 6'-8", 2 doors	Inst	Set	C-18	10.0	55.90	---	33.71	89.61	106.13	C-18	7.5	64.50	---	44.95	109.45	131.47
4'-0" x 6'-8", 4 doors	Inst	Set	C-18	8.0	97.50	---	42.14	139.64	160.28	C-18	6.0	112.50	---	56.18	168.68	196.21
6'-0" x 6'-8", 4 doors	Inst	Set	C-18	8.0	113.10	---	42.14	155.24	175.88	C-18	6.0	130.50	---	56.18	186.68	214.21
8'-0" x 6'-8", 4 doors	Inst	Set	C-18	8.0	161.20	---	42.14	203.34	223.98	C-18	6.0	186.00	---	56.18	242.18	269.71
4'-0" x 8'-0", 4 doors	Inst	Set	C-18	8.0	137.80	---	42.14	179.94	200.58	C-18	6.0	159.00	---	56.18	215.18	242.71
6'-0" x 8'-0", 4 doors	Inst	Set	C-18	8.0	161.20	---	42.14	203.34	223.98	C-18	6.0	186.00	---	56.18	242.18	269.71
8'-0" x 8'-0", 4 doors	Inst	Set	C-18	8.0	226.20	---	42.14	268.34	288.98	C-18	6.0	261.00	---	56.18	317.18	344.71
Sen (ash), flush face, 1-3/8" T hollow core																
2'-0" x 6'-8", 2 doors	Inst	Set	C-18	10.0	70.20	---	33.71	103.91	120.43	C-18	7.5	81.00	---	44.95	125.95	147.97
4'-0" x 6'-8", 4 doors	Inst	Set	C-18	8.0	128.70	---	42.14	170.84	191.48	C-18	6.0	148.50	---	56.18	204.68	232.21
6'-0" x 6'-8", 4 doors	Inst	Set	C-18	8.0	161.20	---	42.14	203.34	223.98	C-18	6.0	186.00	---	56.18	242.18	269.71
8'-0" x 6'-8", 4 doors	Inst	Set	C-18	8.0	224.90	---	42.14	267.04	287.68	C-18	6.0	259.50	---	56.18	315.68	343.21
4'-0" x 8'-0", 4 doors	Inst	Set	C-18	8.0	189.80	---	42.14	231.94	252.58	C-18	6.0	219.00	---	56.18	275.18	302.71
6'-0" x 8'-0", 4 doors	Inst	Set	C-18	8.0	237.90	---	42.14	280.04	300.68	C-18	6.0	274.50	---	56.18	330.68	358.21
8'-0" x 8'-0", 4 doors	Inst	Set	C-18	8.0	301.60	---	42.14	343.74	364.38	C-18	6.0	348.00	---	56.18	404.18	431.71
Prefinished																
Walnut tone, mar resistant finish																
Embossed (distressed wood appearance) lauan, flush face, 1-3/8" T hollow core																
2'-0" x 6'-8", 2 doors	Inst	Set	C-18	10.0	66.30	---	33.71	100.01	116.53	C-18	7.5	76.50	---	44.95	121.45	143.47

Description	Oper	Unit	Costs Based On Large Volume							Costs Based On Small Volume						
			Crew Size	Avg Day Prod	Avg Mat'l Unit Cost	Avg Equip Unit Cost	Avg Labor Unit Cost	Avg Total Unit Cost	Avg Price Incl O&P	Crew Size	Avg Day Prod	Avg Mat'l Unit Cost	Avg Equip Unit Cost	Avg Labor Unit Cost	Avg Total Unit Cost	Avg Price Incl O&P

Doors (continued)

Description	Oper	Unit	Crew Size	Avg Day Prod	Avg Mat'l Unit Cost	Avg Equip Unit Cost	Avg Labor Unit Cost	Avg Total Unit Cost	Avg Price Incl O&P	Crew Size	Avg Day Prod	Avg Mat'l Unit Cost	Avg Equip Unit Cost	Avg Labor Unit Cost	Avg Total Unit Cost	Avg Price Incl O&P
4'-0" x 6'-8", 4 doors	Inst	Set	C-18	8.0	128.70	---	42.14	170.84	191.48	C-18	6.0	148.50	---	56.18	204.68	232.21
6'-0" x 6'-8", 4 doors	Inst	Set	C-18	8.0	171.60	---	42.14	213.74	234.38	C-18	6.0	198.00	---	56.18	254.18	281.71
Lauan, flush face, 1-3/8" T hollow core																
2'-0" x 6'-8", 2 doors	Inst	Set	C-18	10.0	74.10	---	33.71	107.81	124.33	C-18	7.5	85.50	---	44.95	130.45	152.47
4'-0" x 6'-8", 4 doors	Inst	Set	C-18	8.0	144.30	---	42.14	186.44	207.08	C-18	6.0	166.50	---	56.18	222.68	250.21
6'-0" x 6'-8", 4 doors	Inst	Set	C-18	8.0	195.00	---	42.14	237.14	257.78	C-18	6.0	225.00	---	56.18	281.18	308.71
Ponderosa pine, full louver, 1-3/8" T hollow core																
2'-0" x 6'-8", 2 doors	Inst	Set	C-18	10.0	141.70	---	33.71	175.41	191.93	C-18	7.5	163.50	---	44.95	208.45	230.47
4'-0" x 6'-8", 4 doors	Inst	Set	C-18	8.0	274.30	---	42.14	316.44	337.08	C-18	6.0	316.50	---	56.18	372.68	400.21
6'-0" x 6'-8", 4 doors	Inst	Set	C-18	8.0	349.70	---	42.14	391.84	412.48	C-18	6.0	403.50	---	56.18	459.68	487.21
Ponderosa pine, raised louver, 1-3/8" T hollow core																
2'-0" x 6'-8", 2 doors	Inst	Set	C-18	10.0	149.50	---	33.71	183.21	199.73	C-18	7.5	172.50	---	44.95	217.45	239.47
4'-0" x 6'-8", 4 doors	Inst	Set	C-18	8.0	291.20	---	42.14	333.34	353.98	C-18	6.0	336.00	---	56.18	392.18	419.71
6'-0" x 6'-8", 4 doors	Inst	Set	C-18	8.0	349.70	---	42.14	391.84	412.48	C-18	6.0	403.50	---	56.18	459.68	487.21
Sliding or bi-passing unit																
Includes hardware, 4-5/8" jambs, header, and fascia																
Wood inserts; 1-3/8" T hollow core																
Unfinished birch																
4'-0" x 6'-8", 2 doors	Inst	Set	C-18	7.0	117.00	---	48.16	165.16	188.75	C-18	5.3	135.00	---	64.21	199.21	230.67
6'-0" x 6'-8", 2 doors	Inst	Set	C-18	7.0	157.30	---	48.16	205.46	229.05	C-18	5.3	181.50	---	64.21	245.71	277.17
8'-0" x 6'-8", 2 doors	Inst	Set	C-18	7.0	241.80	---	48.16	289.96	313.55	C-18	5.3	279.00	---	64.21	343.21	374.67
4'-0" x 8'-0", 2 doors	Inst	Set	C-18	7.0	176.80	---	48.16	224.96	248.55	C-18	5.3	204.00	---	64.21	268.21	299.67
6'-0" x 8'-0", 2 doors	Inst	Set	C-18	7.0	228.80	---	48.16	276.96	300.55	C-18	5.3	264.00	---	64.21	328.21	359.67
8'-0" x 8'-0", 2 doors	Inst	Set	C-18	7.0	317.20	---	48.16	365.36	388.95	C-18	5.3	366.00	---	64.21	430.21	461.67
10'-0" x 6'-8", 3 doors	Inst	Set	C-18	5.0	328.90	---	67.42	396.32	429.35	C-18	3.8	379.50	---	89.89	469.39	513.44
12'-0" x 6'-8", 3 doors	Inst	Set	C-18	5.0	370.50	---	67.42	437.92	470.95	C-18	3.8	427.50	---	89.89	517.39	561.44
10'-0" x 8'-0", 3 doors	Inst	Set	C-18	5.0	443.30	---	67.42	510.72	543.75	C-18	3.8	511.50	---	89.89	601.39	645.44
12'-0" x 8'-0", 3 doors	Inst	Set	C-18	5.0	490.10	---	67.42	557.52	590.55	C-18	3.8	565.50	---	89.89	655.39	699.44
Unfinished hardboard																
4'-0" x 6'-8", 2 doors	Inst	Set	C-18	7.0	84.50	---	48.16	132.66	156.25	C-18	5.3	97.50	---	64.21	161.71	193.17
6'-0" x 6'-8", 2 doors	Inst	Set	C-18	7.0	104.00	---	48.16	152.16	175.75	C-18	5.3	120.00	---	64.21	184.21	215.67
8'-0" x 6'-8", 2 doors	Inst	Set	C-18	7.0	139.10	---	48.16	187.26	210.85	C-18	5.3	160.50	---	64.21	224.71	256.17
4'-0" x 8'-0", 2 doors	Inst	Set	C-18	7.0	124.80	---	48.16	172.96	196.55	C-18	5.3	144.00	---	64.21	208.21	239.67
6'-0" x 8'-0", 2 doors	Inst	Set	C-18	7.0	141.70	---	48.16	189.86	213.45	C-18	5.3	163.50	---	64.21	227.71	259.17

Doors (continued)

Description	Oper	Unit	Costs Based On Large Volume							Costs Based On Small Volume						
			Crew Size	Avg Day Prod	Avg Mat'l Unit Cost	Avg Equip Unit Cost	Avg Labor Unit Cost	Avg Total Unit Cost	Avg Price Incl O&P	Crew Size	Avg Day Prod	Avg Mat'l Unit Cost	Avg Equip Unit Cost	Avg Labor Unit Cost	Avg Total Unit Cost	Avg Price Incl O&P
8'-0" x 8'-0", 2 doors	Inst	Set	C-18	7.0	179.40	---	48.16	227.56	251.15	C-18	5.3	207.00	---	64.21	271.21	302.67
10'-0" x 6'-8", 3 doors	Inst	Set	C-18	5.0	195.00	---	67.42	262.42	295.45	C-18	3.8	225.00	---	89.89	314.89	358.94
12'-0" x 6'-8", 3 doors	Inst	Set	C-18	5.0	217.10	---	67.42	284.52	317.55	C-18	3.8	250.50	---	89.89	340.39	384.44
10'-0" x 8'-0", 3 doors	Inst	Set	C-18	5.0	256.10	---	67.42	323.52	356.55	C-18	3.8	295.50	---	89.89	385.39	429.44
12'-0" x 8'-0", 3 doors	Inst	Set	C-18	5.0	284.70	---	67.42	352.12	385.15	C-18	3.8	328.50	---	89.89	418.39	462.44
Unfinished lauan																
4'-0" x 6'-8", 2 doors	Inst	Set	C-18	7.0	97.50	---	48.16	145.66	169.25	C-18	5.3	112.50	---	64.21	176.71	208.17
6'-0" x 6'-8", 2 doors	Inst	Set	C-18	7.0	128.70	---	48.16	176.86	200.45	C-18	5.3	148.50	---	64.21	212.71	244.17
8'-0" x 6'-8", 2 doors	Inst	Set	C-18	7.0	197.60	---	48.16	245.76	269.35	C-18	5.3	228.00	---	64.21	292.21	323.67
4'-0" x 8'-0", 2 doors	Inst	Set	C-18	7.0	163.80	---	48.16	211.96	235.55	C-18	5.3	189.00	---	64.21	253.21	284.67
6'-0" x 8'-0", 2 doors	Inst	Set	C-18	7.0	195.00	---	48.16	243.16	266.75	C-18	5.3	225.00	---	64.21	289.21	320.67
8'-0" x 8'-0", 2 doors	Inst	Set	C-18	7.0	262.60	---	48.16	310.76	334.35	C-18	5.3	303.00	---	64.21	367.21	398.67
10'-0" x 6'-8", 3 doors	Inst	Set	C-18	5.0	269.10	---	67.42	336.52	369.55	C-18	3.8	310.50	---	89.89	400.39	444.44
12'-0" x 6'-8", 3 doors	Inst	Set	C-18	5.0	304.20	---	67.42	371.62	404.65	C-18	3.8	351.00	---	89.89	440.89	484.94
10'-0" x 8'-0", 3 doors	Inst	Set	C-18	5.0	373.10	---	67.42	440.52	473.55	C-18	3.8	430.50	---	89.89	520.39	564.44
12'-0" x 8'-0", 3 doors	Inst	Set	C-18	5.0	406.90	---	67.42	474.32	507.35	C-18	3.8	469.50	---	89.89	559.39	603.44
Unfinished red oak																
4'-0" x 6'-8", 2 doors	Inst	Set	C-18	7.0	140.40	---	48.16	188.56	212.15	C-18	5.3	162.00	---	64.21	226.21	257.67
6'-0" x 6'-8", 2 doors	Inst	Set	C-18	7.0	191.10	---	48.16	239.26	262.85	C-18	5.3	220.50	---	64.21	284.71	316.17
8'-0" x 6'-8", 2 doors	Inst	Set	C-18	7.0	305.50	---	48.16	353.66	377.25	C-18	5.3	352.50	---	64.21	416.71	448.17
10'-0" x 6'-8", 3 doors	Inst	Set	C-18	5.0	445.90	---	67.42	513.32	546.35	C-18	3.8	514.50	---	89.89	604.39	648.44
12'-0" x 6'-8", 3 doors	Inst	Set	C-18	5.0	466.70	---	67.42	534.12	567.15	C-18	3.8	538.50	---	89.89	628.39	672.44
Unfinished ash-sen																
4'-0" x 6'-8", 2 doors	Inst	Set	C-18	7.0	118.30	---	48.16	166.46	190.05	C-18	5.3	136.50	---	64.21	200.71	232.17
6'-0" x 6'-8", 2 doors	Inst	Set	C-18	7.0	159.90	---	48.16	208.06	231.65	C-18	5.3	184.50	---	64.21	248.71	280.17
8'-0" x 6'-8", 2 doors	Inst	Set	C-18	7.0	244.40	---	48.16	292.56	316.15	C-18	5.3	282.00	---	64.21	346.21	377.67
4'-0" x 8'-0", 2 doors	Inst	Set	C-18	7.0	183.30	---	48.16	231.46	255.05	C-18	5.3	211.50	---	64.21	275.71	307.17
6'-0" x 8'-0", 2,doors	Inst	Set	C-18	7.0	230.10	---	48.16	278.26	301.85	C-18	5.3	265.50	---	64.21	329.71	361.17
8'-0" x 8'-0", 2 doors	Inst	Set	C-18	7.0	321.10	---	48.16	369.26	392.85	C-18	5.3	370.50	---	64.21	434.71	466.17
10'-0" x 6'-8", 3 doors	Inst	Set	C-18	5.0	332.80	---	67.42	400.22	433.25	C-18	3.8	384.00	---	89.89	473.89	517.94
12'-0" x 6'-8", 3 doors	Inst	Set	C-18	5.0	374.40	---	67.42	441.82	474.85	C-18	3.8	432.00	---	89.89	521.89	565.94
10'-0" x 8'-0", 3 doors	Inst	Set	C-18	5.0	452.40	---	67.42	519.82	552.85	C-18	3.8	522.00	---	89.89	611.89	655.94
12'-0" x 8'-0", 3 doors	Inst	Set	C-18	5.0	496.60	---	67.42	564.02	597.05	C-18	3.8	573.00	---	89.89	662.89	706.94

Description	Oper	Unit	Costs Based On Large Volume							Costs Based On Small Volume						
			Crew Size	Avg Day Prod	Avg Mat'l Unit Cost	Avg Equip Unit Cost	Avg Labor Unit Cost	Avg Total Unit Cost	Avg Price Incl O&P	Crew Size	Avg Day Prod	Avg Mat'l Unit Cost	Avg Equip Unit Cost	Avg Labor Unit Cost	Avg Total Unit Cost	Avg Price Incl O&P

Doors (continued)

Mirror inserts with plate/float glass

Description	Oper	Unit	Crew Size	Avg Day Prod	Avg Mat'l Unit Cost	Avg Equip Unit Cost	Avg Labor Unit Cost	Avg Total Unit Cost	Avg Price Incl O&P	Crew Size	Avg Day Prod	Avg Mat'l Unit Cost	Avg Equip Unit Cost	Avg Labor Unit Cost	Avg Total Unit Cost	Avg Price Incl O&P
3'-11" x 6'-8", 2 doors	Inst	Set	C-18	6.0	204.10	---	56.18	260.28	287.81	C-18	4.5	235.50	---	74.91	310.41	347.12
5'-11" x 6'-8", 2 doors	Inst	Set	C-18	6.0	278.20	---	56.18	334.38	361.91	C-18	4.5	321.00	---	74.91	395.91	432.62
7'-11" x 6'-8", 2 doors	Inst	Set	C-18	6.0	353.60	---	56.18	409.78	437.31	C-18	4.5	408.00	---	74.91	482.91	519.62
3'-11" x 7'-11", 2 doors	Inst	Set	C-18	6.0	260.00	---	56.18	316.18	343.71	C-18	4.5	300.00	---	74.91	374.91	411.62
5-11" x 7'-11", 2 doors	Inst	Set	C-18	6.0	317.20	---	56.18	373.38	400.91	C-18	4.5	366.00	---	74.91	440.91	477.62
7'-11" x 7'-11", 2 doors	Inst	Set	C-18	6.0	414.70	---	56.18	470.88	498.41	C-18	4.5	478.50	---	74.91	553.41	590.12
9'-11" x 6'-8", 3 doors	Inst	Set	C-18	5.0	505.70	---	67.42	573.12	606.15	C-18	3.8	583.50	---	89.89	673.39	717.44
11'-11" x 6'-8", 3 doors	Inst	Set	C-18	5.0	539.50	---	67.42	606.92	639.95	C-18	3.8	622.50	---	89.89	712.39	756.44
9'-11" x 7'-11", 3 doors	Inst	Set	C-18	5.0	555.10	---	67.42	622.52	655.55	C-18	3.8	640.50	---	89.89	730.39	774.44
11'-11" x 7'-11", 3 doors	Inst	Set	C-18	5.0	631.80	---	67.42	699.22	732.25	C-18	3.8	729.00	---	89.89	818.89	862.94
Track and hardware only																
4'-0" x 6'-8", 2 doors	Inst	Set	---	---	20.80	---	---	20.80	20.80	---	---	24.00	---	---	24.00	24.00
6'-0" x 6'-8", 2 doors	Inst	Set	---	---	27.30	---	---	27.30	27.30	---	---	31.50	---	---	31.50	31.50
8'-0" x 6'-8", 2 doors	Inst	Set	---	---	35.10	---	---	35.10	35.10	---	---	40.50	---	---	40.50	40.50
4'-0" x 8'-0", 2 doors	Inst	Set	---	---	20.80	---	---	20.80	20.80	---	---	24.00	---	---	24.00	24.00
6'-0" x 8'-0", 2 doors	Inst	Set	---	---	27.30	---	---	27.30	27.30	---	---	31.50	---	---	31.50	31.50
8'-0" x 8'-0", 2 doors	Inst	Set	---	---	35.10	---	---	35.10	35.10	---	---	40.50	---	---	40.50	40.50
10'-0" x 6'-8", 3 doors	Inst	Set	---	---	48.10	---	---	48.10	48.10	---	---	55.50	---	---	55.50	55.50
12'-0" x 6'-8", 3 doors	Inst	Set	---	---	59.80	---	---	59.80	59.80	---	---	69.00	---	---	69.00	69.00
10'-0" x 8'-0", 3 doors	Inst	Set	---	---	48.10	---	---	48.10	48.10	---	---	55.50	---	---	55.50	55.50
12'-0" x 8'-0", 3 doors	Inst	Set	---	---	59.80	---	---	59.80	59.80	---	---	69.00	---	---	69.00	69.00

Exterior wood door frames

Exterior frame with exterior trim

5/4" x 4-9/16" deep

Description	Oper	Unit	Crew Size	Avg Day Prod	Avg Mat'l Unit Cost	Avg Equip Unit Cost	Avg Labor Unit Cost	Avg Total Unit Cost	Avg Price Incl O&P	Crew Size	Avg Day Prod	Avg Mat'l Unit Cost	Avg Equip Unit Cost	Avg Labor Unit Cost	Avg Total Unit Cost	Avg Price Incl O&P
Pine	Inst	LF	C-18	350	3.00	---	0.96	3.96	4.44	C-18	263	3.60	---	1.28	4.88	5.51
Oak	Inst	LF	C-18	325	4.05	---	1.04	5.09	5.60	C-18	244	4.86	---	1.38	6.24	6.92
Walnut	Inst	LF	C-18	300	5.80	---	1.12	6.92	7.47	C-18	225	6.96	---	1.50	8.46	9.19

5/4" x 5-3/16" deep

Description	Oper	Unit	Crew Size	Avg Day Prod	Avg Mat'l Unit Cost	Avg Equip Unit Cost	Avg Labor Unit Cost	Avg Total Unit Cost	Avg Price Incl O&P	Crew Size	Avg Day Prod	Avg Mat'l Unit Cost	Avg Equip Unit Cost	Avg Labor Unit Cost	Avg Total Unit Cost	Avg Price Incl O&P
Pine	Inst	LF	C-18	350	3.25	---	0.96	4.21	4.69	C-18	263	3.90	---	1.28	5.18	5.81
Oak	Inst	LF	C-18	325	4.20	---	1.04	5.24	5.75	C-18	244	5.04	---	1.38	6.42	7.10
Walnut	Inst	LF	C-18	300	6.60	---	1.12	7.72	8.27	C-18	225	7.92	---	1.50	9.42	10.15

			Costs Based On Large Volume							Costs Based On Small Volume						
Description	Oper	Unit	Crew Size	Avg Day Prod	Avg Mat'l Unit Cost	Avg Equip Unit Cost	Avg Labor Unit Cost	Avg Total Unit Cost	Avg Price Incl O&P	Crew Size	Avg Day Prod	Avg Mat'l Unit Cost	Avg Equip Unit Cost	Avg Labor Unit Cost	Avg Total Unit Cost	Avg Price Incl O&P

Doors (continued)

5/4" x 6-9/16" deep

Pine	Inst	LF	C-18	350	3.75	---	0.96	4.71	5.19	C-18	263	4.50	---	1.28	5.78	6.41
Oak	Inst	LF	C-18	325	5.00	---	1.04	6.04	6.55	C-18	244	6.00	---	1.38	7.38	8.06
Walnut	Inst	LF	C-18	300	7.40	---	1.12	8.52	9.07	C-18	225	8.88	---	1.50	10.38	11.11

Exterior sills

8/4" x 8" deep

No horns	Inst	LF	C-18	80	7.50	---	4.21	11.71	13.78	C-18	60	9.00	---	5.62	14.62	17.37
2" horns	Inst	LF	C-18	80	8.80	---	4.21	13.01	15.08	C-18	60	10.56	---	5.62	16.18	18.93
3" horns	Inst	LF	C-18	80	9.15	---	4.21	13.36	15.43	C-18	60	10.98	---	5.62	16.60	19.35

8/4" x 10" deep

No horns	Inst	LF	C-18	60	9.50	---	5.62	15.12	17.87	C-18	45	11.40	---	7.49	18.89	22.56
2" horns	Inst	LF	C-18	60	10.50	---	5.62	16.12	18.87	C-18	45	12.60	---	7.49	20.09	23.76
3" horns	Inst	LF	C-18	60	11.15	---	5.62	16.77	19.52	C-18	45	13.38	---	7.49	20.87	24.54

Exterior, colonial frame and trim

3' opening, in swing	Inst	EA	C-18	20.0	550.00	---	16.85	566.85	575.11	C-18	15.0	660.00	---	22.47	682.47	693.48
5'4" opening, in/out swing	Inst	EA	C-18	15.0	825.00	---	22.47	847.47	858.48	C-18	11.3	990.00	---	29.96	1019.96	1034.65
6' opening, in/out swing	Inst	EA	C-18	10.0	865.00	---	33.71	898.71	915.23	C-18	7.5	1038.00	---	44.95	1082.95	1104.97

Entrance doors

Stile and rail raised panels

No lites

2 raised panels

2'-8" x 6'-8" x 1-3/4" T	Inst	EA	C-18	12.0	234.00	---	28.09	262.09	275.86	C-18	9.0	270.00	---	37.46	307.46	325.81
3'-0" x 6'-8" x 1-3/4" T	Inst	EA	C-18	11.0	241.80	---	30.65	272.45	287.46	C-18	8.3	279.00	---	40.86	319.86	339.88

4 raised panels

2'-8" x 6'-8" x 1-3/4" T	Inst	EA	C-18	12.0	183.30	---	28.09	211.39	225.16	C-18	9.0	211.50	---	37.46	248.96	267.31
3'-0" x 6'-8" x 1-3/4" T	Inst	EA	C-18	11.0	188.50	---	30.65	219.15	234.16	C-18	8.3	217.50	---	40.86	258.36	278.38

8 raised panels

2'-8" x 6'-8" x 1-3/4" T	Inst	EA	C-18	12.0	195.00	---	28.09	223.09	236.86	C-18	9.0	225.00	---	37.46	262.46	280.81
3'-0" x 6'-8" x 1-3/4" T	Inst	EA	C-18	11.0	221.00	---	30.65	251.65	266.66	C-18	8.3	255.00	---	40.86	295.86	315.88

Lites, tempered clear

2 raised panels, 2 lites

2'-8" x 6'-8" x 1-3/4" T	Inst	EA	C-18	12.0	195.00	---	28.09	223.09	236.86	C-18	9.0	225.00	---	37.46	262.46	280.81
3'-0" x 6'-8" x 1-3/4" T	Inst	EA	C-18	11.0	221.00	---	30.65	251.65	266.66	C-18	8.3	255.00	---	40.86	295.86	315.88

					Costs Based On Large Volume						Costs Based On Small Volume					
Description	Oper	Unit	Crew Size	Avg Day Prod	Avg Mat'l Unit Cost	Avg Equip Unit Cost	Avg Labor Unit Cost	Avg Total Unit Cost	Avg Price Incl O&P	Crew Size	Avg Day Prod	Avg Mat'l Unit Cost	Avg Equip Unit Cost	Avg Labor Unit Cost	Avg Total Unit Cost	Avg Price Incl O&P

Doors (continued)

2 raised panels, 6 lites

2'-8" x 6'-8" x 1-3/4" T | Inst | EA | C-18 | 12.0 | 211.90 | --- | 28.09 | 239.99 | 253.76 | C-18 | 9.0 | 244.50 | --- | 37.46 | 281.96 | 300.31
3'-0" x 6'-8" x 1-3/4" T | Inst | EA | C-18 | 11.0 | 223.60 | --- | 30.65 | 254.25 | 269.26 | C-18 | 8.3 | 258.00 | --- | 40.86 | 298.86 | 318.88

2 raised panels, 9 lites

2'-8" x 6'-8" x 1-3/4" T | Inst | EA | C-18 | 12.0 | 222.30 | --- | 28.09 | 250.39 | 264.16 | C-18 | 9.0 | 256.50 | --- | 37.46 | 293.96 | 312.31
3'-0" x 6'-8" x 1-3/4" T | Inst | EA | C-18 | 11.0 | 234.00 | --- | 30.65 | 264.65 | 279.66 | C-18 | 8.3 | 270.00 | --- | 40.86 | 310.86 | 330.88

Dutch doors

Country style, fir, stile and rail, tempered clear lite
Two raised panels in lower door
1 lite in upper door

2'-8" x 6'-8" x 1-3/4" T | Inst | EA | C-18 | 12.0 | 282.10 | --- | 28.09 | 310.19 | 323.96 | C-18 | 9.0 | 325.50 | --- | 37.46 | 362.96 | 381.31
3'-0" x 6'-8" x 1-3/4" T | Inst | EA | C-18 | 11.0 | 291.20 | --- | 30.65 | 321.85 | 336.86 | C-18 | 8.3 | 336.00 | --- | 40.86 | 376.86 | 396.88

4 lites in upper door

2'-8" x 6'-8" x 1-3/4" T | Inst | EA | C-18 | 12.0 | 328.90 | --- | 28.09 | 356.99 | 370.76 | C-18 | 9.0 | 379.50 | --- | 37.46 | 416.96 | 435.31
3'-0" x 6'-8" x 1-3/4" T | Inst | EA | C-18 | 11.0 | 339.30 | --- | 30.65 | 369.95 | 384.96 | C-18 | 8.3 | 391.50 | --- | 40.86 | 432.36 | 452.38

Four diamond shaped raised panels in lower door
1 lite in upper door

2'-8" x 6'-8" x 1-3/4" T | Inst | EA | C-18 | 12.0 | 284.70 | --- | 28.09 | 312.79 | 326.56 | C-18 | 9.0 | 328.50 | --- | 37.46 | 365.96 | 384.31
3'-0" x 6'-8" x 1-3/4" T | Inst | EA | C-18 | 11.0 | 292.50 | --- | 30.65 | 323.15 | 338.16 | C-18 | 8.3 | 337.50 | --- | 40.86 | 378.36 | 398.38

6 lites in upper door

2'-8" x 6'-8" x 1-3/4" T | Inst | EA | C-18 | 12.0 | 323.70 | --- | 28.09 | 351.79 | 365.56 | C-18 | 9.0 | 373.50 | --- | 37.46 | 410.96 | 429.31
3'-0" x 6'-8" x 1-3/4" T | Inst | EA | C-18 | 11.0 | 332.80 | --- | 30.65 | 363.45 | 378.46 | C-18 | 8.3 | 384.00 | --- | 40.86 | 424.86 | 444.88

French doors

Douglas fir or hemlock 4-1/2" stiles and top rail, 9-1/2" bottom rail, tempered glass with stops
1 lite
1-3/8" T and 1-3/4" T

2'-0" x 6'-8" | Inst | EA | C-18 | 15.0 | 187.20 | --- | 22.47 | 209.67 | 220.68 | C-18 | 11.3 | 216.00 | --- | 29.96 | 245.96 | 260.65
2'-6" x 6'-8" | Inst | EA | C-18 | 13.0 | 187.20 | --- | 25.93 | 213.13 | 225.84 | C-18 | 9.8 | 216.00 | --- | 34.57 | 250.57 | 267.52
3'-0" x 6'-8" | Inst | EA | C-18 | 11.0 | 193.70 | --- | 30.65 | 224.35 | 239.36 | C-18 | 8.3 | 223.50 | --- | 40.86 | 264.36 | 284.38

5 lites, 5 high
1-3/8" T and 1-3/4" T

2'-0" x 6'-8" | Inst | EA | C-18 | 15.0 | 176.80 | --- | 22.47 | 199.27 | 210.28 | C-18 | 11.3 | 204.00 | --- | 29.96 | 233.96 | 248.65
2'-6" x 6'-8" | Inst | EA | C-18 | 13.0 | 176.80 | --- | 25.93 | 202.73 | 215.44 | C-18 | 9.8 | 204.00 | --- | 34.57 | 238.57 | 255.52
3'-0" x 6'-8" | Inst | EA | C-18 | 11.0 | 182.00 | --- | 30.65 | 212.65 | 227.66 | C-18 | 8.3 | 210.00 | --- | 40.86 | 250.86 | 270.88

Description	Oper	Unit	Costs Based On Large Volume							Costs Based On Small Volume						
			Crew Size	Avg Day Prod	Avg Mat'l Unit Cost	Avg Equip Unit Cost	Avg Labor Unit Cost	Avg Total Unit Cost	Avg Price Incl O&P	Crew Size	Avg Day Prod	Avg Mat'l Unit Cost	Avg Equip Unit Cost	Avg Labor Unit Cost	Avg Total Unit Cost	Avg Price Incl O&P

Doors (continued)

10 lites, 5 high
  1-3/8" T and 1-3/4" T

Description	Oper	Unit	Crew Size	Avg Day Prod	Avg Mat'l	Avg Equip	Avg Labor	Avg Total	Avg Price O&P	Crew Size	Avg Day Prod	Avg Mat'l	Avg Equip	Avg Labor	Avg Total	Avg Price O&P
2'-0" x 6'-8"	Inst	EA	C-18	15.0	182.00	---	22.47	204.47	215.48	C-18	11.3	210.00	---	29.96	239.96	254.65
2'-6" x 6'-8"	Inst	EA	C-18	13.0	182.00	---	25.93	207.93	220.64	C-18	9.8	210.00	---	34.57	244.57	261.52
3'-0" x 6'-8"	Inst	EA	C-18	11.0	187.20	---	30.65	217.85	232.86	C-18	8.3	216.00	---	40.86	256.86	276.88

Garden doors
  1 door, "X" unit

2'-6" x 6'-8"	Inst	LS	C-18	12.0	401.70	---	28.09	429.79	443.56	C-18	9.0	463.50	---	37.46	500.96	519.31
3'-0" x 6'-8"	Inst	LS	C-18	10.0	430.30	---	33.71	464.01	480.53	C-18	7.5	496.50	---	44.95	541.45	563.47

  2 doors, "XO/OX" unit

5'-0" x 6'-8"	Inst	LS	C-18	6.6	751.40	---	51.08	802.48	827.50	C-18	5.0	867.00	---	68.10	935.10	968.47
6'-0" x 6'-8"	Inst	LS	C-18	5.5	806.00	---	61.29	867.29	897.32	C-18	4.1	930.00	---	81.72	1011.72	1051.76

  3 doors, "XOX" unit

7'-6" x 6'-8"	Inst	LS	C-18	4.4	1072.50	---	76.61	1149.11	1186.65	C-18	3.3	1237.50	---	102.15	1339.65	1389.70
9'-0" x 6'-8"	Inst	LS	C-18	3.7	1160.90	---	91.11	1252.01	1296.65	C-18	2.8	1339.50	---	121.48	1460.98	1520.50

Fire doors

Natural birch
  One hour rating

2'-6" x 6'-8" x 1-3/4" T	Inst	EA	C-18	13.0	171.60	---	25.93	197.53	210.24	C-18	9.8	198.00	---	34.57	232.57	249.52
3'-0" x 6'-8" x 1-3/4" T	Inst	EA	C-18	11.0	193.70	---	30.65	224.35	239.36	C-18	8.3	223.50	---	40.86	264.36	284.38
2'-6" x 7'-0" x 1-3/4" T	Inst	EA	C-18	13.0	179.40	---	25.93	205.33	218.04	C-18	9.8	207.00	---	34.57	241.57	258.52
3'-0" x 7'-0" x 1-3/4" T	Inst	EA	C-18	11.0	202.80	---	30.65	233.45	248.46	C-18	8.3	234.00	---	40.86	274.86	294.88

  1.5 hour rating

2'-6" x 6'-8" x 1-3/4" T	Inst	EA	C-18	13.0	244.40	---	25.93	270.33	283.04	C-18	9.8	282.00	---	34.57	316.57	333.52
3'-0" x 6'-8" x 1-3/4" T	Inst	EA	C-18	11.0	274.30	---	30.65	304.95	319.96	C-18	8.3	316.50	---	40.86	357.36	377.38
2'-6" x 7'-0" x 1-3/4" T	Inst	EA	C-18	13.0	250.90	---	25.93	276.83	289.54	C-18	9.8	289.50	---	34.57	324.07	341.02
3'-0" x 7'-0" x 1-3/4" T	Inst	EA	C-18	11.0	291.20	---	30.65	321.85	336.86	C-18	8.3	336.00	---	40.86	376.86	396.88

Interior wood door frames

Interior frame
  11/16" x 3-5/8" deep

Pine	Inst	LF	C-18	350	1.20	---	0.96	2.16	2.64	C-18	263	1.44	---	1.28	2.72	3.35
Oak	Inst	LF	C-18	325	1.45	---	1.04	2.49	3.00	C-18	244	1.74	---	1.38	3.12	3.80
Walnut	Inst	LF	C-18	300	2.00	---	1.12	3.12	3.67	C-18	225	2.40	---	1.50	3.90	4.63

Description	Oper	Unit	Costs Based On Large Volume							Costs Based On Small Volume						
			Crew Size	Avg Day Prod	Avg Mat'l Unit Cost	Avg Equip Unit Cost	Avg Labor Unit Cost	Avg Total Unit Cost	Avg Price Incl O&P	Crew Size	Avg Day Prod	Avg Mat'l Unit Cost	Avg Equip Unit Cost	Avg Labor Unit Cost	Avg Total Unit Cost	Avg Price Incl O&P

Doors (continued)

11/16" x 4-9/16" deep

Description	Oper	Unit	Crew Size	Avg Day Prod	Avg Mat'l Unit Cost	Avg Equip Unit Cost	Avg Labor Unit Cost	Avg Total Unit Cost	Avg Price Incl O&P	Crew Size	Avg Day Prod	Avg Mat'l Unit Cost	Avg Equip Unit Cost	Avg Labor Unit Cost	Avg Total Unit Cost	Avg Price Incl O&P
Pine	Inst	LF	C-18	350	2.10	---	0.96	3.06	3.54	C-18	263	2.52	---	1.28	3.80	4.43
Oak	Inst	LF	C-18	325	2.45	---	1.04	3.49	4.00	C-18	244	2.94	---	1.38	4.32	5.00
Walnut	Inst	LF	C-18	300	2.80	---	1.12	3.92	4.47	C-18	225	3.36	---	1.50	4.86	5.59

11/16" x 5-3/16" deep

Description	Oper	Unit	Crew Size	Avg Day Prod	Avg Mat'l Unit Cost	Avg Equip Unit Cost	Avg Labor Unit Cost	Avg Total Unit Cost	Avg Price Incl O&P	Crew Size	Avg Day Prod	Avg Mat'l Unit Cost	Avg Equip Unit Cost	Avg Labor Unit Cost	Avg Total Unit Cost	Avg Price Incl O&P
Pine	Inst	LF	C-18	350	2.40	---	0.96	3.36	3.84	C-18	263	2.88	---	1.28	4.16	4.79
Oak	Inst	LF	C-18	325	3.00	---	1.04	4.04	4.55	C-18	244	3.60	---	1.38	4.98	5.66
Walnut	Inst	LF	C-18	300	3.35	---	1.12	4.47	5.02	C-18	225	4.02	---	1.50	5.52	6.25
Pocket door frame	Inst	EA	C-18	12.0	65.00	---	28.09	93.09	106.86	C-18	9.0	78.00	---	37.46	115.46	133.81

Threshold, oak

Description	Oper	Unit	Crew Size	Avg Day Prod	Avg Mat'l Unit Cost	Avg Equip Unit Cost	Avg Labor Unit Cost	Avg Total Unit Cost	Avg Price Incl O&P	Crew Size	Avg Day Prod	Avg Mat'l Unit Cost	Avg Equip Unit Cost	Avg Labor Unit Cost	Avg Total Unit Cost	Avg Price Incl O&P
5/8" x 3-5/8" deep	Inst	LF	C-18	160	2.00	---	2.11	4.11	5.14	C-18	120	2.40	---	2.81	5.21	6.59
5/8" x 4-5/8" deep	Inst	LF	C-18	150	2.50	---	2.25	4.75	5.85	C-18	113	3.00	---	3.00	6.00	7.46
5/8" x 5-5/8" deep	Inst	LF	C-18	140	3.00	---	2.41	5.41	6.59	C-18	105	3.60	---	3.21	6.81	8.38

Interior doors

Passage doors, flush, hinge hardware included

Hollow core

Hardboard

Description	Oper	Unit	Crew Size	Avg Day Prod	Avg Mat'l Unit Cost	Avg Equip Unit Cost	Avg Labor Unit Cost	Avg Total Unit Cost	Avg Price Incl O&P	Crew Size	Avg Day Prod	Avg Mat'l Unit Cost	Avg Equip Unit Cost	Avg Labor Unit Cost	Avg Total Unit Cost	Avg Price Incl O&P
2'-6" x 6'-8" x 1-3/8" T	Inst	EA	C-18	15.0	32.50	---	22.47	54.97	65.98	C-18	11.3	37.50	---	29.96	67.46	82.15
2'-8" x 6'-8" x 1-3/8" T	Inst	EA	C-18	14.0	36.40	---	24.08	60.48	72.28	C-18	10.5	42.00	---	32.10	74.10	89.84
3'-0" x 6'-8" x 1-3/8" T	Inst	EA	C-18	14.0	39.00	---	24.08	63.08	74.88	C-18	10.5	45.00	---	32.10	77.10	92.84
2'-6" x 7'-0" x 1-3/8" T	Inst	EA	C-18	15.0	41.60	---	22.47	64.07	75.08	C-18	11.3	48.00	---	29.96	77.96	92.65
2'-8" x 7'-0" x 1-3/8" T	Inst	EA	C-18	14.0	44.20	---	24.08	68.28	80.08	C-18	10.5	51.00	---	32.10	83.10	98.84
3'-0" x 7'-0" x 1-3/8" T	Inst	EA	C-18	14.0	45.50	---	24.08	69.58	81.38	C-18	10.5	52.50	---	32.10	84.60	100.34
Lauan																
2'-6" x 6'-8" x 1-3/8" T	Inst	EA	C-18	15.0	37.70	---	22.47	60.17	71.18	C-18	11.3	43.50	---	29.96	73.46	88.15
2'-8" x 6'-8" x 1-3/8" T	Inst	EA	C-18	14.0	46.80	---	24.08	70.88	82.68	C-18	10.5	54.00	---	32.10	86.10	101.84
3'-0" x 6'-8" x 1-3/8" T	Inst	EA	C-18	14.0	50.70	---	24.08	74.78	86.58	C-18	10.5	58.50	---	32.10	90.60	106.34
2'-6" x 7'-0" x 1-3/8" T	Inst	EA	C-18	15.0	52.00	---	22.47	74.47	85.48	C-18	11.3	60.00	---	29.96	89.96	104.65
2'-8" x 7'-0" x 1-3/8" T	Inst	EA	C-18	14.0	54.60	---	24.08	78.68	90.48	C-18	10.5	63.00	---	32.10	95.10	110.84
3'-0" x 7'-0" x 1-3/8" T	Inst	EA	C-18	14.0	58.50	---	24.08	82.58	94.38	C-18	10.5	67.50	---	32.10	99.60	115.34
Birch																
2'-6" x 6'-8" x 1-3/8" T	Inst	EA	C-18	15.0	55.90	---	22.47	78.37	89.38	C-18	11.3	64.50	---	29.96	94.46	109.15
2'-8" x 6'-8" x 1-3/8" T	Inst	EA	C-18	14.0	59.80	---	24.08	83.88	95.68	C-18	10.5	69.00	---	32.10	101.10	116.84
3'-0" x 6'-8" x 1-3/8" T	Inst	EA	C-18	14.0	65.00	---	24.08	89.08	100.88	C-18	10.5	75.00	---	32.10	107.10	122.84

Description	Oper	Unit	Costs Based On Large Volume							Costs Based On Small Volume						
			Crew Size	Avg Day Prod	Avg Mat'l Unit Cost	Avg Equip Unit Cost	Avg Labor Unit Cost	Avg Total Unit Cost	Avg Price Incl O&P	Crew Size	Avg Day Prod	Avg Mat'l Unit Cost	Avg Equip Unit Cost	Avg Labor Unit Cost	Avg Total Unit Cost	Avg Price Incl O&P

Doors (continued)

Description	Oper	Unit	Crew Size	Avg Day Prod	Avg Mat'l Unit Cost	Avg Equip Unit Cost	Avg Labor Unit Cost	Avg Total Unit Cost	Avg Price Incl O&P	Crew Size	Avg Day Prod	Avg Mat'l Unit Cost	Avg Equip Unit Cost	Avg Labor Unit Cost	Avg Total Unit Cost	Avg Price Incl O&P
2'-6" x 7'-0" x 1-3/8" T	Inst	EA	C-18	15.0	63.70	---	22.47	86.17	97.18	C-18	11.3	73.50	---	29.96	103.46	118.15
2'-8" x 7'-0" x 1-3/8" T	Inst	EA	C-18	14.0	68.90	---	24.08	92.98	104.78	C-18	10.5	79.50	---	32.10	111.60	127.34
3'-0" x 7'-0" x 1-3/8" T	Inst	EA	C-18	14.0	72.80	---	24.08	96.88	108.68	C-18	10.5	84.00	---	32.10	116.10	131.84
Solid core																
Hardboard																
2'-6" x 6'-8" x 1-3/8" T	Inst	EA	C-18	14.0	61.10	---	24.08	85.18	96.98	C-18	10.5	70.50	---	32.10	102.60	118.34
2'-8" x 6'-8" x 1-3/8" T	Inst	EA	C-18	13.0	62.40	---	25.93	88.33	101.04	C-18	9.8	72.00	---	34.57	106.57	123.52
3'-0" x 6'-8" x 1-3/8" T	Inst	EA	C-18	13.0	65.00	---	25.93	90.93	103.64	C-18	9.8	75.00	---	34.57	109.57	126.52
2'-6" x 7'-0" x 1-3/8" T	Inst	EA	C-18	14.0	68.90	---	24.08	92.98	104.78	C-18	10.5	79.50	---	32.10	111.60	127.34
2'-8" x 7'-0" x 1-3/8" T	Inst	EA	C-18	13.0	71.50	---	25.93	97.43	110.14	C-18	9.8	82.50	---	34.57	117.07	134.02
3'-0" x 7'-0" x 1-3/8" T	Inst	EA	C-18	13.0	72.80	---	25.93	98.73	111.44	C-18	9.8	84.00	---	34.57	118.57	135.52
Lauan																
2'-6" x 6'-8" x 1-3/8" T	Inst	EA	C-18	14.0	71.50	---	24.08	95.58	107.38	C-18	10.5	82.50	---	32.10	114.60	130.34
2'-8" x 6'-8" x 1-3/8" T	Inst	EA	C-18	13.0	74.10	---	25.93	100.03	112.74	C-18	9.8	85.50	---	34.57	120.07	137.02
3'-0" x 6'-8" x 1-3/8" T	Inst	EA	C-18	13.0	84.50	---	25.93	110.43	123.14	C-18	9.8	97.50	---	34.57	132.07	149.02
2'-6" x 7'-0" x 1-3/8" T	Inst	EA	C-18	14.0	79.30	---	24.08	103.38	115.18	C-18	10.5	91.50	---	32.10	123.60	139.34
2'-8" x 7'-0" x 1-3/8" T	Inst	EA	C-18	13.0	81.90	---	25.93	107.83	120.54	C-18	9.8	94.50	---	34.57	129.07	146.02
3'-0" x 7'-0" x 1-3/8" T	Inst	EA	C-18	13.0	84.50	---	25.93	110.43	123.14	C-18	9.8	97.50	---	34.57	132.07	149.02
Birch																
2'-6" x 6'-8" x 1-3/8" T	Inst	EA	C-18	14.0	83.20	---	24.08	107.28	119.08	C-18	10.5	96.00	---	32.10	128.10	143.84
2'-8" x 6'-8" x 1-3/8" T	Inst	EA	C-18	13.0	87.10	---	25.93	113.03	125.74	C-18	9.8	100.50	---	34.57	135.07	152.02
3'-0" x 6'-8" x 1-3/8" T	Inst	EA	C-18	13.0	92.30	---	25.93	118.23	130.94	C-18	9.8	106.50	---	34.57	141.07	158.02
2'-6" x 7'-0" x 1-3/8" T	Inst	EA	C-18	14.0	91.00	---	24.08	115.08	126.88	C-18	10.5	105.00	---	32.10	137.10	152.84
2'-8" x 7'-0" x 1-3/8" T	Inst	EA	C-18	13.0	96.20	---	25.93	122.13	134.84	C-18	9.8	111.00	---	34.57	145.57	162.52
3'-0" x 7'-0" x 1-3/8" T	Inst	EA	C-18	13.0	100.10	---	25.93	126.03	138.74	C-18	9.8	115.50	---	34.57	150.07	167.02

Pre-hung package door units

Hollow core, flush, pre-hung
Hardboard

Description	Oper	Unit	Crew Size	Avg Day Prod	Avg Mat'l Unit Cost	Avg Equip Unit Cost	Avg Labor Unit Cost	Avg Total Unit Cost	Avg Price Incl O&P	Crew Size	Avg Day Prod	Avg Mat'l Unit Cost	Avg Equip Unit Cost	Avg Labor Unit Cost	Avg Total Unit Cost	Avg Price Incl O&P
2'-6" x 6'-8" x 1-3/8" T	Inst	EA	C-18	12.0	79.30	---	28.09	107.39	121.16	C-18	9.0	91.50	---	37.46	128.96	147.31
2'-8" x 6'-8" x 1-3/8" T	Inst	EA	C-18	12.0	81.90	---	28.09	109.99	123.76	C-18	9.0	94.50	---	37.46	131.96	150.31
3'-0" x 6'-8" x 1-3/8" T	Inst	EA	C-18	12.0	83.20	---	28.09	111.29	125.06	C-18	9.0	96.00	---	37.46	133.46	151.81

Description	Oper	Unit	Costs Based On Large Volume							Costs Based On Small Volume						
			Crew Size	Avg Day Prod	Avg Mat'l Unit Cost	Avg Equip Unit Cost	Avg Labor Unit Cost	Avg Total Unit Cost	Avg Price Incl O&P	Crew Size	Avg Day Prod	Avg Mat'l Unit Cost	Avg Equip Unit Cost	Avg Labor Unit Cost	Avg Total Unit Cost	Avg Price Incl O&P

Doors (continued)

Lauan

Description	Oper	Unit	Crew Size	Avg Day Prod	Avg Mat'l Unit Cost	Avg Equip Unit Cost	Avg Labor Unit Cost	Avg Total Unit Cost	Avg Price Incl O&P	Crew Size	Avg Day Prod	Avg Mat'l Unit Cost	Avg Equip Unit Cost	Avg Labor Unit Cost	Avg Total Unit Cost	Avg Price Incl O&P
2'-6" x 6'-8" x 1-3/8" T	Inst	EA	C-18	12.0	91.00	---	28.09	119.09	132.86	C-18	9.0	105.00	---	37.46	142.46	160.81
2'-8" x 6'-8" x 1-3/8" T	Inst	EA	C-18	12.0	92.30	---	28.09	120.39	134.16	C-18	9.0	106.50	---	37.46	143.96	162.31
3'-0" x 6'-8" x 1-3/8" T	Inst	EA	C-18	12.0	96.20	---	28.09	124.29	138.06	C-18	9.0	111.00	---	37.46	148.46	166.81
Birch																
2'-6" x 6'-8" x 1-3/8" T	Inst	EA	C-18	12.0	102.70	---	28.09	130.79	144.56	C-18	9.0	118.50	---	37.46	155.96	174.31
2'-8" x 6'-8" x 1-3/8" T	Inst	EA	C-18	12.0	106.60	---	28.09	134.69	148.46	C-18	9.0	123.00	---	37.46	160.46	178.81
3'-0" x 6'-8" x 1-3/8" T	Inst	EA	C-18	12.0	111.80	---	28.09	139.89	153.66	C-18	9.0	129.00	---	37.46	166.46	184.81
Ash																
2'-6" x 6'-8" x 1-3/8" T	Inst	EA	C-18	12.0	102.70	---	28.09	130.79	144.56	C-18	9.0	118.50	---	37.46	155.96	174.31
2'-8" x 6'-8" x 1-3/8" T	Inst	EA	C-18	12.0	106.60	---	28.09	134.69	148.46	C-18	9.0	123.00	---	37.46	160.46	178.81
3'-0" x 6'-8" x 1-3/8" T	Inst	EA	C-18	12.0	111.80	---	28.09	139.89	153.66	C-18	9.0	129.00	---	37.46	166.46	184.81

Solid core, flush, pre-hung

Hardboard

Description	Oper	Unit	Crew Size	Avg Day Prod	Avg Mat'l Unit Cost	Avg Equip Unit Cost	Avg Labor Unit Cost	Avg Total Unit Cost	Avg Price Incl O&P	Crew Size	Avg Day Prod	Avg Mat'l Unit Cost	Avg Equip Unit Cost	Avg Labor Unit Cost	Avg Total Unit Cost	Avg Price Incl O&P
2'-6" x 6'-8" x 1-3/8" T	Inst	EA	C-18	11.0	117.00	---	30.65	147.65	162.66	C-18	8.3	135.00	---	40.86	175.86	195.88
2'-8" x 6'-8" x 1-3/8" T	Inst	EA	C-18	11.0	118.30	---	30.65	148.95	163.96	C-18	8.3	136.50	---	40.86	177.36	197.38
3'-0" x 6'-8" x 1-3/8" T	Inst	EA	C-18	11.0	120.90	---	30.65	151.55	166.56	C-18	8.3	139.50	---	40.86	180.36	200.38
Lauan																
2'-6" x 6'-8" x 1-3/8" T	Inst	EA	C-18	11.0	127.40	---	30.65	158.05	173.06	C-18	8.3	147.00	---	40.86	187.86	207.88
2'-8" x 6'-8" x 1-3/8" T	Inst	EA	C-18	11.0	128.70	---	30.65	159.35	174.36	C-18	8.3	148.50	---	40.86	189.36	209.38
3'-0" x 6'-8" x 1-3/8" T	Inst	EA	C-18	11.0	132.60	---	30.65	163.25	178.26	C-18	8.3	153.00	---	40.86	193.86	213.88
Birch																
2'-6" x 6'-8" x 1-3/8" T	Inst	EA	C-18	11.0	139.10	---	30.65	169.75	184.76	C-18	8.3	160.50	---	40.86	201.36	221.38
2'-8" x 6'-8" x 1-3/8" T	Inst	EA	C-18	11.0	141.70	---	30.65	172.35	187.36	C-18	8.3	163.50	---	40.86	204.36	224.38
3'-0" x 6'-8" x 1-3/8" T	Inst	EA	C-18	11.0	146.90	---	30.65	177.55	192.56	C-18	8.3	169.50	---	40.86	210.36	230.38
Ash																
2'-6" x 6'-8" x 1-3/8" T	Inst	EA	C-18	11.0	140.40	---	30.65	171.05	186.06	C-18	8.3	162.00	---	40.86	202.86	222.88
2'-8" x 6'-8" x 1-3/8" T	Inst	EA	C-18	11.0	143.00	---	30.65	173.65	188.66	C-18	8.3	165.00	---	40.86	205.86	225.88
3'-0" x 6'-8" x 1-3/8" T	Inst	EA	C-18	11.0	149.50	---	30.65	180.15	195.16	C-18	8.3	172.50	---	40.86	213.36	233.38

Adjustments

For casing, 2 sets mitered, not applied

Description	Oper	Unit	Crew Size	Avg Day Prod	Avg Mat'l Unit Cost	Avg Equip Unit Cost	Avg Labor Unit Cost	Avg Total Unit Cost	Avg Price Incl O&P	Crew Size	Avg Day Prod	Avg Mat'l Unit Cost	Avg Equip Unit Cost	Avg Labor Unit Cost	Avg Total Unit Cost	Avg Price Incl O&P
7/16" x 1-1/2" finger joint	Inst	EA	---	---	6.50	---	---	6.50	6.50	---	---	7.50	---	---	7.50	7.50
9/16" x 1-1/2" finger joint	Inst	EA	---	---	7.80	---	---	7.80	7.80	---	---	9.00	---	---	9.00	9.00

Description	Oper	Unit	Costs Based On Large Volume							Costs Based On Small Volume						
			Crew Size	Avg Day Prod	Avg Mat'l Unit Cost	Avg Equip Unit Cost	Avg Labor Unit Cost	Avg Total Unit Cost	Avg Price Incl O&P	Crew Size	Avg Day Prod	Avg Mat'l Unit Cost	Avg Equip Unit Cost	Avg Labor Unit Cost	Avg Total Unit Cost	Avg Price Incl O&P

Doors (continued)

Description	Oper	Unit	Crew Size	Avg Day Prod	Avg Mat'l Unit Cost	Avg Equip Unit Cost	Avg Labor Unit Cost	Avg Total Unit Cost	Avg Price Incl O&P	Crew Size	Avg Day Prod	Avg Mat'l Unit Cost	Avg Equip Unit Cost	Avg Labor Unit Cost	Avg Total Unit Cost	Avg Price Incl O&P
5/8" x 1-5/8" solid pine	Inst	EA	---	---	22.10	---	---	22.10	22.10	---	---	25.50	---	---	25.50	25.50
5/8" x 2-1/2" solid pine	Inst	EA	---	---	53.30	---	---	53.30	53.30	---	---	61.50	---	---	61.50	61.50
For solid jamb, 4-9/16" solid pine	Inst	EA	---	---	28.60	---	---	28.60	28.60	---	---	33.00	---	---	33.00	33.00

Screen doors
Aluminum, fiberglass wire, plain grille, with hardware and hinges, closer and latch

Description	Oper	Unit	Crew Size	Avg Day Prod	Avg Mat'l Unit Cost	Avg Equip Unit Cost	Avg Labor Unit Cost	Avg Total Unit Cost	Avg Price Incl O&P	Crew Size	Avg Day Prod	Avg Mat'l Unit Cost	Avg Equip Unit Cost	Avg Labor Unit Cost	Avg Total Unit Cost	Avg Price Incl O&P
3'-0" x 7'-0"	Inst	EA	C-18	16.0	78.00	---	21.07	99.07	109.39	C-18	12.0	90.00	---	28.09	118.09	131.86

Wood screen doors, pine, aluminum wire
1-1/8" T

Description	Oper	Unit	Crew Size	Avg Day Prod	Avg Mat'l Unit Cost	Avg Equip Unit Cost	Avg Labor Unit Cost	Avg Total Unit Cost	Avg Price Incl O&P	Crew Size	Avg Day Prod	Avg Mat'l Unit Cost	Avg Equip Unit Cost	Avg Labor Unit Cost	Avg Total Unit Cost	Avg Price Incl O&P
2'-6" x 6'-9"	Inst	EA	C-18	16.0	107.90	---	21.07	128.97	139.29	C-18	12.0	124.50	---	28.09	152.59	166.36
2'-8" x 6'-9"	Inst	EA	C-18	16.0	107.90	---	21.07	128.97	139.29	C-18	12.0	124.50	---	28.09	152.59	166.36
3'-0" x 6'-9"	Inst	EA	C-18	16.0	115.70	---	21.07	136.77	147.09	C-18	12.0	133.50	---	28.09	161.59	175.36
Half screen																
2'-6" x 6'-9"	Inst	EA	C-18	16.0	140.40	---	21.07	161.47	171.79	C-18	12.0	162.00	---	28.09	190.09	203.86
2'-8" x 6'-9"	Inst	EA	C-18	16.0	140.40	---	21.07	161.47	171.79	C-18	12.0	162.00	---	28.09	190.09	203.86
3'-0" x 6'-9"	Inst	EA	C-18	16.0	154.70	---	21.07	175.77	186.09	C-18	12.0	178.50	---	28.09	206.59	220.36

Special doors
Glass sliding doors, with 5-1/2" anodized aluminum frame, trim, weatherstripping, tempered 3/16" T clear single glazed glass, with screen
6'-8" H

Description	Oper	Unit	Crew Size	Avg Day Prod	Avg Mat'l Unit Cost	Avg Equip Unit Cost	Avg Labor Unit Cost	Avg Total Unit Cost	Avg Price Incl O&P	Crew Size	Avg Day Prod	Avg Mat'l Unit Cost	Avg Equip Unit Cost	Avg Labor Unit Cost	Avg Total Unit Cost	Avg Price Incl O&P
5' W, 2 lites, 1 sliding	Inst	SET	C-18	4.0	336.70	---	84.27	420.97	462.27	C-18	3.0	388.50	---	112.37	500.87	555.92
6' W, 2 lites, 1 sliding	Inst	SET	C-18	4.0	366.60	---	84.27	450.87	492.17	C-18	3.0	423.00	---	112.37	535.37	590.42
7' W, 2 lites, 1 sliding	Inst	SET	C-18	4.0	429.00	---	84.27	513.27	554.57	C-18	3.0	495.00	---	112.37	607.37	662.42
8' W, 2 lites, 1 sliding	Inst	SET	C-18	4.0	429.00	---	84.27	513.27	554.57	C-18	3.0	495.00	---	112.37	607.37	662.42
10' W, 2 lites, 1 sliding	Inst	SET	C-18	4.0	568.10	---	84.27	652.37	693.67	C-18	3.0	655.50	---	112.37	767.87	822.92
9' W, 3 lites, 1 sliding	Inst	SET	C-18	3.0	492.70	---	112.37	605.07	660.12	C-18	2.3	568.50	---	149.82	718.32	791.73
12' W, 3 lites, 1 sliding	Inst	SET	C-18	3.0	581.10	---	112.37	693.47	748.52	C-18	2.3	670.50	---	149.82	820.32	893.73
15' W, 3 lites, 1 sliding	Inst	SET	C-18	3.0	790.40	---	112.37	902.77	957.82	C-18	2.3	912.00	---	149.82	1061.82	1135.23
8'-0" H																
5' W, 2 lites, 1 sliding	Inst	SET	C-18	4.0	383.50	---	84.27	467.77	509.07	C-18	3.0	442.50	---	112.37	554.87	609.92
6' W, 2 lites, 1 sliding	Inst	SET	C-18	4.0	419.90	---	84.27	504.17	545.47	C-18	3.0	484.50	---	112.37	596.87	651.92
7' W, 2 lites, 1 sliding	Inst	SET	C-18	4.0	484.90	---	84.27	569.17	610.47	C-18	3.0	559.50	---	112.37	671.87	726.92
8' W, 2 lites, 1 sliding	Inst	SET	C-18	4.0	484.90	---	84.27	569.17	610.47	C-18	3.0	559.50	---	112.37	671.87	726.92

			Costs Based On Large Volume							Costs Based On Small Volume					

Description	Oper	Unit	Crew Size	Avg Day Prod	Avg Mat'l Unit Cost	Avg Equip Unit Cost	Avg Labor Unit Cost	Avg Total Unit Cost	Avg Price Incl O&P	Crew Size	Avg Day Prod	Avg Mat'l Unit Cost	Avg Equip Unit Cost	Avg Labor Unit Cost	Avg Total Unit Cost	Avg Price Incl O&P

Doors (continued)

Description	Oper	Unit	Crew	Day	Mat'l	Equip	Labor	Total	Price	Crew	Day	Mat'l	Equip	Labor	Total	Price
10' W, 2 lites, 1 sliding	Inst	SET	C-18	4.0	631.80	---	84.27	716.07	757.37	C-18	3.0	729.00	---	112.37	841.37	896.42
9' W, 3 lites, 1 sliding	Inst	SET	C-18	3.0	559.00	---	112.37	671.37	726.42	C-18	2.3	645.00	---	149.82	794.82	868.23
12' W, 3 lites, 1 sliding	Inst	SET	C-18	3.0	657.80	---	112.37	770.17	825.22	C-18	2.3	759.00	---	149.82	908.82	982.23
15' W, 3 lites, 1 sliding	Inst	SET	C-18	3.0	874.90	---	112.37	987.27	1042.32	C-18	2.3	1009.50	---	149.82	1159.32	1232.73

Garage doors
Aluminum frame with plastic skin bonded to polystyrene foam core
Jamb type with hardware and deluxe lock

Description	Oper	Unit	Crew	Day	Mat'l	Equip	Labor	Total	Price	Crew	Day	Mat'l	Equip	Labor	Total	Price
8' x 7', single	Inst	EA	C-18	4.0	354.90	---	84.27	439.17	480.47	C-18	3.0	409.50	---	112.37	521.87	576.92
8' x 8', single	Inst	EA	C-18	4.0	461.50	---	84.27	545.77	587.07	C-18	3.0	532.50	---	112.37	644.87	699.92
9' x 7', single	Inst	EA	C-18	4.0	387.40	---	84.27	471.67	512.97	C-18	3.0	447.00	---	112.37	559.37	614.42
9' x 8', single	Inst	EA	C-18	4.0	496.60	---	84.27	580.87	622.17	C-18	3.0	573.00	---	112.37	685.37	740.42
16' x 7', double	Inst	EA	C-18	3.0	668.20	---	112.37	780.57	835.62	C-18	2.3	771.00	---	149.82	920.82	994.23
16' x 8', double	Inst	EA	C-18	3.0	864.50	---	112.37	976.87	1031.92	C-18	2.3	997.50	---	149.82	1147.32	1220.73

Track type with hardware and deluxe lock

Description	Oper	Unit	Crew	Day	Mat'l	Equip	Labor	Total	Price	Crew	Day	Mat'l	Equip	Labor	Total	Price
8' x 7', single	Inst	EA	C-18	4.0	418.60	---	84.27	502.87	544.17	C-18	3.0	483.00	---	112.37	595.37	650.42
9' x 7', single	Inst	EA	C-18	4.0	460.20	---	84.27	544.47	585.77	C-18	3.0	531.00	---	112.37	643.37	698.42
16' x 7', double	Inst	EA	C-18	3.0	768.30	---	112.37	880.67	935.72	C-18	2.3	886.50	---	149.82	1036.32	1109.73

Sectional type with hardware and key lock

Description	Oper	Unit	Crew	Day	Mat'l	Equip	Labor	Total	Price	Crew	Day	Mat'l	Equip	Labor	Total	Price
8' x 7', single	Inst	EA	C-18	4.0	436.80	---	84.27	521.07	562.37	C-18	3.0	504.00	---	112.37	616.37	671.42
9' x 7', single	Inst	EA	C-18	4.0	464.10	---	84.27	548.37	589.67	C-18	3.0	535.50	---	112.37	647.87	702.92
16' x 7', double	Inst	EA	C-18	3.0	791.70	---	112.37	904.07	959.12	C-18	2.3	913.50	---	149.82	1063.32	1136.73

Fiberglass
Jamb type with hardware and deluxe lock

Description	Oper	Unit	Crew	Day	Mat'l	Equip	Labor	Total	Price	Crew	Day	Mat'l	Equip	Labor	Total	Price
8' x 7', single	Inst	EA	C-18	4.0	312.00	---	84.27	396.27	437.57	C-18	3.0	360.00	---	112.37	472.37	527.42
8' x 8', single	Inst	EA	C-18	4.0	399.10	---	84.27	483.37	524.67	C-18	3.0	460.50	---	112.37	572.87	627.92
9' x 7', single	Inst	EA	C-18	4.0	340.60	---	84.27	424.87	466.17	C-18	3.0	393.00	---	112.37	505.37	560.42
9' x 8', single	Inst	EA	C-18	4.0	443.30	---	84.27	527.57	568.87	C-18	3.0	511.50	---	112.37	623.87	678.92
16' x 7', double	Inst	EA	C-18	3.0	620.10	---	112.37	732.47	787.52	C-18	2.3	715.50	---	149.82	865.32	938.73
16' x 8', double	Inst	EA	C-18	3.0	825.50	---	112.37	937.87	992.92	C-18	2.3	952.50	---	149.82	1102.32	1175.73

Track type with hardware and deluxe lock

Description	Oper	Unit	Crew	Day	Mat'l	Equip	Labor	Total	Price	Crew	Day	Mat'l	Equip	Labor	Total	Price
8' x 7', single	Inst	EA	C-18	4.0	364.00	---	84.27	448.27	489.57	C-18	3.0	420.00	---	112.37	532.37	587.42
9' x 7', single	Inst	EA	C-18	4.0	400.40	---	84.27	484.67	525.97	C-18	3.0	462.00	---	112.37	574.37	629.42
16' x 7', double	Inst	EA	C-18	3.0	704.60	---	112.37	816.97	872.02	C-18	2.3	813.00	---	149.82	962.82	1036.23

			Costs Based On Large Volume							Costs Based On Small Volume						
Description	Oper	Unit	Crew Size	Avg Day Prod	Avg Mat'l Unit Cost	Avg Equip Unit Cost	Avg Labor Unit Cost	Avg Total Unit Cost	Avg Price Incl O&P	Crew Size	Avg Day Prod	Avg Mat'l Unit Cost	Avg Equip Unit Cost	Avg Labor Unit Cost	Avg Total Unit Cost	Avg Price Incl O&P

Doors (continued)

Sectional type with hardware and key lock

Description	Oper	Unit	Crew Size	Avg Day Prod	Avg Mat'l Unit Cost	Avg Equip Unit Cost	Avg Labor Unit Cost	Avg Total Unit Cost	Avg Price Incl O&P	Crew Size	Avg Day Prod	Avg Mat'l Unit Cost	Avg Equip Unit Cost	Avg Labor Unit Cost	Avg Total Unit Cost	Avg Price Incl O&P
8' x 7', single	Inst	EA	C-18	4.0	354.90	---	84.27	439.17	480.47	C-18	3.0	409.50	---	112.37	521.87	576.92
9' x 7', single	Inst	EA	C-18	4.0	396.50	---	84.27	480.77	522.07	C-18	3.0	457.50	---	112.37	569.87	624.92
16' x 7', double	Inst	EA	C-18	3.0	627.90	---	112.37	740.27	795.32	C-18	2.3	724.50	---	149.82	874.32	947.73
Steel																
Jamb type with hardware and deluxe lock																
8' x 7', single	Inst	EA	C-18	4.0	266.50	---	84.27	350.77	392.07	C-18	3.0	307.50	---	112.37	419.87	474.92
8' x 8', single	Inst	EA	C-18	4.0	354.90	---	84.27	439.17	480.47	C-18	3.0	409.50	---	112.37	521.87	576.92
9' x 7', single	Inst	EA	C-18	4.0	292.50	---	84.27	376.77	418.07	C-18	3.0	337.50	---	112.37	449.87	504.92
9' x 8', single	Inst	EA	C-18	4.0	391.30	---	84.27	475.57	516.87	C-18	3.0	451.50	---	112.37	563.87	618.92
16' x 7', double	Inst	EA	C-18	3.0	500.50	---	112.37	612.87	667.92	C-18	2.3	577.50	---	149.82	727.32	800.73
16' x 8', double	Inst	EA	C-18	3.0	705.90	---	112.37	818.27	873.32	C-18	2.3	814.50	---	149.82	964.32	1037.73
Track type with hardware and deluxe lock																
8' x 7', single	Inst	EA	C-18	4.0	304.20	---	84.27	388.47	429.77	C-18	3.0	351.00	---	112.37	463.37	518.42
9' x 7', single	Inst	EA	C-18	4.0	338.00	---	84.27	422.27	463.57	C-18	3.0	390.00	---	112.37	502.37	557.42
16' x 7', double	Inst	EA	C-18	3.0	561.60	---	112.37	673.97	729.02	C-18	2.3	648.00	---	149.82	797.82	871.23
Sectional type with hardware and key lock																
8' x 7', single	Inst	EA	C-18	4.0	327.60	---	84.27	411.87	453.17	C-18	3.0	378.00	---	112.37	490.37	545.42
9' x 7', single	Inst	EA	C-18	4.0	354.90	---	84.27	439.17	480.47	C-18	3.0	409.50	---	112.37	521.87	576.92
16' x 7', double	Inst	EA	C-18	3.0	614.90	---	112.37	727.27	782.32	C-18	2.3	709.50	---	149.82	859.32	932.73
Wood																
Jamb type with hardware and deluxe lock																
8' x 7', single	Inst	EA	C-18	4.0	377.00	---	84.27	461.27	502.57	C-18	3.0	435.00	---	112.37	547.37	602.42
8' x 8', single	Inst	EA	C-18	4.0	464.10	---	84.27	548.37	589.67	C-18	3.0	535.50	---	112.37	647.87	702.92
9' x 7', single	Inst	EA	C-18	4.0	405.60	---	84.27	489.87	531.17	C-18	3.0	468.00	---	112.37	580.37	635.42
9' x 8', single	Inst	EA	C-18	4.0	508.30	---	84.27	592.57	633.87	C-18	3.0	586.50	---	112.37	698.87	753.92
16' x 7', double	Inst	EA	C-18	3.0	685.10	---	112.37	797.47	852.52	C-18	2.3	790.50	---	149.82	940.32	1013.73
16' x 8', double	Inst	EA	C-18	3.0	890.50	---	112.37	1002.87	1057.92	C-18	2.3	1027.50	---	149.82	1177.32	1250.73
Track type with hardware and deluxe lock																
8' x 7', single	Inst	EA	C-18	4.0	429.00	---	84.27	513.27	554.57	C-18	3.0	495.00	---	112.37	607.37	662.42
9' x 7', single	Inst	EA	C-18	4.0	465.40	---	84.27	549.67	590.97	C-18	3.0	537.00	---	112.37	649.37	704.42
16' x 7', double	Inst	EA	C-18	3.0	769.60	---	112.37	881.97	937.02	C-18	2.3	888.00	---	149.82	1037.82	1111.23

			Costs Based On Large Volume							Costs Based On Small Volume						
Description	Oper	Unit	Crew Size	Avg Day Prod	Avg Mat'l Unit Cost	Avg Equip Unit Cost	Avg Labor Unit Cost	Avg Total Unit Cost	Avg Price Incl O&P	Crew Size	Avg Day Prod	Avg Mat'l Unit Cost	Avg Equip Unit Cost	Avg Labor Unit Cost	Avg Total Unit Cost	Avg Price Incl O&P

Doors (continued)

Sectional type with hardware and key lock

8' x 7', single	Inst	EA	C-18	4.0	419.90	---	84.27	504.17	545.47	C-18	3.0	484.50	---	112.37	596.87	651.92
9' x 7', single	Inst	EA	C-18	4.0	461.50	---	84.27	545.77	587.07	C-18	3.0	532.50	---	112.37	644.87	699.92

Drywall. See Sheetrock.

Electrical

General work

Residential service, single phase system. Prices given on a cost per each basis for a unit price system which includes a weathercap, service entrance cable, meter socket, entrance disconnect switch, ground rod, with clamp, ground cable, EMT, and panelboard

Weathercap

100 AMP service	Inst	EA	E-1	8.0	4.50	---	24.49	28.99	40.01	E-1	6.0	5.85	---	32.66	38.51	53.20
200 AMP service	Inst	EA	E-1	5.0	13.10	---	39.19	52.29	69.92	E-1	3.8	17.03	---	52.25	69.28	92.79

Service entrance cable (typical allowance is 20 LF)

100 AMP service	Inst	LF	E-1	65	0.10	---	3.01	3.11	4.47	E-1	49	0.13	---	4.02	4.15	5.96
200 AMP service	Inst	LF	E-1	45	0.22	---	4.35	4.57	6.53	E-1	34	0.29	---	5.81	6.09	8.70

Meter socket

100 AMP service	Inst	EA	E-1	2.00	24.50	---	97.97	122.47	166.55	E-1	1.50	31.85	---	130.62	162.47	221.25
200 AMP service	Inst	EA	E-1	1.25	41.00	---	156.75	197.75	268.28	E-1	0.94	53.30	---	208.99	262.29	356.34

Entrance disconnect switch

100 AMP service	Inst	EA	E-1	1.25	130.00	---	156.75	286.75	357.28	E-1	0.94	169.00	---	208.99	377.99	472.04
200 AMP service	Inst	EA	E-1	0.85	260.00	---	230.51	490.51	594.24	E-1	0.64	338.00	---	307.34	645.34	783.65

Ground rod, with clamp

100 AMP service	Inst	EA	E-1	3.0	13.00	---	65.31	78.31	107.70	E-1	2.25	16.90	---	87.08	103.98	143.17
200 AMP service	Inst	EA	E-1	3.0	25.00	---	65.31	90.31	119.70	E-1	2.25	32.50	---	87.08	119.58	158.77

Ground cable (typical allowance is 10 LF)

100 AMP service	Inst	LF	E-1	100	0.13	---	1.96	2.09	2.97	E-1	75	0.17	---	2.61	2.78	3.96
200 AMP service	Inst	LF	E-1	80	0.21	---	2.45	2.66	3.76	E-1	60	0.27	---	3.27	3.54	5.01

3/4" EMT (typical allowance is 10 LF)

200 AMP service	Inst	LF	E-1	75	0.50	---	2.61	3.11	4.29	E-1	56	0.65	---	3.48	4.13	5.70

			Costs Based On Large Volume							Costs Based On Small Volume						
Description	Oper	Unit	Crew Size	Avg Day Prod	Avg Mat'l Unit Cost	Avg Equip Unit Cost	Avg Labor Unit Cost	Avg Total Unit Cost	Avg Price Incl O&P	Crew Size	Avg Day Prod	Avg Mat'l Unit Cost	Avg Equip Unit Cost	Avg Labor Unit Cost	Avg Total Unit Cost	Avg Price Incl O&P

Electrical (continued)

Panelboard

Description	Oper	Unit	Crew Size	Avg Day Prod	Avg Mat'l Unit Cost	Avg Equip Unit Cost	Avg Labor Unit Cost	Avg Total Unit Cost	Avg Price Incl O&P	Crew Size	Avg Day Prod	Avg Mat'l Unit Cost	Avg Equip Unit Cost	Avg Labor Unit Cost	Avg Total Unit Cost	Avg Price Incl O&P
100 AMP service - 12 circuit	Inst	EA	E-1	0.75	110	---	261	371	489	E-1	0.56	143	---	348	491	648
200 AMP service - 24 circuit	Inst	EA	E-1	0.40	270	---	490	760	980	E-1	0.30	351	---	653	1004	1298

Adjustments for other than normal working situations

Description	Oper	Unit	Crew Size	Avg Day Prod	Avg Mat'l Unit Cost	Avg Equip Unit Cost	Avg Labor Unit Cost	Avg Total Unit Cost	Avg Price Incl O&P	Crew Size	Avg Day Prod	Avg Mat'l Unit Cost	Avg Equip Unit Cost	Avg Labor Unit Cost	Avg Total Unit Cost	Avg Price Incl O&P
Cut and patch, ADD		%	E-1	---	2.0	---	3.0	---	---	E-1	---	5.0	---	9.0	---	---
Dust protection, ADD		%	E-1	---	1.0	---	2.0	---	---	E-1	---	4.0	---	11.0	---	---
Protect existing work, ADD		%	E-1	---	2.0	---	2.0	---	---	E-1	---	5.0	---	7.0	---	---

Wiring per outlet or switch; wall or ceiling

Description	Oper	Unit	Crew Size	Avg Day Prod	Avg Mat'l Unit Cost	Avg Equip Unit Cost	Avg Labor Unit Cost	Avg Total Unit Cost	Avg Price Incl O&P	Crew Size	Avg Day Prod	Avg Mat'l Unit Cost	Avg Equip Unit Cost	Avg Labor Unit Cost	Avg Total Unit Cost	Avg Price Incl O&P
Romex, non-metallic sheathed cable, 600 volt, copper with ground wire	Inst	EA	E-1	16.0	6.00	---	12.25	18.25	23.76	E-1	12.0	7.80	---	16.33	24.13	31.48
BX, flexible armored cable, 600 volt, copper	Inst	EA	E-1	12.0	11.00	---	16.33	27.33	34.68	E-1	9.0	14.30	---	21.77	36.07	45.87
EMT with wire, electric metallic thinwall, 1/2"	Inst	EA	E-1	6.0	14.00	---	32.66	46.66	61.35	E-1	4.5	18.20	---	43.54	61.74	81.33
Rigid with wire, 1/2"	Inst	EA	E-1	4.0	19.00	---	48.98	67.98	90.03	E-1	3.0	24.70	---	65.31	90.01	119.40
Wiring, connection, and installation in closed wall structure, ADD	Inst	EA	E-1	8.0	---	---	24.49	24.49	35.51	E-1	6.0	---	---	32.66	32.66	47.35

Lighting. See Lighting fixtures.

Special systems

Burglary detection systems

Description	Oper	Unit	Crew Size	Avg Day Prod	Avg Mat'l Unit Cost	Avg Equip Unit Cost	Avg Labor Unit Cost	Avg Total Unit Cost	Avg Price Incl O&P	Crew Size	Avg Day Prod	Avg Mat'l Unit Cost	Avg Equip Unit Cost	Avg Labor Unit Cost	Avg Total Unit Cost	Avg Price Incl O&P
Alarm bell	Inst	EA	E-1	4.0	60	---	49	109	131	E-1	3.0	78	---	65	143	173
Burglar alarm, battery operated																
Mechanical trigger	Inst	EA	E-1	4.0	200	---	49	249	271	E-1	3.0	260	---	65	325	355
Electrical trigger	Inst	EA	E-1	4.0	245	---	49	294	316	E-1	3.0	319	---	65	384	413
Adjustments, ADD:																
Outside key control	Inst	EA	---	---	60	---	---	60	60	---	---	60	---	---	60	60
Remote signaling circuitry	Inst	EA	---	---	100	---	---	100	100	---	---	100	---	---	100	100
Card reader																
Standard	Inst	EA	E-1	2.5	650	---	78	728	764	E-1	1.9	845	---	104	949	997
Multi-code	Inst	EA	E-1	2.5	900	---	78	978	1014	E-1	1.9	1170	---	104	1274	1322

| | | | Costs Based On Large Volume | | | | | | Costs Based On Small Volume | | | | | | |
Description	Oper	Unit	Crew Size	Avg Day Prod	Avg Mat'l Unit Cost	Avg Equip Unit Cost	Avg Labor Unit Cost	Avg Total Unit Cost	Avg Price Incl O&P	Crew Size	Avg Day Prod	Avg Mat'l Unit Cost	Avg Equip Unit Cost	Avg Labor Unit Cost	Avg Total Unit Cost	Avg Price Incl O&P

Electrical (continued)

Detectors
Motion

Description	Oper	Unit	Crew Size	Avg Day Prod	Avg Mat'l Unit Cost	Avg Equip Unit Cost	Avg Labor Unit Cost	Avg Total Unit Cost	Avg Price Incl O&P	Crew Size	Avg Day Prod	Avg Mat'l Unit Cost	Avg Equip Unit Cost	Avg Labor Unit Cost	Avg Total Unit Cost	Avg Price Incl O&P
Infrared photoelectric	Inst	EA	E-1	2.0	150	---	98	248	292	E-1	1.5	195	---	131	326	384
Passive infrared	Inst	EA	E-1	2.0	215	---	98	313	357	E-1	1.5	280	---	131	410	469
Ultrasonic, 12 volt	Inst	EA	E-1	2.0	180	---	98	278	322	E-1	1.5	234	---	131	365	423
Microwave																
10' to 200'	Inst	EA	E-1	2.0	500	---	98	598	642	E-1	1.5	650	---	131	781	839
10' to 350'	Inst	EA	E-1	2.0	1400	---	98	1498	1542	E-1	1.5	1820	---	131	1951	2009
Door switches																
Hinge switch	Inst	EA	E-1	5.0	50	---	39	89	107	E-1	3.8	65	---	52	117	141
Magnetic switch	Inst	EA	E-1	5.0	60	---	39	99	117	E-1	3.8	78	---	52	130	154
Exit control locks																
Horn alarm	Inst	EA	E-1	4.0	250	---	49	299	321	E-1	3.0	325	---	65	390	420
Flashing light alarm	Inst	EA	E-1	4.0	275	---	49	324	346	E-1	3.0	358	---	65	423	452
Glass break alarm switch	Inst	EA	E-1	8.0	40	---	24	64	76	E-1	6.0	52	---	33	85	99
Indicating panels																
1 channel	Inst	EA	E-1	2.5	260	---	78	338	374	E-1	1.9	338	---	104	442	490
10 channel	Inst	EA	E-1	1.5	900	---	131	1031	1089	E-1	1.1	1170	---	174	1344	1423
20 channel	Inst	EA	E-1	1.0	1725	---	196	1921	2009	E-1	0.8	2243	---	261	2504	2621
40 channel	Inst	EA	E-1	0.5	3100	---	392	3492	3668	E-1	0.4	4030	---	522	4552	4788
Police connect panel	Inst	EA	E-1	4.0	175	---	49	224	246	E-1	3.0	228	---	65	293	322
Siren	Inst	EA	E-1	4.0	110	---	49	159	181	E-1	3.0	143	---	65	208	238
Switchmats																
30" x 5'	Inst	EA	E-1	5.0	60	---	39	99	117	E-1	3.8	78	---	52	130	154
30" x 25'	Inst	EA	E-1	4.0	150	---	49	199	221	E-1	3.0	195	---	65	260	290
Telephone dialer	Inst	EA	E-1	5.0	275	---	39	314	332	E-1	3.8	358	---	52	410	433

Doorbell systems
includes transformer, button, bell

Description	Oper	Unit	Crew Size	Avg Day Prod	Avg Mat'l Unit Cost	Avg Equip Unit Cost	Avg Labor Unit Cost	Avg Total Unit Cost	Avg Price Incl O&P	Crew Size	Avg Day Prod	Avg Mat'l Unit Cost	Avg Equip Unit Cost	Avg Labor Unit Cost	Avg Total Unit Cost	Avg Price Incl O&P
Door chimes, 2 notes	Inst	EA	E-1	10.0	60	---	20	80	88	E-1	7.5	78	---	26	104	116
Tube type chimes	Inst	EA	E-1	8.0	150	---	24	174	186	E-1	6.0	195	---	33	228	242
Transformer and button only	Inst	EA	E-1	12.0	35	---	16	51	59	E-1	9.0	46	---	22	67	77
Push button only	Inst	EA	E-1	16.0	15	---	12	27	33	E-1	12.0	20	---	16	36	43

Description	Oper	Unit	Costs Based On Large Volume							Costs Based On Small Volume						
			Crew Size	Avg Day Prod	Avg Mat'l Unit Cost	Avg Equip Unit Cost	Avg Labor Unit Cost	Avg Total Unit Cost	Avg Price Incl O&P	Crew Size	Avg Day Prod	Avg Mat'l Unit Cost	Avg Equip Unit Cost	Avg Labor Unit Cost	Avg Total Unit Cost	Avg Price Incl O&P

Electrical (continued)

Fire alarm systems

Description	Oper	Unit	Crew Size	Avg Day Prod	Avg Mat'l	Avg Equip	Avg Labor	Avg Total	Avg Price O&P	Crew Size	Avg Day Prod	Avg Mat'l	Avg Equip	Avg Labor	Avg Total	Avg Price O&P
Battery and rack	Inst	EA	E-1	3.0	575	---	65	640	670	E-1	2.3	748	---	87	835	874
Automatic charger	Inst	EA	E-1	6.0	360	---	33	393	407	E-1	4.5	468	---	44	512	531
Detector																
Fixed temperature	Inst	EA	E-1	6.0	30	---	33	63	77	E-1	4.5	39	---	44	83	102
Rate of rise	Inst	EA	E-1	6.0	40	---	33	73	87	E-1	4.5	52	---	44	96	115
Door holder																
Electro-Magnetic	Inst	EA	E-1	4.0	65	---	49	114	136	E-1	3.0	85	---	65	150	179
Combination holder/closer	Inst	EA	E-1	3.0	350	---	65	415	445	E-1	2.3	455	---	87	542	581
Fire drill switch	Inst	EA	E-1	6.0	75	---	33	108	122	E-1	4.5	98	---	44	141	161
Glass break alarm switch	Inst	EA	E-1	8.0	40	---	24	64	76	E-1	6.0	52	---	33	85	99
Signal bell	Inst	EA	E-1	6.0	45	---	33	78	92	E-1	4.5	59	---	44	102	122
Smoke detector																
Ceiling type	Inst	EA	E-1	6.0	60	---	33	93	107	E-1	4.5	78	---	44	122	141
Duct type	Inst	EA	E-1	3.0	210	---	65	275	305	E-1	2.3	273	---	87	360	399
Intercom systems																
Master stations, with digital AM/FM receiver, up to 20 remote stations, & telephone interface																
Solid state with antenna, 200' wire, 4 speakers	Inst	Set	E-1	2.0	1200	---	98	1298	1342	E-1	1.5	1560	---	131	1691	1749
Solid state with antenna, 1000' wire, 8 speakers	Inst	Set	E-1	1.0	1600	---	196	1796	1884	E-1	0.8	2080	---	261	2341	2459
Room to door intercoms																
Master station, door station, transformer, wire	Inst	Set	E-1	4.0	300	---	49	349	371	E-1	3.0	390	---	65	455	485
Second door station, ADD	Inst	EA	E-1	8.0	40	---	24	64	76	E-1	6.0	52	---	33	85	99
Installation, wiring, and connection in closed wall, ADD	Inst	LS	E-1	5.0	---	---	39	39	57	E-1	3.8	---	---	52	52	76
Telephone, phone-jack wiring																
Pre-wiring, per outlet or jack	Inst	EA	E-1	12.0	20	---	16	36	44	E-1	9.0	26	---	22	48	58
Wiring, connection, and installation in closed wall structure, ADD	Inst	EA	E-1	8.0	---	---	24	24	36	E-1	6.0	---	---	33	33	47

			Costs Based On Large Volume							Costs Based On Small Volume						
Description	Oper	Unit	Crew Size	Avg Day Prod	Avg Mat'l Unit Cost	Avg Equip Unit Cost	Avg Labor Unit Cost	Avg Total Unit Cost	Avg Price Incl O&P	Crew Size	Avg Day Prod	Avg Mat'l Unit Cost	Avg Equip Unit Cost	Avg Labor Unit Cost	Avg Total Unit Cost	Avg Price Incl O&P

Electrical (continued)

Description	Oper	Unit	Crew Size	Avg Day Prod	Avg Mat'l Unit Cost	Avg Equip Unit Cost	Avg Labor Unit Cost	Avg Total Unit Cost	Avg Price Incl O&P	Crew Size	Avg Day Prod	Avg Mat'l Unit Cost	Avg Equip Unit Cost	Avg Labor Unit Cost	Avg Total Unit Cost	Avg Price Incl O&P
Television, antenna outlet 300 ohm Wiring, connection, and installation in closed wall	Inst	EA	E-1	12.0	12	---	16	28	36	E-1	9.0	16	---	22	37	47
structure, ADD	Inst	EA	E-1	8.0	---	---	24	24	36	E-1	6.0	---	---	33	33	47
Thermostat, wiring (heating unit on first floor)																
Thermostat on first floor Wiring, connection, and installation in closed wall	Inst	EA	E-1	8.0	20	---	24	44	56	E-1	6.0	26	---	33	59	73
structure, ADD	Inst	EA	E-1	8.0	---	---	24	24	36	E-1	6.0	---	---	33	33	47
Thermostat on second floor Wiring, connection, and installation in closed wall	Inst	EA	E-1	5.0	25	---	39	64	82	E-1	3.8	33	---	52	85	108
structure, ADD	Inst	EA	E-1	6.0	---	---	33	33	47	E-1	4.5	---	---	44	44	63

Entrances

Description	Oper	Unit	Crew Size	Avg Day Prod	Avg Mat'l Unit Cost	Avg Equip Unit Cost	Avg Labor Unit Cost	Avg Total Unit Cost	Avg Price Incl O&P	Crew Size	Avg Day Prod	Avg Mat'l Unit Cost	Avg Equip Unit Cost	Avg Labor Unit Cost	Avg Total Unit Cost	Avg Price Incl O&P
Single and double door entrances	Demo	EA	L-2	25	---	---	10	10	15	L-2	18	---	---	15	15	22

Colonial design, white pine; includes frames, pediments, and pilasters

Plain carved archway

Single door units

Description	Oper	Unit	Crew Size	Avg Day Prod	Avg Mat'l Unit Cost	Avg Equip Unit Cost	Avg Labor Unit Cost	Avg Total Unit Cost	Avg Price Incl O&P	Crew Size	Avg Day Prod	Avg Mat'l Unit Cost	Avg Equip Unit Cost	Avg Labor Unit Cost	Avg Total Unit Cost	Avg Price Incl O&P
3'-0" W x 6'-8" H	Inst	EA	C-18	8	147	---	42	189	210	C-18	6	170	---	60	230	259

Double door units

Description	Oper	Unit	Crew Size	Avg Day Prod	Avg Mat'l Unit Cost	Avg Equip Unit Cost	Avg Labor Unit Cost	Avg Total Unit Cost	Avg Price Incl O&P	Crew Size	Avg Day Prod	Avg Mat'l Unit Cost	Avg Equip Unit Cost	Avg Labor Unit Cost	Avg Total Unit Cost	Avg Price Incl O&P
Two - 2'-6" W x 6'-8" H	Inst	EA	C-18	6	159	---	56	215	242	C-18	4	183	---	80	263	303
Two - 2'-8" W x 6'-8" H	Inst	EA	C-18	6	159	---	56	215	242	C-18	4	183	---	80	263	303
Two - 3'-0" W x 6'-8" H	Inst	EA	C-18	6	166	---	56	222	250	C-18	4	191	---	80	272	311

Decorative carved archway

Single door units

Description	Oper	Unit	Crew Size	Avg Day Prod	Avg Mat'l Unit Cost	Avg Equip Unit Cost	Avg Labor Unit Cost	Avg Total Unit Cost	Avg Price Incl O&P	Crew Size	Avg Day Prod	Avg Mat'l Unit Cost	Avg Equip Unit Cost	Avg Labor Unit Cost	Avg Total Unit Cost	Avg Price Incl O&P
3'-0" W x 6'-8" H	Inst	EA	C-18	8	221	---	42	264	284	C-18	6	255	---	60	316	345

Double door units

Description	Oper	Unit	Crew Size	Avg Day Prod	Avg Mat'l Unit Cost	Avg Equip Unit Cost	Avg Labor Unit Cost	Avg Total Unit Cost	Avg Price Incl O&P	Crew Size	Avg Day Prod	Avg Mat'l Unit Cost	Avg Equip Unit Cost	Avg Labor Unit Cost	Avg Total Unit Cost	Avg Price Incl O&P
Two - 2'-6" W x 6'-8" H	Inst	EA	C-18	6	254	---	56	310	337	C-18	4	293	---	80	373	412
Two - 2'-8" W x 6'-8" H	Inst	EA	C-18	6	254	---	56	310	337	C-18	4	293	---	80	373	412
Two - 3'-0" W x 6'-8" H	Inst	EA	C-18	6	261	---	56	317	345	C-18	4	301	---	80	382	421

Excavation. See Site Work.

Facebrick. See Masonry.

			Costs Based On Large Volume							Costs Based On Small Volume						
Description	Oper	Unit	Crew Size	Avg Day Prod	Avg Mat'l Unit Cost	Avg Equip Unit Cost	Avg Labor Unit Cost	Avg Total Unit Cost	Avg Price Incl O&P	Crew Size	Avg Day Prod	Avg Mat'l Unit Cost	Avg Equip Unit Cost	Avg Labor Unit Cost	Avg Total Unit Cost	Avg Price Incl O&P

Fans. See HVAC, Ventilating.

Fences

Basketweave

Redwood, preassembled, 8' L panels, includes 4" x 4" line posts, horizontal or vertical weave

Description	Oper	Unit	Crew Size	Avg Day Prod	Avg Mat'l Unit Cost	Avg Equip Unit Cost	Avg Labor Unit Cost	Avg Total Unit Cost	Avg Price Incl O&P	Crew Size	Avg Day Prod	Avg Mat'l Unit Cost	Avg Equip Unit Cost	Avg Labor Unit Cost	Avg Total Unit Cost	Avg Price Incl O&P
5' H	Inst	LF	C-17	240	10.08	---	1.95	12.02	12.98	C-17	192	11.63	---	2.43	14.06	15.25
6' H	Inst	LF	C-17	240	11.70	---	1.95	13.65	14.60	C-17	192	13.50	---	2.43	15.93	17.13
Adjustments																
Corner or end posts, 8' H	Inst	EA	C-1	13	19.50	---	12.97	32.47	38.82	C-1	10	22.50	---	16.21	38.71	46.65
3-1/2" W x 5' H gate, with hardware	Inst	EA	C-1	10	84.50	---	16.85	101.35	109.61	C-1	8	97.50	---	21.07	118.57	128.89

Board

6' H nailed to wood frame; 1 side only; 8' L redwood 4' x 4' (milled) posts set 2' D in
concrete filled holes @ 6' o.c., frame members 2" x 4" (milled) as 2 rails between posts per 6' L
fence section; priced per LF fence

Douglas fir frame members with redwood posts

Milled boards, "dog-eared" one end:

Description	Oper	Unit	Crew Size	Avg Day Prod	Avg Mat'l Unit Cost	Avg Equip Unit Cost	Avg Labor Unit Cost	Avg Total Unit Cost	Avg Price Incl O&P	Crew Size	Avg Day Prod	Avg Mat'l Unit Cost	Avg Equip Unit Cost	Avg Labor Unit Cost	Avg Total Unit Cost	Avg Price Incl O&P
Cedar																
1" x 6" - 6' H	Inst	LF	C-16	60	7.96	---	7.51	15.47	19.15	C-16	48	9.18	---	9.39	18.57	23.18
1" x 8" - 6' H	Inst	LF	C-16	70	7.18	---	6.44	13.62	16.77	C-16	56	8.28	---	8.05	16.33	20.28
1" x 10" - 6' H	Inst	LF	C-16	80	6.38	---	5.64	12.02	14.78	C-16	64	7.37	---	7.05	14.41	17.86
Douglas fir																
1" x 6" - 6' H	Inst	LF	C-16	60	6.49	---	7.51	14.00	17.68	C-16	48	7.49	---	9.39	16.88	21.48
1" x 8" - 6' H	Inst	LF	C-16	70	6.38	---	6.44	12.82	15.98	C-16	56	7.37	---	8.05	15.42	19.36
1" x 10" - 6' H	Inst	LF	C-16	80	7.02	---	5.64	12.66	15.42	C-16	64	8.10	---	7.05	15.15	18.60
Redwood																
1" x 6" - 6' H	Inst	LF	C-16	60	7.68	---	7.51	15.20	18.88	C-16	48	8.87	---	9.39	18.26	22.86
1" x 8" - 6' H	Inst	LF	C-16	70	6.96	---	6.44	13.40	16.55	C-16	56	8.03	---	8.05	16.08	20.02
1" x 10" - 6' H	Inst	LF	C-16	80	6.19	---	5.64	11.82	14.59	C-16	64	7.14	---	7.05	14.19	17.64
Rough boards, both ends squared																
Cedar																
1" x 6" - 6' H	Inst	LF	C-16	65	7.42	---	6.94	14.36	17.76	C-16	52	8.57	---	8.67	17.24	21.48
1" x 8" - 6' H	Inst	LF	C-16	75	6.70	---	6.01	12.71	15.65	C-16	60	7.73	---	7.51	15.24	18.92
1" x 10" - 6' H	Inst	LF	C-16	85	5.95	---	5.30	11.26	13.86	C-16	68	6.87	---	6.63	13.50	16.75

			Costs Based On Large Volume							Costs Based On Small Volume					
Description	Oper	Unit	Crew Size	Avg Day Prod	Avg Mat'l Unit Cost	Avg Equip Unit Cost	Avg Labor Unit Cost	Avg Total Unit Cost	Avg Price Incl O&P	Crew Size	Avg Day Prod	Avg Mat'l Unit Cost	Avg Equip Unit Cost	Avg Labor Unit Cost	Avg Total Unit Cost	Avg Price Incl O&P

Fences (continued)

Douglas fir
1" x 6" - 6' H	Inst	LF	C-16	65	6.10	---	6.94	13.03	16.43	C-16	52	7.04	---	8.67	15.71	19.95
1" x 8" - 6' H	Inst	LF	C-16	75	5.99	---	6.01	12.00	14.95	C-16	60	6.92	---	7.51	14.43	18.11
1" x 10" - 6' H	Inst	LF	C-16	85	6.60	---	5.30	11.91	14.51	C-16	68	7.62	---	6.63	14.25	17.50

Redwood
1" x 6" - 6' H	Inst	LF	C-16	65	7.19	---	6.94	14.13	17.52	C-16	52	8.30	---	8.67	16.97	21.21
1" x 8" - 6' H	Inst	LF	C-16	75	6.51	---	6.01	12.52	15.47	C-16	60	7.52	---	7.51	15.03	18.71
1" x 10" - 6' H	Inst	LF	C-16	85	5.80	---	5.30	11.10	13.70	C-16	68	6.69	---	6.63	13.32	16.57

Redwood frame members with redwood posts
Milled boards, "dog-eared" one end:

Cedar
1" x 6" - 6' H	Inst	LF	C-16	60	8.40	---	7.51	15.91	19.59	C-16	48	9.69	---	9.39	19.08	23.69
1" x 8" - 6' H	Inst	LF	C-16	70	7.58	---	6.44	14.02	17.18	C-16	56	8.75	---	8.05	16.80	20.74
1" x 10" - 6' H	Inst	LF	C-16	80	6.73	---	5.64	12.37	15.13	C-16	64	7.77	---	7.05	14.82	18.27

Douglas fir
1" x 6" - 6' H	Inst	LF	C-16	60	6.93	---	7.51	14.44	18.13	C-16	48	8.00	---	9.39	17.39	21.99
1" x 8" - 6' H	Inst	LF	C-16	70	6.81	---	6.44	13.25	16.41	C-16	56	7.86	---	8.05	15.91	19.86
1" x 10" - 6' H	Inst	LF	C-16	80	7.50	---	5.64	13.14	15.90	C-16	64	8.66	---	7.05	15.70	19.15

Redwood
1" x 6" - 6' H	Inst	LF	C-16	60	8.14	---	7.51	15.65	19.33	C-16	48	9.39	---	9.39	18.78	23.39
1" x 8" - 6' H	Inst	LF	C-16	70	7.37	---	6.44	13.81	16.97	C-16	56	8.51	---	8.05	16.56	20.50
1" x 10" - 6' H	Inst	LF	C-16	80	6.55	---	5.64	12.19	14.95	C-16	64	7.56	---	7.05	14.61	18.06

Rough boards, both ends squared

Cedar
1" x 6" - 6' H	Inst	LF	C-16	65	7.87	---	6.94	14.80	18.20	C-16	52	9.08	---	8.67	17.75	21.99
1" x 8" - 6' H	Inst	LF	C-16	75	7.10	---	6.01	13.11	16.06	C-16	60	8.19	---	7.51	15.70	19.39
1" x 10" - 6' H	Inst	LF	C-16	85	6.31	---	5.30	11.61	14.21	C-16	68	7.28	---	6.63	13.91	17.15

Douglas fir
1" x 6" - 6' H	Inst	LF	C-16	65	6.54	---	6.94	13.48	16.87	C-16	52	7.55	---	8.67	16.22	20.46
1" x 8" - 6' H	Inst	LF	C-16	75	6.44	---	6.01	12.45	15.39	C-16	60	7.43	---	7.51	14.94	18.62
1" x 10" - 6' H	Inst	LF	C-16	85	7.09	---	5.30	12.39	14.99	C-16	68	8.18	---	6.63	14.81	18.05

			Costs Based On Large Volume							Costs Based On Small Volume						
Description	Oper	Unit	Crew Size	Avg Day Prod	Avg Mat'l Unit Cost	Avg Equip Unit Cost	Avg Labor Unit Cost	Avg Total Unit Cost	Avg Price Incl O&P	Crew Size	Avg Day Prod	Avg Mat'l Unit Cost	Avg Equip Unit Cost	Avg Labor Unit Cost	Avg Total Unit Cost	Avg Price Incl O&P

Fences (continued)

Redwood

Description	Oper	Unit	Crew Size	Avg Day Prod	Avg Mat'l Unit Cost	Avg Equip Unit Cost	Avg Labor Unit Cost	Avg Total Unit Cost	Avg Price Incl O&P	Crew Size	Avg Day Prod	Avg Mat'l Unit Cost	Avg Equip Unit Cost	Avg Labor Unit Cost	Avg Total Unit Cost	Avg Price Incl O&P
1" x 6" - 6' H	Inst	LF	C-16	65	7.63	---	6.94	14.57	17.97	C-16	52	8.81	---	8.67	17.48	21.72
1" x 8" - 6' H	Inst	LF	C-16	75	6.92	---	6.01	12.93	15.87	C-16	60	7.98	---	7.51	15.49	19.18
1" x 10" - 6' H	Inst	LF	C-16	85	6.15	---	5.30	11.45	14.05	C-16	68	7.10	---	6.63	13.73	16.97

Chain link

9 gauge galvanized steel, includes top rail (1-5/8" o.d.), line posts (2" o.d.) @10', sleeves

Description	Oper	Unit	Crew Size	Avg Day Prod	Avg Mat'l Unit Cost	Avg Equip Unit Cost	Avg Labor Unit Cost	Avg Total Unit Cost	Avg Price Incl O&P	Crew Size	Avg Day Prod	Avg Mat'l Unit Cost	Avg Equip Unit Cost	Avg Labor Unit Cost	Avg Total Unit Cost	Avg Price Incl O&P
36" H	Inst	LF	H-2	220	4.28	---	2.15	6.42	7.43	H-2	176	4.94	---	2.68	7.62	8.88
42" H	Inst	LF	H-2	210	4.67	---	2.25	6.92	7.97	H-2	168	5.39	---	2.81	8.20	9.52
48" H	Inst	LF	H-2	200	5.06	---	2.36	7.42	8.53	H-2	160	5.84	---	2.95	8.79	10.17
60" H	Inst	LF	H-2	180	5.84	---	2.62	8.46	9.69	H-2	144	6.74	---	3.28	10.01	11.56
72" H	Inst	LF	H-2	160	6.75	---	2.95	9.70	11.09	H-2	128	7.79	---	3.69	11.47	13.21

Adjustments

Description	Oper	Unit	Crew Size	Avg Day Prod	Avg Mat'l Unit Cost	Avg Equip Unit Cost	Avg Labor Unit Cost	Avg Total Unit Cost	Avg Price Incl O&P	Crew Size	Avg Day Prod	Avg Mat'l Unit Cost	Avg Equip Unit Cost	Avg Labor Unit Cost	Avg Total Unit Cost	Avg Price Incl O&P
11-1/2" gauge galvanized steel fabric, DEDUCT		%	---	27.0	---	---	---	---	---	---	---	27.0	---	---	---	---
12 gauge galvanized steel fabric, DEDUCT		%	---	29.0	---	---	---	---	---	---	---	29.0	---	---	---	---
9 gauge green vinyl-coated fabric, DEDUCT		%	---	4.0	---	---	---	---	---	---	---	4.0	---	---	---	---
11 gauge green vinyl-coated fabric, DEDUCT		%	---	3.8	---	---	---	---	---	---	---	3.8	---	---	---	---

Filler strips, ADD
Aluminum, baked on enamel finish
Diagonal, 1-7/8" W

Description	Oper	Unit	Crew Size	Avg Day Prod	Avg Mat'l Unit Cost	Avg Equip Unit Cost	Avg Labor Unit Cost	Avg Total Unit Cost	Avg Price Incl O&P	Crew Size	Avg Day Prod	Avg Mat'l Unit Cost	Avg Equip Unit Cost	Avg Labor Unit Cost	Avg Total Unit Cost	Avg Price Incl O&P
48" H	Inst	LF	L-2	100	3.38	---	2.60	5.98	7.23	L-2	80	3.90	---	3.25	7.15	8.71
60" H	Inst	LF	L-2	100	4.10	---	2.60	6.70	7.94	L-2	80	4.73	---	3.25	7.98	9.54
72" H	Inst	LF	L-2	100	4.80	---	2.60	7.40	8.65	L-2	80	5.54	---	3.25	8.79	10.35

Vertical, 1-1/4" W

Description	Oper	Unit	Crew Size	Avg Day Prod	Avg Mat'l Unit Cost	Avg Equip Unit Cost	Avg Labor Unit Cost	Avg Total Unit Cost	Avg Price Incl O&P	Crew Size	Avg Day Prod	Avg Mat'l Unit Cost	Avg Equip Unit Cost	Avg Labor Unit Cost	Avg Total Unit Cost	Avg Price Incl O&P
48" H	Inst	LF	L-2	100	3.38	---	2.60	5.98	7.23	L-2	80	3.90	---	3.25	7.15	8.71
60" H	Inst	LF	L-2	100	3.98	---	2.60	6.58	7.83	L-2	80	4.59	---	3.25	7.84	9.40
72" H	Inst	LF	L-2	100	4.68	---	2.60	7.28	8.53	L-2	80	5.40	---	3.25	8.65	10.21

Wood, redwood stain
Vertical, 1-1/4" W

Description	Oper	Unit	Crew Size	Avg Day Prod	Avg Mat'l Unit Cost	Avg Equip Unit Cost	Avg Labor Unit Cost	Avg Total Unit Cost	Avg Price Incl O&P	Crew Size	Avg Day Prod	Avg Mat'l Unit Cost	Avg Equip Unit Cost	Avg Labor Unit Cost	Avg Total Unit Cost	Avg Price Incl O&P
48" H	Inst	LF	L-2	100	3.38	---	2.60	5.98	7.23	L-2	80	3.90	---	3.25	7.15	8.71
60" H	Inst	LF	L-2	100	4.10	---	2.60	6.70	7.94	L-2	80	4.73	---	3.25	7.98	9.54
72" H	Inst	LF	L-2	100	4.80	---	2.60	7.40	8.65	L-2	80	5.54	---	3.25	8.79	10.35

			Costs Based On Large Volume							Costs Based On Small Volume						
Description	Oper	Unit	Crew Size	Avg Day Prod	Avg Mat'l Unit Cost	Avg Equip Unit Cost	Avg Labor Unit Cost	Avg Total Unit Cost	Avg Price Incl O&P	Crew Size	Avg Day Prod	Avg Mat'l Unit Cost	Avg Equip Unit Cost	Avg Labor Unit Cost	Avg Total Unit Cost	Avg Price Incl O&P

Fences (continued)

Corner posts (2-1/2" o.d.), installed, heavyweight

Description	Oper	Unit	Crew Size	Avg Day Prod	Avg Mat'l Unit Cost	Avg Equip Unit Cost	Avg Labor Unit Cost	Avg Total Unit Cost	Avg Price Incl O&P	Crew Size	Avg Day Prod	Avg Mat'l Unit Cost	Avg Equip Unit Cost	Avg Labor Unit Cost	Avg Total Unit Cost	Avg Price Incl O&P
36" H	Inst	EA	H-1	28	19.50	---	6.11	25.61	28.48	H-1	22	22.50	---	7.64	30.14	33.73
42" H	Inst	EA	H-1	27	22.10	---	6.34	28.44	31.42	H-1	22	25.50	---	7.92	33.42	37.14
48" H	Inst	EA	H-1	25	23.40	---	6.84	30.24	33.46	H-1	20	27.00	---	8.56	35.56	39.58
60" H	Inst	EA	H-1	22	27.95	---	7.78	35.73	39.38	H-1	18	32.25	---	9.72	41.97	46.54
72" H	Inst	EA	H-1	20	32.50	---	8.56	41.06	45.08	H-1	16	37.50	---	10.69	48.19	53.22

End or gate posts (2-1/2" o.d.), installed, heavyweight

Description	Oper	Unit	Crew Size	Avg Day Prod	Avg Mat'l Unit Cost	Avg Equip Unit Cost	Avg Labor Unit Cost	Avg Total Unit Cost	Avg Price Incl O&P	Crew Size	Avg Day Prod	Avg Mat'l Unit Cost	Avg Equip Unit Cost	Avg Labor Unit Cost	Avg Total Unit Cost	Avg Price Incl O&P
36" H	Inst	EA	H-1	28	14.95	---	6.11	21.06	23.93	H-1	22	17.25	---	7.64	24.89	28.48
42" H	Inst	EA	H-1	27	16.25	---	6.34	22.59	25.57	H-1	22	18.75	---	7.92	26.67	30.39
48" H	Inst	EA	H-1	25	18.20	---	6.84	25.04	28.26	H-1	20	21.00	---	8.56	29.56	33.58
60" H	Inst	EA	H-1	22	21.45	---	7.78	29.23	32.88	H-1	18	24.75	---	9.72	34.47	39.04
72" H	Inst	EA	H-1	20	24.70	---	8.56	33.26	37.28	H-1	16	28.50	---	10.69	39.19	44.22

Gates, square corner frame 9 gauge wire, installed

3' wide walk gates

Description	Oper	Unit	Crew Size	Avg Day Prod	Avg Mat'l Unit Cost	Avg Equip Unit Cost	Avg Labor Unit Cost	Avg Total Unit Cost	Avg Price Incl O&P	Crew Size	Avg Day Prod	Avg Mat'l Unit Cost	Avg Equip Unit Cost	Avg Labor Unit Cost	Avg Total Unit Cost	Avg Price Incl O&P
36" H	Inst	EA	H-1	13	54.60	---	13.16	67.76	73.95	H-1	10	63.00	---	16.45	79.45	87.18
42" H	Inst	EA	H-1	13	57.20	---	13.16	70.36	76.55	H-1	10	66.00	---	16.45	82.45	90.18
48" H	Inst	EA	H-1	12	58.50	---	14.26	72.76	79.46	H-1	10	67.50	---	17.82	85.32	93.70
60" H	Inst	EA	H-1	12	70.20	---	14.26	84.46	91.16	H-1	10	81.00	---	17.82	98.82	107.20
72" H	Inst	EA	H-1	11	81.90	---	15.55	97.45	104.77	H-1	9	94.50	---	19.44	113.94	123.08

12' wide driveway double gates

Description	Oper	Unit	Crew Size	Avg Day Prod	Avg Mat'l Unit Cost	Avg Equip Unit Cost	Avg Labor Unit Cost	Avg Total Unit Cost	Avg Price Incl O&P	Crew Size	Avg Day Prod	Avg Mat'l Unit Cost	Avg Equip Unit Cost	Avg Labor Unit Cost	Avg Total Unit Cost	Avg Price Incl O&P
36" H	Inst	EA	H-1	6	144.30	---	28.52	172.82	186.22	H-1	5	166.50	---	35.65	202.15	218.90
42" H	Inst	EA	H-1	6	152.10	---	28.52	180.62	194.02	H-1	5	175.50	---	35.65	211.15	227.90
48" H	Inst	EA	H-1	5	157.30	---	34.22	191.52	207.60	H-1	4	181.50	---	42.78	224.28	244.38
60" H	Inst	EA	H-1	5	188.50	---	34.22	222.72	238.80	H-1	4	217.50	---	42.78	260.28	280.38
72" H	Inst	EA	H-1	4	208.00	---	42.78	250.78	270.88	H-1	3	240.00	---	53.47	293.47	318.60

Split rail, red cedar, 10' long sectional spans

Description	Oper	Unit	Crew Size	Avg Day Prod	Avg Mat'l Unit Cost	Avg Equip Unit Cost	Avg Labor Unit Cost	Avg Total Unit Cost	Avg Price Incl O&P	Crew Size	Avg Day Prod	Avg Mat'l Unit Cost	Avg Equip Unit Cost	Avg Labor Unit Cost	Avg Total Unit Cost	Avg Price Incl O&P
Split fence, rail only	---	EA	---	---	8.84	---	---	8.84	8.84	---	---	10.20	---	---	10.20	10.20

Bored 2 rail posts

Description	Oper	Unit	Crew Size	Avg Day Prod	Avg Mat'l Unit Cost	Avg Equip Unit Cost	Avg Labor Unit Cost	Avg Total Unit Cost	Avg Price Incl O&P	Crew Size	Avg Day Prod	Avg Mat'l Unit Cost	Avg Equip Unit Cost	Avg Labor Unit Cost	Avg Total Unit Cost	Avg Price Incl O&P
5'-6" line or end posts	Inst	EA	C-1	13	10.40	---	12.97	23.37	29.72	C-1	10	12.00	---	16.21	28.21	36.15
5'-6" corner posts	Inst	EA	C-1	13	11.70	---	12.97	24.67	31.02	C-1	10	13.50	---	16.21	29.71	37.65

Bored 3 rail posts

Description	Oper	Unit	Crew Size	Avg Day Prod	Avg Mat'l Unit Cost	Avg Equip Unit Cost	Avg Labor Unit Cost	Avg Total Unit Cost	Avg Price Incl O&P	Crew Size	Avg Day Prod	Avg Mat'l Unit Cost	Avg Equip Unit Cost	Avg Labor Unit Cost	Avg Total Unit Cost	Avg Price Incl O&P
6'-6" line or end posts	Inst	EA	C-1	13	13.00	---	12.97	25.97	32.32	C-1	10	15.00	---	16.21	31.21	39.15
6'-6" corner posts	Inst	EA	C-1	13	14.30	---	12.97	27.27	33.62	C-1	10	16.50	---	16.21	32.71	40.65

			Costs Based On Large Volume							Costs Based On Small Volume						
Description	Oper	Unit	Crew Size	Avg Day Prod	Avg Mat'l Unit Cost	Avg Equip Unit Cost	Avg Labor Unit Cost	Avg Total Unit Cost	Avg Price Incl O&P	Crew Size	Avg Day Prod	Avg Mat'l Unit Cost	Avg Equip Unit Cost	Avg Labor Unit Cost	Avg Total Unit Cost	Avg Price Incl O&P

Fences (continued)

Complete fence estimate (not including gates)

Description	Oper	Unit	Crew Size	Avg Day Prod	Avg Mat'l Unit Cost	Avg Equip Unit Cost	Avg Labor Unit Cost	Avg Total Unit Cost	Avg Price Incl O&P	Crew Size	Avg Day Prod	Avg Mat'l Unit Cost	Avg Equip Unit Cost	Avg Labor Unit Cost	Avg Total Unit Cost	Avg Price Incl O&P
2 rail, 36" H, 5'-6" post	Inst	EA	C-9	380	2.94	---	0.79	3.72	4.11	C-9	304	3.39	---	0.98	4.37	4.85
3 rail, 48" H, 6'-6" post	Inst	EA	C-9	330	3.82	---	0.90	4.73	5.17	C-9	264	4.41	---	1.13	5.54	6.10

Gate
2 rail

Description	Oper	Unit	Crew Size	Avg Day Prod	Avg Mat'l Unit Cost	Avg Equip Unit Cost	Avg Labor Unit Cost	Avg Total Unit Cost	Avg Price Incl O&P	Crew Size	Avg Day Prod	Avg Mat'l Unit Cost	Avg Equip Unit Cost	Avg Labor Unit Cost	Avg Total Unit Cost	Avg Price Incl O&P
3-1/2' W	Inst	EA	C-1	13	52.00	---	12.97	64.97	71.32	C-1	10	60.00	---	16.21	76.21	84.15
5' W	Inst	EA	C-1	10	67.60	---	16.85	84.45	92.71	C-1	8	78.00	---	21.07	99.07	109.39

3 rail

Description	Oper	Unit	Crew Size	Avg Day Prod	Avg Mat'l Unit Cost	Avg Equip Unit Cost	Avg Labor Unit Cost	Avg Total Unit Cost	Avg Price Incl O&P	Crew Size	Avg Day Prod	Avg Mat'l Unit Cost	Avg Equip Unit Cost	Avg Labor Unit Cost	Avg Total Unit Cost	Avg Price Incl O&P
3-1/2' W	Inst	EA	C-1	13	65.00	---	12.97	77.97	84.32	C-1	10	75.00	---	16.21	91.21	99.15
5' W	Inst	EA	C-1	10	75.40	---	16.85	92.25	100.51	C-1	8	87.00	---	21.07	108.07	118.39

Fiberglass panels

Corrugated

8', 10', or 12' L panels; 2-1/2" W x 1/2"
Corrugation; nailed on wood frame

Description	Oper	Unit	Crew Size	Avg Day Prod	Avg Mat'l Unit Cost	Avg Equip Unit Cost	Avg Labor Unit Cost	Avg Total Unit Cost	Avg Price Incl O&P	Crew Size	Avg Day Prod	Avg Mat'l Unit Cost	Avg Equip Unit Cost	Avg Labor Unit Cost	Avg Total Unit Cost	Avg Price Incl O&P
4 oz., .03" T, 26" W	Inst	SF	C-1	200	0.66	---	0.84	1.51	1.92	C-1	160	0.77	---	1.05	1.82	2.33
5 oz., .037" T, 26" W	Inst	SF	C-1	200	0.77	---	0.84	1.61	2.02	C-1	160	0.89	---	1.05	1.94	2.45
6 oz., .045" T, 26" W	Inst	SF	C-1	200	0.88	---	0.84	1.73	2.14	C-1	160	1.02	---	1.05	2.07	2.59
8 oz., .06" T, 26" W	Inst	SF	C-1	200	1.17	---	0.84	2.01	2.43	C-1	160	1.35	---	1.05	2.40	2.92

Flat panels

8', 10', or 12' L panels; clear, green, and white

Description	Oper	Unit	Crew Size	Avg Day Prod	Avg Mat'l Unit Cost	Avg Equip Unit Cost	Avg Labor Unit Cost	Avg Total Unit Cost	Avg Price Incl O&P	Crew Size	Avg Day Prod	Avg Mat'l Unit Cost	Avg Equip Unit Cost	Avg Labor Unit Cost	Avg Total Unit Cost	Avg Price Incl O&P
5 oz., 48" w	Inst	SF	C-1	200	0.91	---	0.84	1.75	2.17	C-1	160	1.05	---	1.05	2.10	2.62
6 oz., 48" W	Inst	SF	C-1	200	1.05	---	0.84	1.90	2.31	C-1	160	1.22	---	1.05	2.27	2.78
8 oz., 48" W	Inst	SF	C-1	200	1.39	---	0.84	2.23	2.65	C-1	160	1.61	---	1.05	2.66	3.17

Solar block

8', 10', 12' L panels; 2-1/2" W x 1/2" D corrugation; nailed on wood frame

Description	Oper	Unit	Crew Size	Avg Day Prod	Avg Mat'l Unit Cost	Avg Equip Unit Cost	Avg Labor Unit Cost	Avg Total Unit Cost	Avg Price Incl O&P	Crew Size	Avg Day Prod	Avg Mat'l Unit Cost	Avg Equip Unit Cost	Avg Labor Unit Cost	Avg Total Unit Cost	Avg Price Incl O&P
5 oz., 26" W	Inst	SF	C-1	200	0.77	---	0.84	1.61	2.02	C-1	160	0.89	---	1.05	1.94	2.45

Accessories
Wood corrugated

Description	Oper	Unit	Crew Size	Avg Day Prod	Avg Mat'l Unit Cost	Avg Equip Unit Cost	Avg Labor Unit Cost	Avg Total Unit Cost	Avg Price Incl O&P	Crew Size	Avg Day Prod	Avg Mat'l Unit Cost	Avg Equip Unit Cost	Avg Labor Unit Cost	Avg Total Unit Cost	Avg Price Incl O&P
2-1/2" W x 1-1/2" D x 6' L	Inst	EA	---	---	2.17	---	---	---	2.17	---	---	2.51	---	---	---	2.51
2-1/2" W x 1-1/2" D x 8' L	Inst	EA	---	---	2.87	---	---	---	2.87	---	---	3.32	---	---	---	3.32
2-1/2" W x 3/4" D x 6' L	Inst	EA	---	---	1.09	---	---	---	1.09	---	---	1.26	---	---	---	1.26
2-1/2" W x 3/4" D x 8' L	Inst	EA	---	---	1.40	---	---	---	1.40	---	---	1.62	---	---	---	1.62

			Costs Based On Large Volume							Costs Based On Small Volume					
Description	Oper	Unit	Crew Size	Avg Day Prod	Avg Mat'l Unit Cost	Avg Equip Unit Cost	Avg Labor Unit Cost	Avg Total Unit Cost	Avg Price Incl O&P	Crew Size	Avg Day Prod	Avg Mat'l Unit Cost	Avg Equip Unit Cost	Avg Labor Unit Cost	Avg Total Unit Cost	Avg Price Incl O&P

Fiberglass panels (continued)

Rubber corrugated

Description	Oper	Unit	Crew Size	Avg Day Prod	Avg Mat'l Unit Cost	Avg Equip Unit Cost	Avg Labor Unit Cost	Avg Total Unit Cost	Avg Price Incl O&P	Crew Size	Avg Day Prod	Avg Mat'l Unit Cost	Avg Equip Unit Cost	Avg Labor Unit Cost	Avg Total Unit Cost	Avg Price Incl O&P
1" x 3"	Inst	EA	---	---	1.09	---	---	---	1.09	---	---	1.26	---	---	---	1.26
Polyfoam corrugated																
1" x 3"	Inst	EA	---	---	0.77	---	---	---	0.77	---	---	0.89	---	---	---	0.89
Vertical crown moulding																
Wood																
1-1/2" x 6' L	Inst	EA	---	---	2.17	---	---	---	2.17	---	---	2.51	---	---	---	2.51
1-1/2" x 8' L	Inst	EA	---	---	2.87	---	---	---	2.87	---	---	3.32	---	---	---	3.32
Polyfoam, 1" x 1" x 3' L	Inst	EA	---	---	1.05	---	---	---	1.05	---	---	1.22	---	---	---	1.22
Rubber, 1" x 1" x 3' L	Inst	EA	---	---	1.99	---	---	---	1.99	---	---	2.30	---	---	---	2.30

Fireplaces

Woodburning, prefabricated; no masonry support required, installs directly on floor; ceramic backed firebox with black vitreous enamel side panels. No finish plastering or brick hearthwork included. Fire screen, 9" (i.d.) factory-built insulated chimneys with flue, lining, damper, and flashing with rain cap included. Chimney height from floor to where chimney exits through roof

36" W fireplace unit with:

Description	Oper	Unit	Crew Size	Avg Day Prod	Avg Mat'l Unit Cost	Avg Equip Unit Cost	Avg Labor Unit Cost	Avg Total Unit Cost	Avg Price Incl O&P	Crew Size	Avg Day Prod	Avg Mat'l Unit Cost	Avg Equip Unit Cost	Avg Labor Unit Cost	Avg Total Unit Cost	Avg Price Incl O&P
Up to 9'-0" chimney height	Inst	LS	C-9	1.3	845	---	230	1075	1188	C-9	0.9	976	---	185	1161	1252
9'-3" to 12'-2" chimney	Inst	LS	C-9	1.2	889	---	249	1138	1260	C-9	0.8	1026	---	201	1227	1325
12'-3" to 15'-1" chimney	Inst	LS	C-9	1.1	933	---	271	1205	1338	C-9	0.8	1077	---	219	1296	1403
15'-2" to 18'-0" chimney	Inst	LS	C-9	1.0	977	---	299	1275	1422	C-9	0.7	1127	---	241	1368	1486
18'-1" to 20'-11" chimney	Inst	LS	C-9	0.9	1021	---	332	1352	1515	C-9	0.6	1178	---	268	1445	1576
21'-3" to 23'-10" chimney	Inst	LS	C-9	0.8	1065	---	373	1438	1621	C-9	0.6	1228	---	301	1529	1677
23'-11" to 24'-9" chimney	Inst	LS	C-9	0.7	1108	---	427	1535	1744	C-9	0.5	1279	---	344	1623	1791
42" W fireplace unit with:																
Up to 9'-0" chimney height	Inst	LS	C-9	1.3	956	---	230	1186	1299	C-9	0.9	1103	---	185	1289	1379
9'-3" to 12'-2" chimney	Inst	LS	C-9	1.2	1006	---	249	1255	1377	C-9	0.8	1161	---	201	1362	1460
12'-3" to 15'-1" chimney	Inst	LS	C-9	1.1	1057	---	271	1328	1461	C-9	0.8	1219	---	219	1438	1545
15'-2" to 18'-0" chimney	Inst	LS	C-9	1.0	1107	---	299	1405	1552	C-9	0.7	1277	---	241	1518	1636
18'-1" to 20'-11" chimney	Inst	LS	C-9	0.9	1157	---	332	1489	1651	C-9	0.6	1335	---	268	1602	1733
21'-3" to 23'-10" chimney	Inst	LS	C-9	0.8	1207	---	373	1580	1763	C-9	0.6	1393	---	301	1694	1841
23'-11" to 24'-9" chimney	Inst	LS	C-9	0.7	1257	---	427	1684	1893	C-9	0.5	1450	---	344	1794	1963

			Costs Based On Large Volume							Costs Based On Small Volume					
Description	Oper	Unit	Crew Size	Avg Day Prod	Avg Mat'l Unit Cost	Avg Equip Unit Cost	Avg Labor Unit Cost	Avg Total Unit Cost	Avg Price Incl O&P	Crew Size	Avg Day Prod	Avg Mat'l Unit Cost	Avg Equip Unit Cost	Avg Labor Unit Cost	Avg Total Unit Cost	Avg Price Incl O&P

Fireplaces (continued)

Accessories

Description	Oper	Unit	Crew Size	Avg Day Prod	Avg Mat'l Unit Cost	Avg Equip Unit Cost	Avg Labor Unit Cost	Avg Total Unit Cost	Avg Price Incl O&P	Crew Size	Avg Day Prod	Avg Mat'l Unit Cost	Avg Equip Unit Cost	Avg Labor Unit Cost	Avg Total Unit Cost	Avg Price Incl O&P
Log lighter with gas valve (straight or angle pattern)	Inst	EA	S-1	6.0	28.60	---	49.77	78.37	102.75	S-1	4.2	33.00	---	40.13	73.13	92.79
Log lighter, less gas valve (straight, angle, tee pattern)	Inst	EA	S-1	12.0	10.40	---	24.88	35.28	47.48	S-1	8.4	12.00	---	---	12.00	12.00
Gas valve for log lighter	Inst	EA	S-1	12.0	18.85	---	24.88	43.73	55.93	S-1	8.4	21.75	---	---	21.75	21.75
Spare parts																
Gas valve key	Inst	EA	---	---	1.30	---	---	1.30	1.30	---	---	1.50	---	---	1.50	1.50
Stem extender	Inst	EA	---	---	2.28	---	---	2.28	2.28	---	---	2.63	---	---	2.63	2.63
Extra long	Inst	EA	---	---	3.25	---	---	3.25	3.25	---	---	3.75	---	---	3.75	3.75
Valve floor plate	Inst	EA	---	---	2.73	---	---	2.73	2.73	---	---	3.15	---	---	3.15	3.15
Lighter burner tube (12" to 17")	Inst	EA	---	---	5.53	---	---	5.53	5.53	---	---	6.38	---	---	6.38	6.38

Mantels. See Mantels, fireplace

Flashing. See Sheet metal.

Floor finishes. See individual items.

Floor joists. See Framing.

Food center

Includes wiring, connection and installation in exposed drainboard only.
Built-in models, 1/4 hp, 4-1/4" x 6-3/4" x

Description	Oper	Unit	Crew Size	Avg Day Prod	Avg Mat'l Unit Cost	Avg Equip Unit Cost	Avg Labor Unit Cost	Avg Total Unit Cost	Avg Price Incl O&P	Crew Size	Avg Day Prod	Avg Mat'l Unit Cost	Avg Equip Unit Cost	Avg Labor Unit Cost	Avg Total Unit Cost	Avg Price Incl O&P
10" rough cut, 110 volts, 6 speed	Inst	EA	E-1	3.00	439.28	---	65.31	504.59	534.64	E-1	2.00	506.87	---	97.97	604.83	649.90
Blender	---	EA	---	---	47.19	---	---	---	47.19	---	---	54.45	---	---	---	54.45
Fruit juicer	---	EA	---	---	23.11	---	---	---	23.11	---	---	26.67	---	---	---	26.67
Ice crusher	---	EA	---	---	74.93	---	---	---	74.93	---	---	86.46	---	---	---	86.46
Knife sharpener	---	EA	---	---	49.19	---	---	---	49.19	---	---	56.76	---	---	---	56.76
Meat grinder, shredder/slicer with power post	---	EA	---	---	286.00	---	---	---	286.00	---	---	330.00	---	---	---	330.00
Mixer	---	EA	---	---	134.28	---	---	---	134.28	---	---	154.94	---	---	---	154.94
Food processor	---	EA	---	---	238.81	---	---	---	238.81	---	---	275.55	---	---	---	275.55

			Costs Based On Large Volume							Costs Based On Small Volume						
Description	Oper	Unit	Crew Size	Avg Day Prod	Avg Mat'l Unit Cost	Avg Equip Unit Cost	Avg Labor Unit Cost	Avg Total Unit Cost	Avg Price Incl O&P	Crew Size	Avg Day Prod	Avg Mat'l Unit Cost	Avg Equip Unit Cost	Avg Labor Unit Cost	Avg Total Unit Cost	Avg Price Incl O&P

Footings. See Concrete, Cast-In-Place.

Formica. See Countertops.

Forming. See Concrete, Cast-In-Place.

Foundations. See Concrete, Masonry.

Framing. Rough Carpentry

Dimension lumber

Beams
set on steel columns, not wood columns
  Built-up:

Description	Oper	Unit	Crew Size	Avg Day Prod	Avg Mat'l Unit Cost	Avg Equip Unit Cost	Avg Labor Unit Cost	Avg Total Unit Cost	Avg Price Incl O&P	Crew Size	Avg Day Prod	Avg Mat'l Unit Cost	Avg Equip Unit Cost	Avg Labor Unit Cost	Avg Total Unit Cost	Avg Price Incl O&P
4" x 6" - 10' (2 pieces)	Demo	LF	L-2	661	---	---	0.39	0.39	0.58	L-2	595	---	---	0.44	0.44	0.65
4" x 6" - 10' (2 pieces)	Inst	LF	C-18	377	0.71	0.12	0.89	1.73	2.17	C-18	339	0.85	0.14	0.99	1.98	2.47
4" x 8" - 10' (2 pieces)	Demo	LF	L-2	595	---	---	0.44	0.44	0.65	L-2	536	---	---	0.49	0.49	0.72
4" x 8" - 10' (2 pieces)	Inst	LF	C-18	350	0.99	0.13	0.96	2.09	2.56	C-18	315	1.19	0.15	1.07	2.41	2.93
4" x 10" - 10' (2 pieces)	Demo	LF	L-2	540	---	---	0.48	0.48	0.71	L-2	486	---	---	0.54	0.54	0.79
4" x 10" - 10' (2 pieces)	Inst	LF	C-18	325	1.33	0.14	1.04	2.51	3.02	C-18	293	1.60	0.16	1.15	2.91	3.48
4" x 12" - 12' (2 pieces)	Demo	LF	L-2	495	---	---	0.53	0.53	0.78	L-2	446	---	---	0.58	0.58	0.86
4" x 12" - 12' (2 pieces)	Inst	LF	C-18	304	1.63	0.15	1.11	2.90	3.44	C-18	274	1.96	0.17	1.23	3.36	3.97
6" x 8" - 10' (2 pieces)	Demo	LF	L-4	744	---	---	0.80	0.80	1.19	L-4	670	---	---	0.89	0.89	1.32
6" x 8" - 10' (2 pieces)	Inst	LF	C-21	435	1.72	0.11	1.37	3.20	3.87	C-21	392	2.07	0.12	1.53	3.71	4.46
6" x 10" - 10' (2 pieces)	Demo	LF	L-4	676	---	---	0.88	0.88	1.31	L-4	608	---	---	0.98	0.98	1.45
6" x 10" - 10' (2 pieces)	Inst	LF	C-21	405	2.20	0.12	1.47	3.78	4.51	C-21	365	2.63	0.13	1.64	4.40	5.20
6" x 12" - 12' (2 pieces)	Demo	LF	L-4	619	---	---	0.96	0.96	1.43	L-4	557	---	---	1.07	1.07	1.59
6" x 12" - 12' (2 pieces)	Inst	LF	C-21	379	2.69	0.12	1.58	4.39	5.16	C-21	341	3.23	0.14	1.75	5.12	5.97
6" x 8" - 10' (3 pieces)	Demo	LF	L-4	733	---	---	0.81	0.81	1.21	L-4	660	---	---	0.91	0.91	1.34
6" x 8" - 10' (3 pieces)	Inst	LF	C-21	425	1.51	0.11	1.41	3.02	3.71	C-21	383	1.81	0.12	1.56	3.49	4.26
6" x 10" - 10' (3 pieces)	Demo	LF	L-4	669	---	---	0.89	0.89	1.32	L-4	602	---	---	0.99	0.99	1.47
6" x 10" - 10' (3 pieces)	Inst	LF	C-21	397	2.03	0.12	1.50	3.65	4.38	C-21	357	2.43	0.13	1.67	4.23	5.05
6" x 12" - 12' (3 pieces)	Demo	LF	L-4	609	---	---	0.98	0.98	1.45	L-4	548	---	---	1.09	1.09	1.61
6" x 12" - 12' (3 pieces)	Inst	LF	C-21	368	2.48	0.13	1.62	4.23	5.02	C-21	331	2.97	0.14	1.80	4.92	5.80

Framing

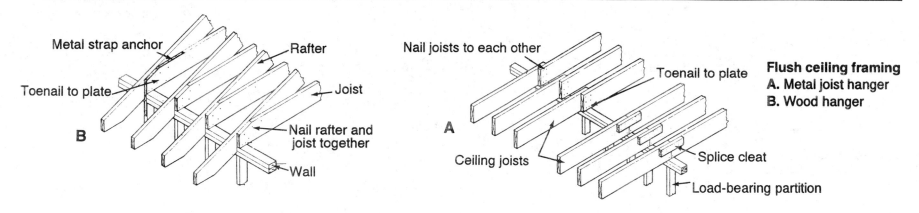

Metal strap anchor

Toenail to plate

B

Rafter

Joist

Nail rafter and joist together

Wall

Nail joists to each other

Toenail to plate

A

Ceiling joists

Splice cleat

Load-bearing partition

Flush ceiling framing
A. Metal joist hanger
B. Wood hanger

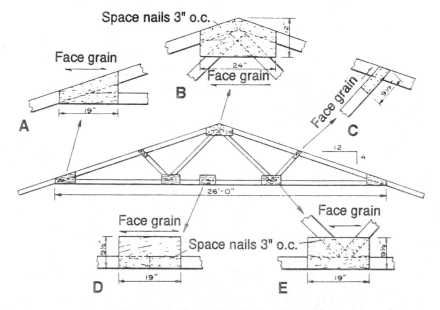

Space nails 3" o.c.

Face grain

A

19"

Face grain

B

24"

2

Face grain

C

9½

Face grain

D

19"

9½

Space nails 3" o.c.

Face grain

E

19"

9½

12
4

26'-0"

Construction of a 26 foot W truss:
A. Bevel-heel gusset
B. Peak gusset
C. Upper chord intermediate gusset
D. Splice of lower chord
E. Lower chord intermediate gusset

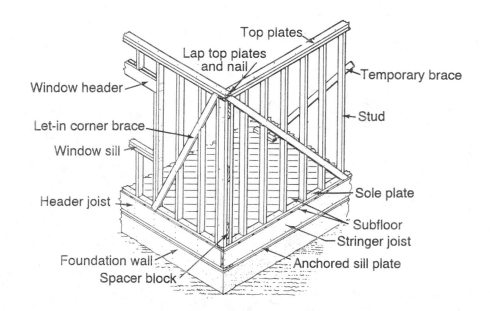

Top plates

Lap top plates and nail

Window header

Let-in corner brace

Window sill

Header joist

Foundation wall

Spacer block

Temporary brace

Stud

Sole plate

Subfloor

Stringer joist

Anchored sill plate

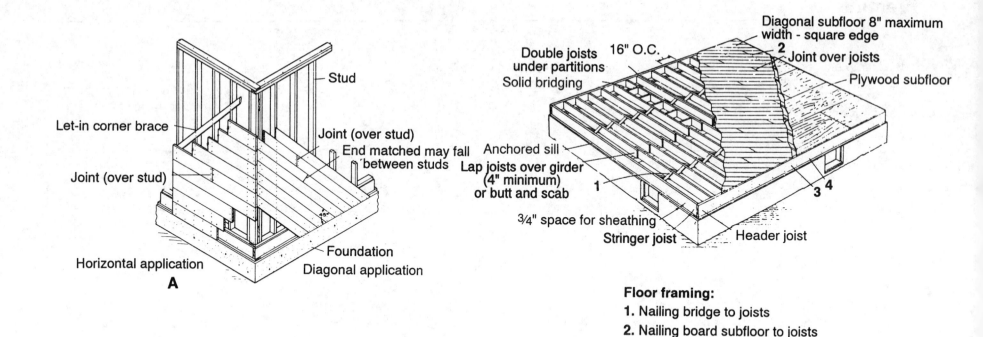

Stud

Let-in corner brace

Joint (over stud)

Joint (over stud)

End matched may fall between studs

Horizontal application

Foundation

Diagonal application

A

Double joists under partitions

Solid bridging

16" O.C.

Diagonal subfloor 8" maximum width - square edge

Joint over joists

Plywood subfloor

Anchored sill

Lap joists over girder (4" minimum) or butt and scab

¾" space for sheathing

Stringer joist

Header joist

Floor framing:

1. Nailing bridge to joists
2. Nailing board subfloor to joists
3. Nailing header to joists
4. Toenailing header to sill

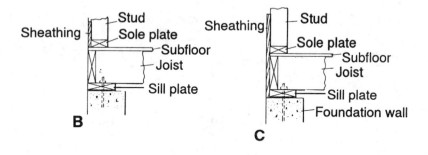

Sheathing

Stud

Sole plate

Subfloor

Joist

Sill plate

B

Sheathing

Stud

Sole plate

Subfloor

Joist

Sill plate

Foundation wall

C

Application of wood sheathing:

A. Horizontal and diagonal
B. Started at subfloor
C. Started at foundation wall

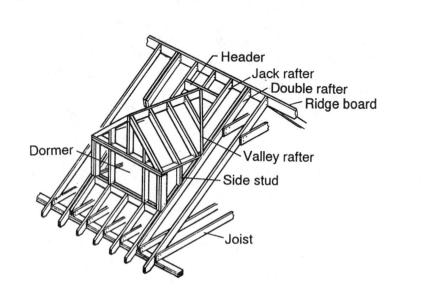

Header

Jack rafter

Double rafter

Ridge board

Dormer

Valley rafter

Side stud

Joist

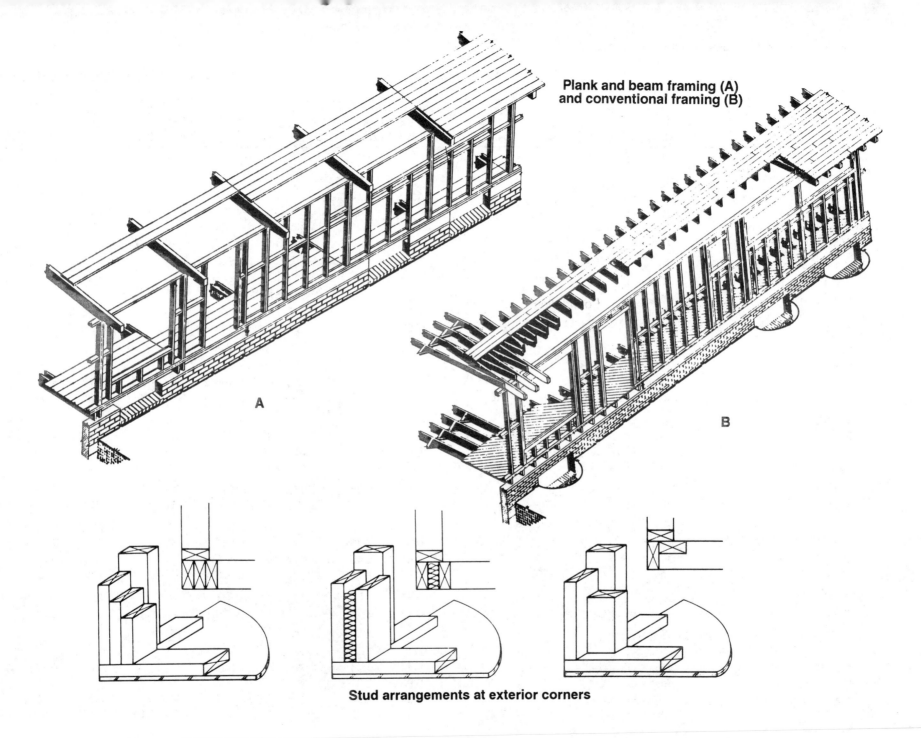

Plank and beam framing (A) and conventional framing (B)

A

B

Stud arrangements at exterior corners

				Costs Based On Large Volume							Costs Based On Small Volume					
Description	Oper	Unit	Crew Size	Avg Day Prod	Avg Mat'l Unit Cost	Avg Equip Unit Cost	Avg Labor Unit Cost	Avg Total Unit Cost	Avg Price Incl O&P	Crew Size	Avg Day Prod	Avg Mat'l Unit Cost	Avg Equip Unit Cost	Avg Labor Unit Cost	Avg Total Unit Cost	Avg Price Incl O&P

Framing (continued)

Description	Oper	Unit	Crew Size	Avg Day Prod	Avg Mat'l Unit Cost	Avg Equip Unit Cost	Avg Labor Unit Cost	Avg Total Unit Cost	Avg Price Incl O&P	Crew Size	Avg Day Prod	Avg Mat'l Unit Cost	Avg Equip Unit Cost	Avg Labor Unit Cost	Avg Total Unit Cost	Avg Price Incl O&P
9" x 10" - 12' (3 pieces)	Demo	LF	L-4	541	---	---	1.10	1.10	1.63	L-4	487	---	---	1.23	1.23	1.82
9" x 10" - 12' (3 pieces)	Inst	LF	C-21	338	2.99	0.14	1.77	4.89	5.76	C-21	304	3.59	0.15	1.96	5.70	6.66
9" x 12" - 12' (3 pieces)	Demo	LF	L-4	493	---	---	1.21	1.21	1.79	L-4	444	---	---	1.35	1.35	1.99
9" x 12" - 12' (3 pieces)	Inst	LF	C-21	315	3.66	0.15	1.90	5.71	6.64	C-21	284	4.40	0.16	2.11	6.67	7.70
Single member:																
2" x 6"	Demo	LF	L-2	947	---	---	0.27	0.27	0.41	L-2	852	---	---	0.31	0.31	0.45
2" x 6"	Inst	LF	C-18	520	0.43	0.09	0.65	1.17	1.49	C-18	468	0.52	0.10	0.72	1.34	1.69
2" x 8"	Demo	LF	L-2	874	---	---	0.30	0.30	0.44	L-2	787	---	---	0.33	0.33	0.49
2" x 8"	Inst	LF	C-18	491	0.57	0.10	0.69	1.35	1.69	C-18	442	0.68	0.11	0.76	1.55	1.92
2" x 10"	Demo	LF	L-2	811	---	---	0.32	0.32	0.47	L-2	730	---	---	0.36	0.36	0.53
2" x 10"	Inst	LF	C-18	465	0.74	0.10	0.72	1.57	1.92	C-18	419	0.89	0.11	0.81	1.81	2.20
2" x 12"	Demo	LF	L-2	756	---	---	0.34	0.34	0.51	L-2	680	---	---	0.38	0.38	0.57
2" x 12"	Inst	LF	C-18	440	0.89	0.11	0.77	1.76	2.14	C-18	396	1.07	0.12	0.85	2.04	2.46
3" x 6"	Demo	LF	L-2	832	---	---	0.31	0.31	0.46	L-2	749	---	---	0.35	0.35	0.51
3" x 6"	Inst	LF	C-18	472	0.69	0.10	0.71	1.51	1.86	C-18	425	0.83	0.11	0.79	1.74	2.13
3" x 8"	Demo	LF	L-2	756	---	---	0.34	0.34	0.51	L-2	680	---	---	0.38	0.38	0.57
3" x 8"	Inst	LF	C-18	440	0.94	0.11	0.77	1.81	2.18	C-18	396	1.12	0.12	0.85	2.09	2.51
3" x 10"	Demo	LF	L-2	680	---	---	0.38	0.38	0.57	L-2	612	---	---	0.42	0.42	0.63
3" x 10"	Inst	LF	C-18	408	1.17	0.11	0.83	2.11	2.52	C-18	367	1.41	0.13	0.92	2.45	2.90
3" x 12"	Demo	LF	L-2	623	---	---	0.42	0.42	0.62	L-2	561	---	---	0.46	0.46	0.69
3" x 12"	Inst	LF	C-18	381	1.42	0.12	0.88	2.43	2.86	C-18	343	1.70	0.14	0.98	2.82	3.30
3" x 14"	Demo	LF	L-2	575	---	---	0.45	0.45	0.67	L-2	518	---	---	0.50	0.50	0.74
3" x 14"	Inst	LF	C-18	359	1.68	0.13	0.94	2.75	3.21	C-18	323	2.01	0.14	1.04	3.20	3.71
4" x 6"	Demo	LF	L-2	745	---	---	0.35	0.35	0.52	L-2	671	---	---	0.39	0.39	0.57
4" x 6"	Inst	LF	C-18	433	1.02	0.11	0.78	1.91	2.29	C-18	390	1.23	0.12	0.87	2.21	2.64
4" x 8"	Demo	LF	L-2	653	---	---	0.40	0.40	0.59	L-2	588	---	---	0.44	0.44	0.65
4" x 8"	Inst	LF	C-18	394	1.39	0.12	0.86	2.37	2.79	C-18	355	1.67	0.13	0.95	2.75	3.22
4" x 10"	Demo	LF	L-2	586	---	---	0.44	0.44	0.66	L-2	527	---	---	0.49	0.49	0.73
4" x 10"	Inst	LF	C-18	363	1.78	0.13	0.93	2.84	3.30	C-18	327	2.14	0.14	1.03	3.32	3.82
4" x 12"	Demo	LF	L-2	522	---	---	0.50	0.50	0.74	L-2	470	---	---	0.55	0.55	0.82
4" x 12"	Inst	LF	C-18	333	2.21	0.14	1.01	3.36	3.86	C-18	300	2.65	0.16	1.12	3.93	4.49

					Costs Based On Large Volume						Costs Based On Small Volume					
Description	Oper	Unit	Crew Size	Avg Day Prod	Avg Mat'l Unit Cost	Avg Equip Unit Cost	Avg Labor Unit Cost	Avg Total Unit Cost	Avg Price Incl O&P	Crew Size	Avg Day Prod	Avg Mat'l Unit Cost	Avg Equip Unit Cost	Avg Labor Unit Cost	Avg Total Unit Cost	Avg Price Incl O&P

Framing (continued)

Description	Oper	Unit	Crew	Prod	Mat'l	Equip	Labor	Total	O&P	Crew	Prod	Mat'l	Equip	Labor	Total	O&P
4" x 14"	Demo	LF	L-8	642	---	---	0.61	0.61	0.90	L-8	578	---	---	0.68	0.68	1.00
4" x 14"	Inst	LF	C-17	404	2.67	0.12	1.16	3.94	4.51	C-17	364	3.20	0.13	1.28	4.62	5.24
4" x 16"	Demo	LF	L-8	594	---	---	0.66	0.66	0.97	L-8	535	---	---	0.73	0.73	1.08
4" x 16"	Inst	LF	C-17	379	3.15	0.12	1.23	4.50	5.11	C-17	341	3.78	0.14	1.37	5.28	5.96
6" x 8"	Demo	LF	L-9	782	---	---	0.67	0.67	0.98	L-9	704	---	---	0.74	0.74	1.09
6" x 8"	Inst	LF	C-21	466	2.08	0.10	1.28	3.46	4.09	C-21	419	2.50	0.11	1.42	4.03	4.73
6" x 10"	Demo	LF	L-9	694	---	---	0.75	0.75	1.11	L-9	625	---	---	0.83	0.83	1.23
6" x 10"	Inst	LF	C-21	427	2.74	0.11	1.40	4.25	4.93	C-21	384	3.29	0.12	1.55	4.96	5.72
6" x 12"	Demo	LF	L-9	631	---	---	0.82	0.82	1.22	L-9	568	---	---	0.92	0.92	1.36
6" x 12"	Inst	LF	C-21	396	3.47	0.12	1.51	5.09	5.83	C-21	356	4.16	0.13	1.68	5.97	6.79
6" x 14"	Demo	LF	L-9	567	---	---	0.92	0.92	1.36	L-9	510	---	---	1.02	1.02	1.51
6" x 14"	Inst	LF	C-21	365	4.26	0.13	1.64	6.02	6.82	C-21	329	5.11	0.14	1.82	7.07	7.96
6" x 16"	Demo	LF	L-9	523	---	---	0.99	0.99	1.47	L-9	471	---	---	1.11	1.11	1.64
6" x 16"	Inst	LF	C-21	342	5.12	0.14	1.75	7.00	7.85	C-21	308	6.14	0.15	1.94	8.23	9.18
8" x 8"	Demo	LF	L-9	668	---	---	0.78	0.78	1.15	L-9	601	---	---	0.87	0.87	1.28
8" x 8"	Inst	LF	C-21	413	2.91	0.11	1.45	4.47	5.18	C-21	372	3.50	0.13	1.61	5.23	6.02
8" x 10"	Demo	LF	L-9	556	---	---	0.94	0.94	1.38	L-9	500	---	---	1.04	1.04	1.54
8" x 10"	Inst	LF	C-21	363	3.84	0.13	1.65	5.62	6.42	C-21	327	4.61	0.14	1.83	6.58	7.48
8" x 12"	Demo	LF	L-9	515	---	---	1.01	1.01	1.49	L-9	464	---	---	1.12	1.12	1.66
8" x 12"	Inst	LF	C-21	341	4.85	0.14	1.75	6.74	7.60	C-21	307	5.82	0.15	1.95	7.92	8.87
8" x 14"	Demo	LF	L-9	468	---	---	1.11	1.11	1.64	L-9	421	---	---	1.23	1.23	1.83
8" x 14"	Inst	LF	C-21	316	5.95	0.15	1.89	7.99	8.91	C-21	284	7.14	0.16	2.10	9.40	10.43
8" x 16"	Demo	LF	L-9	423	---	---	1.23	1.23	1.82	L-9	381	---	---	1.37	1.37	2.02
8" x 16"	Inst	LF	C-21	292	7.14	0.16	2.05	9.35	10.35	C-21	263	8.57	0.18	2.27	11.02	12.13

Blocking
Horizontal, for studs

Description	Oper	Unit	Crew	Prod	Mat'l	Equip	Labor	Total	O&P	Crew	Prod	Mat'l	Equip	Labor	Total	O&P
2" x 4" - 12"	Demo	LF	L-1	710	---	---	0.18	0.18	0.27	L-1	625	---	---	0.21	0.21	0.31
2" x 4" - 12"	Inst	LF	C-1	224	0.22	0.21	0.75	1.18	1.55	C-1	197	0.26	0.24	0.86	1.35	1.77
2" x 4" - 16"	Demo	LF	L-1	761	---	---	0.17	0.17	0.25	L-1	670	---	---	0.19	0.19	0.29
2" x 4" - 16"	Inst	LF	C-1	274	0.22	0.17	0.62	1.00	1.30	C-1	241	0.26	0.19	0.70	1.15	1.49

Description	Oper	Unit	Costs Based On Large Volume							Costs Based On Small Volume						
			Crew Size	Avg Day Prod	Avg Mat'l Unit Cost	Avg Equip Unit Cost	Avg Labor Unit Cost	Avg Total Unit Cost	Avg Price Incl O&P	Crew Size	Avg Day Prod	Avg Mat'l Unit Cost	Avg Equip Unit Cost	Avg Labor Unit Cost	Avg Total Unit Cost	Avg Price Incl O&P

Framing (continued)

Description	Oper	Unit	Crew Size	Avg Day Prod	Avg Mat'l Unit Cost	Avg Equip Unit Cost	Avg Labor Unit Cost	Avg Total Unit Cost	Avg Price Incl O&P	Crew Size	Avg Day Prod	Avg Mat'l Unit Cost	Avg Equip Unit Cost	Avg Labor Unit Cost	Avg Total Unit Cost	Avg Price Incl O&P
2" x 4" - 24"	Demo	LF	L-1	810	---	---	0.16	0.16	0.24	L-1	713	---	---	0.18	0.18	0.27
2" x 4" - 24"	Inst	LF	C-1	350	0.22	0.13	0.48	0.83	1.07	C-1	308	0.26	0.15	0.55	0.96	1.23
2" x 6" - 12"	Demo	LF	L-1	608	---	---	0.21	0.21	0.32	L-1	535	---	---	0.24	0.24	0.36
2" x 6" - 12"	Inst	LF	C-1	207	0.34	0.23	0.81	1.38	1.78	C-1	182	0.41	0.26	0.93	1.59	2.04
2" x 6" - 16"	Demo	LF	L-1	647	---	---	0.20	0.20	0.30	L-1	569	---	---	0.23	0.23	0.34
2" x 6" - 16"	Inst	LF	C-1	252	0.34	0.19	0.67	1.19	1.52	C-1	222	0.41	0.21	0.76	1.38	1.75
2" x 6" - 24"	Demo	LF	L-1	683	---	---	0.19	0.19	0.28	L-1	601	---	---	0.22	0.22	0.32
2" x 6" - 24"	Inst	LF	C-1	323	0.34	0.14	0.52	1.01	1.26	C-1	284	0.41	0.16	0.59	1.17	1.46
2" x 8" - 12"	Demo	LF	L-1	532	---	---	0.24	0.24	0.36	L-1	468	---	---	0.28	0.28	0.41
2" x 8" - 12"	Inst	LF	C-1	193	0.48	0.24	0.87	1.59	2.02	C-1	170	0.57	0.27	0.99	1.84	2.33
2" x 8" - 16"	Demo	LF	L-1	564	---	---	0.23	0.23	0.34	L-1	496	---	---	0.26	0.26	0.39
2" x 8" - 16"	Inst	LF	C-1	235	0.48	0.20	0.72	1.39	1.75	C-1	207	0.57	0.23	0.82	1.61	2.01
2" x 8" - 24"	Demo	LF	L-1	592	---	---	0.22	0.22	0.33	L-1	521	---	---	0.25	0.25	0.37
2" x 8" - 24"	Inst	LF	C-1	299	0.48	0.16	0.56	1.20	1.47	C-1	263	0.57	0.18	0.64	1.39	1.71
Bracing																
Diagonal let-ins																
Studs, 12" o.c.																
1" x 6"	Demo	Set	L-1	714	---	---	0.18	0.18	0.27	L-1	628	---	---	0.21	0.21	0.31
1" x 6"	Inst	Set	C-1	143	0.20	0.33	1.18	1.70	2.28	C-1	126	0.24	0.37	1.34	1.95	2.60
Studs, 16" o.c.																
1" x 6"	Demo	Set	L-1	770	---	---	0.17	0.17	0.25	L-1	678	---	---	0.19	0.19	0.28
1" x 6"	Inst	Set	C-1	184	0.20	0.25	0.92	1.37	1.82	C-1	162	0.24	0.29	1.04	1.57	2.08
Studs, 24" o.c.																
1" x 6"	Demo	Set	L-1	825	---	---	0.16	0.16	0.23	L-1	726.00	---	---	0.18	0.18	0.27
1" x 6"	Inst	Set	C-1	242	0.20	0.19	0.70	1.09	1.43	C-1	212.96	0.24	0.22	0.79	1.25	1.63
Bridging																
Bridging, "X" type																
For joists 12" o.c.																
1" x 3"	Demo	Set	L-1	550	---	---	0.24	0.24	0.35	L-1	484	---	---	0.27	0.27	0.40
1" x 3"	Inst	Set	C-1	140	0.33	0.33	1.20	1.86	2.45	C-1	123	0.39	0.38	1.37	2.14	2.81

Costs Based On Large Volume | Costs Based On Small Volume

Description	Oper	Unit	Crew Size	Avg Day Prod	Avg Mat'l Unit Cost	Avg Equip Unit Cost	Avg Labor Unit Cost	Avg Total Unit Cost	Avg Price Incl O&P	Crew Size	Avg Day Prod	Avg Mat'l Unit Cost	Avg Equip Unit Cost	Avg Labor Unit Cost	Avg Total Unit Cost	Avg Price Incl O&P
Framing (continued)																
2" x 2"	Demo	Set	L-1	536	---	---	0.24	0.24	0.36	L-1	472	---	---	0.28	0.28	0.41
2" x 2"	Inst	Set	C-1	139	0.40	0.34	1.21	1.95	2.55	C-1	122	0.49	0.38	1.38	2.24	2.92
For joists 16" o.c.																
1" x 3"	Demo	Set	L-1	530	---	---	0.25	0.25	0.36	L-1	466	---	---	0.28	0.28	0.41
1" x 3"	Inst	Set	C-1	137	0.35	0.34	1.23	1.92	2.52	C-1	121	0.41	0.39	1.40	2.20	2.88
2" x 2"	Demo	Set	L-1	515	---	---	0.25	0.25	0.37	L-1	453	---	---	0.29	0.29	0.42
2" x 2"	Inst	Set	C-1	136	0.43	0.34	1.24	2.01	2.62	C-1	120	0.51	0.39	1.41	2.31	3.00
For joists 24" o.c.																
1" x 3"	Demo	Set	L-1	507	---	---	0.26	0.26	0.38	L-1	446	---	---	0.29	0.29	0.43
1" x 3"	Inst	Set	C-1	492	0.38	0.09	0.34	0.82	0.98	C-1	433	0.45	0.11	0.39	0.95	1.14
2" x 2"	Demo	Set	L-1	134	---	---	0.97	0.97	1.44	L-1	118	---	---	1.10	1.10	1.63
2" x 2"	Inst	Set	C-1	133	0.47	0.35	1.27	2.09	2.71	C-1	117	0.56	0.40	1.44	2.40	3.11
Bridging, solid, between joists																
2" x 6" - 12"	Demo	Set	L-1	590	---	---	0.22	0.22	0.33	L-1	519	---	---	0.25	0.25	0.37
2" x 6" - 12"	Inst	Set	C-1	199	0.34	0.23	0.85	1.42	1.84	C-1	175	0.41	0.27	0.96	1.64	2.11
2" x 8" - 12"	Demo	Set	L-1	517	---	---	0.25	0.25	0.37	L-1	455	---	---	0.29	0.29	0.42
2" x 8" - 12"	Inst	Set	C-1	185	0.48	0.25	0.91	1.64	2.09	C-1	163	0.57	0.29	1.04	1.90	2.40
2" x 10" - 12"	Demo	Set	L-1	455	---	---	0.29	0.29	0.42	L-1	400	---	---	0.32	0.32	0.48
2" x 10" - 12"	Inst	Set	C-1	174	0.65	0.27	0.97	1.89	2.36	C-1	153	0.78	0.30	1.10	2.19	2.73
2" x 12" - 12"	Demo	Set	L-1	412	---	---	0.32	0.32	0.47	L-1	363	---	---	0.36	0.36	0.53
2" x 12" - 12"	Inst	Set	C-1	164	0.80	0.28	1.03	2.11	2.62	C-1	144	0.96	0.32	1.17	2.45	3.03
2" x 6" - 16"	Demo	Set	L-1	628	---	---	0.21	0.21	0.31	L-1	553	---	---	0.24	0.24	0.35
2" x 6" - 16"	Inst	Set	C-1	243	0.34	0.19	0.69	1.23	1.57	C-1	214	0.41	0.22	0.79	1.41	1.80
2" x 8" - 16"	Demo	Set	L-1	548	---	---	0.24	0.24	0.35	L-1	482	---	---	0.27	0.27	0.40
2" x 8" - 16"	Inst	Set	C-1	226	0.48	0.21	0.75	1.43	1.80	C-1	199	0.57	0.23	0.85	1.66	2.07
2" x 10" - 16"	Demo	Set	L-1	481	---	---	0.27	0.27	0.40	L-1	423	---	---	0.31	0.31	0.45
2" x 10" - 16"	Inst	Set	C-1	211	0.65	0.22	0.80	1.67	2.06	C-1	186	0.78	0.25	0.91	1.94	2.39
2" x 12" - 16"	Demo	Set	L-1	434	---	---	0.30	0.30	0.44	L-1	382	---	---	0.34	0.34	0.50
2" x 12" - 16"	Inst	Set	C-1	198	0.80	0.24	0.85	1.89	2.31	C-1	174	0.96	0.27	0.97	2.20	2.67

Description	Oper	Unit	Costs Based On Large Volume							Costs Based On Small Volume						
			Crew Size	Avg Day Prod	Avg Mat'l Unit Cost	Avg Equip Unit Cost	Avg Labor Unit Cost	Avg Total Unit Cost	Avg Price Incl O&P	Crew Size	Avg Day Prod	Avg Mat'l Unit Cost	Avg Equip Unit Cost	Avg Labor Unit Cost	Avg Total Unit Cost	Avg Price Incl O&P

Framing (continued)

Description	Oper	Unit	Crew Size	Avg Day Prod	Avg Mat'l Unit Cost	Avg Equip Unit Cost	Avg Labor Unit Cost	Avg Total Unit Cost	Avg Price Incl O&P	Crew Size	Avg Day Prod	Avg Mat'l Unit Cost	Avg Equip Unit Cost	Avg Labor Unit Cost	Avg Total Unit Cost	Avg Price Incl O&P
2" x 6" - 24"	Demo	Set	L-1	664	---	---	0.20	0.20	0.29	L-1	584	---	---	0.22	0.22	0.33
2" x 6" - 24"	Inst	Set	C-1	312	0.34	0.15	0.54	1.03	1.29	C-1	275	0.41	0.17	0.61	1.19	1.49
2" x 8" - 24"	Demo	Set	L-1	577	---	---	0.23	0.23	0.33	L-1	508	---	---	0.26	0.26	0.38
2" x 8" - 24"	Inst	Set	C-1	289	0.48	0.16	0.58	1.22	1.51	C-1	254	0.57	0.18	0.66	1.42	1.74
2" x 10" - 24"	Demo	Set	L-1	504	---	---	0.26	0.26	0.38	L-1	444	---	---	0.29	0.29	0.43
2" x 10" - 24"	Inst	Set	C-1	270	0.65	0.17	0.62	1.45	1.76	C-1	238	0.78	0.20	0.71	1.69	2.04
2" x 12" - 24"	Demo	Set	L-1	453	---	---	0.29	0.29	0.42	L-1	399	---	---	0.33	0.33	0.48
2" x 12" - 24"	Inst	Set	C-1	252	0.80	0.19	0.67	1.66	1.98	C-1	222	0.96	0.21	0.76	1.93	2.31

Columns or posts
Without base or cap, hardware, or chamfer corners

Description	Oper	Unit	Crew Size	Avg Day Prod	Avg Mat'l Unit Cost	Avg Equip Unit Cost	Avg Labor Unit Cost	Avg Total Unit Cost	Avg Price Incl O&P	Crew Size	Avg Day Prod	Avg Mat'l Unit Cost	Avg Equip Unit Cost	Avg Labor Unit Cost	Avg Total Unit Cost	Avg Price Incl O&P
4" x 4" - 8'	Demo	LF	L-2	584	---	---	0.45	0.45	0.66	L-2	537	---	---	0.48	0.48	0.72
4" x 4" - 8'	Inst	LF	C-18	315	0.61	0.15	1.07	1.83	2.35	C-18	290	0.73	0.16	1.16	2.06	2.63
4" x 6" - 8'	Demo	LF	L-2	529	---	---	0.49	0.49	0.73	L-2	487	---	---	0.53	0.53	0.79
4" x 6" - 8'	Inst	LF	C-18	294	0.95	0.16	1.15	2.25	2.82	C-18	270	1.14	0.17	1.25	2.56	3.17
4" x 8" - 8'	Demo	LF	L-2	479	---	---	0.54	0.54	0.80	L-2	441	---	---	0.59	0.59	0.87
4" x 8" - 8'	Inst	LF	C-18	274	1.32	0.17	1.23	2.72	3.32	C-18	252	1.58	0.19	1.34	3.10	3.76
6" x 6" - 8'	Demo	LF	L-2	432	---	---	0.60	0.60	0.89	L-2	397	---	---	0.65	0.65	0.97
6" x 6" - 8'	Inst	LF	C-18	248	1.41	0.19	1.36	2.96	3.62	C-18	228	1.69	0.20	1.48	3.38	4.10
6" x 8" - 8'	Demo	LF	L-2	385	---	---	0.68	0.68	1.00	L-2	354	---	---	0.73	0.73	1.09
6" x 8" - 8'	Inst	LF	C-18	229	2.01	0.20	1.47	3.68	4.40	C-18	211	2.41	0.22	1.60	4.23	5.01
6" x 10" - 8'	Demo	LF	L-2	345	---	---	0.75	0.75	1.12	L-2	317	---	---	0.82	0.82	1.21
6" x 10" - 8'	Inst	LF	C-18	212	2.67	0.22	1.59	4.48	5.25	C-18	195	3.20	0.24	1.73	5.17	6.01
8" x 8" - 8'	Demo	LF	L-2	319	---	---	0.82	0.82	1.21	L-2	293	---	---	0.89	0.89	1.31
8" x 8" - 8'	Inst	LF	C-18	196	2.84	0.24	1.72	4.80	5.64	C-18	180	3.41	0.26	1.87	5.54	6.45
8" x 10" - 8'	Demo	LF	L-2	269	---	---	0.97	0.97	1.43	L-2	247	---	---	1.05	1.05	1.56
8" x 10" - 8'	Inst	LF	C-18	174	3.77	0.27	1.94	5.97	6.92	C-18	160	4.52	0.29	2.11	6.92	7.95

Fascia

Description	Oper	Unit	Crew Size	Avg Day Prod	Avg Mat'l Unit Cost	Avg Equip Unit Cost	Avg Labor Unit Cost	Avg Total Unit Cost	Avg Price Incl O&P	Crew Size	Avg Day Prod	Avg Mat'l Unit Cost	Avg Equip Unit Cost	Avg Labor Unit Cost	Avg Total Unit Cost	Avg Price Incl O&P
1" x 4" - 12'	Demo	LF	L-2	1315	---	---	0.20	0.20	0.29	L-2	1210	---	---	0.21	0.21	0.32
1" x 4" - 12'	Inst	LF	C-10	386	0.14	0.12	0.92	1.17	1.62	C-10	355	0.16	0.13	1.00	1.29	1.78

Description	Oper	Unit	Costs Based On Large Volume							Costs Based On Small Volume						
			Crew Size	Avg Day Prod	Avg Mat'l Unit Cost	Avg Equip Unit Cost	Avg Labor Unit Cost	Avg Total Unit Cost	Avg Price Incl O&P	Crew Size	Avg Day Prod	Avg Mat'l Unit Cost	Avg Equip Unit Cost	Avg Labor Unit Cost	Avg Total Unit Cost	Avg Price Incl O&P

Framing (continued)

Furring strips on ceilings

Description	Oper	Unit	Crew Size	Avg Day Prod	Avg Mat'l Unit Cost	Avg Equip Unit Cost	Avg Labor Unit Cost	Avg Total Unit Cost	Avg Price Incl O&P	Crew Size	Avg Day Prod	Avg Mat'l Unit Cost	Avg Equip Unit Cost	Avg Labor Unit Cost	Avg Total Unit Cost	Avg Price Incl O&P
1" x 3" on wood	Demo	LF	L-2	625	---	---	0.42	0.42	0.62	L-2	575	---	---	0.45	0.45	0.67
1" x 3" on wood	Inst	LF	C-18	574	0.09	0.08	0.59	0.76	1.05	C-18	528	0.11	0.09	0.64	0.84	1.15

Furring strips on walls

Description	Oper	Unit	Crew Size	Avg Day Prod	Avg Mat'l Unit Cost	Avg Equip Unit Cost	Avg Labor Unit Cost	Avg Total Unit Cost	Avg Price Incl O&P	Crew Size	Avg Day Prod	Avg Mat'l Unit Cost	Avg Equip Unit Cost	Avg Labor Unit Cost	Avg Total Unit Cost	Avg Price Incl O&P
1" x 3" on wood	Demo	LF	L-2	805	---	---	0.32	0.32	0.48	L-2	741	---	---	0.35	0.35	0.52
1" x 3" on wood	Inst	LF	C-18	741	0.09	0.06	0.45	0.61	0.83	C-18	682	0.11	0.07	0.49	0.67	0.92
1" x 3" on masonry	Demo	LF	L-2	625	---	---	0.42	0.42	0.62	L-2	575	---	---	0.45	0.45	0.67
1" x 3" on masonry	Inst	LF	C-18	574	0.09	0.08	0.59	0.76	1.05	C-18	528	0.11	0.09	0.64	0.83	1.15
1" x 3" on concrete	Demo	LF	L-2	456	---	---	0.57	0.57	0.84	L-2	420	---	---	0.62	0.62	0.92
1" x 3" on concrete	Inst	LF	C-18	418	0.09	0.11	0.81	1.01	1.41	C-18	385	0.11	0.12	0.88	1.11	1.54

Headers or lintels, over openings

Four feet wide:

Description	Oper	Unit	Crew Size	Avg Day Prod	Avg Mat'l Unit Cost	Avg Equip Unit Cost	Avg Labor Unit Cost	Avg Total Unit Cost	Avg Price Incl O&P	Crew Size	Avg Day Prod	Avg Mat'l Unit Cost	Avg Equip Unit Cost	Avg Labor Unit Cost	Avg Total Unit Cost	Avg Price Incl O&P
4" x 6"	Demo	LF	L-1	338	---	---	0.38	0.38	0.57	L-1	304	---	---	0.43	0.43	0.63
4" x 6"	Inst	LF	C-1	198	0.94	0.24	0.85	2.03	2.45	C-1	178	1.13	0.26	0.95	2.34	2.80
4" x 8"	Demo	LF	L-1	302	---	---	0.43	0.43	0.64	L-1	272	---	---	0.48	0.48	0.71
4" x 8"	Inst	LF	C-1	186	1.31	0.25	0.91	2.47	2.91	C-1	167	1.57	0.28	1.01	2.86	3.35
4" x 12"	Demo	LF	L-1	250	---	---	0.52	0.52	0.77	L-1	225	---	---	0.58	0.58	0.86
4" x 12"	Inst	LF	C-1	164	2.13	0.28	1.03	3.44	3.95	C-1	148	2.56	0.32	1.14	4.02	4.58
4" x 14"	Demo	LF	L-1	236	---	---	0.55	0.55	0.82	L-1	212	---	---	0.61	0.61	0.91
4" x 14"	Inst	LF	C-1	158	2.59	0.30	1.07	3.95	4.47	C-1	142	3.11	0.33	1.19	4.62	5.20

Six feet wide:

Description	Oper	Unit	Crew Size	Avg Day Prod	Avg Mat'l Unit Cost	Avg Equip Unit Cost	Avg Labor Unit Cost	Avg Total Unit Cost	Avg Price Incl O&P	Crew Size	Avg Day Prod	Avg Mat'l Unit Cost	Avg Equip Unit Cost	Avg Labor Unit Cost	Avg Total Unit Cost	Avg Price Incl O&P
4" x 12"	Demo	LF	L-2	583	---	---	0.45	0.45	0.66	L-2	525	---	---	0.50	0.50	0.73
4" x 12"	Inst	LF	C-18	392	2.13	0.12	0.86	3.11	3.53	C-18	353	2.56	0.13	0.96	3.65	4.11

Eight feet wide:

Description	Oper	Unit	Crew Size	Avg Day Prod	Avg Mat'l Unit Cost	Avg Equip Unit Cost	Avg Labor Unit Cost	Avg Total Unit Cost	Avg Price Incl O&P	Crew Size	Avg Day Prod	Avg Mat'l Unit Cost	Avg Equip Unit Cost	Avg Labor Unit Cost	Avg Total Unit Cost	Avg Price Incl O&P
4" x 12"	Demo	LF	L-2	655	---	---	0.40	0.40	0.59	L-2	590	---	---	0.44	0.44	0.65
4" x 12"	Inst	LF	C-18	453	2.13	0.10	0.74	2.98	3.34	C-18	408	2.56	0.11	0.83	3.50	3.90

Ten feet wide:

Description	Oper	Unit	Crew Size	Avg Day Prod	Avg Mat'l Unit Cost	Avg Equip Unit Cost	Avg Labor Unit Cost	Avg Total Unit Cost	Avg Price Incl O&P	Crew Size	Avg Day Prod	Avg Mat'l Unit Cost	Avg Equip Unit Cost	Avg Labor Unit Cost	Avg Total Unit Cost	Avg Price Incl O&P
4" x 12"	Demo	LF	L-2	743	---	---	0.35	0.35	0.52	L-2	669	---	---	0.39	0.39	0.58
4" x 12"	Inst	LF	C-18	535	2.13	0.09	0.63	2.85	3.16	C-18	482	2.56	0.10	0.70	3.36	3.70
4" x 14"	Demo	LF	L-2	627	---	---	0.41	0.41	0.61	L-2	564	---	---	0.46	0.46	0.68
4" x 14"	Inst	LF	C-18	455	2.59	0.10	0.74	3.43	3.79	C-18	410	3.11	0.11	0.82	4.04	4.45

			Costs Based On Large Volume							Costs Based On Small Volume						

Framing (continued)

Description	Oper	Unit	Crew Size	Avg Day Prod	Avg Mat'l Unit Cost	Avg Equip Unit Cost	Avg Labor Unit Cost	Avg Total Unit Cost	Avg Price Incl O&P	Crew Size	Avg Day Prod	Avg Mat'l Unit Cost	Avg Equip Unit Cost	Avg Labor Unit Cost	Avg Total Unit Cost	Avg Price Incl O&P
Twelve feet wide:																
4" x 14"	Demo	LF	L-2	667	---	---	0.39	0.39	0.58	L-2	600	---	---	0.43	0.43	0.64
4" x 14"	Inst	LF	C-18	491	2.59	0.10	0.69	3.37	3.71	C-18	442	3.11	0.11	0.76	3.97	4.35
4" x 16"	Demo	LF	L-2	562	---	---	0.46	0.46	0.68	L-2	506	---	---	0.51	0.51	0.76
4" x 16"	Inst	LF	C-18	429	3.07	0.11	0.79	3.96	4.35	C-18	386	3.68	0.12	0.87	4.68	5.10
Fourteen feet wide:																
4" x 16"	Demo	LF	L-2	588	---	---	0.44	0.44	0.65	L-2	529	---	---	0.49	0.49	0.73
4" x 16"	Inst	LF	C-18	453	3.07	0.10	0.74	3.92	4.28	C-18	408	3.68	0.11	0.83	4.62	5.03
Sixteen feet wide:																
4" x 16"	Demo	LF	L-2	611	---	---	0.43	0.43	0.63	L-2	550	---	---	0.47	0.47	0.70
4" x 16"	Inst	LF	C-18	475	3.07	0.10	0.71	3.88	4.22	C-18	428	3.68	0.11	0.79	4.58	4.97
Eighteen feet wide:																
4" x 16"	Demo	LF	L-2	632	---	---	0.41	0.41	0.61	L-2	569	---	---	0.46	0.46	0.68
4" x 16"	Inst	LF	C-18	495	3.07	0.09	0.68	3.84	4.18	C-18	446	3.68	0.10	0.76	4.54	4.91
Joists																
Ceiling/floor, per LF of stick																
2" x 4" -6'	Demo	LF	L-2	1024	---	---	0.25	0.25	0.38	L-2	901	---	---	0.29	0.29	0.43
2" x 4" -6'	Inst	LF	C-18	844	0.21	0.06	0.40	0.67	0.86	C-18	743	0.26	0.06	0.45	0.77	1.00
2" x 4" -8'	Demo	LF	L-2	1213	---	---	0.21	0.21	0.32	L-2	1067	---	---	0.24	0.24	0.36
2" x 4" -8'	Inst	LF	C-18	1004	0.21	0.05	0.34	0.60	0.76	C-18	884	0.26	0.05	0.38	0.69	0.88
2" x 4" -10'	Demo	LF	L-2	1367	---	---	0.19	0.19	0.28	L-2	1203	---	---	0.22	0.22	0.32
2" x 4" -10'	Inst	LF	C-18	1134	0.21	0.04	0.30	0.55	0.70	C-18	998	0.26	0.05	0.34	0.64	0.81
2" x 4" -12'	Demo	LF	L-2	1498	---	---	0.17	0.17	0.26	L-2	1318	---	---	0.20	0.20	0.29
2" x 4" -12'	Inst	LF	C-18	1245	0.21	0.04	0.27	0.52	0.66	C-18	1096	0.26	0.04	0.31	0.61	0.76
2" x 6" - 8'	Demo	LF	L-2	1064	---	---	0.24	0.24	0.36	L-2	936	---	---	0.28	0.28	0.41
2" x 6" - 8'	Inst	LF	C-18	891	0.34	0.05	0.38	0.77	0.95	C-18	784	0.41	0.06	0.43	0.90	1.11
2" x 6" - 10'	Demo	LF	L-2	1195	---	---	0.22	0.22	0.32	L-2	1052	---	---	0.25	0.25	0.37
2" x 6" - 10'	Inst	LF	C-18	1005	0.34	0.05	0.34	0.72	0.88	C-18	884	0.41	0.05	0.38	0.84	1.03

			Costs Based On Large Volume							Costs Based On Small Volume						
Description	Oper	Unit	Crew Size	Avg Day Prod	Avg Mat'l Unit Cost	Avg Equip Unit Cost	Avg Labor Unit Cost	Avg Total Unit Cost	Avg Price Incl O&P	Crew Size	Avg Day Prod	Avg Mat'l Unit Cost	Avg Equip Unit Cost	Avg Labor Unit Cost	Avg Total Unit Cost	Avg Price Incl O&P

Framing (continued)

Description	Oper	Unit	Crew Size	Avg Day Prod	Avg Mat'l Unit Cost	Avg Equip Unit Cost	Avg Labor Unit Cost	Avg Total Unit Cost	Avg Price Incl O&P	Crew Size	Avg Day Prod	Avg Mat'l Unit Cost	Avg Equip Unit Cost	Avg Labor Unit Cost	Avg Total Unit Cost	Avg Price Incl O&P
2" x 6" - 12'	Demo	LF	L-2	1307	---	---	0.20	0.20	0.29	L-2	1150	---	---	0.23	0.23	0.33
2" x 6" - 12'	Inst	LF	C-18	1102	0.34	0.04	0.31	0.69	0.84	C-18	970	0.41	0.05	0.35	0.80	0.97
2" x 6" - 14'	Demo	LF	L-2	1405	---	---	0.19	0.19	0.27	L-2	1236	---	---	0.21	0.21	0.31
2" x 6" - 14'	Inst	LF	C-18	1187	0.34	0.04	0.28	0.66	0.80	C-18	1045	0.41	0.04	0.32	0.77	0.93
2" x 8" - 10'	Demo	LF	L-2	1060	---	---	0.25	0.25	0.36	L-2	933	---	---	0.28	0.28	0.41
2" x 8" - 10'	Inst	LF	C-18	901	0.48	0.05	0.37	0.90	1.09	C-18	793	0.57	0.06	0.43	1.06	1.26
2" x 8" - 12'	Demo	LF	L-2	1156	---	---	0.22	0.22	0.33	L-2	1017	---	---	0.26	0.26	0.38
2" x 8" - 12'	Inst	LF	C-18	986	0.48	0.05	0.34	0.87	1.03	C-18	868	0.57	0.05	0.39	1.01	1.20
2" x 8" - 14'	Demo	LF	L-2	1239	---	---	0.21	0.21	0.31	L-2	1090	---	---	0.24	0.24	0.35
2" x 8" - 14'	Inst	LF	C-18	1059	0.48	0.04	0.32	0.84	0.99	C-18	932	0.57	0.05	0.36	0.98	1.16
2" x 8" - 16'	Demo	LF	L-2	1313	---	---	0.20	0.20	0.29	L-2	1155	---	---	0.23	0.23	0.33
2" x 8" - 16'	Inst	LF	C-18	1124	0.48	0.04	0.30	0.82	0.96	C-18	989	0.57	0.05	0.34	0.96	1.13
2" x 10" - 12'	Demo	LF	L-2	1034	---	---	0.25	0.25	0.37	L-2	910	---	---	0.29	0.29	0.42
2" x 10" - 12'	Inst	LF	C-18	890	0.65	0.05	0.38	1.08	1.27	C-18	783	0.78	0.06	0.43	1.27	1.48
2" x 10" - 14'	Demo	LF	L-2	1106	---	---	0.24	0.24	0.35	L-2	973	---	---	0.27	0.27	0.40
2" x 10" - 14'	Inst	LF	C-18	955	0.65	0.05	0.35	1.05	1.23	C-18	840	0.78	0.06	0.40	1.24	1.43
2" x 10" - 16'	Demo	LF	L-2	1171	---	---	0.22	0.22	0.33	L-2	1030	---	---	0.25	0.25	0.37
2" x 10" - 16'	Inst	LF	C-18	1013	0.65	0.05	0.33	1.03	1.19	C-18	891	0.78	0.05	0.38	1.21	1.40
2" x 10" - 18'	Demo	LF	L-2	1229	---	---	0.21	0.21	0.31	L-2	1082	---	---	0.24	0.24	0.36
2" x 10" - 18'	Inst	LF	C-18	1065	0.65	0.04	0.32	1.01	1.17	C-18	937	0.78	0.05	0.36	1.19	1.37
2" x 12" - 14'	Demo	LF	L-2	998	---	---	0.26	0.26	0.39	L-2	878	---	---	0.30	0.30	0.44
2" x 12" - 14'	Inst	LF	C-18	869	0.80	0.05	0.39	1.24	1.43	C-18	765	0.96	0.06	0.44	1.46	1.68
2" x 12" - 16'	Demo	LF	L-2	1054	---	---	0.25	0.25	0.37	L-2	928	---	---	0.28	0.28	0.42
2" x 12" - 16'	Inst	LF	C-18	920	0.80	0.05	0.37	1.22	1.40	C-18	810	0.96	0.06	0.42	1.43	1.64
2" x 12" - 18'	Demo	LF	L-2	1106	---	---	0.24	0.24	0.35	L-2	973	---	---	0.27	0.27	0.40
2" x 12" - 18'	Inst	LF	C-18	967	0.80	0.05	0.35	1.20	1.37	C-18	851	0.96	0.05	0.40	1.41	1.61
2" x 12" - 20'	Demo	LF	L-2	1151	---	---	0.23	0.23	0.33	L-2	1013	---	---	0.26	0.26	0.38
2" x 12" - 20'	Inst	LF	C-18	1009	0.80	0.05	0.33	1.18	1.34	C-18	888	0.96	0.05	0.38	1.39	1.58

| | | | Costs Based On Large Volume | | | | | | | Costs Based On Small Volume | | | | | | |
Description	Oper	Unit	Crew Size	Avg Day Prod	Avg Mat'l Unit Cost	Avg Equip Unit Cost	Avg Labor Unit Cost	Avg Total Unit Cost	Avg Price Incl O&P	Crew Size	Avg Day Prod	Avg Mat'l Unit Cost	Avg Equip Unit Cost	Avg Labor Unit Cost	Avg Total Unit Cost	Avg Price Incl O&P

Framing (continued)

Description	Oper	Unit	Crew Size	Avg Day Prod	Avg Mat'l Unit Cost	Avg Equip Unit Cost	Avg Labor Unit Cost	Avg Total Unit Cost	Avg Price Incl O&P	Crew Size	Avg Day Prod	Avg Mat'l Unit Cost	Avg Equip Unit Cost	Avg Labor Unit Cost	Avg Total Unit Cost	Avg Price Incl O&P
3" x 8" - 12'	Demo	LF	L-2	942	---	---	0.28	0.28	0.41	L-2	829	---	---	0.31	0.31	0.46
3" x 8" - 12'	Inst	LF	C-18	816	0.84	0.06	0.41	1.31	1.52	C-18	718	1.01	0.06	0.47	1.55	1.78
3" x 8" - 14'	Demo	LF	L-2	1007	---	---	0.26	0.26	0.38	L-2	886	---	---	0.29	0.29	0.43
3" x 8" - 14'	Inst	LF	C-18	876	0.84	0.05	0.38	1.28	1.47	C-18	771	1.01	0.06	0.44	1.51	1.72
3" x 8" - 16'	Demo	LF	L-2	1065	---	---	0.24	0.24	0.36	L-2	937	---	---	0.28	0.28	0.41
3" x 8" - 16'	Inst	LF	C-18	929	0.84	0.05	0.36	1.26	1.43	C-18	818	1.01	0.06	0.41	1.48	1.68
3" x 8" - 18'	Demo	LF	L-2	1117	---	---	0.23	0.23	0.34	L-2	983	---	---	0.26	0.26	0.39
3" x 8" - 18'	Inst	LF	C-18	976	0.84	0.05	0.35	1.24	1.41	C-18	859	1.01	0.05	0.39	1.46	1.65
3" x 10" - 16'	Demo	LF	L-2	918	---	---	0.28	0.28	0.42	L-2	808	---	---	0.32	0.32	0.48
3" x 10" - 16'	Inst	LF	C-18	811	1.08	0.06	0.42	1.55	1.76	C-18	714	1.30	0.07	0.47	1.83	2.07
3" x 10" - 18'	Demo	LF	L-2	961	---	---	0.27	0.27	0.40	L-2	846	---	---	0.31	0.31	0.46
3" x 10" - 18'	Inst	LF	C-18	851	1.08	0.05	0.40	1.53	1.73	C-18	749	1.30	0.06	0.45	1.81	2.03
3" x 10" - 20'	Demo	LF	L-2	1001	---	---	0.26	0.26	0.38	L-2	881	---	---	0.30	0.30	0.44
3" x 10" - 20'	Inst	LF	C-18	888	1.08	0.05	0.38	1.51	1.70	C-18	781	1.30	0.06	0.43	1.79	2.00
3" x 12" - 16'	Demo	LF	L-2	821	---	---	0.32	0.32	0.47	L-2	722	---	---	0.36	0.36	0.53
3" x 12" - 16'	Inst	LF	C-18	732	1.08	0.06	0.46	1.60	1.83	C-18	644	1.30	0.07	0.52	1.89	2.15
3" x 12" - 18'	Demo	LF	L-2	873	---	---	0.30	0.30	0.44	L-2	768	---	---	0.34	0.34	0.50
3" x 12" - 18'	Inst	LF	C-18	781	1.33	0.06	0.43	1.82	2.03	C-18	687	1.59	0.07	0.49	2.15	2.39
3" x 12" - 20'	Demo	LF	L-2	898	---	---	0.29	0.29	0.43	L-2	790	---	---	0.33	0.33	0.49
3" x 12" - 20'	Inst	LF	C-18	805	1.33	0.06	0.42	1.80	2.01	C-18	708	1.59	0.07	0.48	2.14	2.37
3" x 12" - 22'	Demo	LF	L-2	925	---	---	0.28	0.28	0.42	L-2	814	---	---	0.32	0.32	0.47
3" x 12" - 22'	Inst	LF	C-18	830	1.33	0.06	0.41	1.79	1.99	C-18	730	1.59	0.06	0.46	2.12	2.35
Ceiling/floor, per SF of area																
2" x 4" - 6', 12" o.c.	Demo	SF	L-2	998	---	---	0.26	0.26	0.39	L-2	878	---	---	0.30	0.30	0.44
2" x 4" - 6', 12" o.c.	Inst	SF	C-18	823	0.22	0.06	0.41	0.69	0.89	C-18	724	0.26	0.06	0.47	0.79	1.02
2" x 4" - 6', 16" o.c.	Demo	SF	L-2	1188	---	---	0.22	0.22	0.32	L-2	1045	---	---	0.25	0.25	0.37
2" x 4" - 6', 16" o.c.	Inst	SF	C-18	979	0.17	0.05	0.34	0.56	0.73	C-18	862	0.20	0.05	0.39	0.65	0.84
2" x 4" - 6', 24" o.c.	Demo	SF	L-2	1559	---	---	0.17	0.17	0.25	L-2	1372	---	---	0.19	0.19	0.28
2" x 4" - 6', 24" o.c.	Inst	SF	C-18	1285	0.12	0.04	0.26	0.41	0.54	C-18	1131	0.14	0.04	0.30	0.48	0.62

			Costs Based On Large Volume							Costs Based On Small Volume						
Description	Oper	Unit	Crew Size	Avg Day Prod	Avg Mat'l Unit Cost	Avg Equip Unit Cost	Avg Labor Unit Cost	Avg Total Unit Cost	Avg Price Incl O&P	Crew Size	Avg Day Prod	Avg Mat'l Unit Cost	Avg Equip Unit Cost	Avg Labor Unit Cost	Avg Total Unit Cost	Avg Price Incl O&P

Framing (continued)

Description	Oper	Unit	Crew Size	Avg Day Prod	Avg Mat'l Unit Cost	Avg Equip Unit Cost	Avg Labor Unit Cost	Avg Total Unit Cost	Avg Price Incl O&P	Crew Size	Avg Day Prod	Avg Mat'l Unit Cost	Avg Equip Unit Cost	Avg Labor Unit Cost	Avg Total Unit Cost	Avg Price Incl O&P
2" x 6" - 8', 12" o.c.	Demo	SF	L-2	1037	---	---	0.25	0.25	0.37	L-2	913	---	---	0.29	0.29	0.42
2" x 6" - 8', 12" o.c.	Inst	SF	C-18	869	0.35	0.05	0.39	0.79	0.98	C-18	765	0.42	0.06	0.44	0.92	1.14
2" x 6" - 8', 16" o.c.	Demo	SF	L-2	1235	---	---	0.21	0.21	0.31	L-2	1087	---	---	0.24	0.24	0.35
2" x 6" - 8', 16" o.c.	Inst	SF	C-18	1034	0.27	0.05	0.33	0.64	0.80	C-18	910	0.32	0.05	0.37	0.74	0.92
2" x 6" - 8', 24" o.c.	Demo	SF	L-2	1622	---	---	0.16	0.16	0.24	L-2	1427	---	---	0.18	0.18	0.27
2" x 6" - 8', 24" o.c.	Inst	SF	C-18	1359	0.18	0.03	0.25	0.47	0.59	C-18	1196	0.22	0.04	0.28	0.54	0.68
2" x 8" - 10', 12" o.c.	Demo	SF	L-2	1033	---	---	0.25	0.25	0.37	L-2	909	---	---	0.29	0.29	0.42
2" x 8" - 10', 12" o.c.	Inst	SF	C-18	878	0.49	0.05	0.38	0.93	1.12	C-18	773	0.59	0.06	0.44	1.08	1.30
2" x 8" - 10', 16" o.c.	Demo	SF	L-2	1231	---	---	0.21	0.21	0.31	L-2	1083	---	---	0.24	0.24	0.36
2" x 8" - 10', 16" o.c.	Inst	SF	C-18	1046	0.37	0.04	0.32	0.74	0.90	C-18	920	0.44	0.05	0.37	0.86	1.04
2" x 8" - 10', 24" o.c.	Demo	SF	L-2	1614	---	---	0.16	0.16	0.24	L-2	1420	---	---	0.18	0.18	0.27
2" x 8" - 10', 24" o.c.	Inst	SF	C-18	1372	0.25	0.03	0.25	0.53	0.65	C-18	1207	0.31	0.04	0.28	0.62	0.76
2" x 10" - 12', 12" o.c.	Demo	SF	L-2	1008	---	---	0.26	0.26	0.38	L-2	887	---	---	0.29	0.29	0.43
2" x 10" - 12', 12" o.c.	Inst	SF	C-18	868	0.67	0.05	0.39	1.11	1.30	C-18	764	0.80	0.06	0.44	1.30	1.52
2" x 10" - 12', 16" o.c.	Demo	SF	L-2	1200	---	---	0.22	0.22	0.32	L-2	1056	---	---	0.25	0.25	0.36
2" x 10" - 12', 16" o.c.	Inst	SF	C-18	1033	0.50	0.05	0.33	0.88	1.04	C-18	909	0.61	0.05	0.37	1.03	1.21
2" x 10" - 12', 24" o.c.	Demo	SF	L-2	1575	---	---	0.17	0.17	0.24	L-2	1386	---	---	0.19	0.19	0.28
2" x 10" - 12', 24" o.c.	Inst	SF	C-18	1356	0.35	0.03	0.25	0.63	0.75	C-18	1193	0.42	0.04	0.28	0.74	0.88
2" x 12" - 14', 12" o.c.	Demo	SF	L-2	974	---	---	0.27	0.27	0.40	L-2	857	---	---	0.30	0.30	0.45
2" x 12" - 14', 12" o.c.	Inst	SF	C-18	848	0.82	0.06	0.40	1.27	1.47	C-18	746	0.98	0.06	0.45	1.50	1.72
2" x 12" - 14', 16" o.c.	Demo	SF	L-2	1159	---	---	0.22	0.22	0.33	L-2	1020	---	---	0.26	0.26	0.38
2" x 12" - 14', 16" o.c.	Inst	SF	C-18	1009	0.62	0.05	0.33	1.00	1.17	C-18	888	0.75	0.05	0.38	1.18	1.36
2" x 12" - 14', 24" o.c.	Demo	SF	L-2	1521	---	---	0.17	0.17	0.25	L-2	1338	---	---	0.19	0.19	0.29
2" x 12" - 14', 24" o.c.	Inst	SF	C-18	1324	0.42	0.04	0.25	0.71	0.84	C-18	1165	0.51	0.04	0.29	0.84	0.98
3" x 8" - 14', 12" o.c.	Demo	SF	L-2	983	---	---	0.26	0.26	0.39	L-2	865	---	---	0.30	0.30	0.44
3" x 8" - 14', 12" o.c.	Inst	SF	C-18	855	0.86	0.05	0.39	1.31	1.51	C-18	752	1.04	0.06	0.45	1.55	1.77
3" x 8" - 14', 16" o.c.	Demo	SF	L-2	1170	---	---	0.22	0.22	0.33	L-2	1030	---	---	0.25	0.25	0.37
3" x 8" - 14', 16" o.c.	Inst	SF	C-18	1017	0.66	0.05	0.33	1.03	1.20	C-18	895	0.79	0.05	0.38	1.22	1.40

			Costs Based On Large Volume							Costs Based On Small Volume						
Description	Oper	Unit	Crew Size	Avg Day Prod	Avg Mat'l Unit Cost	Avg Equip Unit Cost	Avg Labor Unit Cost	Avg Total Unit Cost	Avg Price Incl O&P	Crew Size	Avg Day Prod	Avg Mat'l Unit Cost	Avg Equip Unit Cost	Avg Labor Unit Cost	Avg Total Unit Cost	Avg Price Incl O&P

Framing (continued)

Description	Oper	Unit	Crew Size	Avg Day Prod	Avg Mat'l Unit Cost	Avg Equip Unit Cost	Avg Labor Unit Cost	Avg Total Unit Cost	Avg Price Incl O&P	Crew Size	Avg Day Prod	Avg Mat'l Unit Cost	Avg Equip Unit Cost	Avg Labor Unit Cost	Avg Total Unit Cost	Avg Price Incl O&P
3" x 8" - 14', 24" o.c.	Demo	SF	L-2	1535	---	---	0.17	0.17	0.25	L-2	1351	---	---	0.19	0.19	0.28
3" x 8" - 14', 24" o.c.	Inst	SF	C-18	1335	0.45	0.03	0.25	0.73	0.86	C-18	1175	0.54	0.04	0.29	0.86	1.00
3" x 10" - 16', 12" o.c.	Demo	SF	L-2	896	---	---	0.29	0.29	0.43	L-2	788	---	---	0.33	0.33	0.49
3" x 10" - 16', 12" o.c.	Inst	SF	C-18	792	1.11	0.06	0.43	1.59	1.80	C-18	697	1.33	0.07	0.48	1.88	2.12
3" x 10" - 16', 16" o.c.	Demo	SF	L-2	1066	---	---	0.24	0.24	0.36	L-2	938	---	---	0.28	0.28	0.41
3" x 10" - 16', 16" o.c.	Inst	SF	C-18	943	0.84	0.05	0.36	1.25	1.42	C-18	830	1.01	0.06	0.41	1.47	1.67
3" x 10" - 16', 24" o.c.	Demo	SF	L-2	1399	---	---	0.19	0.19	0.28	L-2	1231	---	---	0.21	0.21	0.31
3" x 10" - 16', 24" o.c.	Inst	SF	C-18	1236	0.57	0.04	0.27	0.88	1.01	C-18	1088	0.68	0.04	0.31	1.04	1.19
3" x 12" - 18', 12" o.c.	Demo	SF	L-2	851	---	---	0.31	0.31	0.45	L-2	749	---	---	0.35	0.35	0.51
3" x 12" - 18', 12" o.c.	Inst	SF	C-18	762	1.36	0.06	0.44	1.87	2.08	C-18	671	1.64	0.07	0.50	2.21	2.45
3" x 12" - 18', 16" o.c.	Demo	SF	L-2	1013	---	---	0.26	0.26	0.38	L-2	891	---	---	0.29	0.29	0.43
3" x 12" - 18', 16" o.c.	Inst	SF	C-18	907	1.03	0.05	0.37	1.46	1.64	C-18	798	1.24	0.06	0.42	1.72	1.93
3" x 12" - 18', 24" o.c.	Demo	SF	L-2	1330	---	---	0.20	0.20	0.29	L-2	1170	---	---	0.22	0.22	0.33
3" x 12" - 18', 24" o.c.	Inst	SF	C-18	1190	0.70	0.04	0.28	1.03	1.16	C-18	1047	0.84	0.04	0.32	1.21	1.37

Ledgers
Nailed:

Description	Oper	Unit	Crew Size	Avg Day Prod	Avg Mat'l Unit Cost	Avg Equip Unit Cost	Avg Labor Unit Cost	Avg Total Unit Cost	Avg Price Incl O&P	Crew Size	Avg Day Prod	Avg Mat'l Unit Cost	Avg Equip Unit Cost	Avg Labor Unit Cost	Avg Total Unit Cost	Avg Price Incl O&P
2" x 4" - 12'	Demo	LF	L-2	1220	---	---	0.21	0.21	0.32	L-2	1122	---	---	0.23	0.23	0.34
2" x 4" - 12'	Inst	LF	C-18	857	0.26	0.05	0.39	0.71	0.90	C-18	788	0.31	0.06	0.43	0.80	1.01
2" x 6" - 12'	Demo	LF	L-2	1082	---	---	0.24	0.24	0.36	L-2	995	---	---	0.26	0.26	0.39
2" x 6" - 12'	Inst	LF	C-18	774	0.39	0.06	0.44	0.88	1.09	C-18	712	0.46	0.07	0.47	1.00	1.23
2" x 8" - 12'	Demo	LF	L-2	973	---	---	0.27	0.27	0.40	L-2	895	---	---	0.29	0.29	0.43
2" x 8" - 12'	Inst	LF	C-18	706	0.52	0.07	0.48	1.07	1.30	C-18	650	0.63	0.07	0.52	1.22	1.47

Bolted; bolts pre-embedded, not included in labor/material costs; securing ledger included in labor cost.

Description	Oper	Unit	Crew Size	Avg Day Prod	Avg Mat'l Unit Cost	Avg Equip Unit Cost	Avg Labor Unit Cost	Avg Total Unit Cost	Avg Price Incl O&P	Crew Size	Avg Day Prod	Avg Mat'l Unit Cost	Avg Equip Unit Cost	Avg Labor Unit Cost	Avg Total Unit Cost	Avg Price Incl O&P
3" x 6" - 12'	Demo	LF	L-2	718	---	---	0.36	0.36	0.54	L-2	661	---	---	0.39	0.39	0.58
3" x 6" - 12'	Inst	LF	C-18	513	0.59	0.09	0.66	1.34	1.66	C-18	472	0.71	0.10	0.71	1.53	1.88
3" x 8" - 12'	Demo	LF	L-2	654	---	---	0.40	0.40	0.59	L-2	602	---	---	0.43	0.43	0.64
3" x 8" - 12'	Inst	LF	C-18	475	0.84	0.10	0.71	1.64	1.99	C-18	437	1.00	0.11	0.77	1.88	2.26

			Costs Based On Large Volume							Costs Based On Small Volume					
Description	Oper	Unit	Crew Size	Avg Day Prod	Avg Mat'l Unit Cost	Avg Equip Unit Cost	Avg Labor Unit Cost	Avg Total Unit Cost	Avg Price Incl O&P	Crew Size	Avg Day Prod	Avg Mat'l Unit Cost	Avg Equip Unit Cost	Avg Labor Unit Cost	Avg Total Unit Cost	Avg Price Incl O&P

Framing (continued)

3" x 10" - 12'	Demo	LF	L-2	593	---	---	0.44	0.44	0.65	L-2	546	---	---	0.48	0.48	0.71
3" x 10" - 12'	Inst	LF	C-18	438	1.07	0.11	0.77	1.95	2.33	C-18	403	1.29	0.12	0.84	2.24	2.65
3" x 12" - 12'	Demo	LF	L-2	547	---	---	0.48	0.48	0.70	L-2	503	---	---	0.52	0.52	0.76
3" x 12" - 12'	Inst	LF	C-18	408	1.32	0.11	0.83	2.26	2.67	C-18	375	1.58	0.12	0.90	2.61	3.05

Patio Framing

Wood deck

4" x 4" rough sawn beams (4'-0" o.c.) leveled 1/16" to 1/8" and nailed to pre-set concrete piers with wood block on top; 2" T x 4" W decking (S4S) with 1/2" spacing, double-nailed beam junctures

Fir	Demo	SF	L-2	380	---	---	0.68	0.68	1.01	L-2	350	---	---	0.74	0.74	1.10
Fir	Inst	SF	C-12	330	2.83	0.14	1.17	4.14	4.71	C-12	304	3.40	0.15	1.27	4.82	5.44
Redwood	Demo	SF	L-2	380	---	---	0.68	0.68	1.01	L-2	350	---	---	0.74	0.74	1.10
Redwood	Inst	SF	C-12	330	2.99	0.14	1.17	4.30	4.87	C-12	304	3.41	0.15	1.27	4.84	5.46

Wood awning

4" x 4" columns (10'-0" o.c.) nailed to wood; 2" x 6" beams nailed horizontally to either side of columns; 2" x 6" ledger nailed to wall studs; 2" x 6" joists (4'-0" o.c.) nailed to ledger and toe-nailed on top of beams; 2" x 2" (4" o.c.) nailed to joists for sunscreen

Redwood, rough sawn	Demo	SF	L-2	290	---	---	0.90	0.90	1.33	L-2	267	---	---	0.97	0.97	1.44
Redwood, rough sawn	Inst	SF	C-17	250	0.66	0.19	1.87	2.71	3.63	C-17	230	0.79	0.20	2.03	3.02	4.02

Plates; joined with studs before setting

Double Top, nailed:

2" x 4" - 8'	Demo	LF	L-2	820	---	---	0.32	0.32	0.47	L-2	722	---	---	0.36	0.36	0.53
2" x 4" - 8'	Inst	LF	C-11	410	0.48	0.11	0.90	1.50	1.94	C-11	361	0.58	0.13	1.02	1.73	2.23
2" x 6" - 8'	Demo	LF	L-2	820	---	---	0.32	0.32	0.47	L-2	722	---	---	0.36	0.36	0.53
2" x 6" - 8'	Inst	LF	C-11	410	0.72	0.11	0.90	1.73	2.17	C-11	361	0.86	0.13	1.02	2.01	2.51

Single Bottom, nailed:

2" x 4" - 8'	Demo	LF	L-2	1360	---	---	0.19	0.19	0.28	L-2	1197	---	---	0.22	0.22	0.32
2" x 4" - 8'	Inst	LF	C-11	815	0.24	0.06	0.45	0.75	0.97	C-11	717	0.29	0.07	0.52	0.87	1.12
2" x 6" - 8'	Demo	LF	L-2	1360	---	---	0.19	0.19	0.28	L-2	1197	---	---	0.22	0.22	0.32
2" x 6" - 8'	Inst	LF	C-11	815	0.36	0.06	0.45	0.87	1.09	C-11	717	0.43	0.07	0.52	1.01	1.26

			Costs Based On Large Volume							Costs Based On Small Volume						
Description	Oper	Unit	Crew Size	Avg Day Prod	Avg Mat'l Unit Cost	Avg Equip Unit Cost	Avg Labor Unit Cost	Avg Total Unit Cost	Avg Price Incl O&P	Crew Size	Avg Day Prod	Avg Mat'l Unit Cost	Avg Equip Unit Cost	Avg Labor Unit Cost	Avg Total Unit Cost	Avg Price Incl O&P

Framing (continued)

Sill or Bottom, bolted; bolts pre-embedded and not included in labor/material costs;
securing plate included in labor cost:

Description	Oper	Unit	Crew Size	Avg Day Prod	Avg Mat'l	Avg Equip	Avg Labor	Avg Total	Avg Price	Crew Size	Avg Day Prod	Avg Mat'l	Avg Equip	Avg Labor	Avg Total	Avg Price
2" x 4" - 8'	Demo	LF	L-2	685	---	---	0.38	0.38	0.56	L-2	630	---	---	0.41	0.41	0.61
2" x 4" - 8'	Inst	LF	C-11	560	0.20	0.08	0.66	0.94	1.26	C-11	515	0.24	0.09	0.72	1.04	1.40
2" x 6" - 8'	Demo	LF	L-2	685	---	---	0.38	0.38	0.56	L-2	630	---	---	0.41	0.41	0.61
2" x 6" - 8'	Inst	LF	C-11	560	0.32	0.08	0.66	1.06	1.38	C-11	515	0.38	0.09	0.72	1.19	1.54

Rafters, per LF of stick
Common, gable or hip, to 1/3 pitch:

Description	Oper	Unit	Crew Size	Avg Day Prod	Avg Mat'l	Avg Equip	Avg Labor	Avg Total	Avg Price	Crew Size	Avg Day Prod	Avg Mat'l	Avg Equip	Avg Labor	Avg Total	Avg Price
2" x 4" - 8' Avg.	Demo	LF	L-2	744	---	---	0.35	0.35	0.52	L-2	670	---	---	0.39	0.39	0.57
2" x 4" - 8' Avg.	Inst	LF	C-18	683	0.21	0.07	0.49	0.78	1.02	C-18	615	0.26	0.08	0.55	0.88	1.15
2" x 4" - 10' Avg.	Demo	LF	L-2	861	---	---	0.30	0.30	0.45	L-2	775	---	---	0.34	0.34	0.50
2" x 4" - 10' Avg.	Inst	LF	C-18	793	0.21	0.06	0.43	0.70	0.91	C-18	714	0.26	0.07	0.47	0.79	1.03
2" x 4" - 12' Avg.	Demo	LF	L-2	963	---	---	0.27	0.27	0.40	L-2	867	---	---	0.30	0.30	0.44
2" x 4" - 12' Avg.	Inst	LF	C-18	887	0.21	0.05	0.38	0.65	0.83	C-18	798	0.26	0.06	0.42	0.74	0.94
2" x 4" - 14' Avg.	Demo	LF	L-2	1053	---	---	0.25	0.25	0.37	L-2	948	---	---	0.27	0.27	0.41
2" x 4" - 14' Avg.	Inst	LF	C-18	971	0.21	0.05	0.35	0.61	0.78	C-18	874	0.26	0.05	0.39	0.70	0.88
2" x 4" - 16' Avg.	Demo	LF	L-2	1134	---	---	0.23	0.23	0.34	L-2	1021	---	---	0.25	0.25	0.38
2" x 4" - 16' Avg.	Inst	LF	C-18	1047	0.21	0.04	0.32	0.58	0.74	C-18	942	0.26	0.05	0.36	0.66	0.84
2" x 6" - 10' Avg.	Demo	LF	L-2	748	---	---	0.35	0.35	0.51	L-2	673	---	---	0.39	0.39	0.57
2" x 6" - 10' Avg.	Inst	LF	C-18	691	0.34	0.07	0.49	0.89	1.13	C-18	622	0.40	0.08	0.54	1.02	1.29
2" x 6" - 12' Avg.	Demo	LF	L-2	843	---	---	0.31	0.31	0.46	L-2	759	---	---	0.34	0.34	0.51
2" x 6" - 12' Avg.	Inst	LF	C-18	779	0.34	0.06	0.43	0.83	1.04	C-18	701	0.40	0.07	0.48	0.95	1.19
2" x 6" - 14' Avg.	Demo	LF	L-2	929	---	---	0.28	0.28	0.41	L-2	836	---	---	0.31	0.31	0.46
2" x 6" - 14' Avg.	Inst	LF	C-18	860	0.34	0.05	0.39	0.78	0.98	C-18	774	0.40	0.06	0.44	0.90	1.11
2" x 6" - 16' Avg.	Demo	LF	L-2	1008	---	---	0.26	0.26	0.38	L-2	907	---	---	0.29	0.29	0.42
2" x 6" - 16' Avg.	Inst	LF	C-18	933	0.34	0.05	0.36	0.75	0.93	C-18	840	0.40	0.06	0.40	0.86	1.06
2" x 6" - 18' Avg.	Demo	LF	L-2	1081	---	---	0.24	0.24	0.36	L-2	973	---	---	0.27	0.27	0.40
2" x 6" - 18' Avg.	Inst	LF	C-18	1001	0.34	0.05	0.34	0.72	0.89	C-18	901	0.40	0.05	0.37	0.83	1.01
2" x 8" - 12' Avg.	Demo	LF	L-2	737	---	---	0.35	0.35	0.52	L-2	663	---	---	0.39	0.39	0.58
2" x 8" - 12' Avg.	Inst	LF	C-18	684	0.48	0.07	0.49	1.04	1.28	C-18	616	0.57	0.08	0.55	1.19	1.46

			Costs Based On Large Volume							Costs Based On Small Volume						
Description	Oper	Unit	Crew Size	Avg Day Prod	Avg Mat'l Unit Cost	Avg Equip Unit Cost	Avg Labor Unit Cost	Avg Total Unit Cost	Avg Price Incl O&P	Crew Size	Avg Day Prod	Avg Mat'l Unit Cost	Avg Equip Unit Cost	Avg Labor Unit Cost	Avg Total Unit Cost	Avg Price Incl O&P

Framing (continued)

Description	Oper	Unit	Crew Size	Avg Day Prod	Avg Mat'l Unit Cost	Avg Equip Unit Cost	Avg Labor Unit Cost	Avg Total Unit Cost	Avg Price Incl O&P	Crew Size	Avg Day Prod	Avg Mat'l Unit Cost	Avg Equip Unit Cost	Avg Labor Unit Cost	Avg Total Unit Cost	Avg Price Incl O&P
2" x 8" - 14' Avg.	Demo	LF	L-2	804	---	---	0.32	0.32	0.48	L-2	724	---	---	0.36	0.36	0.53
2" x 8" - 14' Avg.	Inst	LF	C-18	748	0.48	0.06	0.45	0.99	1.21	C-18	673	0.57	0.07	0.50	1.14	1.39
2" x 8" - 16' Avg.	Demo	LF	L-2	864	---	---	0.30	0.30	0.45	L-2	778	---	---	0.33	0.33	0.50
2" x 8" - 16' Avg.	Inst	LF	C-18	805	0.48	0.06	0.42	0.95	1.16	C-18	725	0.57	0.06	0.47	1.10	1.33
2" x 8" - 18' Avg.	Demo	LF	L-2	918	---	---	0.28	0.28	0.42	L-2	826	---	---	0.31	0.31	0.47
2" x 8" - 18' Avg.	Inst	LF	C-18	856	0.48	0.05	0.39	0.92	1.12	C-18	770	0.57	0.06	0.44	1.07	1.28
2" x 8" - 20' Avg.	Demo	LF	L-2	967	---	---	0.27	0.27	0.40	L-2	870	---	---	0.30	0.30	0.44
2" x 8" - 20' Avg.	Inst	LF	C-18	902	0.48	0.05	0.37	0.90	1.08	C-18	812	0.57	0.06	0.42	1.04	1.25
Common, gable or hip, to 3/8-1/2 pitch:																
2" x 4" - 8' Avg.	Demo	LF	L-2	632	---	---	0.41	0.41	0.61	L-2	569	---	---	0.46	0.46	0.68
2" x 4" - 8' Avg.	Inst	LF	C-18	580	0.21	0.08	0.58	0.88	1.16	C-18	522	0.26	0.09	0.65	0.99	1.31
2" x 4" - 10' Avg.	Demo	LF	L-2	740	---	---	0.35	0.35	0.52	L-2	666	---	---	0.39	0.39	0.58
2" x 4" - 10' Avg.	Inst	LF	C-18	680	0.21	0.07	0.50	0.78	1.02	C-18	612	0.26	0.08	0.55	0.88	1.15
2" x 4" - 12' Avg.	Demo	LF	L-2	836	---	---	0.31	0.31	0.46	L-2	752	---	---	0.35	0.35	0.51
2" x 4" - 12' Avg.	Inst	LF	C-18	769	0.21	0.06	0.44	0.71	0.93	C-18	692	0.26	0.07	0.49	0.81	1.05
2" x 4" - 14' Avg.	Demo	LF	L-2	922	---	---	0.28	0.28	0.42	L-2	830	---	---	0.31	0.31	0.46
2" x 4" - 14' Avg.	Inst	LF	C-18	849	0.21	0.05	0.40	0.67	0.86	C-18	764	0.26	0.06	0.44	0.76	0.97
2" x 4" - 16' Avg.	Demo	LF	L-2	999	---	---	0.26	0.26	0.39	L-2	899	---	---	0.29	0.29	0.43
2" x 4" - 16' Avg.	Inst	LF	C-18	921	0.21	0.05	0.37	0.63	0.81	C-18	829	0.26	0.06	0.41	0.72	0.92
2" x 6" - 10' Avg.	Demo	LF	L-2	655	---	---	0.40	0.40	0.59	L-2	590	---	---	0.44	0.44	0.65
2" x 6" - 10' Avg.	Inst	LF	C-18	604	0.34	0.08	0.56	0.97	1.25	C-18	544	0.40	0.09	0.62	1.11	1.41
2" x 6" - 12' Avg.	Demo	LF	L-2	743	---	---	0.35	0.35	0.52	L-2	669	---	---	0.39	0.39	0.58
2" x 6" - 12' Avg.	Inst	LF	C-18	686	0.34	0.07	0.49	0.90	1.14	C-18	617	0.40	0.08	0.55	1.03	1.29
2" x 6" - 14' Avg.	Demo	LF	L-2	825	---	---	0.32	0.32	0.47	L-2	743	---	---	0.35	0.35	0.52
2" x 6" - 14' Avg.	Inst	LF	C-18	762	0.34	0.06	0.44	0.84	1.06	C-18	686	0.40	0.07	0.49	0.96	1.20
2" x 6" - 16' Avg.	Demo	LF	L-2	901	---	---	0.29	0.29	0.43	L-2	811	---	---	0.32	0.32	0.47
2" x 6" - 16' Avg.	Inst	LF	C-18	833	0.34	0.06	0.40	0.80	1.00	C-18	750	0.40	0.06	0.45	0.92	1.14
2" x 6" - 18' Avg.	Demo	LF	L-2	971	---	---	0.27	0.27	0.40	L-2	874	---	---	0.30	0.30	0.44
2" x 6" - 18' Avg.	Inst	LF	C-18	898	0.34	0.05	0.38	0.76	0.95	C-18	808	0.40	0.06	0.42	0.88	1.08
2" x 8" - 12' Avg.	Demo	LF	L-2	660	---	---	0.39	0.39	0.58	L-2	594	---	---	0.44	0.44	0.65
2" x 8" - 12' Avg.	Inst	LF	C-18	612	0.48	0.08	0.55	1.10	1.37	C-18	551	0.57	0.08	0.61	1.27	1.57

| | | | Costs Based On Large Volume | | | | | | | Costs Based On Small Volume | | | | | | |
Description	Oper	Unit	Crew Size	Avg Day Prod	Avg Mat'l Unit Cost	Avg Equip Unit Cost	Avg Labor Unit Cost	Avg Total Unit Cost	Avg Price Incl O&P	Crew Size	Avg Day Prod	Avg Mat'l Unit Cost	Avg Equip Unit Cost	Avg Labor Unit Cost	Avg Total Unit Cost	Avg Price Incl O&P

Framing (continued)

Description	Oper	Unit	Crew	Prod	Mat'l	Equip	Labor	Total	O&P	Crew	Prod	Mat'l	Equip	Labor	Total	O&P
2" x 8" - 14' Avg.	Demo	LF	L-2	725	---	---	0.36	0.36	0.53	L-2	653	---	---	0.40	0.40	0.59
2" x 8" - 14' Avg.	Inst	LF	C-18	673	0.48	0.07	0.50	1.05	1.29	C-18	606	0.57	0.08	0.56	1.20	1.48
2" x 8" - 16' Avg.	Demo	LF	L-2	784	---	---	0.33	0.33	0.49	L-2	706	---	---	0.37	0.37	0.55
2" x 8" - 16' Avg.	Inst	LF	C-18	728	0.48	0.06	0.46	1.00	1.23	C-18	655	0.57	0.07	0.51	1.16	1.41
2" x 8" - 18' Avg.	Demo	LF	L-2	837	---	---	0.31	0.31	0.46	L-2	753	---	---	0.35	0.35	0.51
2" x 8" - 18' Avg.	Inst	LF	C-18	779	0.48	0.06	0.43	0.97	1.18	C-18	701	0.57	0.07	0.48	1.12	1.35
2" x 8" - 20' Avg.	Demo	LF	L-2	885	---	---	0.29	0.29	0.43	L-2	797	---	---	0.33	0.33	0.48
2" x 8" - 20' Avg.	Inst	LF	C-18	825	0.48	0.06	0.41	0.94	1.14	C-18	743	0.57	0.06	0.45	1.09	1.31
Common, cut-up roofs, to 1/3 pitch:																
2" x 4" - 8' Avg.	Demo	LF	L-2	588	---	---	0.44	0.44	0.65	L-2	529	---	---	0.49	0.49	0.73
2" x 4" - 8' Avg.	Inst	LF	C-18	539	0.22	0.09	0.63	0.93	1.24	C-18	485	0.26	0.10	0.69	1.05	1.39
2" x 4" - 10' Avg.	Demo	LF	L-2	692	---	---	0.38	0.38	0.56	L-2	623	---	---	0.42	0.42	0.62
2" x 4" - 10' Avg.	Inst	LF	C-18	635	0.22	0.07	0.53	0.82	1.08	C-18	572	0.26	0.08	0.59	0.93	1.22
2" x 4" - 12' Avg.	Demo	LF	L-2	783	---	---	0.33	0.33	0.49	L-2	705	---	---	0.37	0.37	0.55
2" x 4" - 12' Avg.	Inst	LF	C-18	720	0.22	0.06	0.47	0.75	0.98	C-18	648	0.26	0.07	0.52	0.85	1.11
2" x 4" - 14' Avg.	Demo	LF	L-2	868	---	---	0.30	0.30	0.44	L-2	781	---	---	0.33	0.33	0.49
2" x 4" - 14' Avg.	Inst	LF	C-18	799	0.22	0.06	0.42	0.70	0.90	C-18	719	0.26	0.06	0.47	0.79	1.02
2" x 4" - 16' Avg.	Demo	LF	L-2	945	---	---	0.28	0.28	0.41	L-2	851	---	---	0.31	0.31	0.45
2" x 4" - 16' Avg.	Inst	LF	C-18	870	0.22	0.05	0.39	0.66	0.85	C-18	783	0.26	0.06	0.43	0.75	0.96
2" x 6" - 10' Avg.	Demo	LF	L-2	617	---	---	0.42	0.42	0.62	L-2	555	---	---	0.47	0.47	0.69
2" x 6" - 10' Avg.	Inst	LF	C-18	569	0.34	0.08	0.59	1.02	1.31	C-18	512	0.41	0.09	0.66	1.16	1.48
2" x 6" - 12' Avg.	Demo	LF	L-2	703	---	---	0.37	0.37	0.55	L-2	633	---	---	0.41	0.41	0.61
2" x 6" - 12' Avg.	Inst	LF	C-18	648	0.34	0.07	0.52	0.94	1.19	C-18	583	0.41	0.08	0.58	1.07	1.35
2" x 6" - 14' Avg.	Demo	LF	L-2	781	---	---	0.33	0.33	0.49	L-2	703	---	---	0.37	0.37	0.55
2" x 6" - 14' Avg.	Inst	LF	C-18	721	0.34	0.06	0.47	0.88	1.10	C-18	649	0.41	0.07	0.52	1.00	1.26
2" x 6" - 16' Avg.	Demo	LF	L-2	855	---	---	0.30	0.30	0.45	L-2	770	---	---	0.34	0.34	0.50
2" x 6" - 16' Avg.	Inst	LF	C-18	789	0.34	0.06	0.43	0.83	1.04	C-18	710	0.41	0.07	0.47	0.95	1.18
2" x 6" - 18' Avg.	Demo	LF	L-2	923	---	---	0.28	0.28	0.42	L-2	831	---	---	0.31	0.31	0.46
2" x 6" - 18' Avg.	Inst	LF	C-18	853	0.34	0.05	0.40	0.79	0.99	C-18	768	0.41	0.06	0.44	0.91	1.13
2" x 8" - 12' Avg.	Demo	LF	L-2	627	---	---	0.41	0.41	0.61	L-2	564	---	---	0.46	0.46	0.68
2" x 8" - 12' Avg.	Inst	LF	C-18	580	0.48	0.08	0.58	1.15	1.43	C-18	522	0.58	0.09	0.65	1.32	1.63

Framing (continued)

			Costs Based On Large Volume							Costs Based On Small Volume						
Description	Oper	Unit	Crew Size	Avg Day Prod	Avg Mat'l Unit Cost	Avg Equip Unit Cost	Avg Labor Unit Cost	Avg Total Unit Cost	Avg Price Incl O&P	Crew Size	Avg Day Prod	Avg Mat'l Unit Cost	Avg Equip Unit Cost	Avg Labor Unit Cost	Avg Total Unit Cost	Avg Price Incl O&P
2" x 8" - 14' Avg.	Demo	LF	L-2	691	---	---	0.38	0.38	0.56	L-2	622	---	---	0.42	0.42	0.62
2" x 8" - 14' Avg.	Inst	LF	C-18	641	0.48	0.07	0.53	1.08	1.34	C-18	577	0.58	0.08	0.58	1.25	1.53
2" x 8" - 16' Avg.	Demo	LF	L-2	749	---	---	0.35	0.35	0.51	L-2	674	---	---	0.39	0.39	0.57
2" x 8" - 16' Avg.	Inst	LF	C-18	695	0.48	0.07	0.49	1.04	1.27	C-18	626	0.58	0.07	0.54	1.19	1.46
2" x 8" - 18' Avg.	Demo	LF	L-2	802	---	---	0.32	0.32	0.48	L-2	722	---	---	0.36	0.36	0.53
2" x 8" - 18' Avg.	Inst	LF	C-18	745	0.48	0.06	0.45	1.00	1.22	C-18	671	0.58	0.07	0.50	1.15	1.40
2" x 8" - 20' Avg.	Demo	LF	L-2	850	---	---	0.31	0.31	0.45	L-2	765	---	---	0.34	0.34	0.50
2" x 8" - 20' Avg.	Inst	LF	C-18	791	0.48	0.06	0.43	0.97	1.18	C-18	712	0.58	0.07	0.47	1.12	1.35
Common, cut-up roofs, to 3/8-1/2 pitch:																
2" x 4" - 8' Avg.	Demo	LF	L-2	516	---	---	0.50	0.50	0.75	L-2	464	---	---	0.56	0.56	0.83
2" x 4" - 8' Avg.	Inst	LF	C-18	473	0.22	0.10	0.71	1.03	1.38	C-18	426	0.26	0.11	0.79	1.16	1.55
2" x 4" - 10' Avg.	Demo	LF	L-2	612	---	---	0.42	0.42	0.63	L-2	551	---	---	0.47	0.47	0.70
2" x 4" - 10' Avg.	Inst	LF	C-18	562	0.22	0.08	0.60	0.90	1.19	C-18	506	0.26	0.09	0.67	1.02	1.35
2" x 4" - 12' Avg.	Demo	LF	L-2	697	---	---	0.37	0.37	0.55	L-2	627	---	---	0.41	0.41	0.61
2" x 4" - 12' Avg.	Inst	LF	C-18	640	0.22	0.07	0.53	0.82	1.07	C-18	576	0.26	0.08	0.59	0.93	1.21
2" x 4" - 14' Avg.	Demo	LF	L-2	777	---	---	0.33	0.33	0.50	L-2	699	---	---	0.37	0.37	0.55
2" x 4" - 14' Avg.	Inst	LF	C-18	714	0.22	0.07	0.47	0.75	0.99	C-18	643	0.26	0.07	0.52	0.86	1.11
2" x 4" - 16' Avg.	Demo	LF	L-2	849	---	---	0.31	0.31	0.45	L-2	764	---	---	0.34	0.34	0.50
2" x 4" - 16' Avg.	Inst	LF	C-18	780	0.22	0.06	0.43	0.71	0.92	C-18	702	0.26	0.07	0.48	0.81	1.04
2" x 6" - 10' Avg.	Demo	LF	L-2	552	---	---	0.47	0.47	0.70	L-2	497	---	---	0.52	0.52	0.77
2" x 6" - 10' Avg.	Inst	LF	C-18	508	0.34	0.09	0.66	1.10	1.42	C-18	457	0.41	0.10	0.74	1.25	1.61
2" x 6" - 12' Avg.	Demo	LF	L-2	633	---	---	0.41	0.41	0.61	L-2	570	---	---	0.46	0.46	0.68
2" x 6" - 12' Avg.	Inst	LF	C-18	583	0.34	0.08	0.58	1.00	1.28	C-18	525	0.41	0.09	0.64	1.14	1.46
2" x 6" - 14' Avg.	Demo	LF	L-2	706	---	---	0.37	0.37	0.55	L-2	635	---	---	0.41	0.41	0.61
2" x 6" - 14' Avg.	Inst	LF	C-18	651	0.34	0.07	0.52	0.93	1.19	C-18	586	0.41	0.08	0.58	1.07	1.35
2" x 6" - 16' Avg.	Demo	LF	L-2	776	---	---	0.34	0.34	0.50	L-2	698	---	---	0.37	0.37	0.55
2" x 6" - 16' Avg.	Inst	LF	C-18	715	0.34	0.07	0.47	0.88	1.11	C-18	644	0.41	0.07	0.52	1.01	1.26
2" x 6" - 18' Avg.	Demo	LF	L-2	842	---	---	0.31	0.31	0.46	L-2	758	---	---	0.34	0.34	0.51
2" x 6" - 18' Avg.	Inst	LF	C-18	777	0.34	0.06	0.43	0.84	1.05	C-18	699	0.41	0.07	0.48	0.96	1.20
2" x 8" - 12' Avg.	Demo	LF	L-2	570	---	---	0.46	0.46	0.68	L-2	513	---	---	0.51	0.51	0.75
2" x 8" - 12' Avg.	Inst	LF	C-18	527	0.48	0.09	0.64	1.21	1.53	C-18	474	0.58	0.10	0.71	1.39	1.74

			Costs Based On Large Volume							Costs Based On Small Volume						
Description	Oper	Unit	Crew Size	Avg Day Prod	Avg Mat'l Unit Cost	Avg Equip Unit Cost	Avg Labor Unit Cost	Avg Total Unit Cost	Avg Price Incl O&P	Crew Size	Avg Day Prod	Avg Mat'l Unit Cost	Avg Equip Unit Cost	Avg Labor Unit Cost	Avg Total Unit Cost	Avg Price Incl O&P

Framing (continued)

Description	Oper	Unit	Crew Size	Avg Day Prod	Avg Mat'l Unit Cost	Avg Equip Unit Cost	Avg Labor Unit Cost	Avg Total Unit Cost	Avg Price Incl O&P	Crew Size	Avg Day Prod	Avg Mat'l Unit Cost	Avg Equip Unit Cost	Avg Labor Unit Cost	Avg Total Unit Cost	Avg Price Incl O&P
2" x 8" - 14' Avg.	Demo	LF	L-2	632	---	---	0.41	0.41	0.61	L-2	569	---	---	0.46	0.46	0.68
2" x 8" - 14' Avg.	Inst	LF	C-18	585	0.48	0.08	0.58	1.14	1.42	C-18	527	0.58	0.09	0.64	1.31	1.62
2" x 8" - 16' Avg.	Demo	LF	L-2	688	---	---	0.38	0.38	0.56	L-2	619	---	---	0.42	0.42	0.62
2" x 8" - 16' Avg.	Inst	LF	C-18	638	0.48	0.07	0.53	1.09	1.34	C-18	574	0.58	0.08	0.59	1.25	1.54
2" x 8" - 18' Avg.	Demo	LF	L-2	739	---	---	0.35	0.35	0.52	L-2	665	---	---	0.39	0.39	0.58
2" x 8" - 18' Avg.	Inst	LF	C-18	686	0.48	0.07	0.49	1.04	1.28	C-18	617	0.58	0.08	0.55	1.20	1.47
2" x 8" - 20' Avg.	Demo	LF	L-2	787	---	---	0.33	0.33	0.49	L-2	708	---	---	0.37	0.37	0.54
2" x 8" - 20' Avg.	Inst	LF	C-18	731	0.48	0.06	0.46	1.01	1.23	C-18	658	0.58	0.07	0.51	1.16	1.41

Rafters, per SF of area
Gable or hip, to 1/3 pitch:

Description	Oper	Unit	Crew Size	Avg Day Prod	Avg Mat'l Unit Cost	Avg Equip Unit Cost	Avg Labor Unit Cost	Avg Total Unit Cost	Avg Price Incl O&P	Crew Size	Avg Day Prod	Avg Mat'l Unit Cost	Avg Equip Unit Cost	Avg Labor Unit Cost	Avg Total Unit Cost	Avg Price Incl O&P
2" x 4" - 12" o.c.	Demo	SF	L-2	840	---	---	0.31	0.31	0.46	L-2	756	---	---	0.34	0.34	0.51
2" x 4" - 12" o.c.	Inst	SF	C-18	774	0.26	0.06	0.44	0.76	0.97	C-18	697	0.32	0.07	0.48	0.87	1.10
2" x 4" - 16" o.c.	Demo	SF	L-2	1000	---	---	0.26	0.26	0.38	L-2	900	---	---	0.29	0.29	0.43
2" x 4" - 16" o.c.	Inst	SF	C-18	921	0.21	0.05	0.37	0.63	0.81	C-18	829	0.25	0.06	0.41	0.72	0.91
2" x 4" - 24" o.c.	Demo	SF	L-2	1312	---	---	0.20	0.20	0.29	L-2	1181	---	---	0.22	0.22	0.33
2" x 4" - 24" o.c.	Inst	SF	C-18	1208	0.16	0.04	0.28	0.48	0.61	C-18	1087	0.19	0.04	0.31	0.54	0.69
2" x 6" - 12" o.c.	Demo	SF	L-2	821	---	---	0.32	0.32	0.47	L-2	739	---	---	0.35	0.35	0.52
2" x 6" - 12" o.c.	Inst	SF	C-18	760	0.39	0.06	0.44	0.89	1.11	C-18	684	0.47	0.07	0.49	1.03	1.27
2" x 6" - 16" o.c.	Demo	SF	L-2	978	---	---	0.27	0.27	0.39	L-2	880	---	---	0.30	0.30	0.44
2" x 6" - 16" o.c.	Inst	SF	C-18	905	0.31	0.05	0.37	0.73	0.91	C-18	815	0.37	0.06	0.41	0.84	1.04
2" x 6" - 24" o.c.	Demo	SF	L-2	1284	---	---	0.20	0.20	0.30	L-2	1156	---	---	0.23	0.23	0.33
2" x 6" - 24" o.c.	Inst	SF	C-18	1187	0.22	0.04	0.28	0.55	0.69	C-18	1068	0.27	0.04	0.32	0.63	0.78
2" x 8" - 12" o.c.	Demo	SF	L-2	785	---	---	0.33	0.33	0.49	L-2	707	---	---	0.37	0.37	0.54
2" x 8" - 12" o.c.	Inst	SF	C-18	730	0.53	0.06	0.46	1.06	1.28	C-18	657	0.64	0.07	0.51	1.22	1.47
2" x 8" - 16" o.c.	Demo	SF	L-2	933	---	---	0.28	0.28	0.41	L-2	840	---	---	0.31	0.31	0.46
2" x 8" - 16" o.c.	Inst	SF	C-18	868	0.41	0.05	0.39	0.85	1.04	C-18	781	0.50	0.06	0.43	0.99	1.20
2" x 8" - 24" o.c.	Demo	SF	L-2	1225	---	---	0.21	0.21	0.31	L-2	1103	---	---	0.24	0.24	0.35
2" x 8" - 24" o.c.	Inst	SF	C-18	1140	0.30	0.04	0.30	0.63	0.78	C-18	1026	0.36	0.05	0.33	0.73	0.89

			Costs Based On Large Volume							Costs Based On Small Volume						
Description	Oper	Unit	Crew Size	Avg Day Prod	Avg Mat'l Unit Cost	Avg Equip Unit Cost	Avg Labor Unit Cost	Avg Total Unit Cost	Avg Price Incl O&P	Crew Size	Avg Day Prod	Avg Mat'l Unit Cost	Avg Equip Unit Cost	Avg Labor Unit Cost	Avg Total Unit Cost	Avg Price Incl O&P

Framing (continued)

Gable or hip, to 3/8-1/2 pitch:

Description	Oper	Unit	Crew Size	Avg Day Prod	Avg Mat'l Unit Cost	Avg Equip Unit Cost	Avg Labor Unit Cost	Avg Total Unit Cost	Avg Price Incl O&P	Crew Size	Avg Day Prod	Avg Mat'l Unit Cost	Avg Equip Unit Cost	Avg Labor Unit Cost	Avg Total Unit Cost	Avg Price Incl O&P
2" x 4" - 12" o.c.	Demo	SF	L-2	723	---	---	0.36	0.36	0.53	L-2	651	---	---	0.40	0.40	0.59
2" x 4" - 12" o.c.	Inst	SF	C-18	664	0.26	0.07	0.51	0.84	1.09	C-18	598	0.32	0.08	0.56	0.96	1.23
2" x 4" - 16" o.c.	Demo	SF	L-2	859	---	---	0.30	0.30	0.45	L-2	773	---	---	0.34	0.34	0.50
2" x 4" - 16" o.c.	Inst	SF	C-18	789	0.21	0.06	0.43	0.70	0.91	C-18	710	0.25	0.07	0.47	0.79	1.03
2" x 4" - 24" o.c.	Demo	SF	L-2	1127	---	---	0.23	0.23	0.34	L-2	1014	---	---	0.26	0.26	0.38
2" x 4" - 24" o.c.	Inst	SF	C-18	1036	0.16	0.05	0.33	0.53	0.69	C-18	932	0.19	0.05	0.36	0.60	0.78
2" x 6" - 12" o.c.	Demo	SF	L-2	725	---	---	0.36	0.36	0.53	L-2	653	---	---	0.40	0.40	0.59
2" x 6" - 12" o.c.	Inst	SF	C-18	669	0.39	0.07	0.50	0.96	1.21	C-18	602	0.47	0.08	0.56	1.11	1.38
2" x 6" - 16" o.c.	Demo	SF	L-2	862	---	---	0.30	0.30	0.45	L-2	776	---	---	0.34	0.34	0.50
2" x 6" - 16" o.c.	Inst	SF	C-18	796	0.31	0.06	0.42	0.79	1.00	C-18	716	0.37	0.07	0.47	0.90	1.14
2" x 6" - 24" o.c.	Demo	SF	L-2	1132	---	---	0.23	0.23	0.34	L-2	1019	---	---	0.26	0.26	0.38
2" x 6" - 24" o.c.	Inst	SF	C-18	1045	0.22	0.04	0.32	0.59	0.75	C-18	941	0.27	0.05	0.36	0.68	0.85
2" x 8" - 12" o.c.	Demo	SF	L-2	708	---	---	0.37	0.37	0.54	L-2	637	---	---	0.41	0.41	0.60
2" x 8" - 12" o.c.	Inst	SF	C-18	657	0.53	0.07	0.51	1.12	1.37	C-18	591	0.64	0.08	0.57	1.29	1.57
2" x 8" - 16" o.c.	Demo	SF	L-2	841	---	---	0.31	0.31	0.46	L-2	757	---	---	0.34	0.34	0.51
2" x 8" - 16" o.c.	Inst	SF	C-18	781	0.41	0.06	0.43	0.90	1.12	C-18	703	0.50	0.07	0.48	1.04	1.28
2" x 8" - 24" o.c.	Demo	SF	L-2	1351	---	---	0.19	0.19	0.28	L-2	1216	---	---	0.21	0.21	0.32
2" x 8" - 24" o.c.	Inst	SF	C-18	1260	0.30	0.04	0.27	0.60	0.73	C-18	1134	0.36	0.04	0.30	0.69	0.84

Cut-up roofs, to 1/3 pitch:

Description	Oper	Unit	Crew Size	Avg Day Prod	Avg Mat'l Unit Cost	Avg Equip Unit Cost	Avg Labor Unit Cost	Avg Total Unit Cost	Avg Price Incl O&P	Crew Size	Avg Day Prod	Avg Mat'l Unit Cost	Avg Equip Unit Cost	Avg Labor Unit Cost	Avg Total Unit Cost	Avg Price Incl O&P
2" x 4" - 12" o.c.	Demo	SF	L-2	674	---	---	0.39	0.39	0.57	L-2	607	---	---	0.43	0.43	0.63
2" x 4" - 12" o.c.	Inst	SF	C-18	619	0.27	0.08	0.54	0.89	1.15	C-18	557	0.32	0.08	0.61	1.01	1.31
2" x 4" - 16" o.c.	Demo	SF	L-2	804	---	---	0.32	0.32	0.48	L-2	724	---	---	0.36	0.36	0.53
2" x 4" - 16" o.c.	Inst	SF	C-18	738	0.21	0.06	0.46	0.73	0.96	C-18	664	0.26	0.07	0.51	0.83	1.08
2" x 4" - 24" o.c.	Demo	SF	L-2	1055	---	---	0.25	0.25	0.36	L-2	950	---	---	0.27	0.27	0.41
2" x 4" - 24" o.c.	Inst	SF	C-18	968	0.16	0.05	0.35	0.56	0.73	C-18	871	0.19	0.05	0.39	0.63	0.82
2" x 6" - 12" o.c.	Demo	SF	L-2	686	---	---	0.38	0.38	0.56	L-2	617	---	---	0.42	0.42	0.62
2" x 6" - 12" o.c.	Inst	SF	C-18	633	0.40	0.07	0.53	1.00	1.26	C-18	570	0.48	0.08	0.59	1.15	1.44

			Costs Based On Large Volume							Costs Based On Small Volume					
Description	Oper	Unit	Crew Size	Avg Day Prod	Avg Mat'l Unit Cost	Avg Equip Unit Cost	Avg Labor Unit Cost	Avg Total Unit Cost	Avg Price Incl O&P	Crew Size	Avg Day Prod	Avg Mat'l Unit Cost	Avg Equip Unit Cost	Avg Labor Unit Cost	Avg Total Unit Cost	Avg Price Incl O&P

Framing (continued)

Description	Oper	Unit	Crew Size	Avg Day Prod	Avg Mat'l Unit Cost	Avg Equip Unit Cost	Avg Labor Unit Cost	Avg Total Unit Cost	Avg Price Incl O&P	Crew Size	Avg Day Prod	Avg Mat'l Unit Cost	Avg Equip Unit Cost	Avg Labor Unit Cost	Avg Total Unit Cost	Avg Price Incl O&P
2" x 6" - 16" o.c.	Demo	SF	L-2	816	---	---	0.32	0.32	0.47	L-2	734	---	---	0.35	0.35	0.52
2" x 6" - 16" o.c.	Inst	SF	C-18	752	0.31	0.06	0.45	0.82	1.04	C-18	677	0.37	0.07	0.50	0.94	1.19
2" x 6" - 24" o.c.	Demo	SF	L-2	1070	---	---	0.24	0.24	0.36	L-2	963	---	---	0.27	0.27	0.40
2" x 6" - 24" o.c.	Inst	SF	C-18	987	0.23	0.05	0.34	0.62	0.78	C-18	888	0.27	0.05	0.38	0.71	0.89
2" x 8" - 12" o.c.	Demo	SF	L-2	674	---	---	0.39	0.39	0.57	L-2	607	---	---	0.43	0.43	0.63
2" x 8" - 12" o.c.	Inst	SF	C-18	625	0.54	0.07	0.54	1.16	1.42	C-18	563	0.65	0.08	0.60	1.33	1.63
2" x 8" - 16" o.c.	Demo	SF	L-2	802	---	---	0.32	0.32	0.48	L-2	722	---	---	0.36	0.36	0.53
2" x 8" - 16" o.c.	Inst	SF	C-18	744	0.42	0.06	0.45	0.93	1.16	C-18	670	0.50	0.07	0.50	1.08	1.32
2" x 8" - 24" o.c.	Demo	SF	L-2	1053	---	---	0.25	0.25	0.37	L-2	948	---	---	0.27	0.27	0.41
2" x 8" - 24" o.c.	Inst	SF	C-18	976	0.30	0.05	0.35	0.69	0.86	C-18	878	0.36	0.05	0.38	0.80	0.99

Cut-up roofs, to 3/8-1/2 pitch:

Description	Oper	Unit	Crew Size	Avg Day Prod	Avg Mat'l Unit Cost	Avg Equip Unit Cost	Avg Labor Unit Cost	Avg Total Unit Cost	Avg Price Incl O&P	Crew Size	Avg Day Prod	Avg Mat'l Unit Cost	Avg Equip Unit Cost	Avg Labor Unit Cost	Avg Total Unit Cost	Avg Price Incl O&P
2" x 4" - 12" o.c.	Demo	SF	L-2	596	---	---	0.44	0.44	0.65	L-2	536	---	---	0.48	0.48	0.72
2" x 4" - 12" o.c.	Inst	SF	C-18	547	0.27	0.09	0.62	0.97	1.27	C-18	492	0.32	0.09	0.68	1.10	1.43
2" x 4" - 16" o.c.	Demo	SF	L-2	712	---	---	0.37	0.37	0.54	L-2	641	---	---	0.41	0.41	0.60
2" x 4" - 16" o.c.	Inst	SF	C-18	653	0.21	0.07	0.52	0.80	1.05	C-18	588	0.26	0.08	0.57	0.91	1.19
2" x 4" - 24" o.c.	Demo	SF	L-2	932	---	---	0.28	0.28	0.41	L-2	839	---	---	0.31	0.31	0.46
2" x 4" - 24" o.c.	Inst	SF	C-18	855	0.16	0.05	0.39	0.61	0.80	C-18	770	0.19	0.06	0.44	0.69	0.91
2" x 6" - 12" o.c.	Demo	SF	L-2	618	---	---	0.42	0.42	0.62	L-2	556	---	---	0.47	0.47	0.69
2" x 6" - 12" o.c.	Inst	SF	C-18	569	0.40	0.08	0.59	1.07	1.36	C-18	512	0.48	0.09	0.66	1.22	1.55
2" x 6" - 16" o.c.	Demo	SF	L-2	735	---	---	0.35	0.35	0.52	L-2	662	---	---	0.39	0.39	0.58
2" x 6" - 16" o.c.	Inst	SF	C-18	677	0.31	0.07	0.50	0.88	1.12	C-18	609	0.37	0.08	0.55	1.00	1.28
2" x 6" - 24" o.c.	Demo	SF	L-2	963	---	---	0.27	0.27	0.40	L-2	867	---	---	0.30	0.30	0.44
2" x 6" - 24" o.c.	Inst	SF	C-18	887	0.23	0.05	0.38	0.66	0.85	C-18	798	0.27	0.06	0.42	0.75	0.96
2" x 8" - 12" o.c.	Demo	SF	L-2	616	---	---	0.42	0.42	0.62	L-2	554	---	---	0.47	0.47	0.69
2" x 8" - 12" o.c.	Inst	SF	C-18	570	0.54	0.08	0.59	1.21	1.50	C-18	513	0.65	0.09	0.66	1.40	1.72
2" x 8" - 16" o.c.	Demo	SF	L-2	734	---	---	0.35	0.35	0.52	L-2	661	---	---	0.39	0.39	0.58
2" x 8" - 16" o.c.	Inst	SF	C-18	679	0.42	0.07	0.50	0.98	1.23	C-18	611	0.50	0.08	0.55	1.13	1.40

Description	Oper	Unit	Costs Based On Large Volume							Costs Based On Small Volume						
			Crew Size	Avg Day Prod	Avg Mat'l Unit Cost	Avg Equip Unit Cost	Avg Labor Unit Cost	Avg Total Unit Cost	Avg Price Incl O&P	Crew Size	Avg Day Prod	Avg Mat'l Unit Cost	Avg Equip Unit Cost	Avg Labor Unit Cost	Avg Total Unit Cost	Avg Price Incl O&P

Framing (continued)

Description	Oper	Unit	Crew Size	Avg Day Prod	Avg Mat'l Unit Cost	Avg Equip Unit Cost	Avg Labor Unit Cost	Avg Total Unit Cost	Avg Price Incl O&P	Crew Size	Avg Day Prod	Avg Mat'l Unit Cost	Avg Equip Unit Cost	Avg Labor Unit Cost	Avg Total Unit Cost	Avg Price Incl O&P
2" x 8" - 24" o.c.	Demo	SF	L-2	962	---	---	0.27	0.27	0.40	L-2	866	---	---	0.30	0.30	0.44
2" x 8" - 24" o.c.	Inst	SF	C-18	891	0.30	0.05	0.38	0.73	0.92	C-18	802	0.36	0.06	0.42	0.84	1.05

Roof Decking, solid, T&G, dry for Plank-and-Beam construction

Description	Oper	Unit	Crew Size	Avg Day Prod	Avg Mat'l Unit Cost	Avg Equip Unit Cost	Avg Labor Unit Cost	Avg Total Unit Cost	Avg Price Incl O&P	Crew Size	Avg Day Prod	Avg Mat'l Unit Cost	Avg Equip Unit Cost	Avg Labor Unit Cost	Avg Total Unit Cost	Avg Price Incl O&P
2" x 6" - 12'	Demo	SF	L-2	645	---	---	0.40	0.40	0.60	L-2	581	---	---	0.45	0.45	0.66
2" x 6" - 12'	Inst	SF	C-11	480	0.10	0.10	0.70	0.90	1.24	C-11	432	0.12	0.11	0.78	1.01	1.39
2" x 8" - 12'	Demo	SF	L-2	900	---	---	0.29	0.29	0.43	L-2	810	---	---	0.32	0.32	0.48
2" x 8" - 12'	Inst	SF	C-11	670	0.07	0.07	0.50	0.64	0.89	C-11	603	0.09	0.08	0.56	0.72	1.00

Studs/Plates, per LF of stick
Walls or Partitions:

Description	Oper	Unit	Crew Size	Avg Day Prod	Avg Mat'l Unit Cost	Avg Equip Unit Cost	Avg Labor Unit Cost	Avg Total Unit Cost	Avg Price Incl O&P	Crew Size	Avg Day Prod	Avg Mat'l Unit Cost	Avg Equip Unit Cost	Avg Labor Unit Cost	Avg Total Unit Cost	Avg Price Incl O&P
2" x 4" - 8' Avg.	Demo	LF	L-2	1031	---	---	0.25	0.25	0.37	L-2	907	---	---	0.29	0.29	0.42
2" x 4" - 8' Avg.	Inst	LF	C-18	848	0.21	0.06	0.40	0.66	0.86	C-18	746	0.25	0.06	0.45	0.77	0.99
2" x 4" - 10' Avg.	Demo	LF	L-2	1172	---	---	0.22	0.22	0.33	L-2	1031	---	---	0.25	0.25	0.37
2" x 4" - 10' Avg.	Inst	LF	C-18	969	0.21	0.05	0.35	0.61	0.78	C-18	853	0.25	0.05	0.40	0.70	0.90
2" x 4" - 12' Avg.	Demo	LF	L-2	1308	---	---	0.20	0.20	0.29	L-2	1151	---	---	0.23	0.23	0.33
2" x 4" - 12' Avg.	Inst	LF	C-18	1082	0.21	0.04	0.31	0.56	0.72	C-18	952	0.25	0.05	0.35	0.66	0.83
2" x 6" - 8' Avg.	Demo	LF	L-2	904	---	---	0.29	0.29	0.43	L-2	796	---	---	0.33	0.33	0.48
2" x 6" - 8' Avg.	Inst	LF	C-18	752	0.33	0.06	0.45	0.84	1.06	C-18	662	0.39	0.07	0.51	0.97	1.22
2" x 6" - 10' Avg.	Demo	LF	L-2	1032	---	---	0.25	0.25	0.37	L-2	908	---	---	0.29	0.29	0.42
2" x 6" - 10' Avg.	Inst	LF	C-18	861	0.33	0.05	0.39	0.77	0.97	C-18	758	0.39	0.06	0.44	0.90	1.12
2" x 6" - 12' Avg.	Demo	LF	L-2	1142	---	---	0.23	0.23	0.34	L-2	1005	---	---	0.26	0.26	0.38
2" x 6" - 12' Avg.	Inst	LF	C-18	956	0.33	0.05	0.35	0.73	0.90	C-18	841	0.39	0.06	0.40	0.85	1.05
2" x 8" - 8' Avg.	Demo	LF	L-2	802	---	---	0.32	0.32	0.48	L-2	706	---	---	0.37	0.37	0.55
2" x 8" - 8' Avg.	Inst	LF	C-18	673	0.46	0.07	0.50	1.03	1.28	C-18	592	0.55	0.08	0.57	1.20	1.48
2" x 8" - 10' Avg.	Demo	LF	L-2	915	---	---	0.28	0.28	0.42	L-2	805	---	---	0.32	0.32	0.48
2" x 8" - 10' Avg.	Inst	LF	C-18	771	0.46	0.06	0.44	0.96	1.17	C-18	678	0.55	0.07	0.50	1.12	1.36
2" x 8" - 12' Avg.	Demo	LF	L-2	1010	---	---	0.26	0.26	0.38	L-2	889	---	---	0.29	0.29	0.43
2" x 8" - 12' Avg.	Inst	LF	C-18	855	0.46	0.05	0.39	0.91	1.10	C-18	752	0.55	0.06	0.45	1.06	1.28

Gable Ends:

Description	Oper	Unit	Crew Size	Avg Day Prod	Avg Mat'l Unit Cost	Avg Equip Unit Cost	Avg Labor Unit Cost	Avg Total Unit Cost	Avg Price Incl O&P	Crew Size	Avg Day Prod	Avg Mat'l Unit Cost	Avg Equip Unit Cost	Avg Labor Unit Cost	Avg Total Unit Cost	Avg Price Incl O&P
2" x 4" - 3' Avg.	Demo	LF	L-2	660	---	---	0.39	0.39	0.58	L-2	581	---	---	0.45	0.45	0.66
2" x 4" - 3' Avg.	Inst	LF	C-18	344	0.22	0.14	0.98	1.33	1.82	C-18	303	0.26	0.15	1.11	1.53	2.08

Framing (continued)

Description	Oper	Unit	Crew Size	Avg Day Prod	Avg Mat'l Unit Cost	Avg Equip Unit Cost	Avg Labor Unit Cost	Avg Total Unit Cost	Avg Price Incl O&P	Crew Size	Avg Day Prod	Avg Mat'l Unit Cost	Avg Equip Unit Cost	Avg Labor Unit Cost	Avg Total Unit Cost	Avg Price Incl O&P
					Costs Based On Large Volume								Costs Based On Small Volume				
2" x 4" - 4' Avg.	Demo	LF	L-2	818	---	---	0.32	0.32	0.47	L-2	720	---	---	0.36	0.36	0.53
2" x 4" - 4' Avg.	Inst	LF	C-18	431	0.22	0.11	0.78	1.11	1.49	C-18	379	0.26	0.12	0.89	1.28	1.71
2" x 4" - 5' Avg.	Demo	LF	L-2	960	---	---	0.27	0.27	0.40	L-2	845	---	---	0.31	0.31	0.46
2" x 4" - 5' Avg.	Inst	LF	C-18	511	0.22	0.09	0.66	0.97	1.29	C-18	450	0.26	0.10	0.75	1.12	1.48
2" x 4" - 6' Avg.	Demo	LF	L-2	1082	---	---	0.24	0.24	0.36	L-2	952	---	---	0.27	0.27	0.40
2" x 4" - 6' Avg.	Inst	LF	C-18	580	0.22	0.08	0.58	0.88	1.17	C-18	510	0.26	0.09	0.66	1.02	1.34
2" x 6" - 3' Avg.	Demo	LF	L-2	634	---	---	0.41	0.41	0.61	L-2	558	---	---	0.47	0.47	0.69
2" x 6" - 3' Avg.	Inst	LF	C-18	337	0.34	0.14	1.00	1.48	1.97	C-18	297	0.41	0.16	1.14	1.71	2.26
2" x 6" - 4' Avg.	Demo	LF	L-2	778	---	---	0.33	0.33	0.49	L-2	685	---	---	0.38	0.38	0.56
2" x 6" - 4' Avg.	Inst	LF	C-18	420	0.34	0.11	0.80	1.26	1.65	C-18	370	0.41	0.13	0.91	1.45	1.90
2" x 6" - 5' Avg.	Demo	LF	L-2	906	---	---	0.29	0.29	0.42	L-2	797	---	---	0.33	0.33	0.48
2" x 6" - 5' Avg.	Inst	LF	C-18	495	0.34	0.09	0.68	1.12	1.45	C-18	436	0.41	0.11	0.77	1.29	1.67
2" x 6" - 6' Avg.	Demo	LF	L-2	1013	---	---	0.26	0.26	0.38	L-2	891	---	---	0.29	0.29	0.43
2" x 6" - 6' Avg.	Inst	LF	C-18	560	0.34	0.08	0.60	1.03	1.32	C-18	493	0.41	0.09	0.68	1.19	1.53
2" x 8" - 3' Avg.	Demo	LF	L-2	610	---	---	0.43	0.43	0.63	L-2	537	---	---	0.48	0.48	0.72
2" x 8" - 3' Avg.	Inst	LF	C-18	330	0.48	0.14	1.02	1.64	2.14	C-18	290	0.58	0.16	1.16	1.90	2.47
2" x 8" - 4' Avg.	Demo	LF	L-2	742	---	---	0.35	0.35	0.52	L-2	653	---	---	0.40	0.40	0.59
2" x 8" - 4' Avg.	Inst	LF	C-18	409	0.48	0.11	0.82	1.42	1.82	C-18	360	0.58	0.13	0.94	1.64	2.10
2" x 8" - 5' Avg.	Demo	LF	L-2	857	---	---	0.30	0.30	0.45	L-2	754	---	---	0.34	0.34	0.51
2" x 8" - 5' Avg.	Inst	LF	C-18	480	0.48	0.10	0.70	1.28	1.62	C-18	422	0.58	0.11	0.80	1.49	1.88
2" x 8" - 6' Avg.	Demo	LF	L-2	953	---	---	0.27	0.27	0.40	L-2	839	---	---	0.31	0.31	0.46
2" x 8" - 6' Avg.	Inst	LF	C-18	541	0.48	0.09	0.62	1.19	1.50	C-18	476	0.58	0.10	0.71	1.38	1.73

Studs/Plates, per SF of area
Walls or Partitions:

Description	Oper	Unit	Crew Size	Avg Day Prod	Avg Mat'l Unit Cost	Avg Equip Unit Cost	Avg Labor Unit Cost	Avg Total Unit Cost	Avg Price Incl O&P	Crew Size	Avg Day Prod	Avg Mat'l Unit Cost	Avg Equip Unit Cost	Avg Labor Unit Cost	Avg Total Unit Cost	Avg Price Incl O&P
2" x 4" - 8', 12" o.c.	Demo	SF	L-2	601	---	---	0.43	0.43	0.64	L-2	529	---	---	0.49	0.49	0.73
2" x 4" - 8', 12" o.c.	Inst	SF	C-18	495	0.37	0.09	0.68	1.14	1.48	C-18	436	0.44	0.11	0.77	1.32	1.70
2" x 4" - 8', 16" o.c.	Demo	SF	L-2	674	---	---	0.39	0.39	0.57	L-2	593	---	---	0.44	0.44	0.65
2" x 4" - 8', 16" o.c.	Inst	SF	C-18	555	0.30	0.08	0.61	0.99	1.29	C-18	488	0.36	0.10	0.69	1.15	1.49
2" x 4" - 8', 24" o.c.	Demo	SF	L-2	793	---	---	0.33	0.33	0.49	L-2	698	---	---	0.37	0.37	0.55
2" x 4" - 8', 24" o.c.	Inst	SF	C-18	652	0.24	0.07	0.52	0.82	1.08	C-18	574	0.28	0.08	0.59	0.95	1.24

			Costs Based On Large Volume							Costs Based On Small Volume						
Description	Oper	Unit	Crew Size	Avg Day Prod	Avg Mat'l Unit Cost	Avg Equip Unit Cost	Avg Labor Unit Cost	Avg Total Unit Cost	Avg Price Incl O&P	Crew Size	Avg Day Prod	Avg Mat'l Unit Cost	Avg Equip Unit Cost	Avg Labor Unit Cost	Avg Total Unit Cost	Avg Price Incl O&P

Framing (continued)

Description	Oper	Unit	Crew Size	Avg Day Prod	Mat'l	Equip	Labor	Total	Price O&P	Crew Size	Avg Day Prod	Mat'l	Equip	Labor	Total	Price O&P
2" x 4" - 10', 12" o.c.	Demo	SF	L-2	723	---	---	0.36	0.36	0.53	L-2	636	---	---	0.41	0.41	0.60
2" x 4" - 10', 12" o.c.	Inst	SF	C-18	597	0.35	0.08	0.56	1.00	1.27	C-18	525	0.42	0.09	0.64	1.15	1.47
2" x 4" - 10', 16" o.c.	Demo	SF	L-2	816	---	---	0.32	0.32	0.47	L-2	718	---	---	0.36	0.36	0.54
2" x 4" - 10', 16" o.c.	Inst	SF	C-18	674	0.29	0.07	0.50	0.86	1.10	C-18	593	0.35	0.08	0.57	0.99	1.27
2" x 4" - 10', 24" o.c.	Demo	SF	L-2	973	---	---	0.27	0.27	0.40	L-2	856	---	---	0.30	0.30	0.45
2" x 4" - 10', 24" o.c.	Inst	SF	C-18	803	0.22	0.06	0.42	0.70	0.91	C-18	707	0.27	0.07	0.48	0.81	1.05
2" x 4" - 12', 12" o.c.	Demo	SF	L-2	826	---	---	0.31	0.31	0.47	L-2	727	---	---	0.36	0.36	0.53
2" x 4" - 12', 12" o.c.	Inst	SF	C-18	683	0.34	0.07	0.49	0.91	1.15	C-18	601	0.41	0.08	0.56	1.05	1.33
2" x 4" - 12', 16" o.c.	Demo	SF	L-2	941	---	---	0.28	0.28	0.41	L-2	828	---	---	0.31	0.31	0.46
2" x 4" - 12', 16" o.c.	Inst	SF	C-18	778	0.28	0.06	0.43	0.77	0.98	C-18	685	0.33	0.07	0.49	0.89	1.13
2" x 4" - 12', 24" o.c.	Demo	SF	L-2	1142	---	---	0.23	0.23	0.34	L-2	1005	---	---	0.26	0.26	0.38
2" x 4" - 12', 24" o.c.	Inst	SF	C-18	944	0.21	0.05	0.36	0.62	0.79	C-18	831	0.25	0.06	0.41	0.72	0.92
2" x 6" - 8', 12" o.c.	Demo	SF	L-2	528	---	---	0.49	0.49	0.73	L-2	465	---	---	0.56	0.56	0.83
2" x 6" - 8', 12" o.c.	Inst	SF	C-18	439	0.57	0.11	0.77	1.45	1.82	C-18	386	0.69	0.12	0.87	1.68	2.11
2" x 6" - 8', 16" o.c.	Demo	SF	L-2	591	---	---	0.44	0.44	0.65	L-2	520	---	---	0.50	0.50	0.74
2" x 6" - 8', 16" o.c.	Inst	SF	C-18	491	0.47	0.10	0.69	1.25	1.59	C-18	432	0.56	0.11	0.78	1.45	1.83
2" x 6" - 8', 24" o.c.	Demo	SF	L-2	696	---	---	0.37	0.37	0.55	L-2	612	---	---	0.42	0.42	0.63
2" x 6" - 8', 24" o.c.	Inst	SF	C-18	579	0.36	0.08	0.58	1.02	1.31	C-18	510	0.43	0.09	0.66	1.19	1.51
2" x 6" - 10', 12" o.c.	Demo	SF	L-2	632	---	---	0.41	0.41	0.61	L-2	556	---	---	0.47	0.47	0.69
2" x 6" - 10', 12" o.c.	Inst	SF	C-11	528	0.55	0.09	0.70	1.33	1.68	C-11	465	0.66	0.10	0.80	1.55	1.94
2" x 6" - 10', 16" o.c.	Demo	SF	L-2	714	---	---	0.36	0.36	0.54	L-2	628	---	---	0.41	0.41	0.61
2" x 6" - 10', 16" o.c.	Inst	SF	C-11	596	0.44	0.08	0.62	1.14	1.44	C-11	524	0.53	0.09	0.70	1.32	1.67
2" x 6" - 10', 24" o.c.	Demo	SF	L-2	852	---	---	0.31	0.31	0.45	L-2	750	---	---	0.35	0.35	0.51
2" x 6" - 10', 24" o.c.	Inst	SF	C-18	711	0.34	0.07	0.47	0.88	1.11	C-18	626	0.41	0.07	0.54	1.02	1.28
2" x 6" - 12', 12" o.c.	Demo	SF	L-2	721	---	---	0.36	0.36	0.53	L-2	634	---	---	0.41	0.41	0.61
2" x 6" - 12', 12" o.c.	Inst	SF	C-18	603	0.53	0.08	0.56	1.17	1.44	C-18	531	0.64	0.09	0.64	1.36	1.67
2" x 6" - 12', 16" o.c.	Demo	SF	L-2	822	---	---	0.32	0.32	0.47	L-2	723	---	---	0.36	0.36	0.53
2" x 6" - 12', 16" o.c.	Inst	SF	C-18	688	0.43	0.07	0.49	0.98	1.22	C-18	605	0.51	0.08	0.56	1.15	1.42

			Costs Based On Large Volume							Costs Based On Small Volume						
Description	Oper	Unit	Crew Size	Avg Day Prod	Avg Mat'l Unit Cost	Avg Equip Unit Cost	Avg Labor Unit Cost	Avg Total Unit Cost	Avg Price Incl O&P	Crew Size	Avg Day Prod	Avg Mat'l Unit Cost	Avg Equip Unit Cost	Avg Labor Unit Cost	Avg Total Unit Cost	Avg Price Incl O&P

Framing (continued)

2" x 6" - 12', 24" o.c.	Demo	SF	L-2	996	---	---	0.26	0.26	0.39	L-2	876	---	---	0.30	0.30	0.44
2" x 6" - 12', 24" o.c.	Inst	SF	C-18	834	0.32	0.06	0.40	0.78	0.98	C-18	734	0.39	0.06	0.46	0.91	1.14
2" x 8" - 8', 12" o.c.	Demo	SF	L-2	468	---	---	0.56	0.56	0.82	L-2	412	---	---	0.63	0.63	0.93
2" x 8" - 8', 12" o.c.	Inst	SF	C-18	392	0.80	0.12	0.86	1.78	2.20	C-18	345	0.96	0.14	0.98	2.07	2.55
2" x 8" - 8', 16" o.c.	Demo	SF	L-2	525	---	---	0.50	0.50	0.73	L-2	462	---	---	0.56	0.56	0.83
2" x 8" - 8', 16" o.c.	Inst	SF	C-18	441	0.65	0.11	0.76	1.52	1.89	C-18	388	0.78	0.12	0.87	1.77	2.19
2" x 8" - 8', 24" o.c.	Demo	SF	L-2	618	---	---	0.42	0.42	0.62	L-2	544	---	---	0.48	0.48	0.71
2" x 8" - 8', 24" o.c.	Inst	SF	C-18	518	0.50	0.09	0.65	1.24	1.56	C-18	456	0.60	0.10	0.74	1.44	1.80
2" x 8" - 10', 12" o.c.	Demo	SF	L-2	560	---	---	0.46	0.46	0.69	L-2	493	---	---	0.53	0.53	0.78
2" x 8" - 10', 12" o.c.	Inst	SF	C-18	472	0.76	0.10	0.71	1.58	1.93	C-18	415	0.91	0.11	0.81	1.84	2.24
2" x 8" - 10', 16" o.c.	Demo	SF	L-2	633	---	---	0.41	0.41	0.61	L-2	557	---	---	0.47	0.47	0.69
2" x 8" - 10', 16" o.c.	Inst	SF	C-18	533	0.61	0.09	0.63	1.33	1.64	C-18	469	0.74	0.10	0.72	1.56	1.91
2" x 8" - 10', 24" o.c.	Demo	SF	L-2	754	---	---	0.34	0.34	0.51	L-2	664	---	---	0.39	0.39	0.58
2" x 8" - 10', 24" o.c.	Inst	SF	C-18	636	0.47	0.07	0.53	1.07	1.33	C-18	560	0.56	0.08	0.60	1.25	1.54
2" x 8" - 12', 12" o.c.	Demo	SF	L-2	638	---	---	0.41	0.41	0.60	L-2	561	---	---	0.46	0.46	0.69
2" x 8" - 12', 12" o.c.	Inst	SF	C-18	540	0.74	0.09	0.62	1.45	1.76	C-18	475	0.89	0.10	0.71	1.70	2.05
2" x 8" - 12', 16" o.c.	Demo	SF	L-2	728	---	---	0.36	0.36	0.53	L-2	641	---	---	0.41	0.41	0.60
2" x 8" - 12', 16" o.c.	Inst	SF	C-18	615	0.59	0.08	0.55	1.22	1.49	C-18	541	0.71	0.09	0.62	1.42	1.73
2" x 8" - 12', 24" o.c.	Demo	SF	L-2	881	---	---	0.30	0.30	0.44	L-2	775	---	---	0.34	0.34	0.50
2" x 8" - 12', 24" o.c.	Inst	SF	C-18	745	0.44	0.06	0.45	0.96	1.18	C-18	656	0.53	0.07	0.51	1.12	1.37
Gable Ends:																
2" x 4" - 3' Avg., 12" o.c.	Demo	SF	L-2	216	---	---	1.20	1.20	1.78	L-2	190	---	---	1.37	1.37	2.03
2" x 4" - 3' Avg., 12" o.c.	Inst	SF	C-18	113	0.66	0.41	2.98	4.06	5.52	C-18	99	0.79	0.47	3.39	4.65	6.31
2" x 4" - 3' Avg., 16" o.c.	Demo	SF	L-2	231	---	---	1.13	1.13	1.67	L-2	203	---	---	1.28	1.28	1.89
2" x 4" - 3' Avg., 16" o.c.	Inst	SF	C-18	121	0.62	0.39	2.79	3.79	5.16	C-18	106	0.75	0.44	3.17	4.35	5.90
2" x 4" - 3' Avg., 24" o.c.	Demo	SF	L-2	253	---	---	1.03	1.03	1.52	L-2	223	---	---	1.17	1.17	1.73
2" x 4" - 3' Avg., 24" o.c.	Inst	SF	C-18	132	0.57	0.35	2.55	3.48	4.73	C-18	116	0.69	0.40	2.90	3.99	5.41

			Costs Based On Large Volume							Costs Based On Small Volume						
Description	Oper	Unit	Crew Size	Avg Day Prod	Avg Mat'l Unit Cost	Avg Equip Unit Cost	Avg Labor Unit Cost	Avg Total Unit Cost	Avg Price Incl O&P	Crew Size	Avg Day Prod	Avg Mat'l Unit Cost	Avg Equip Unit Cost	Avg Labor Unit Cost	Avg Total Unit Cost	Avg Price Incl O&P

Framing (continued)

Description	Oper	Unit	Crew Size	Avg Day Prod	Avg Mat'l Unit Cost	Avg Equip Unit Cost	Avg Labor Unit Cost	Avg Total Unit Cost	Avg Price Incl O&P	Crew Size	Avg Day Prod	Avg Mat'l Unit Cost	Avg Equip Unit Cost	Avg Labor Unit Cost	Avg Total Unit Cost	Avg Price Incl O&P
2" x 4" - 4' Avg., 12" o.c.	Demo	SF	L-2	323	---	---	0.81	0.81	1.19	L-2	284	---	---	0.92	0.92	1.35
2" x 4" - 4' Avg., 12" o.c.	Inst	SF	C-18	170	0.55	0.27	1.98	2.81	3.78	C-18	150	0.66	0.31	2.25	3.23	4.33
2" x 4" - 4' Avg., 16" o.c.	Demo	SF	L-2	348	---	---	0.75	0.75	1.11	L-2	306	---	---	0.85	0.85	1.26
2" x 4" - 4' Avg., 16" o.c.	Inst	SF	C-18	183	0.52	0.26	1.84	2.61	3.52	C-18	161	0.62	0.29	2.09	3.00	4.03
2" x 4" - 4' Avg., 24" o.c.	Demo	SF	L-2	384	---	---	0.68	0.68	1.00	L-2	338	---	---	0.77	0.77	1.14
2" x 4" - 4' Avg., 24" o.c.	Inst	SF	C-18	202	0.47	0.23	1.67	2.37	3.18	C-18	178	0.56	0.26	1.90	2.72	3.65
2" x 4" - 5' Avg., 12" o.c.	Demo	SF	L-2	412	---	---	0.63	0.63	0.93	L-2	363	---	---	0.72	0.72	1.06
2" x 4" - 5' Avg., 12" o.c.	Inst	SF	C-18	219	0.51	0.21	1.54	2.26	3.02	C-18	193	0.61	0.24	1.75	2.60	3.46
2" x 4" - 5' Avg., 16" o.c.	Demo	SF	L-2	454	---	---	0.57	0.57	0.85	L-2	400	---	---	0.65	0.65	0.96
2" x 4" - 5' Avg., 16" o.c.	Inst	SF	C-18	241	0.47	0.19	1.40	2.06	2.75	C-18	212	0.56	0.22	1.59	2.37	3.15
2" x 4" - 5' Avg., 24" o.c.	Demo	SF	L-2	514	---	---	0.51	0.51	0.75	L-2	452	---	---	0.58	0.58	0.85
2" x 4" - 5' Avg., 24" o.c.	Inst	SF	C-18	273	0.42	0.17	1.23	1.82	2.43	C-18	240	0.50	0.19	1.40	2.10	2.79
2" x 4" - 6' Avg., 12" o.c.	Demo	SF	L-2	501	---	---	0.52	0.52	0.77	L-2	441	---	---	0.59	0.59	0.87
2" x 4" - 6' Avg., 12" o.c.	Inst	SF	C-18	269	0.48	0.17	1.25	1.90	2.52	C-18	237	0.57	0.20	1.42	2.19	2.89
2" x 4" - 6' Avg., 16" o.c.	Demo	SF	L-2	564	---	---	0.46	0.46	0.68	L-2	496	---	---	0.52	0.52	0.78
2" x 4" - 6' Avg., 16" o.c.	Inst	SF	C-18	302	0.42	0.15	1.12	1.69	2.24	C-18	266	0.51	0.18	1.27	1.95	2.57
2" x 4" - 6' Avg., 24" o.c.	Demo	SF	L-2	650	---	---	0.40	0.40	0.59	L-2	572	---	---	0.45	0.45	0.67
2" x 4" - 6' Avg., 24" o.c.	Inst	SF	C-18	349	0.37	0.13	0.97	1.47	1.95	C-18	307	0.45	0.15	1.10	1.70	2.24
2" x 6" - 3' Avg., 12" o.c.	Demo	SF	L-2	208	---	---	1.25	1.25	1.85	L-2	183	---	---	1.42	1.42	2.10
2" x 6" - 3' Avg., 12" o.c.	Inst	SF	C-18	110	1.04	0.42	3.06	4.53	6.03	C-18	97	1.25	0.48	3.48	5.22	6.92
2" x 6" - 3' Avg., 16" o.c.	Demo	SF	L-2	222	---	---	1.17	1.17	1.73	L-2	195	---	---	1.33	1.33	1.97
2" x 6" - 3' Avg., 16" o.c.	Inst	SF	C-18	118	0.98	0.40	2.86	4.23	5.63	C-18	104	1.18	0.45	3.25	4.87	6.46
2" x 6" - 3' Avg., 24" o.c.	Demo	SF	L-2	243	---	---	1.07	1.07	1.58	L-2	214	---	---	1.22	1.22	1.80
2" x 6" - 3' Avg., 24" o.c.	Inst	SF	C-18	129	0.90	0.36	2.61	3.88	5.16	C-18	114	1.08	0.41	2.97	4.46	5.92
2" x 6" - 4' Avg., 12" o.c.	Demo	SF	L-2	307	---	---	0.85	0.85	1.25	L-2	270	---	---	0.96	0.96	1.42
2" x 6" - 4' Avg., 12" o.c.	Inst	SF	C-18	166	0.87	0.28	2.03	3.18	4.18	C-18	146	1.05	0.32	2.31	3.67	4.80
2" x 6" - 4' Avg., 16" o.c.	Demo	SF	L-2	331	---	---	0.79	0.79	1.16	L-2	291	---	---	0.89	0.89	1.32
2" x 6" - 4' Avg., 16" o.c.	Inst	SF	C-18	179	0.81	0.26	1.88	2.95	3.88	C-18	158	0.97	0.30	2.14	3.41	4.46

Framing (continued)

			Costs Based On Large Volume							Costs Based On Small Volume						
Description	Oper	Unit	Crew Size	Avg Day Prod	Avg Mat'l Unit Cost	Avg Equip Unit Cost	Avg Labor Unit Cost	Avg Total Unit Cost	Avg Price Incl O&P	Crew Size	Avg Day Prod	Avg Mat'l Unit Cost	Avg Equip Unit Cost	Avg Labor Unit Cost	Avg Total Unit Cost	Avg Price Incl O&P
2" x 6" - 4' Avg., 24" o.c.	Demo	SF	L-2	366	---	---	0.71	0.71	1.05	L-2	322	---	---	0.81	0.81	1.20
2" x 6" - 4' Avg., 24" o.c.	Inst	SF	C-18	197	0.73	0.24	1.71	2.68	3.52	C-18	173	0.88	0.27	1.94	3.09	4.05
2" x 6" - 5' Avg., 12" o.c.	Demo	SF	L-2	389	---	---	0.67	0.67	0.99	L-2	342	---	---	0.76	0.76	1.12
2" x 6" - 5' Avg., 12" o.c.	Inst	SF	C-18	212	0.80	0.22	1.59	2.61	3.39	C-18	187	0.96	0.25	1.81	3.02	3.90
2" x 6" - 5' Avg., 16" o.c.	Demo	SF	L-2	428	---	---	0.61	0.61	0.90	L-2	377	---	---	0.69	0.69	1.02
2" x 6" - 5' Avg., 16" o.c.	Inst	SF	C-18	234	0.73	0.20	1.44	2.37	3.08	C-18	206	0.88	0.23	1.64	2.74	3.54
2" x 6" - 5' Avg., 24" o.c.	Demo	SF	L-2	484	---	---	0.54	0.54	0.80	L-2	426	---	---	0.61	0.61	0.90
2" x 6" - 5' Avg., 24" o.c.	Inst	SF	C-18	265	0.65	0.18	1.27	2.10	2.72	C-18	233	0.78	0.20	1.45	2.43	3.13
2" x 6" - 6' Avg., 12" o.c.	Demo	SF	L-2	469	---	---	0.55	0.55	0.82	L-2	413	---	---	0.63	0.63	0.93
2" x 6" - 6' Avg., 12" o.c.	Inst	SF	C-18	259	0.75	0.18	1.30	2.23	2.87	C-18	228	0.89	0.20	1.48	2.58	3.30
2" x 6" - 6' Avg., 16" o.c.	Demo	SF	L-2	528	---	---	0.49	0.49	0.73	L-2	465	---	---	0.56	0.56	0.83
2" x 6" - 6' Avg., 16" o.c.	Inst	SF	C-18	292	0.66	0.16	1.15	1.98	2.54	C-18	257	0.80	0.18	1.31	2.29	2.93
2" x 6" - 6' Avg., 24" o.c.	Demo	SF	L-2	609	---	---	0.43	0.43	0.63	L-2	536	---	---	0.49	0.49	0.72
2" x 6" - 6' Avg., 24" o.c.	Inst	SF	C-18	336	0.58	0.14	1.00	1.72	2.21	C-18	296	0.70	0.16	1.14	1.99	2.55
2" x 8" - 3' Avg., 12" o.c.	Demo	SF	L-2	200	---	---	1.30	1.30	1.92	L-2	176	---	---	1.48	1.48	2.19
2" x 8" - 3' Avg., 12" o.c.	Inst	SF	C-18	108	1.47	0.43	3.12	5.02	6.55	C-18	95	1.76	0.49	3.55	5.80	7.53
2" x 8" - 3' Avg., 16" o.c.	Demo	SF	L-2	213	---	---	1.22	1.22	1.81	L-2	187	---	---	1.39	1.39	2.05
2" x 8" - 3' Avg., 16" o.c.	Inst	SF	C-18	116	1.38	0.40	2.91	4.69	6.11	C-18	102	1.65	0.46	3.30	5.41	7.03
2" x 8" - 3' Avg., 24" o.c.	Demo	SF	L-2	234	---	---	1.11	1.11	1.64	L-2	206	---	---	1.26	1.26	1.87
2" x 8" - 3' Avg., 24" o.c.	Inst	SF	C-18	127	1.26	0.37	2.65	4.28	5.58	C-18	112	1.51	0.42	3.02	4.95	6.43
2" x 8" - 4' Avg., 12" o.c.	Demo	SF	L-2	293	---	---	0.89	0.89	1.31	L-2	258	---	---	1.01	1.01	1.49
2" x 8" - 4' Avg., 12" o.c.	Inst	SF	C-18	161	1.22	0.29	2.09	3.61	4.63	C-18	142	1.47	0.33	2.38	4.18	5.34
2" x 8" - 4' Avg., 16" o.c.	Demo	SF	L-2	316	---	---	0.82	0.82	1.22	L-2	278	---	---	0.94	0.94	1.38
2" x 8" - 4' Avg., 16" o.c.	Inst	SF	C-18	174	1.13	0.27	1.94	3.34	4.29	C-18	153	1.36	0.30	2.20	3.87	4.95
2" x 8" - 4' Avg., 24" o.c.	Demo	SF	L-2	349	---	---	0.75	0.75	1.10	L-2	307	---	---	0.85	0.85	1.25
2" x 8" - 4' Avg., 24" o.c.	Inst	SF	C-18	192	1.03	0.24	1.76	3.02	3.88	C-18	169	1.23	0.28	2.00	3.50	4.48
2" x 8" - 5' Avg., 12" o.c.	Demo	SF	L-2	368	---	---	0.71	0.71	1.05	L-2	324	---	---	0.80	0.80	1.19
2" x 8" - 5' Avg., 12" o.c.	Inst	SF	C-18	206	1.12	0.23	1.64	2.98	3.79	C-18	181	1.34	0.26	1.86	3.46	4.37

Description	Oper	Unit	Costs Based On Large Volume							Costs Based On Small Volume						
			Crew Size	Avg Day Prod	Avg Mat'l Unit Cost	Avg Equip Unit Cost	Avg Labor Unit Cost	Avg Total Unit Cost	Avg Price Incl O&P	Crew Size	Avg Day Prod	Avg Mat'l Unit Cost	Avg Equip Unit Cost	Avg Labor Unit Cost	Avg Total Unit Cost	Avg Price Incl O&P

Framing (continued)

Description	Oper	Unit	Crew Size	Avg Day Prod	Avg Mat'l Unit Cost	Avg Equip Unit Cost	Avg Labor Unit Cost	Avg Total Unit Cost	Avg Price Incl O&P	Crew Size	Avg Day Prod	Avg Mat'l Unit Cost	Avg Equip Unit Cost	Avg Labor Unit Cost	Avg Total Unit Cost	Avg Price Incl O&P
2" x 8" - 5' Avg., 16" o.c.	Demo	SF	L-2	405	---	---	0.64	0.64	0.95	L-2	356	---	---	0.73	0.73	1.08
2" x 8" - 5' Avg., 16" o.c.	Inst	SF	C-18	227	1.00	0.21	1.49	2.70	3.42	C-18	200	1.21	0.23	1.69	3.13	3.95
2" x 8" - 5' Avg., 24" o.c.	Demo	SF	L-2	458	---	---	0.57	0.57	0.84	L-2	403	---	---	0.65	0.65	0.96
2" x 8" - 5' Avg., 24" o.c.	Inst	SF	C-18	257	0.89	0.18	1.31	2.38	3.02	C-18	226	1.07	0.21	1.49	2.76	3.49
2" x 8" - 6' Avg., 12" o.c.	Demo	SF	L-2	441	---	---	0.59	0.59	0.87	L-2	388	---	---	0.67	0.67	0.99
2" x 8" - 6' Avg., 12" o.c.	Inst	SF	C-18	251	1.04	0.19	1.34	2.57	3.23	C-18	221	1.25	0.21	1.53	2.99	3.74
2" x 8" - 6' Avg., 16" o.c.	Demo	SF	L-2	497	---	---	0.52	0.52	0.77	L-2	437	---	---	0.59	0.59	0.88
2" x 8" - 6' Avg., 16" o.c.	Inst	SF	C-18	282	0.93	0.17	1.20	2.29	2.87	C-18	248	1.11	0.19	1.36	2.66	3.32
2" x 8" - 6' Avg., 24" o.c.	Demo	SF	L-2	573	---	---	0.45	0.45	0.67	L-2	504	---	---	0.52	0.52	0.76
2" x 8" - 6' Avg., 24" o.c.	Inst	SF	C-18	325	0.81	0.14	1.04	1.99	2.50	C-18	286	0.97	0.16	1.18	2.31	2.89

Studs/Plates, per LF of wall or partition
Walls or Partitions

Description	Oper	Unit	Crew Size	Avg Day Prod	Avg Mat'l Unit Cost	Avg Equip Unit Cost	Avg Labor Unit Cost	Avg Total Unit Cost	Avg Price Incl O&P	Crew Size	Avg Day Prod	Avg Mat'l Unit Cost	Avg Equip Unit Cost	Avg Labor Unit Cost	Avg Total Unit Cost	Avg Price Incl O&P
2" x 4" - 8', 12" o.c.	Demo	LF	L-2	75	---	---	3.47	3.47	5.13	L-2	66	---	---	3.94	3.94	5.83
2" x 4" - 8', 12" o.c.	Inst	LF	C-18	62	2.71	0.75	5.44	8.90	11.57	C-18	55	3.26	0.86	6.18	10.29	13.32
2" x 4" - 10', 12" o.c.	Demo	LF	L-2	72	---	---	3.61	3.61	5.35	L-2	63	---	---	4.10	4.10	6.08
2" x 4" - 10', 12" o.c.	Inst	LF	C-18	59	3.24	0.79	5.71	9.74	12.54	C-18	52	3.89	0.90	6.49	11.28	14.46
2" x 4" - 12', 12" o.c.	Demo	LF	L-2	69	---	---	3.77	3.77	5.58	L-2	61	---	---	4.28	4.28	6.34
2" x 4" - 12', 12" o.c.	Inst	LF	C-18	57	3.77	0.82	5.91	10.51	13.40	C-18	50	4.53	0.93	6.72	12.18	15.47
2" x 4" - 8', 16" o.c.	Demo	LF	L-2	85	---	---	3.06	3.06	4.53	L-2	75	---	---	3.48	3.48	5.15
2" x 4" - 8', 16" o.c.	Inst	LF	C-18	70	2.20	0.67	4.82	7.68	10.04	C-18	62	2.64	0.76	5.47	8.87	11.55
2" x 4" - 10', 16" o.c.	Demo	LF	L-2	81	---	---	3.21	3.21	4.75	L-2	71	---	---	3.65	3.65	5.40
2" x 4" - 10', 16" o.c.	Inst	LF	C-18	67	2.59	0.70	5.03	8.32	10.78	C-18	59	3.11	0.79	5.72	9.62	12.42
2" x 4" - 12', 16" o.c.	Demo	LF	L-2	78	---	---	3.33	3.33	4.93	L-2	69	---	---	3.79	3.79	5.61
2" x 4" - 12', 16" o.c.	Inst	LF	C-18	65	2.96	0.72	5.19	8.87	11.41	C-18	57	3.55	0.82	5.89	10.26	13.15
2" x 4" - 8', 24" o.c.	Demo	LF	L-2	99	---	---	2.63	2.63	3.89	L-2	87	---	---	2.99	2.99	4.42
2" x 4" - 8', 24" o.c.	Inst	LF	C-18	81	1.66	0.58	4.16	6.39	8.43	C-18	71	1.99	0.65	4.73	7.37	9.69
2" x 4" - 10', 24" o.c.	Demo	LF	L-2	97	---	---	2.68	2.68	3.97	L-2	85	---	---	3.05	3.05	4.51
2" x 4" - 10', 24" o.c.	Inst	LF	C-18	80	1.94	0.58	4.21	6.74	8.81	C-18	70	2.33	0.66	4.79	7.78	10.13
2" x 4" - 12', 24" o.c.	Demo	LF	L-2	95	---	---	2.74	2.74	4.05	L-2	84	---	---	3.11	3.11	4.60
2" x 4" - 12', 24" o.c.	Inst	LF	C-18	79	2.19	0.59	4.27	7.04	9.13	C-18	70	2.62	0.67	4.85	8.14	10.52

			Costs Based On Large Volume							Costs Based On Small Volume					
Description	Oper	Unit	Crew Size	Avg Day Prod	Avg Mat'l Unit Cost	Avg Equip Unit Cost	Avg Labor Unit Cost	Avg Total Unit Cost	Avg Price Incl O&P	Crew Size	Avg Day Prod	Avg Mat'l Unit Cost	Avg Equip Unit Cost	Avg Labor Unit Cost	Avg Total Unit Cost	Avg Price Incl O&P

Framing (continued)

Description	Oper	Unit	Crew Size	Avg Day Prod	Avg Mat'l Unit Cost	Avg Equip Unit Cost	Avg Labor Unit Cost	Avg Total Unit Cost	Avg Price Incl O&P	Crew Size	Avg Day Prod	Avg Mat'l Unit Cost	Avg Equip Unit Cost	Avg Labor Unit Cost	Avg Total Unit Cost	Avg Price Incl O&P
2" x 6" - 8', 12" o.c.	Demo	LF	L-2	66	---	---	3.94	3.94	5.83	L-2	58	---	---	4.48	4.48	6.63
2" x 6" - 8', 12" o.c.	Inst	LF	C-18	55	4.34	0.85	6.13	11.32	14.32	C-18	48	5.21	0.96	6.96	13.14	16.55
2" x 6" - 10', 12" o.c.	Demo	LF	L-2	63	---	---	4.13	4.13	6.11	L-2	55	---	---	4.69	4.69	6.94
2" x 6" - 10', 12" o.c.	Inst	LF	C-18	53	5.17	0.88	6.36	12.41	15.52	C-18	47	6.20	1.00	7.23	14.43	17.97
2" x 6" - 12', 12" o.c.	Demo	LF	L-2	60	---	---	4.33	4.33	6.42	L-2	53	---	---	4.93	4.93	7.29
2" x 6" - 12', 12" o.c.	Inst	LF	C-18	50	6.01	0.93	6.74	13.68	16.98	C-18	44	7.21	1.06	7.66	15.93	19.68
2" x 6" - 8', 16" o.c.	Demo	LF	L-2	74	---	---	3.51	3.51	5.20	L-2	65	---	---	3.99	3.99	5.91
2" x 6" - 8', 16" o.c.	Inst	LF	C-11	61	3.51	0.77	5.53	9.80	12.51	C-11	54	4.21	0.87	6.28	11.36	14.44
2" x 6" - 10', 16" o.c.	Demo	LF	L-2	71	---	---	3.66	3.66	5.42	L-2	62	---	---	4.16	4.16	6.16
2" x 6" - 10', 16" o.c.	Inst	LF	C-11	59	4.13	0.79	5.71	10.63	13.43	C-11	52	4.95	0.90	6.49	12.35	15.53
2" x 6" - 12', 16" o.c.	Demo	LF	L-2	69	---	---	3.77	3.77	5.58	L-2	61	---	---	4.28	4.28	6.34
2" x 6" - 12', 16" o.c.	Inst	LF	C-18	58	4.76	0.80	5.81	11.37	14.22	C-18	51	5.71	0.91	6.60	13.23	16.46
2" x 6" - 8', 24" o.c.	Demo	LF	L-2	87	---	---	2.99	2.99	4.42	L-2	77	---	---	3.40	3.40	5.03
2" x 6" - 8', 24" o.c.	Inst	LF	C-18	72	2.65	0.65	4.68	7.98	10.28	C-18	63	3.18	0.74	5.32	9.24	11.85
2" x 6" - 10', 24" o.c.	Demo	LF	L-2	85	---	---	3.06	3.06	4.53	L-2	75	---	---	3.48	3.48	5.15
2" x 6" - 10', 24" o.c.	Inst	LF	C-18	71	3.09	0.66	4.75	8.49	10.82	C-18	62	3.71	0.75	5.40	9.85	12.49
2" x 6" - 12', 24" o.c.	Demo	LF	L-2	84	---	---	3.10	3.10	4.58	L-2	74	---	---	3.52	3.52	5.21
2" x 6" - 12', 24" o.c.	Inst	LF	C-18	70	3.51	0.67	4.82	8.99	11.35	C-18	62	4.21	0.76	5.47	10.44	13.12
2" x 8" - 8', 12" o.c.	Demo	LF	L-2	59	---	---	4.41	4.41	6.52	L-2	52	---	---	5.01	5.01	7.41
2" x 8" - 8', 12" o.c.	Inst	LF	C-18	49	6.16	0.95	6.88	13.99	17.36	C-18	43	7.39	1.08	7.82	16.29	20.12
2" x 8" - 10', 12" o.c.	Demo	LF	L-2	56	---	---	4.64	4.64	6.87	L-2	49	---	---	5.28	5.28	7.81
2" x 8" - 10', 12" o.c.	Inst	LF	C-18	47	7.32	0.99	7.17	15.49	19.00	C-18	41	8.79	1.13	8.15	18.07	22.06
2" x 8" - 12', 12" o.c.	Demo	LF	L-2	53	---	---	4.91	4.91	7.26	L-2	47	---	---	5.58	5.58	8.25
2" x 8" - 12', 12" o.c.	Inst	LF	C-18	45	8.54	1.04	7.49	17.07	20.74	C-18	40	10.25	1.18	8.51	19.94	24.11
2" x 8" - 8', 16" o.c.	Demo	LF	L-2	66	---	---	3.94	3.94	5.83	L-2	58	---	---	4.48	4.48	6.63
2" x 8" - 8', 16" o.c.	Inst	LF	C-18	55	4.95	0.85	6.13	11.93	14.93	C-18	48	5.94	0.96	6.96	13.87	17.28
2" x 8" - 10', 16" o.c.	Demo	LF	L-2	63	---	---	4.13	4.13	6.11	L-2	55	---	---	4.69	4.69	6.94
2" x 8" - 10', 16" o.c.	Inst	LF	C-18	53	5.85	0.88	6.36	13.09	16.20	C-18	47	7.01	1.00	7.23	15.24	18.78
2" x 8" - 12', 16" o.c.	Demo	LF	L-2	61	---	---	4.26	4.26	6.31	L-2	54	---	---	4.85	4.85	7.17
2" x 8" - 12', 16" o.c.	Inst	LF	C-18	52	6.77	0.90	6.48	14.15	17.32	C-18	46	8.12	1.02	7.37	16.51	20.12

Description	Oper	Unit	Costs Based On Large Volume							Costs Based On Small Volume						
			Crew Size	Avg Day Prod	Avg Mat'l Unit Cost	Avg Equip Unit Cost	Avg Labor Unit Cost	Avg Total Unit Cost	Avg Price Incl O&P	Crew Size	Avg Day Prod	Avg Mat'l Unit Cost	Avg Equip Unit Cost	Avg Labor Unit Cost	Avg Total Unit Cost	Avg Price Incl O&P

Framing (continued)

Description	Oper	Unit	Crew Size	Avg Day Prod	Avg Mat'l Unit Cost	Avg Equip Unit Cost	Avg Labor Unit Cost	Avg Total Unit Cost	Avg Price Incl O&P	Crew Size	Avg Day Prod	Avg Mat'l Unit Cost	Avg Equip Unit Cost	Avg Labor Unit Cost	Avg Total Unit Cost	Avg Price Incl O&P
2" x 8" - 8', 24" o.c.	Demo	LF	L-2	78	---	---	3.33	3.33	4.93	L-2	69	---	---	3.79	3.79	5.61
2" x 8" - 8', 24" o.c.	Inst	LF	C-18	65	3.77	0.72	5.19	9.67	12.21	C-18	57	4.52	0.82	5.89	11.23	14.12
2" x 8" - 10', 24" o.c.	Demo	LF	L-2	76	---	---	3.42	3.42	5.06	L-2	67	---	---	3.89	3.89	5.76
2" x 8" - 10', 24" o.c.	Inst	LF	C-18	64	4.37	0.73	5.27	10.36	12.94	C-18	56	5.24	0.83	5.99	12.05	14.99
2" x 8" - 12', 24" o.c.	Demo	LF	L-2	74	---	---	3.51	3.51	5.20	L-2	65	---	---	3.99	3.99	5.91
2" x 8" - 12', 24" o.c.	Inst	LF	C-18	62	4.95	0.75	5.44	11.14	13.81	C-18	55	5.94	0.86	6.18	12.98	16.00

Sheathing
Walls
Boards, 1" x 8"

Description	Oper	Unit	Crew Size	Avg Day Prod	Avg Mat'l Unit Cost	Avg Equip Unit Cost	Avg Labor Unit Cost	Avg Total Unit Cost	Avg Price Incl O&P	Crew Size	Avg Day Prod	Avg Mat'l Unit Cost	Avg Equip Unit Cost	Avg Labor Unit Cost	Avg Total Unit Cost	Avg Price Incl O&P
Horizontal	Demo	SF	L-2	755	---	---	0.34	0.34	0.51	L-2	680	---	---	0.38	0.38	0.57
Horizontal	Inst	SF	C-18	695	0.43	0.07	0.49	0.98	1.22	C-18	626	0.52	0.07	0.54	1.13	1.40
Diagonal	Demo	SF	L-2	672	---	---	0.39	0.39	0.57	L-2	605	---	---	0.43	0.43	0.64
Diagonal	Inst	SF	C-18	618	0.47	0.08	0.55	1.09	1.36	C-18	556	0.56	0.08	0.61	1.25	1.55
Plywood:																
3/8"	Demo	SF	L-2	1145	---	---	0.23	0.23	0.34	L-2	1031	---	---	0.25	0.25	0.37
3/8"	Inst	SF	C-18	1059	0.36	0.04	0.32	0.72	0.88	C-18	953	0.43	0.05	0.35	0.83	1.01
1/2"	Demo	SF	L-2	1145	---	---	0.23	0.23	0.34	L-2	1031	---	---	0.25	0.25	0.37
1/2"	Inst	SF	C-18	1059	0.45	0.04	0.32	0.81	0.97	C-18	953	0.54	0.05	0.35	0.94	1.11
5/8"	Demo	SF	L-2	1145	---	---	0.23	0.23	0.34	L-2	1031	---	---	0.25	0.25	0.37
5/8"	Inst	SF	C-18	1059	0.49	0.04	0.32	0.85	1.01	C-18	953	0.59	0.05	0.35	0.99	1.17
Gypsum board, 1/2"	Demo	SF	L-2	1145	---	---	0.23	0.23	0.34	L-2	1031	---	---	0.25	0.25	0.37
Gypsum board, 1/2"	Inst	SF	C-18	1059	0.34	0.04	0.32	0.70	0.86	C-18	953	0.40	0.05	0.35	0.81	0.98
Wood fiber board, 1/2"																
With vapor barrier	Demo	SF	L-2	1145	---	---	0.23	0.23	0.34	L-2	1031	---	---	0.25	0.25	0.37
With vapor barrier	Inst	SF	C-18	1059	0.54	0.04	0.32	0.90	1.05	C-18	953	0.64	0.05	0.35	1.05	1.22
Without vapor barrier	Demo	SF	L-2	1145	---	---	0.23	0.23	0.34	L-2	1031	---	---	0.25	0.25	0.37
Without vapor barrier	Inst	SF	C-18	1059	0.43	0.04	0.32	0.79	0.94	C-18	953	0.51	0.05	0.35	0.91	1.09
Asphalt impregnated	Demo	SF	L-2	1145	---	---	0.23	0.23	0.34	L-2	1031	---	---	0.25	0.25	0.37
Asphalt impregnated	Inst	SF	C-18	1059	0.25	0.04	0.32	0.61	0.77	C-18	953	0.30	0.05	0.35	0.70	0.88

			Costs Based On Large Volume							Costs Based On Small Volume					
Description	Oper	Unit	Crew Size	Avg Day Prod	Avg Mat'l Unit Cost	Avg Equip Unit Cost	Avg Labor Unit Cost	Avg Total Unit Cost	Avg Price Incl O&P	Crew Size	Avg Day Prod	Avg Mat'l Unit Cost	Avg Equip Unit Cost	Avg Labor Unit Cost	Avg Total Unit Cost	Avg Price Incl O&P

Framing (continued)

Roof

Boards, 1" x 8"

Description	Oper	Unit	Crew Size	Avg Day Prod	Avg Mat'l	Avg Equip	Avg Labor	Avg Total	Avg Price O&P	Crew Size	Avg Day Prod	Avg Mat'l	Avg Equip	Avg Labor	Avg Total	Avg Price O&P
Horizontal	Demo	SF	L-2	884	---	---	0.29	0.29	0.44	L-2	796	---	---	0.33	0.33	0.48
Horizontal	Inst	SF	C-18	819	0.43	0.06	0.41	0.90	1.10	C-18	737	0.52	0.06	0.46	1.04	1.26
Diagonal	Demo	SF	L-2	775	---	---	0.34	0.34	0.50	L-2	698	---	---	0.37	0.37	0.55
Diagonal	Inst	SF	C-18	715	0.47	0.07	0.47	1.01	1.24	C-18	644	0.56	0.07	0.52	1.16	1.42

Plywood:

Description	Oper	Unit	Crew Size	Avg Day Prod	Avg Mat'l	Avg Equip	Avg Labor	Avg Total	Avg Price O&P	Crew Size	Avg Day Prod	Avg Mat'l	Avg Equip	Avg Labor	Avg Total	Avg Price O&P
1/2"	Demo	SF	L-2	1277	---	---	0.20	0.20	0.30	L-2	1149	---	---	0.23	0.23	0.33
1/2"	Inst	SF	C-18	1187	0.43	0.04	0.28	0.75	0.89	C-18	1068	0.51	0.04	0.32	0.87	1.03
5/8"	Demo	SF	L-2	1277	---	---	0.20	0.20	0.30	L-2	1149	---	---	0.23	0.23	0.33
5/8"	Inst	SF	C-18	1187	0.47	0.04	0.28	0.79	0.93	C-18	1068	0.56	0.04	0.32	0.92	1.08
3/4"	Demo	SF	L-2	1187	---	---	0.22	0.22	0.32	L-2	1068	---	---	0.24	0.24	0.36
3/4"	Inst	SF	C-18	1102	0.56	0.04	0.31	0.91	1.06	C-18	992	0.68	0.05	0.34	1.06	1.23

Sub-floor

Boards, 1" x 8"

Description	Oper	Unit	Crew Size	Avg Day Prod	Avg Mat'l	Avg Equip	Avg Labor	Avg Total	Avg Price O&P	Crew Size	Avg Day Prod	Avg Mat'l	Avg Equip	Avg Labor	Avg Total	Avg Price O&P
Horizontal	Demo	SF	L-2	828	---	---	0.31	0.31	0.46	L-2	745	---	---	0.35	0.35	0.52
Horizontal	Inst	SF	C-18	764	0.44	0.06	0.44	0.94	1.16	C-18	688	0.53	0.07	0.49	1.09	1.33
Diagonal	Demo	SF	L-2	729	---	---	0.36	0.36	0.53	L-2	656	---	---	0.40	0.40	0.59
Diagonal	Inst	SF	C-18	672	0.48	0.07	0.50	1.05	1.30	C-18	605	0.58	0.08	0.56	1.21	1.49

Plywood:

Description	Oper	Unit	Crew Size	Avg Day Prod	Avg Mat'l	Avg Equip	Avg Labor	Avg Total	Avg Price O&P	Crew Size	Avg Day Prod	Avg Mat'l	Avg Equip	Avg Labor	Avg Total	Avg Price O&P
5/8"	Demo	SF	L-2	1216	---	---	0.21	0.21	0.32	L-2	1094	---	---	0.24	0.24	0.35
5/8"	Inst	SF	C-18	1127	0.49	0.04	0.30	0.83	0.98	C-18	1014	0.49	0.05	0.33	0.87	1.03
3/4"	Demo	SF	L-2	1134	---	---	0.23	0.23	0.34	L-2	1021	---	---	0.25	0.25	0.38
3/4"	Inst	SF	C-18	1050	0.59	0.04	0.32	0.96	1.11	C-18	945	0.59	0.05	0.36	1.00	1.17
1-1/8"	Demo	SF	L-2	854	---	---	0.30	0.30	0.45	L-2	769	---	---	0.34	0.34	0.50
1-1/8"	Inst	SF	C-18	798	1.05	0.06	0.42	1.53	1.74	C-18	718	1.05	0.06	0.47	1.59	1.82

Particleboard

Description	Oper	Unit	Crew Size	Avg Day Prod	Avg Mat'l	Avg Equip	Avg Labor	Avg Total	Avg Price O&P	Crew Size	Avg Day Prod	Avg Mat'l	Avg Equip	Avg Labor	Avg Total	Avg Price O&P
5/8"	Demo	SF	L-2	1216	---	---	0.21	0.21	0.32	L-2	1094	---	---	0.24	0.24	0.35
5/8"	Inst	SF	C-18	1127	0.34	0.04	0.30	0.68	0.82	C-18	1014	0.34	0.05	0.33	0.72	0.88

			Costs Based On Large Volume							Costs Based On Small Volume						
Description	Oper	Unit	Crew Size	Avg Day Prod	Avg Mat'l Unit Cost	Avg Equip Unit Cost	Avg Labor Unit Cost	Avg Total Unit Cost	Avg Price Incl O&P	Crew Size	Avg Day Prod	Avg Mat'l Unit Cost	Avg Equip Unit Cost	Avg Labor Unit Cost	Avg Total Unit Cost	Avg Price Incl O&P

Framing (continued)

3/4"	Demo	SF	L-2	1134	---	---	0.23	0.23	0.34	L-2	1021	---	---	0.25	0.25	0.38
3/4"	Inst	SF	C-18	1050	0.43	0.04	0.32	0.79	0.95	C-18	945	0.43	0.05	0.36	0.83	1.01
Underlayment																
Plywood, 3/8"	Demo	SF	L-2	1216	---	---	0.21	0.21	0.32	L-2	1094	---	---	0.24	0.24	0.35
Plywood, 3/8"	Inst	SF	C-18	1127	0.44	0.04	0.30	0.78	0.92	C-18	1014	0.44	0.05	0.33	0.81	0.98
Hardboard, 0.215"	Demo	SF	L-2	1216	---	---	0.21	0.21	0.32	L-2	1094	---	---	0.24	0.24	0.35
Hardboard, 0.215"	Inst	SF	C-18	1127	0.34	0.04	0.30	0.68	0.82	C-18	1014	0.34	0.05	0.33	0.72	0.88

Trusses
Shop fabricated, wood "W" type 1/8 pitch
3" rise in 12" run:

20' span	Demo	EA	L-9	37	---	---	14.06	14.06	20.81	L-9	33	---	---	15.62	15.62	23.12
20' span	Inst	EA	C-22	34	21.00	1.37	19.83	42.20	51.91	C-22	31	25.20	1.53	22.03	48.75	59.55
22' span	Demo	EA	L-9	37	---	---	14.06	14.06	20.81	L-9	33	---	---	15.62	15.62	23.12
22' span	Inst	EA	C-22	34	23.38	1.37	19.83	44.58	54.30	C-22	31	28.05	1.53	22.03	51.61	62.41
24' span	Demo	EA	L-9	37	---	---	14.06	14.06	20.81	L-9	33	---	---	15.62	15.62	23.12
24' span	Inst	EA	C-22	34	24.53	1.37	19.83	45.73	55.44	C-22	31	29.43	1.53	22.03	52.99	63.78
26' span	Demo	EA	L-9	37	---	---	14.06	14.06	20.81	L-9	33	---	---	15.62	15.62	23.12
26' span	Inst	EA	C-22	34	26.90	1.37	19.83	48.10	57.82	C-22	31	32.28	1.53	22.03	55.84	66.63
28' span	Demo	EA	L-9	37	---	---	14.06	14.06	20.81	L-9	33	---	---	15.62	15.62	23.12
28' span	Inst	EA	C-22	34	28.06	1.37	19.83	49.26	58.97	C-22	31	33.67	1.53	22.03	57.22	68.02
30' span	Demo	EA	L-9	35	---	---	14.86	14.86	22.00	L-9	32	---	---	16.51	16.51	24.44
30' span	Inst	EA	C-22	32	30.43	1.46	21.07	52.95	63.28	C-22	29	36.51	1.62	23.41	61.54	73.01
32' span	Demo	EA	L-9	35	---	---	14.86	14.86	22.00	L-9	32	---	---	16.51	16.51	24.44
32' span	Inst	EA	C-22	32	31.27	1.46	21.07	53.80	64.12	C-22	29	37.52	1.62	23.41	62.55	74.02
34' span	Demo	EA	L-9	35	---	---	14.86	14.86	22.00	L-9	32	---	---	16.51	16.51	24.44
34' span	Inst	EA	C-22	32	32.73	1.46	21.07	55.26	65.58	C-22	29	39.28	1.62	23.41	64.31	75.78
36' span	Demo	EA	L-9	33	---	---	15.76	15.76	23.33	L-9	30	---	---	17.51	17.51	25.92
36' span	Inst	EA	C-22	31	35.11	1.51	21.75	58.36	69.02	C-22	28	42.13	1.67	24.16	67.96	79.81

			Costs Based On Large Volume							Costs Based On Small Volume						
Description	Oper	Unit	Crew Size	Avg Day Prod	Avg Mat'l Unit Cost	Avg Equip Unit Cost	Avg Labor Unit Cost	Avg Total Unit Cost	Avg Price Incl O&P	Crew Size	Avg Day Prod	Avg Mat'l Unit Cost	Avg Equip Unit Cost	Avg Labor Unit Cost	Avg Total Unit Cost	Avg Price Incl O&P

Framing (continued)

Shop fabricated, wood "W" type 5/24 pitch

5" rise in 12" run:

Description	Oper	Unit	Crew Size	Avg Day Prod	Avg Mat'l Unit Cost	Avg Equip Unit Cost	Avg Labor Unit Cost	Avg Total Unit Cost	Avg Price Incl O&P	Crew Size	Avg Day Prod	Avg Mat'l Unit Cost	Avg Equip Unit Cost	Avg Labor Unit Cost	Avg Total Unit Cost	Avg Price Incl O&P
20' span	Demo	EA	L-9	37	---	---	14.06	14.06	20.81	L-9	33	---	---	15.62	15.62	23.12
20' span	Inst	EA	C-22	34	16.66	1.37	19.83	37.87	47.58	C-22	31	20.00	1.53	22.03	43.55	54.35
22' span	Demo	EA	L-9	37	---	---	14.06	14.06	20.81	L-9	33	---	---	15.62	15.62	23.12
22' span	Inst	EA	C-22	34	18.66	1.37	19.83	39.86	49.58	C-22	31	22.39	1.53	22.03	45.95	56.75
24' span	Demo	EA	L-9	37	---	---	14.06	14.06	20.81	L-9	33	---	---	15.62	15.62	23.12
24' span	Inst	EA	C-22	34	19.75	1.37	19.83	40.95	50.67	C-22	31	23.70	1.53	22.03	47.26	58.05
26' span	Demo	EA	L-9	37	---	---	14.06	14.06	20.81	L-9	33	---	---	15.62	15.62	23.12
26' span	Inst	EA	C-22	34	22.06	1.37	19.83	43.26	52.98	C-22	31	26.47	1.53	22.03	50.03	60.83
28' span	Demo	EA	L-9	37	---	---	14.06	14.06	20.81	L-9	33	---	---	15.62	15.62	23.12
28' span	Inst	EA	C-22	34	26.20	1.37	19.83	47.40	57.12	C-22	31	31.44	1.53	22.03	55.00	65.79
30' span	Demo	EA	L-9	35	---	---	14.86	14.86	22.00	L-9	32	---	---	16.51	16.51	24.44
30' span	Inst	EA	C-22	32	28.20	1.46	21.07	50.73	61.05	C-22	29	33.84	1.62	23.41	58.87	70.35
32' span	Demo	EA	L-9	35	---	---	14.86	14.86	22.00	L-9	32	---	---	16.51	16.51	24.44
32' span	Inst	EA	C-22	32	28.99	1.46	21.07	51.51	61.84	C-22	29	34.78	1.62	23.41	59.81	71.28
34' span	Demo	EA	L-9	35	---	---	14.86	14.86	22.00	L-9	32	---	---	16.51	16.51	24.44
34' span	Inst	EA	C-22	32	31.90	1.46	21.07	54.43	64.75	C-22	29	38.28	1.62	23.41	63.31	74.78
36' span	Demo	EA	L-9	33	---	---	15.76	15.76	23.33	L-9	30	---	---	17.51	17.51	25.92
36' span	Inst	EA	C-22	31	32.99	1.51	21.75	56.24	66.90	C-22	28	39.59	1.67	24.16	65.42	77.26

Shop fabricated, wood "W" type 1/4 pitch

6" rise in 12" run:

Description	Oper	Unit	Crew Size	Avg Day Prod	Avg Mat'l Unit Cost	Avg Equip Unit Cost	Avg Labor Unit Cost	Avg Total Unit Cost	Avg Price Incl O&P	Crew Size	Avg Day Prod	Avg Mat'l Unit Cost	Avg Equip Unit Cost	Avg Labor Unit Cost	Avg Total Unit Cost	Avg Price Incl O&P
20' span	Demo	EA	L-9	37	---	---	14.06	14.06	20.81	L-9	33	---	---	15.62	15.62	23.12
20' span	Inst	EA	C-22	34	1.98	1.37	19.83	23.18	32.89	C-22	31	1.98	1.53	22.03	25.53	36.33
22' span	Demo	EA	L-9	37	---	---	14.06	14.06	20.81	L-9	33	---	---	15.62	15.62	23.12
22' span	Inst	EA	C-22	34	2.14	1.37	19.83	23.34	33.05	C-22	31	2.14	1.53	22.03	25.69	36.49
24' span	Demo	EA	L-9	37	---	---	14.06	14.06	20.81	L-9	33	---	---	15.62	15.62	23.12
24' span	Inst	EA	C-22	34	2.31	1.37	19.83	23.51	33.22	C-22	31	2.31	1.53	22.03	25.86	36.66

			Costs Based On Large Volume							Costs Based On Small Volume						
Description	Oper	Unit	Crew Size	Avg Day Prod	Avg Mat'l Unit Cost	Avg Equip Unit Cost	Avg Labor Unit Cost	Avg Total Unit Cost	Avg Price Incl O&P	Crew Size	Avg Day Prod	Avg Mat'l Unit Cost	Avg Equip Unit Cost	Avg Labor Unit Cost	Avg Total Unit Cost	Avg Price Incl O&P

Framing (continued)

Description	Oper	Unit	Crew Size	Avg Day Prod	Avg Mat'l Unit Cost	Avg Equip Unit Cost	Avg Labor Unit Cost	Avg Total Unit Cost	Avg Price Incl O&P	Crew Size	Avg Day Prod	Avg Mat'l Unit Cost	Avg Equip Unit Cost	Avg Labor Unit Cost	Avg Total Unit Cost	Avg Price Incl O&P
26' span	Demo	EA	L-9	37	---	---	14.06	14.06	20.81	L-9	33	---	---	15.62	15.62	23.12
26' span	Inst	EA	C-22	34	2.48	1.37	19.83	23.68	33.39	C-22	31	2.48	1.53	22.03	26.03	36.83
28' span	Demo	EA	L-9	37	---	---	14.06	14.06	20.81	L-9	33	---	---	15.62	15.62	23.12
28' span	Inst	EA	C-22	34	2.64	1.37	19.83	23.84	33.55	C-22	31	2.64	1.53	22.03	26.19	36.99
30' span	Demo	EA	L-9	35	---	---	14.86	14.86	22.00	L-9	32	---	---	16.51	16.51	24.44
30' span	Inst	EA	C-22	32	2.81	1.46	21.07	25.33	35.66	C-22	29	2.81	1.62	23.41	27.84	39.31
32' span	Demo	EA	L-9	35	---	---	14.86	14.86	22.00	L-9	32	---	---	16.51	16.51	24.44
32' span	Inst	EA	C-22	32	2.98	1.46	21.07	25.50	35.83	C-22	29	2.98	1.62	23.41	28.01	39.48
34' span	Demo	EA	L-9	35	---	---	14.86	14.86	22.00	L-9	32	---	---	16.51	16.51	24.44
34' span	Inst	EA	C-22	32	3.14	1.46	21.07	25.66	35.99	C-22	29	3.14	1.62	23.41	28.17	39.64
36' span	Demo	EA	L-9	33	---	---	15.76	15.76	23.33	L-9	30	---	---	17.51	17.51	25.92
36' span	Inst	EA	C-22	31	3.31	1.51	21.75	26.56	37.22	C-22	28	3.31	1.67	24.16	29.14	40.98

Garage door operators

Radio controlled for single or double doors. Labor includes wiring, connection, and installation

Chain drive, 1/4 hp, with receiver

Description	Oper	Unit	Crew Size	Avg Day Prod	Avg Mat'l Unit Cost	Avg Equip Unit Cost	Avg Labor Unit Cost	Avg Total Unit Cost	Avg Price Incl O&P	Crew Size	Avg Day Prod	Avg Mat'l Unit Cost	Avg Equip Unit Cost	Avg Labor Unit Cost	Avg Total Unit Cost	Avg Price Incl O&P
and one transmitter	Inst	LS	E-4	3.00	166.40	---	86.69	253.09	292.11	E-4	2.10	192.00	---	123.85	315.85	371.58

Screw-worm drive, 1/3 hp, with

Description	Oper	Unit	Crew Size	Avg Day Prod	Avg Mat'l Unit Cost	Avg Equip Unit Cost	Avg Labor Unit Cost	Avg Total Unit Cost	Avg Price Incl O&P	Crew Size	Avg Day Prod	Avg Mat'l Unit Cost	Avg Equip Unit Cost	Avg Labor Unit Cost	Avg Total Unit Cost	Avg Price Incl O&P
receiver and one transmitter	Inst	LS	E-4	3.00	227.50	---	86.69	314.19	353.21	E-4	2.10	262.50	---	123.85	386.35	442.08

Deluxe models, 1/2 hp, with receiver, transmitter, and time delay light

Description	Oper	Unit	Crew Size	Avg Day Prod	Avg Mat'l Unit Cost	Avg Equip Unit Cost	Avg Labor Unit Cost	Avg Total Unit Cost	Avg Price Incl O&P	Crew Size	Avg Day Prod	Avg Mat'l Unit Cost	Avg Equip Unit Cost	Avg Labor Unit Cost	Avg Total Unit Cost	Avg Price Incl O&P
Chain drive, not for vault-type garages	Inst	LS	E-4	3.00	331.50	---	86.69	418.19	457.21	E-4	2.10	382.50	---	123.85	506.35	562.08
Screw drive with threaded worm screw	Inst	LS	E-4	3.00	357.50	---	86.69	444.19	483.21	E-4	2.10	412.50	---	123.85	536.35	592.08

Adjustments:

Description	Oper	Unit	Crew Size	Avg Day Prod	Avg Mat'l Unit Cost	Avg Equip Unit Cost	Avg Labor Unit Cost	Avg Total Unit Cost	Avg Price Incl O&P	Crew Size	Avg Day Prod	Avg Mat'l Unit Cost	Avg Equip Unit Cost	Avg Labor Unit Cost	Avg Total Unit Cost	Avg Price Incl O&P
Additional transmitters, ADD Exterior key switch	Inst	EA	E-1	12.00	13.00	---	21.67	34.67	44.43	E-1	8.40	15.00	---	30.96	45.96	59.90
To remove and replace unit and receiver	Inst	LS	C-18	6.00	---	---	43.35	43.35	64.59	C-18	4.20	---	---	61.92	61.92	92.27

Garage doors. See Doors.

Garbage disposers. See Plumbing.

Description	Oper	Unit	Costs Based On Large Volume							Costs Based On Small Volume						
			Crew Size	Avg Day Prod	Avg Mat'l Unit Cost	Avg Equip Unit Cost	Avg Labor Unit Cost	Avg Total Unit Cost	Avg Price Incl O&P	Crew Size	Avg Day Prod	Avg Mat'l Unit Cost	Avg Equip Unit Cost	Avg Labor Unit Cost	Avg Total Unit Cost	Avg Price Incl O&P

Girders. See Framing.

Glass and glazing

3/16" T float with putty in wood sash

Description	Oper	Unit	Crew Size	Avg Day Prod	Avg Mat'l Unit Cost	Avg Equip Unit Cost	Avg Labor Unit Cost	Avg Total Unit Cost	Avg Price Incl O&P	Crew Size	Avg Day Prod	Avg Mat'l Unit Cost	Avg Equip Unit Cost	Avg Labor Unit Cost	Avg Total Unit Cost	Avg Price Incl O&P
8" x 12"	Demo	SF	G-1	25	---	---	6.66	6.66	9.79	G-1	15	---	---	11.10	11.10	16.31
8" x 12"	Inst	SF	G-1	30	5.11	---	5.55	10.66	13.27	G-1	18	6.18	---	9.25	15.43	19.77
12" x 16"	Demo	SF	G-1	45	---	---	3.70	3.70	5.44	G-1	27	---	---	6.17	6.17	9.06
12" x 16"	Inst	SF	G-1	55	4.19	---	3.03	7.21	8.64	G-1	33	5.06	---	5.04	10.10	12.48
14" x 20"	Demo	SF	G-1	55	---	---	3.03	3.03	4.45	G-1	33	---	---	5.04	5.04	7.42
14" x 20"	Inst	SF	G-1	70	3.82	---	2.38	6.19	7.31	G-1	42	4.61	---	3.96	8.57	10.44
16" x 24"	Demo	SF	G-1	70	---	---	2.38	2.38	3.50	G-1	42	---	---	3.96	3.96	5.83
16" x 24"	Inst	SF	G-1	90	3.56	---	1.85	5.41	6.28	G-1	54	4.31	---	3.08	7.39	8.84
24" x 26"	Demo	SF	G-1	95	---	---	1.75	1.75	2.58	G-1	57	---	---	2.92	2.92	4.29
24" x 26"	Inst	SF	G-1	120	3.20	---	1.39	4.59	5.24	G-1	72	3.87	---	2.31	6.18	7.27
36" x 24"	Demo	SF	G-1	125	---	---	1.33	1.33	1.96	G-1	75	---	---	2.22	2.22	3.26
36" x 24"	Inst	SF	G-1	155	3.05	---	1.07	4.12	4.63	G-1	93	3.68	---	1.79	5.47	6.31

1/8" T float with putty in steel sash

Description	Oper	Unit	Crew Size	Avg Day Prod	Avg Mat'l Unit Cost	Avg Equip Unit Cost	Avg Labor Unit Cost	Avg Total Unit Cost	Avg Price Incl O&P	Crew Size	Avg Day Prod	Avg Mat'l Unit Cost	Avg Equip Unit Cost	Avg Labor Unit Cost	Avg Total Unit Cost	Avg Price Incl O&P
12" x 16"	Demo	SF	G-1	45	---	---	3.70	3.70	5.44	G-1	27	---	---	6.17	6.17	9.06
12" x 16"	Inst	SF	G-1	55	5.89	---	3.03	8.92	10.34	G-1	33	7.12	---	5.04	12.16	14.53
16" x 20"	Demo	SF	G-1	65	---	---	2.56	2.56	3.76	G-1	39	---	---	4.27	4.27	6.27
16" x 20"	Inst	SF	G-1	80	5.09	---	2.08	7.17	8.15	G-1	48	6.15	---	3.47	9.62	11.25
16" x 24"	Demo	SF	G-1	70	---	---	2.38	2.38	3.50	G-1	42	---	---	3.96	3.96	5.83
16" x 24"	Inst	SF	G-1	90	4.90	---	1.85	6.75	7.61	G-1	54	5.92	---	3.08	9.00	10.45
24" x 24"	Demo	SF	G-1	95	---	---	1.75	1.75	2.58	G-1	57	---	---	2.92	2.92	4.29
24" x 24"	Inst	SF	G-1	120	4.32	---	1.39	5.71	6.36	G-1	72	5.22	---	2.31	7.53	8.62
28" x 32"	Demo	SF	G-1	130	---	---	1.28	1.28	1.88	G-1	78	---	---	2.13	2.13	3.14
28" x 32"	Inst	SF	G-1	160	4.01	---	1.04	5.05	5.54	G-1	96	4.84	---	1.73	6.58	7.39
36" x 36"	Demo	SF	G-1	160	---	---	1.04	1.04	1.53	G-1	96	---	---	1.73	1.73	2.55
36" x 36"	Inst	SF	G-1	205	3.73	---	0.81	4.54	4.93	G-1	123	4.51	---	1.35	5.86	6.50
36" x 48"	Demo	SF	G-1	190	---	---	0.88	0.88	1.29	G-1	114	---	---	1.46	1.46	2.15
36" x 48"	Inst	SF	G-1	250	3.55	---	0.67	4.22	4.53	G-1	150	4.29	---	1.11	5.40	5.92

1/4" T float

With putty and points in wood sash

Description	Oper	Unit	Crew Size	Avg Day Prod	Avg Mat'l Unit Cost	Avg Equip Unit Cost	Avg Labor Unit Cost	Avg Total Unit Cost	Avg Price Incl O&P	Crew Size	Avg Day Prod	Avg Mat'l Unit Cost	Avg Equip Unit Cost	Avg Labor Unit Cost	Avg Total Unit Cost	Avg Price Incl O&P
72" x 48"	Demo	SF	G-1	185	---	---	0.90	0.90	1.32	G-1	111	---	---	1.50	1.50	2.20
72" x 48"	Inst	SF	G-2	305	3.53	---	1.09	4.62	5.13	G-2	183	4.26	---	1.82	6.08	6.94

			Costs Based On Large Volume							Costs Based On Small Volume					
Description	Oper	Unit	Crew Size	Avg Day Prod	Avg Mat'l Unit Cost	Avg Equip Unit Cost	Avg Labor Unit Cost	Avg Total Unit Cost	Avg Price Incl O&P	Crew Size	Avg Day Prod	Avg Mat'l Unit Cost	Avg Equip Unit Cost	Avg Labor Unit Cost	Avg Total Unit Cost	Avg Price Incl O&P

Glass and Glazing (continued)

With aluminum channel and rigid neoprene rubber in aluminum sash

48" x 96"	Demo	SF	G-1	175	---	---	0.95	0.95	1.40	G-1	105	---	---	1.59	1.59	2.33
48" x 96"	Inst	SF	G-2	350	3.13	---	0.95	4.08	4.53	G-2	210	3.78	---	1.59	5.37	6.12
96" x 96"	Demo	SF	G-1	180	---	---	0.92	0.92	1.36	G-1	108	---	---	1.54	1.54	2.27
96" x 96"	Inst	SF	G-3	645	3.10	---	0.77	3.87	4.23	G-3	387	3.74	---	1.29	5.03	5.64

1" T insulating glass (2 pieces 1/4" T float with 1/2" air space) with putty and points in wood sash

To 6.0 SF	Demo	SF	G-1	50	---	---	3.33	3.33	4.89	G-1	30	---	---	5.55	5.55	8.16
To 6.0 SF	Inst	SF	G-1	60	8.50	---	2.77	11.27	12.57	G-1	36	10.27	---	4.62	14.89	17.06
6.1 SF to 12.0 SF	Demo	SF	G-1	110	---	---	1.51	1.51	2.22	G-1	66	---	---	2.52	2.52	3.71
6.1 SF to 12.0 SF	Inst	SF	G-1	135	7.66	---	1.23	8.89	9.47	G-1	81	9.25	---	2.06	11.31	12.27
12.1 SF to 18.0 SF	Demo	SF	G-1	150	---	---	1.11	1.11	1.63	G-1	90	---	---	1.85	1.85	2.72
12.1 SF to 18.0 SF	Inst	SF	G-1	180	7.31	---	0.92	8.23	8.67	G-1	108	8.83	---	1.54	10.37	11.10
18.1 SF to 24.0 SF	Demo	SF	G-1	145	---	---	1.15	1.15	1.69	G-1	87	---	---	1.91	1.91	2.81
18.1 SF to 24.0 SF	Inst	SF	G-2	250	7.07	---	1.33	8.40	9.03	G-2	150	8.54	---	2.22	10.76	11.80

Aluminum sliding door glass with aluminum channel and rigid neoprene rubber

3/16" T tempered glass

34" W x 76" H	Demo	SF	G-1	195	---	---	0.85	0.85	1.25	G-1	117	---	---	1.42	1.42	2.09
34" W x 76" H	Inst	SF	G-1	245	2.58	---	0.68	3.26	3.58	G-1	147	3.12	---	1.13	4.25	4.78
46" W x 76" H	Demo	SF	G-1	245	---	---	0.68	0.68	1.00	G-1	147	---	---	1.13	1.13	1.66
46" W x 76" H	Inst	SF	G-2	330	2.56	---	1.01	3.56	4.04	G-2	198	3.09	---	1.68	4.77	5.56

5/8" T insulating glass (2 pieces 3/16" T tempered with 1/4" T air space)

34" W x 76" H	Demo	SF	G-1	170	---	---	0.98	0.98	1.44	G-1	102	---	---	1.63	1.63	2.40
34" W x 76" H	Inst	SF	G-1	230	5.17	---	0.72	5.90	6.24	G-1	138	6.25	---	1.21	7.46	8.02
46" W x 76" H	Demo	SF	G-1	215	---	---	0.77	0.77	1.14	G-1	129	---	---	1.29	1.29	1.90
46" W x 76" H	Inst	SF	G-2	310	5.10	---	1.07	6.17	6.68	G-2	186	6.16	---	1.79	7.95	8.79

Mirrors

Decorator oval mirrors, antique gold

16" x 24"	Inst	EA	C-1	50	45.60	---	3.37	48.97	50.62	C-1	38	57.00	---	4.49	61.49	63.70
16" x 32"	Inst	EA	C-1	50	54.00	---	3.37	57.37	59.02	C-1	38	67.50	---	4.49	71.99	74.20

Grading. See Site Work.

			Costs Based On Large Volume							Costs Based On Small Volume					
Description	Oper	Unit	Crew Size	Avg Day Prod	Avg Mat'l Unit Cost	Avg Equip Unit Cost	Avg Labor Unit Cost	Avg Total Unit Cost	Avg Price Incl O&P	Crew Size	Avg Day Prod	Avg Mat'l Unit Cost	Avg Equip Unit Cost	Avg Labor Unit Cost	Avg Total Unit Cost	Avg Price Incl O&P

Gutters and Downspouts

Aluminum

Baked on painted finish (white or brown)

Gutter, 5" box type

Description	Oper	Unit	Crew Size	Avg Day Prod	Avg Mat'l Unit Cost	Avg Equip Unit Cost	Avg Labor Unit Cost	Avg Total Unit Cost	Avg Price Incl O&P	Crew Size	Avg Day Prod	Avg Mat'l Unit Cost	Avg Equip Unit Cost	Avg Labor Unit Cost	Avg Total Unit Cost	Avg Price Incl O&P
Heavyweight gauge	Inst	LF	U-2	215	1.11	---	1.77	2.87	3.71	U-2	129	1.28	---	2.95	4.22	5.61
Standard weight gauge	Inst	LF	U-2	215	0.88	---	1.77	2.65	3.49	U-2	129	1.02	---	2.95	3.97	5.36
End caps	Inst	EA	U-2	---	0.64	---	---	---	0.64	U-2	---	0.74	---	---	---	0.74
Drop outlet for downspouts	Inst	EA	U-2	---	3.63	---	---	---	3.63	U-2	---	4.19	---	---	---	4.19
Inside/outside corner	Inst	EA	U-2	---	3.89	---	---	---	3.89	U-2	---	4.49	---	---	---	4.49
Joint connector (4 ea/pkg)	Inst	Pkg	U-2	---	4.93	---	---	---	4.93	U-2	---	5.69	---	---	---	5.69
Strap hanger (10 ea/pkg)	Inst	Pkg	U-2	---	7.14	---	---	---	7.14	U-2	---	8.24	---	---	---	8.24
Fascia bracket (4 ea/pkg)	Inst	Pkg	U-2	---	7.14	---	---	---	7.14	U-2	---	8.24	---	---	---	8.24
Spike and ferrule (10 ea/pkg)	Inst	Pkg	U-2	---	5.84	---	---	---	5.84	U-2	---	6.74	---	---	---	6.74

Downspout, corrugated square

Description	Oper	Unit	Crew Size	Avg Day Prod	Avg Mat'l Unit Cost	Avg Equip Unit Cost	Avg Labor Unit Cost	Avg Total Unit Cost	Avg Price Incl O&P	Crew Size	Avg Day Prod	Avg Mat'l Unit Cost	Avg Equip Unit Cost	Avg Labor Unit Cost	Avg Total Unit Cost	Avg Price Incl O&P
Standard weight gauge	Inst	LF	U-2	250	0.91	---	---	---	0.91	U-2	150	1.05	---	---	---	1.05
Regular/side elbow	Inst	EA	U-2	---	1.55	---	---	---	1.55	U-2	---	1.79	---	---	---	1.79
Downspout holder (4 ea/pkg)	Inst	Pkg	U-2	---	4.54	---	---	---	4.54	U-2	---	5.24	---	---	---	5.24

Steel

Natural finish

Gutter, 4" box type

Description	Oper	Unit	Crew Size	Avg Day Prod	Avg Mat'l Unit Cost	Avg Equip Unit Cost	Avg Labor Unit Cost	Avg Total Unit Cost	Avg Price Incl O&P	Crew Size	Avg Day Prod	Avg Mat'l Unit Cost	Avg Equip Unit Cost	Avg Labor Unit Cost	Avg Total Unit Cost	Avg Price Incl O&P
Heavyweight gauge	Inst	LF	U-2	215	0.59	---	---	---	0.59	U-2	129	0.68	---	---	---	0.68
End caps	Inst	EA	U-2	---	0.64	---	---	---	0.64	U-2	---	0.74	---	---	---	0.74
Drop outlet for downspouts	Inst	EA	U-2	---	2.33	---	---	---	2.33	U-2	---	2.69	---	---	---	2.69
Inside/outside corner	Inst	EA	U-2	---	3.50	---	---	---	3.50	U-2	---	4.04	---	---	---	4.04
Joint connector (4 ea/pkg)	Inst	Pkg	U-2	---	3.63	---	---	---	3.63	U-2	---	4.19	---	---	---	4.19
Strap hanger (10 ea/pkg)	Inst	Pkg	U-2	---	4.93	---	---	---	4.93	U-2	---	5.69	---	---	---	5.69
Fascia bracket (4 ea/pkg)	Inst	Pkg	U-2	---	3.89	---	---	---	3.89	U-2	---	4.49	---	---	---	4.49
Spike and ferrule (10 ea/pkg)	Inst	Pkg	U-2	---	4.28	---	---	---	4.28	U-2	---	4.94	---	---	---	4.94

Downspout, 30 gauge galvanized steel

Description	Oper	Unit	Crew Size	Avg Day Prod	Avg Mat'l Unit Cost	Avg Equip Unit Cost	Avg Labor Unit Cost	Avg Total Unit Cost	Avg Price Incl O&P	Crew Size	Avg Day Prod	Avg Mat'l Unit Cost	Avg Equip Unit Cost	Avg Labor Unit Cost	Avg Total Unit Cost	Avg Price Incl O&P
Standard weight gauge	Inst	LF	U-2	---	0.98	---	---	---	0.98	U-2	---	1.13	---	---	---	1.13
Regular/side elbow	Inst	EA	U-2	---	1.55	---	---	---	1.55	U-2	---	1.79	---	---	---	1.79
Downspout holder (4 ea/pkg)	Inst	Pkg	U-2	---	4.15	---	---	---	4.15	U-2	---	4.79	---	---	---	4.79

Costs Based On Large Volume   Costs Based On Small Volume

Description	Oper	Unit	Crew Size	Avg Day Prod	Avg Mat'l Unit Cost	Avg Equip Unit Cost	Avg Labor Unit Cost	Avg Total Unit Cost	Avg Price Incl O&P	Crew Size	Avg Day Prod	Avg Mat'l Unit Cost	Avg Equip Unit Cost	Avg Labor Unit Cost	Avg Total Unit Cost	Avg Price Incl O&P

Gutters and Downspouts (continued)

Vinyl
Extruded 5" PVC

White

Description	Oper	Unit	Crew	Day	Mat'l				O&P	Crew	Day	Mat'l				O&P
Gutter	Inst	LF	U-2	215	0.85	---	---	---	0.85	U-2	129	0.98	---	---	---	0.98
End caps	Inst	EA	U-2	---	1.55	---	---	---	1.55	U-2	---	1.79	---	---	---	1.79
Drop outlet for downspouts	Inst	EA	U-2	---	6.49	---	---	---	6.49	U-2	---	7.49	---	---	---	7.49
Inside/outside corner	Inst	EA	U-2	---	6.49	---	---	---	6.49	U-2	---	7.49	---	---	---	7.49
Joint connector	Inst	EA	U-2	---	1.94	---	---	---	1.94	U-2	---	2.24	---	---	---	2.24
Fascia bracket (4 ea/pkg)	Inst	Pkg	U-2	---	8.44	---	---	---	8.44	U-2	---	9.74	---	---	---	9.74
Downspout	Inst	LF	U-2	250	1.24	---	---	---	1.24	U-2	150	1.43	---	---	---	1.43
Downspout driplet	Inst	EA	U-2	---	8.44	---	---	---	8.44	U-2	---	9.74	---	---	---	9.74
Downspout joiner	Inst	EA	U-2	---	2.59	---	---	---	2.59	U-2	---	2.99	---	---	---	2.99
Downspout holder (2 ea/pkg)	Inst	Pkg	U-2	---	8.44	---	---	---	8.44	U-2	---	9.74	---	---	---	9.74
Regular elbow	Inst	EA	U-2	---	2.85	---	---	---	2.85	U-2	---	3.29	---	---	---	3.29
Well cap	Inst	EA	U-2	---	3.63	---	---	---	3.63	U-2	---	4.19	---	---	---	4.19
Well outlet	Inst	EA	U-2	---	4.93	---	---	---	4.93	U-2	---	5.69	---	---	---	5.69
Expansion joint connector	Inst	EA	U-2	---	7.79	---	---	---	7.79	U-2	---	8.99	---	---	---	8.99
Rafter adapter (4 ea/pkg)	Inst	Pkg	U-2	---	5.19	---	---	---	5.19	U-2	---	5.99	---	---	---	5.99

Brown

Description	Oper	Unit	Crew	Day	Mat'l				O&P	Crew	Day	Mat'l				O&P
Gutter	Inst	LF	U-2	215	1.04	---	---	---	1.04	U-2	129	1.20	---	---	---	1.20
End caps	Inst	EA	U-2	---	1.94	---	---	---	1.94	U-2	---	2.24	---	---	---	2.24
Drop outlet for downspouts	Inst	EA	U-2	---	7.79	---	---	---	7.79	U-2	---	8.99	---	---	---	8.99
Inside/outside corner	Inst	EA	U-2	---	8.44	---	---	---	8.44	U-2	---	9.74	---	---	---	9.74
Joint connector	Inst	EA	U-2	---	2.59	---	---	---	2.59	U-2	---	2.99	---	---	---	2.99
Fascia bracket (4 ea/pkg)	Inst	Pkg	U-2	---	9.74	---	---	---	9.74	U-2	---	11.24	---	---	---	11.24
Downspout	Inst	LF	U-2	250	1.37	---	---	---	1.37	U-2	150	1.58	---	---	---	1.58
Downspout driplet	Inst	EA	U-2	---	9.74	---	---	---	9.74	U-2	---	11.24	---	---	---	11.24
Downspout joiner	Inst	EA	U-2	---	3.11	---	---	---	3.11	U-2	---	3.59	---	---	---	3.59
Downspout holder (2 ea/pkg)	Inst	Pkg	U-2	---	10.39	---	---	---	10.39	U-2	---	11.99	---	---	---	11.99
Regular elbow	Inst	EA	U-2	---	3.63	---	---	---	3.63	U-2	---	4.19	---	---	---	4.19
Well cap	Inst	EA	U-2	---	4.28	---	---	---	4.28	U-2	---	4.94	---	---	---	4.94
Well outlet	Inst	EA	U-2	---	6.23	---	---	---	6.23	U-2	---	7.19	---	---	---	7.19
Expansion joint connector	Inst	EA	U-2	---	9.74	---	---	---	9.74	U-2	---	11.24	---	---	---	11.24
Rafter adapter (4 ea/pkg)	Inst	Pkg	U-2	---	5.19	---	---	---	5.19	U-2	---	5.99	---	---	---	5.99

Description	Oper	Unit	Costs Based On Large Volume							Costs Based On Small Volume						
			Crew Size	Avg Day Prod	Avg Mat'l Unit Cost	Avg Equip Unit Cost	Avg Labor Unit Cost	Avg Total Unit Cost	Avg Price Incl O&P	Crew Size	Avg Day Prod	Avg Mat'l Unit Cost	Avg Equip Unit Cost	Avg Labor Unit Cost	Avg Total Unit Cost	Avg Price Incl O&P

Hardboard. See Paneling.

Hardwood flooring

Includes waste and nails

Strip

Installed over wood subfloor

Prefinished oak, prime

25/32" x 3-1/4"

Lay floor	Inst	SF	F-4	400	7.26	---	1.01	8.27	8.73	F-4	280	9.07	---	1.44	10.52	11.18
Wax, polish, machine buff	Inst	SF	F-3	860	0.03	---	0.22	0.25	0.34	F-3	602	0.04	---	0.31	0.35	0.49

25/32" x 2-1/4"

Lay floor	Inst	SF	F-4	300	6.49	---	1.34	7.83	8.45	F-4	210	8.11	---	1.92	10.03	10.91
Wax, polish, machine buff	Inst	SF	F-3	860	0.03	---	0.22	0.25	0.34	F-3	602	0.04	---	0.31	0.35	0.49

Unfinished

25/32" x 3-1/4", lay floor only

Fir

Vertical grain	Inst	SF	F-4	480	2.20	---	0.84	3.04	3.42	F-4	336	2.75	---	1.20	3.95	4.50
Flat grain	Inst	SF	F-4	480	2.45	---	0.84	3.29	3.67	F-4	336	3.06	---	1.20	4.26	4.81
Yellow pine	Inst	SF	F-4	480	2.07	---	0.84	2.91	3.30	F-4	336	2.59	---	1.20	3.79	4.34

25/32" x 2-1/4", lay floor only

Maple	Inst	SF	F-4	320	3.30	---	1.26	4.56	5.14	F-4	224	4.12	---	1.80	5.92	6.75
Oak	Inst	SF	F-4	320	7.15	---	1.26	8.41	8.99	F-4	224	8.94	---	1.80	10.74	11.57
Yellow pine	Inst	SF	F-4	385	2.17	---	1.05	3.21	3.69	F-4	270	2.71	---	1.50	4.20	4.89

Related materials and operations

Machine sand, fill and finish

New floors		SF	F-3	430	0.07	---	0.43	0.50	0.70	F-3	301	0.08	---	0.62	0.70	0.98
Damaged floors		SF	F-3	285	0.10	---	0.65	0.75	1.05	F-3	200	0.12	---	0.93	1.05	1.48
Wax polish and machine buff		SF	F-3	860	0.03	---	0.22	0.25	0.34	F-3	602	0.04	---	0.31	0.35	0.49

Block

Laid in mastic over wood subfloor covered with felt

Oak, 5/16" x 12" x 12"

Lay floor only

Prefinished	Inst	SF	F-4	520	8.46	---	0.78	9.23	9.59	F-4	364	10.57	---	1.11	11.68	12.19
Unfinished	Inst	SF	F-4	550	4.78	---	0.73	5.51	5.85	F-4	385	5.98	---	1.05	7.02	7.50

Hardwood flooring

1. **Strip** flooring is nailed into place over wood subflooring or over wood sleeper strips. Using 3-1/4"W strips leaves 25% cutting and fitting waste; 2-1/4"W strips leaves 33% waste. Nails and the respective cutting and fitting waste have been included in the unit costs.

2. **Block** flooring is laid in mastic applied to felt-covered wood subfloor. Mastic, 5% block waste and felt are included in material unit costs for block or parquet flooring.

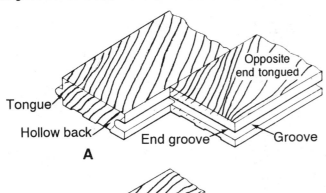

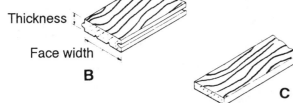

Strip flooring, A, side and end matched; B, side matched; C, square edged

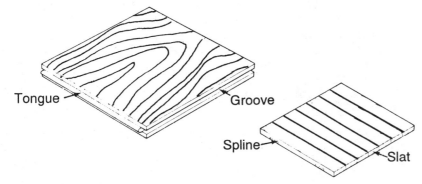

Two types of wood block flooring

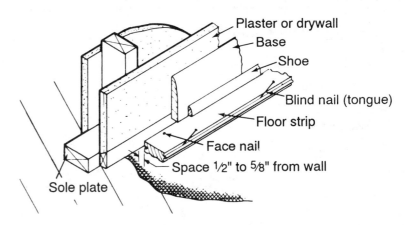

Installation of first strip of flooring

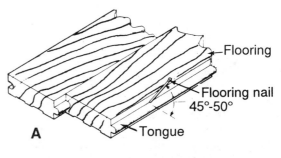

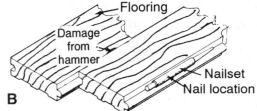

Nailing of flooring; A, angle of nailing; B, setting the nail without damage to the flooring

			Costs Based On Large Volume							Costs Based On Small Volume					
Description	Oper	Unit	Crew Size	Avg Day Prod	Avg Mat'l Unit Cost	Avg Equip Unit Cost	Avg Labor Unit Cost	Avg Total Unit Cost	Avg Price Incl O&P	Crew Size	Avg Day Prod	Avg Mat'l Unit Cost	Avg Equip Unit Cost	Avg Labor Unit Cost	Avg Total Unit Cost	Avg Price Incl O&P

Hardwood flooring (continued)

Teak, 5/16" x 12" x 12"
Lay floor only

Description	Oper	Unit	Crew Size	Avg Day Prod	Avg Mat'l Unit Cost	Avg Equip Unit Cost	Avg Labor Unit Cost	Avg Total Unit Cost	Avg Price Incl O&P	Crew Size	Avg Day Prod	Avg Mat'l Unit Cost	Avg Equip Unit Cost	Avg Labor Unit Cost	Avg Total Unit Cost	Avg Price Incl O&P
Prefinished	Inst	SF	F-4	520	10.03	---	0.78	10.81	11.16	F-4	364	12.54	---	1.11	13.65	14.16
Unfinished	Inst	SF	F-4	550	6.36	---	0.73	7.09	7.43	F-4	385	7.94	---	1.05	8.99	9.47

Oak, 13/16" x 12" x 12"
Lay floor only

Description	Oper	Unit	Crew Size	Avg Day Prod	Avg Mat'l Unit Cost	Avg Equip Unit Cost	Avg Labor Unit Cost	Avg Total Unit Cost	Avg Price Incl O&P	Crew Size	Avg Day Prod	Avg Mat'l Unit Cost	Avg Equip Unit Cost	Avg Labor Unit Cost	Avg Total Unit Cost	Avg Price Incl O&P
Prefinished	Inst	SF	F-4	520	14.76	---	0.78	15.53	15.89	F-4	364	18.44	---	1.11	19.55	20.06
Unfinished	Inst	SF	F-4	500	11.08	---	0.81	11.89	12.26	F-4	350	13.85	---	1.15	15.00	15.53

Machine sand, fill and finish

Description	Oper	Unit	Crew Size	Avg Day Prod	Avg Mat'l Unit Cost	Avg Equip Unit Cost	Avg Labor Unit Cost	Avg Total Unit Cost	Avg Price Incl O&P	Crew Size	Avg Day Prod	Avg Mat'l Unit Cost	Avg Equip Unit Cost	Avg Labor Unit Cost	Avg Total Unit Cost	Avg Price Incl O&P
New floors	Inst	SF	F-3	430	0.07	---	0.43	0.50	0.70	F-3	301	0.08	---	0.62	0.70	0.98
Damaged floors	Inst	SF	F-3	285	0.10	---	0.65	0.75	1.05	F-3	200	0.12	---	0.93	1.05	1.48
Wax polish and machine buff	Inst	SF	F-3	860	0.03	---	0.22	0.25	0.34	F-3	602	0.04	---	0.31	0.35	0.49

Parquetry, 5/16" x 9" x 9"

Lay floor only

Description	Oper	Unit	Crew Size	Avg Day Prod	Avg Mat'l Unit Cost	Avg Equip Unit Cost	Avg Labor Unit Cost	Avg Total Unit Cost	Avg Price Incl O&P	Crew Size	Avg Day Prod	Avg Mat'l Unit Cost	Avg Equip Unit Cost	Avg Labor Unit Cost	Avg Total Unit Cost	Avg Price Incl O&P
Oak	Inst	SF	F-4	500	4.78	---	0.81	5.59	5.96	F-4	350	5.98	---	1.15	7.13	7.66
Walnut	Inst	SF	F-4	500	11.45	---	0.81	12.25	12.63	F-4	350	14.31	---	1.15	15.46	15.99
Teak	Inst	SF	F-4	500	6.36	---	0.81	7.16	7.53	F-4	350	7.94	---	1.15	9.10	9.63

Machine sand, fill and finish

Description	Oper	Unit	Crew Size	Avg Day Prod	Avg Mat'l Unit Cost	Avg Equip Unit Cost	Avg Labor Unit Cost	Avg Total Unit Cost	Avg Price Incl O&P	Crew Size	Avg Day Prod	Avg Mat'l Unit Cost	Avg Equip Unit Cost	Avg Labor Unit Cost	Avg Total Unit Cost	Avg Price Incl O&P
New floors	Inst	SF	F-3	430	0.07	---	0.43	0.50	0.70	F-3	301	0.08	---	0.62	0.70	0.98
Damaged floors	Inst	SF	F-3	285	0.10	---	0.65	0.75	1.05	F-3	200	0.12	---	0.93	1.05	1.48
Wax polish and machine buff	Inst	SF	F-3	860	0.03	---	0.22	0.25	0.34	F-3	602	0.04	---	0.31	0.35	0.49

Acrylic wood parquet blocks

Description	Oper	Unit	Crew Size	Avg Day Prod	Avg Mat'l Unit Cost	Avg Equip Unit Cost	Avg Labor Unit Cost	Avg Total Unit Cost	Avg Price Incl O&P	Crew Size	Avg Day Prod	Avg Mat'l Unit Cost	Avg Equip Unit Cost	Avg Labor Unit Cost	Avg Total Unit Cost	Avg Price Incl O&P
5/16" x 12" x 12" set in epoxy	Inst	SF	F-4	500	5.60	---	0.81	6.41	6.78	F-4	350	7.00	---	1.15	8.15	8.68
Wax polish and machine buff	Inst	SF	F-3	860	0.03	---	0.22	0.25	0.34	F-3	602	0.04	---	0.31	0.35	0.49

Heaters. See HVAC, Heating.

HVAC. Heating, Ventilating, Air Conditioning

Heating
Boilers
Electric fired heaters,
includes standard controls and trim, ASME.
Hot Water

Description	Oper	Unit	Crew Size	Avg Day Prod	Avg Mat'l Unit Cost	Avg Equip Unit Cost	Avg Labor Unit Cost	Avg Total Unit Cost	Avg Price Incl O&P	Crew Size	Avg Day Prod	Avg Mat'l Unit Cost	Avg Equip Unit Cost	Avg Labor Unit Cost	Avg Total Unit Cost	Avg Price Incl O&P
12 KW, 40 MBH	Inst	EA	S-4	1.20	2400	---	437	2837	3039	S-4	0.96	2760	---	547	3307	3558
24 KW, 85 MBH	Inst	EA	S-4	1.10	2600	---	477	3077	3297	S-4	0.88	2990	---	596	3586	3861

			Costs Based On Large Volume							Costs Based On Small Volume					
Description	Oper	Unit	Crew Size	Avg Day Prod	Avg Mat'l Unit Cost	Avg Equip Unit Cost	Avg Labor Unit Cost	Avg Total Unit Cost	Avg Price Incl O&P	Crew Size	Avg Day Prod	Avg Mat'l Unit Cost	Avg Equip Unit Cost	Avg Labor Unit Cost	Avg Total Unit Cost	Avg Price Incl O&P

HVAC (continued)

Description	Oper	Unit	Crew Size	Avg Day Prod	Avg Mat'l Unit Cost	Avg Equip Unit Cost	Avg Labor Unit Cost	Avg Total Unit Cost	Avg Price Incl O&P	Crew Size	Avg Day Prod	Avg Mat'l Unit Cost	Avg Equip Unit Cost	Avg Labor Unit Cost	Avg Total Unit Cost	Avg Price Incl O&P
30 KW, 105 MBH	Inst	EA	S-4	1.00	2650	---	525	3175	3416	S-4	0.80	3048	---	656	3704	4005
36 KW, 120 MBH	Inst	EA	S-4	1.00	2800	---	525	3325	3566	S-4	0.80	3220	---	656	3876	4178
45 KW, 155 MBH	Inst	EA	S-4	0.95	2900	---	553	3453	3707	S-4	0.76	3335	---	691	4026	4343
60 KW, 205 MBH	Inst	EA	S-4	0.90	3300	---	583	3883	4152	S-4	0.72	3795	---	729	4524	4859
Steam																
6 KW, 20 MBH	Inst	EA	S-4	1.10	2900	---	477	3377	3597	S-4	0.88	3335	---	596	3931	4206
60 KW, 205 MBH	Inst	EA	S-4	0.90	4100	---	583	4683	4951	S-4	0.72	4715	---	729	5444	5779
240 KW, 820 MBH	Inst	EA	S-4	0.40	13500	---	1312	14812	15416	S-4	0.32	15525	---	1640	17165	17920
600 KW, 2050 MBH	Inst	EA	S-5	0.30	18500	---	2413	20913	22023	S-5	0.24	21275	---	3016	24291	25678
Minimum Job Charge	Inst	Job	S-4		---	---	262	262	383	S-4		---	---	328	328	479

Gas fired heaters,
natural or propane, heaters; includes standard controls; MBH gross output.

Cast iron with insulated jacket.

Description	Oper	Unit	Crew Size	Avg Day Prod	Avg Mat'l Unit Cost	Avg Equip Unit Cost	Avg Labor Unit Cost	Avg Total Unit Cost	Avg Price Incl O&P	Crew Size	Avg Day Prod	Avg Mat'l Unit Cost	Avg Equip Unit Cost	Avg Labor Unit Cost	Avg Total Unit Cost	Avg Price Incl O&P
Hot Water																
80 MBH	Inst	EA	S-6	1.40	950	---	377	1327	1501	S-6	1.12	1093	---	471	1564	1781
100 MBH	Inst	EA	S-6	1.25	1100	---	422	1522	1717	S-6	1.00	1265	---	528	1793	2036
120 MBH	Inst	EA	S-6	1.00	1200	---	528	1728	1971	S-6	0.80	1380	---	660	2040	2343
160 MBH	Inst	EA	S-6	0.95	1500	---	556	2056	2311	S-6	0.76	1725	---	695	2420	2739
440 MBH	Inst	EA	S-6	0.60	3100	---	880	3980	4385	S-6	0.48	3565	---	1100	4665	5171
2000 MBH	Inst	EA	S-6	0.35	13000	---	1508	14508	15202	S-6	0.28	14950	---	1885	16835	17703
Steam																
80 MBH	Inst	EA	S-6	1.30	1050	---	406	1456	1643	S-6	1.04	1208	---	508	1715	1949
160 MBH	Inst	EA	S-6	0.85	1600	---	621	2221	2507	S-6	0.68	1840	---	776	2616	2973
440 MBH	Inst	EA	S-6	0.50	3250	---	1056	4306	4792	S-6	0.40	3738	---	1320	5057	5664
1875 MBH	Inst	EA	S-6	0.25	11500	---	2112	13612	14583	S-6	0.20	13225	---	2640	15865	17079

Steel with insulated jacket; includes burner and one zone valve

Description	Oper	Unit	Crew Size	Avg Day Prod	Avg Mat'l Unit Cost	Avg Equip Unit Cost	Avg Labor Unit Cost	Avg Total Unit Cost	Avg Price Incl O&P	Crew Size	Avg Day Prod	Avg Mat'l Unit Cost	Avg Equip Unit Cost	Avg Labor Unit Cost	Avg Total Unit Cost	Avg Price Incl O&P
Hot Water																
50 MBH	Inst	EA	S-6	1.75	1100	---	302	1402	1540	S-6	1.40	1265	---	377	1642	1816
70 MBH	Inst	EA	S-6	1.60	1400	---	330	1730	1882	S-6	1.28	1610	---	412	2022	2212
90 MBH	Inst	EA	S-6	1.50	1425	---	352	1777	1939	S-6	1.20	1639	---	440	2079	2281
105 MBH	Inst	EA	S-6	1.25	1600	---	422	2022	2217	S-6	1.00	1840	---	528	2368	2611
130 MBH	Inst	EA	S-6	1.20	1800	---	440	2240	2442	S-6	0.96	2070	---	550	2620	2873
150 MBH	Inst	EA	S-6	1.15	2100	---	459	2559	2770	S-6	0.92	2415	---	574	2989	3253
185 MBH	Inst	EA	S-6	1.10	2500	---	480	2980	3201	S-6	0.88	2875	---	600	3475	3751

			Costs Based On Large Volume							Costs Based On Small Volume						
Description	Oper	Unit	Crew Size	Avg Day Prod	Avg Mat'l Unit Cost	Avg Equip Unit Cost	Avg Labor Unit Cost	Avg Total Unit Cost	Avg Price Incl O&P	Crew Size	Avg Day Prod	Avg Mat'l Unit Cost	Avg Equip Unit Cost	Avg Labor Unit Cost	Avg Total Unit Cost	Avg Price Incl O&P

HVAC (continued)

Description	Oper	Unit	Crew Size	Avg Day Prod	Avg Mat'l Unit Cost	Avg Equip Unit Cost	Avg Labor Unit Cost	Avg Total Unit Cost	Avg Price Incl O&P	Crew Size	Avg Day Prod	Avg Mat'l Unit Cost	Avg Equip Unit Cost	Avg Labor Unit Cost	Avg Total Unit Cost	Avg Price Incl O&P
235 MBH	Inst	EA	S-6	1.00	3200	---	528	3728	3971	S-6	0.80	3680	---	660	4340	4643
290 MBH	Inst	EA	S-6	0.90	3600	---	587	4187	4456	S-6	0.72	4140	---	733	4873	5210
480 MBH	Inst	EA	S-6	0.55	5200	---	960	6160	6601	S-6	0.44	5980	---	1200	7180	7732
640 MBH	Inst	EA	S-6	0.50	6100	---	1056	7156	7642	S-6	0.40	7015	---	1320	8335	8942
800 MBH	Inst	EA	S-6	0.45	7200	---	1173	8373	8913	S-6	0.36	8280	---	1466	9746	10421
960 MBH	Inst	EA	S-6	0.40	8900	---	1320	10220	10827	S-6	0.32	10235	---	1650	11885	12644
Minimum Job Charge	Inst	Job	S-6		---	---	264	264	385	S-6		---	---	330	330	482

Oil fired heaters,
 includes standard controls; flame retention burner; MBH gross output

Cast Iron with insulated jacket
 Hot water

Description	Oper	Unit	Crew Size	Avg Day Prod	Avg Mat'l Unit Cost	Avg Equip Unit Cost	Avg Labor Unit Cost	Avg Total Unit Cost	Avg Price Incl O&P	Crew Size	Avg Day Prod	Avg Mat'l Unit Cost	Avg Equip Unit Cost	Avg Labor Unit Cost	Avg Total Unit Cost	Avg Price Incl O&P
90 MBH	Inst	EA	S-6	1.00	1500	---	528	2028	2271	S-6	0.80	1725	---	660	2385	2688
190 MBH	Inst	EA	S-6	0.70	2100	---	754	2854	3201	S-6	0.56	2415	---	943	3358	3791
1000 MBH	Inst	EA	S-6	0.30	6900	---	1760	8660	9469	S-6	0.24	7935	---	2200	10135	11146
1320 MBH	Inst	EA	S-6	0.25	8100	---	2112	10212	11183	S-6	0.20	9315	---	2640	11955	13169
2100 MBH	Inst	EA	S-6	0.20	12500	---	2640	15140	16354	S-6	0.16	14375	---	3299	17674	19192

Steam

Description	Oper	Unit	Crew Size	Avg Day Prod	Avg Mat'l Unit Cost	Avg Equip Unit Cost	Avg Labor Unit Cost	Avg Total Unit Cost	Avg Price Incl O&P	Crew Size	Avg Day Prod	Avg Mat'l Unit Cost	Avg Equip Unit Cost	Avg Labor Unit Cost	Avg Total Unit Cost	Avg Price Incl O&P
110 MBH	Inst	EA	S-6	1.10	1400	---	480	1880	2101	S-6	0.88	1610	---	600	2210	2486
205 MBH	Inst	EA	S-6	0.80	1900	---	660	2560	2863	S-6	0.64	2185	---	825	3010	3389
1085 MBH	Inst	EA	S-6	0.40	6670	---	1320	7990	8597	S-6	0.32	7671	---	1650	9320	10079
1360 MBH	Inst	EA	S-6	0.33	7900	---	1600	9500	10236	S-6	0.26	9085	---	2000	11085	12005
2175 MBH	Inst	EA	S-6	0.25	12000	---	2112	14112	15083	S-6	0.20	13800	---	2640	16440	17654

Steel insulated jacket burner
 Hot water

Description	Oper	Unit	Crew Size	Avg Day Prod	Avg Mat'l Unit Cost	Avg Equip Unit Cost	Avg Labor Unit Cost	Avg Total Unit Cost	Avg Price Incl O&P	Crew Size	Avg Day Prod	Avg Mat'l Unit Cost	Avg Equip Unit Cost	Avg Labor Unit Cost	Avg Total Unit Cost	Avg Price Incl O&P
105 MBH	Inst	EA	S-6	1.75	1650	---	302	1952	2090	S-6	1.40	1898	---	377	2275	2448
120 MBH	Inst	EA	S-6	1.60	1700	---	330	2030	2182	S-6	1.28	1955	---	412	2367	2557
140 MBH	Inst	EA	S-6	1.40	1750	---	377	2127	2301	S-6	1.12	2013	---	471	2484	2701
170 MBH	Inst	EA	S-6	1.20	2100	---	440	2540	2742	S-6	0.96	2415	---	550	2965	3218
225 MBH	Inst	EA	S-6	0.90	2450	---	587	3037	3306	S-6	0.72	2818	---	733	3551	3888
315 MBH	Inst	EA	S-6	0.80	3600	---	660	4260	4563	S-6	0.64	4140	---	825	4965	5344
420 MBH	Inst	EA	S-6	0.70	4100	---	754	4854	5201	S-6	0.56	4715	---	943	5658	6091
Minimum Job Charge	Inst	Job	S-6		---	---	264	264	385	S-6		---	---	330	330	482

			Costs Based On Large Volume							Costs Based On Small Volume					
Description	Oper	Unit	Crew Size	Avg Day Prod	Avg Mat'l Unit Cost	Avg Equip Unit Cost	Avg Labor Unit Cost	Avg Total Unit Cost	Avg Price Incl O&P	Crew Size	Avg Day Prod	Avg Mat'l Unit Cost	Avg Equip Unit Cost	Avg Labor Unit Cost	Avg Total Unit Cost	Avg Price Incl O&P

HVAC (continued)

Boiler accessories

Burners

Conversion, gas fired, LP or natural. Residential, gun type

Description	Oper	Unit	Crew Size	Avg Day Prod	Avg Mat'l Unit Cost	Avg Equip Unit Cost	Avg Labor Unit Cost	Avg Total Unit Cost	Avg Price Incl O&P	Crew Size	Avg Day Prod	Avg Mat'l Unit Cost	Avg Equip Unit Cost	Avg Labor Unit Cost	Avg Total Unit Cost	Avg Price Incl O&P
35 to 180 MBH	Inst	EA	S-2	2.00	175	---	164	339	415	S-2	1.60	201	---	206	407	501
50 to 240 MBH	Inst	EA	S-2	1.75	200	---	188	388	474	S-2	1.40	230	---	235	465	573
200 to 400 MBH	Inst	EA	S-2	1.50	375	---	219	594	695	S-2	1.20	431	---	274	705	832
Flame retention, oil fired assembly.																
2.0 to 5.0 GPH	Inst	EA	S-2	1.75	425	---	188	613	699	S-2	1.40	489	---	235	724	832

Forced warm air systems

Duct furnaces

Furnace includes burner, controls, stainless steel heat exchanger. Gas fired with an electric ignition.

Outdoor installation, includes vent cap

Description	Oper	Unit	Crew Size	Avg Day Prod	Avg Mat'l Unit Cost	Avg Equip Unit Cost	Avg Labor Unit Cost	Avg Total Unit Cost	Avg Price Incl O&P	Crew Size	Avg Day Prod	Avg Mat'l Unit Cost	Avg Equip Unit Cost	Avg Labor Unit Cost	Avg Total Unit Cost	Avg Price Incl O&P
225 MBH output	Inst	EA	S-2	2.00	2450	---	164	2614	2690	S-2	1.60	2818	---	206	3023	3118
375 MBH output	Inst	EA	S-2	1.50	3600	---	219	3819	3920	S-2	1.20	4140	---	274	4414	4540
450 MBH output	Inst	EA	S-2	1.25	3800	---	263	4063	4184	S-2	1.00	4370	---	329	4699	4850

Furnaces, hot air heating with blowers and standard controls; gas, oil, or flue piping not included (see below)

Electric fired, UL listed, heat staging, 240 volt run and connection

Description	Oper	Unit	Crew Size	Avg Day Prod	Avg Mat'l Unit Cost	Avg Equip Unit Cost	Avg Labor Unit Cost	Avg Total Unit Cost	Avg Price Incl O&P	Crew Size	Avg Day Prod	Avg Mat'l Unit Cost	Avg Equip Unit Cost	Avg Labor Unit Cost	Avg Total Unit Cost	Avg Price Incl O&P
30 MBH	Inst	EA	U-5	3.60	325	---	116	441	496	U-5	2.88	374	---	145	519	587
75 MBH	Inst	EA	U-5	3.50	550	---	119	669	726	U-5	2.80	633	---	149	782	852
85 MBH	Inst	EA	U-5	3.20	610	---	131	741	802	U-5	2.56	702	---	163	865	942
90 MBH	Inst	EA	U-5	3.00	775	---	139	914	980	U-5	2.40	891	---	174	1066	1147
Minimum Job Charge	Inst	Job	U-4		---	---	160	160	235	U-4		---	---	200	200	294

Gas fired, AGA certified, direct drive models

Description	Oper	Unit	Crew Size	Avg Day Prod	Avg Mat'l Unit Cost	Avg Equip Unit Cost	Avg Labor Unit Cost	Avg Total Unit Cost	Avg Price Incl O&P	Crew Size	Avg Day Prod	Avg Mat'l Unit Cost	Avg Equip Unit Cost	Avg Labor Unit Cost	Avg Total Unit Cost	Avg Price Incl O&P
40 MBH	Inst	EA	U-4	3.60	425	---	89	514	556	U-4	2.88	489	---	111	600	652
65 MBH	Inst	EA	U-4	3.50	450	---	92	542	585	U-4	2.80	518	---	114	632	686
80 MBH	Inst	EA	U-4	3.40	500	---	94	594	638	U-4	2.72	575	---	118	693	748
85 MBH	Inst	EA	U-4	3.20	550	---	100	650	697	U-4	2.56	633	---	125	758	816
105 MBH	Inst	EA	U-4	3.00	600	---	107	707	757	U-4	2.40	690	---	133	823	886
125 MBH	Inst	EA	U-4	2.80	725	---	114	839	893	U-4	2.24	834	---	143	977	1044

			Costs Based On Large Volume							Costs Based On Small Volume						
Description	Oper	Unit	Crew Size	Avg Day Prod	Avg Mat'l Unit Cost	Avg Equip Unit Cost	Avg Labor Unit Cost	Avg Total Unit Cost	Avg Price Incl O&P	Crew Size	Avg Day Prod	Avg Mat'l Unit Cost	Avg Equip Unit Cost	Avg Labor Unit Cost	Avg Total Unit Cost	Avg Price Incl O&P

HVAC (continued)

Description	Oper	Unit	Crew Size	Avg Day Prod	Avg Mat'l Unit Cost	Avg Equip Unit Cost	Avg Labor Unit Cost	Avg Total Unit Cost	Avg Price Incl O&P	Crew Size	Avg Day Prod	Avg Mat'l Unit Cost	Avg Equip Unit Cost	Avg Labor Unit Cost	Avg Total Unit Cost	Avg Price Incl O&P
160 MBH	Inst	EA	U-4	2.60	1150	---	123	1273	1331	U-4	2.08	1323	---	154	1476	1549
200 MBH	Inst	EA	U-4	2.40	2100	---	133	2233	2296	U-4	1.92	2415	---	167	2582	2660
Minimum Job Charge	Inst	Job	U-4		---	---	160	160	235	U-4		---	---	200	200	294
Gas line with couplings	Inst	LF	S-2	45.00	6.00	---	7.31	13.31	16.67	S-2	36.00	6.90	---	9.14	16.04	20.24
Oil fired, UL listed, gun type burner																
55 MBH	Inst	EA	U-4	3.50	750	---	92	842	885	U-4	2.80	863	---	114	977	1031
100 MBH	Inst	EA	U-4	3.00	830	---	107	937	987	U-4	2.40	955	---	133	1088	1151
125 MBH	Inst	EA	U-4	2.80	940	---	114	1054	1108	U-4	2.24	1081	---	143	1224	1291
150 MBH	Inst	EA	U-4	2.60	1200	---	123	1323	1381	U-4	2.08	1380	---	154	1534	1606
200 MBH	Inst	EA	U-4	2.40	1850	---	133	1983	2046	U-4	1.92	2128	---	167	2294	2373
Minimum Job Charge	Inst	Job	U-4		---	---	160	160	235	U-4		---	---	200	200	294
Oil line with couplings	Inst	LF	S-2	80.00	3.00	---	4.11	7.11	9.00	S-2	64.00	3.45	---	5.14	8.59	10.95
Combo fired (wood, coal, oil combination) complete with burner																
115 MBH (based on oil)	Inst	EA	U-4	2.90	3450	---	110	3560	3612	U-4	2.32	3968	---	138	4106	4170
140 MBH (based on oil)	Inst	EA	U-4	2.70	3500	---	119	3619	3674	U-4	2.16	4025	---	148	4173	4243
150 MBH (based on oil)	Inst	EA	U-4	2.60	3550	---	123	3673	3731	U-4	2.08	4083	---	154	4236	4309
170 MBH (based on oil)	Inst	EA	U-4	2.50	3600	---	128	3728	3788	U-4	2.00	4140	---	160	4300	4375
Minimum Job Charge	Inst	Job	U-4		---	---	160	160	235	U-4		---	---	200	200	294
Oil line with couplings	Inst	LF	S-2	80.00	3.00	---	4.11	7.11	9.00	S-2	64.00	3.45	---	5.14	8.59	10.95
Space heaters																
Gas Fired																
Unit includes cabinet, grilles, fan, controls, burner and thermostat; no flue piping included (see below)																
Floor mounted																
60 MBH	Inst	EA	S-2	8.00	500	---	41	541	560	S-2	6.40	575	---	51	626	650
180 MBH	Inst	EA	S-2	4.00	700	---	82	782	820	S-2	3.20	805	---	103	908	955
Suspension mounted, propeller fan																
36 MBH	Inst	EA	S-2	6.00	400	---	55	455	480	S-2	4.80	460	---	69	529	560
60 MBH	Inst	EA	S-2	5.00	500	---	66	566	596	S-2	4.00	575	---	82	657	695
120 MBH	Inst	EA	S-2	4.00	650	---	82	732	770	S-2	3.20	748	---	103	850	898
320 MBH	Inst	EA	S-2	2.00	1350	---	164	1514	1590	S-2	1.60	1553	---	206	1758	1853
Powered venter, adapter	ADD	EA	S-2	8.00	225	---	41	266	285	S-2	6.40	259	---	51	310	334

			Costs Based On Large Volume							Costs Based On Small Volume						
Description	Oper	Unit	Crew Size	Avg Day Prod	Avg Mat'l Unit Cost	Avg Equip Unit Cost	Avg Labor Unit Cost	Avg Total Unit Cost	Avg Price Incl O&P	Crew Size	Avg Day Prod	Avg Mat'l Unit Cost	Avg Equip Unit Cost	Avg Labor Unit Cost	Avg Total Unit Cost	Avg Price Incl O&P

HVAC (continued)

Wall furnace, self-contained thermostat

Single capacity, recessed or surface mounted, 1-speed fan

Description	Oper	Unit	Crew Size	Avg Day Prod	Avg Mat'l Unit Cost	Avg Equip Unit Cost	Avg Labor Unit Cost	Avg Total Unit Cost	Avg Price Incl O&P	Crew Size	Avg Day Prod	Avg Mat'l Unit Cost	Avg Equip Unit Cost	Avg Labor Unit Cost	Avg Total Unit Cost	Avg Price Incl O&P
15 MBH	Inst	EA	S-2	5.00	300	---	66	366	396	S-2	4.00	345	---	82	427	465
25 MBH	Inst	EA	S-2	4.00	325	---	82	407	445	S-2	3.20	374	---	103	477	524
35 MBH	Inst	EA	S-2	3.00	400	---	110	510	560	S-2	2.40	460	---	137	597	660

Dual capacity, recessed or surface mounted, 2 speed blowers

Description	Oper	Unit	Crew Size	Avg Day Prod	Avg Mat'l Unit Cost	Avg Equip Unit Cost	Avg Labor Unit Cost	Avg Total Unit Cost	Avg Price Incl O&P	Crew Size	Avg Day Prod	Avg Mat'l Unit Cost	Avg Equip Unit Cost	Avg Labor Unit Cost	Avg Total Unit Cost	Avg Price Incl O&P
50 MBH (direct vent)	Inst	EA	S-2	2.00	500	---	164	664	740	S-2	1.60	575	---	206	781	875
60 MBH (up vent)	Inst	EA	S-2	2.00	450	---	164	614	690	S-2	1.60	518	---	206	723	818

Register kit for circulating

Description	Oper	Unit	Crew Size	Avg Day Prod	Avg Mat'l Unit Cost	Avg Equip Unit Cost	Avg Labor Unit Cost	Avg Total Unit Cost	Avg Price Incl O&P	Crew Size	Avg Day Prod	Avg Mat'l Unit Cost	Avg Equip Unit Cost	Avg Labor Unit Cost	Avg Total Unit Cost	Avg Price Incl O&P
heat to second room	Inst	EA	S-2	8.00	35		41	76	95	S-2	6.40	40	---	51	92	115
Minimum Job Charge	Inst	Job	S-2		---	---	164	164	240	S-2		---	---	206	206	300

Bathroom heaters

Electric Fired

Ceiling Heat-A-Ventlite, includes grille, blower, 4" round duct

13" x 7" dia., 3-way switch

Description	Oper	Unit	Crew Size	Avg Day Prod	Avg Mat'l Unit Cost	Avg Equip Unit Cost	Avg Labor Unit Cost	Avg Total Unit Cost	Avg Price Incl O&P	Crew Size	Avg Day Prod	Avg Mat'l Unit Cost	Avg Equip Unit Cost	Avg Labor Unit Cost	Avg Total Unit Cost	Avg Price Incl O&P
1500 watt	Inst	EA	E-1	2.00	200	---	98	298	342	E-1	1.60	230	---	122	352	408
1800 watt	Inst	EA	E-1	2.00	225	---	98	323	367	E-1	1.60	259	---	122	381	436

Ceiling Heat-A-Lite, includes grille, airotor wheel, 13" x 7"

Description	Oper	Unit	Crew Size	Avg Day Prod	Avg Mat'l Unit Cost	Avg Equip Unit Cost	Avg Labor Unit Cost	Avg Total Unit Cost	Avg Price Incl O&P	Crew Size	Avg Day Prod	Avg Mat'l Unit Cost	Avg Equip Unit Cost	Avg Labor Unit Cost	Avg Total Unit Cost	Avg Price Incl O&P
dia., 2 way switch, 1500 watt	Inst	EA	E-1	2.00	150	---	98	248	292	E-1	1.60	173	---	122	295	350

Ceiling radiant heating using infrared lamps

Recessed, Heat-A-Lamp

Description	Oper	Unit	Crew Size	Avg Day Prod	Avg Mat'l Unit Cost	Avg Equip Unit Cost	Avg Labor Unit Cost	Avg Total Unit Cost	Avg Price Incl O&P	Crew Size	Avg Day Prod	Avg Mat'l Unit Cost	Avg Equip Unit Cost	Avg Labor Unit Cost	Avg Total Unit Cost	Avg Price Incl O&P
One bulb, 250 watt lamp	Inst	EA	E-1	2.40	35	---	82	117	153	E-1	1.92	40	---	102	142	188
Two bulb, 500 watt lamp	Inst	EA	E-1	2.40	60	---	82	142	178	E-1	1.92	69	---	102	171	217
Three bulb, 750 watt lamp	Inst	EA	E-1	2.40	110	---	82	192	228	E-1	1.92	127	---	102	229	274

Recessed, Heat-A-Vent

Description	Oper	Unit	Crew Size	Avg Day Prod	Avg Mat'l Unit Cost	Avg Equip Unit Cost	Avg Labor Unit Cost	Avg Total Unit Cost	Avg Price Incl O&P	Crew Size	Avg Day Prod	Avg Mat'l Unit Cost	Avg Equip Unit Cost	Avg Labor Unit Cost	Avg Total Unit Cost	Avg Price Incl O&P
One bulb, 250 watt lamp	Inst	EA	E-1	2.40	75	---	82	157	193	E-1	1.92	86	---	102	188	234
Two bulb, 500 watt lamp	Inst	EA	E-1	2.40	90	---	82	172	208	E-1	1.92	104	---	102	206	251

Wall Heaters, recessed

Fan forced

Description	Oper	Unit	Crew Size	Avg Day Prod	Avg Mat'l Unit Cost	Avg Equip Unit Cost	Avg Labor Unit Cost	Avg Total Unit Cost	Avg Price Incl O&P	Crew Size	Avg Day Prod	Avg Mat'l Unit Cost	Avg Equip Unit Cost	Avg Labor Unit Cost	Avg Total Unit Cost	Avg Price Incl O&P
1250 watt heating element	Inst	EA	E-1	3.20	100	---	61	161	189	E-1	2.56	115	---	77	192	226

Radiant heating

Description	Oper	Unit	Crew Size	Avg Day Prod	Avg Mat'l Unit Cost	Avg Equip Unit Cost	Avg Labor Unit Cost	Avg Total Unit Cost	Avg Price Incl O&P	Crew Size	Avg Day Prod	Avg Mat'l Unit Cost	Avg Equip Unit Cost	Avg Labor Unit Cost	Avg Total Unit Cost	Avg Price Incl O&P
1200 watt heating element	Inst	EA	E-1	3.20	80	---	61	141	169	E-1	2.56	92	---	77	169	203
1500 watt heating element	Inst	EA	E-1	3.20	85	---	61	146	174	E-1	2.56	98	---	77	174	209

			Costs Based On Large Volume							Costs Based On Small Volume						
Description	Oper	Unit	Crew Size	Avg Day Prod	Avg Mat'l Unit Cost	Avg Equip Unit Cost	Avg Labor Unit Cost	Avg Total Unit Cost	Avg Price Incl O&P	Crew Size	Avg Day Prod	Avg Mat'l Unit Cost	Avg Equip Unit Cost	Avg Labor Unit Cost	Avg Total Unit Cost	Avg Price Incl O&P

HVAC (continued)

Wiring, connection, and installation in closed wall or ceiling structure

Description	Oper	Unit	Crew Size	Avg Day Prod	Avg Mat'l Unit Cost	Avg Equip Unit Cost	Avg Labor Unit Cost	Avg Total Unit Cost	Avg Price Incl O&P	Crew Size	Avg Day Prod	Avg Mat'l Unit Cost	Avg Equip Unit Cost	Avg Labor Unit Cost	Avg Total Unit Cost	Avg Price Incl O&P
ADD	Inst	EA	E-1		--	---	---	59	59	86	E-1			---	---	74	74	108

Flue piping or vent chimney
Prefabricated metal, UL listed
Gas, double wall, galv. steel

Description	Oper	Unit	Crew Size	Avg Day Prod	Avg Mat'l Unit Cost	Avg Equip Unit Cost	Avg Labor Unit Cost	Avg Total Unit Cost	Avg Price Incl O&P	Crew Size	Avg Day Prod	Avg Mat'l Unit Cost	Avg Equip Unit Cost	Avg Labor Unit Cost	Avg Total Unit Cost	Avg Price Incl O&P
3" diameter	Inst	VLF	U-2	70	2.90	---	5.43	8.33	10.89	U-2	56	3.33	---	6.79	10.13	13.32
4" diameter	Inst	VLF	U-2	66	3.50	---	5.76	9.26	11.97	U-2	53	4.03	---	7.20	11.23	14.62
5" diameter	Inst	VLF	U-2	62	4.00	---	6.14	10.14	13.02	U-2	50	4.60	---	7.67	12.27	15.87
6" diameter	Inst	VLF	U-2	58	4.50	---	6.56	11.06	14.14	U-2	46	5.18	---	8.20	13.37	17.23
7" diameter	Inst	VLF	U-2	54	6.25	---	7.04	13.29	16.61	U-2	43	7.19	---	8.81	15.99	20.13
8" diameter	Inst	VLF	U-2	50	7.00	---	7.61	14.61	18.18	U-2	40	8.05	---	9.51	17.56	22.03
10" diameter	Inst	VLF	U-2	46	14.50	---	8.27	22.77	26.66	U-2	37	16.68	---	10.34	27.01	31.87
12" diameter	Inst	VLF	U-2	42	20.00	---	9.06	29.06	33.31	U-2	34	23.00	---	11.32	34.32	39.64
16" diameter	Inst	VLF	U-2	38	42.00	---	10.01	52.01	56.72	U-2	30	48.30	---	12.51	60.81	66.70
20" diameter	Inst	VLF	U-2	34	65.00	---	11.19	76.19	81.45	U-2	27	74.75	---	13.99	88.74	95.31
24" diameter	Inst	VLF	U-2	30	100.00	---	12.68	112.68	118.64	U-2	24	115.00	---	15.85	130.85	138.30
Vent damper, bi-metal, 6" flue																
Gas, auto., electric	Inst	EA	U-2	6	125.00	---	63.40	188.40	218.20	U-2	5	143.75	---	79.25	223.00	260.25
Oil, auto., electric	Inst	EA	U-2	6	150.00	---	63.40	213.40	243.20	U-2	5	172.50	---	79.25	251.75	289.00
Any fuel, double wall, stainless steel																
6" diameter	Inst	VLF	U-2	60	17.75	---	6.34	24.09	27.07	U-2	48	20.41	---	7.93	28.34	32.06
7" diameter	Inst	VLF	U-2	56	22.50	---	6.79	29.29	32.49	U-2	45	25.88	---	8.49	34.37	38.36
8" diameter	Inst	VLF	U-2	52	26.50	---	7.32	33.82	37.25	U-2	42	30.48	---	9.14	39.62	43.92
10" diameter	Inst	VLF	U-2	48	38.00	---	7.93	45.93	49.65	U-2	38	43.70	---	9.91	53.61	58.26
12" diameter	Inst	VLF	U-2	44	51.00	---	8.65	59.65	63.71	U-2	35	58.65	---	10.81	69.46	74.54
14" diameter	Inst	VLF	U-2	40	66.00	---	9.51	75.51	79.98	U-2	32	75.90	---	11.89	87.79	93.38
Any fuel, double wall, stainless steel fittings																
Roof support																
6" diameter	Inst	EA	U-2	30	45.00	---	12.68	57.68	63.64	U-2	24	51.75	---	15.85	67.60	75.05
7" diameter	Inst	EA	U-2	28	51.00	---	13.59	64.59	70.97	U-2	22	58.65	---	16.98	75.63	83.62
8" diameter	Inst	EA	U-2	26	54.00	---	14.63	68.63	75.51	U-2	21	62.10	---	18.29	80.39	88.99
10" diameter	Inst	EA	U-2	24	71.00	---	15.85	86.85	94.30	U-2	19	81.65	---	19.81	101.46	110.78
12" diameter	Inst	EA	U-2	22	85.00	---	17.29	102.29	110.42	U-2	18	97.75	---	21.61	119.36	129.52
14" diameter	Inst	EA	U-2	20	109.00	---	19.02	128.02	136.96	U-2	16	125.35	---	23.78	149.13	160.30

			Costs Based On Large Volume							Costs Based On Small Volume					
Description	Oper	Unit	Crew Size	Avg Day Prod	Avg Mat'l Unit Cost	Avg Equip Unit Cost	Avg Labor Unit Cost	Avg Total Unit Cost	Avg Price Incl O&P	Crew Size	Avg Day Prod	Avg Mat'l Unit Cost	Avg Equip Unit Cost	Avg Labor Unit Cost	Avg Total Unit Cost	Avg Price Incl O&P

HVAC (continued)

Elbow 15 degree

Description	Oper	Unit	Crew Size	Day Prod	Mat'l	Equip	Labor	Total	Price O&P	Crew Size	Day Prod	Mat'l	Equip	Labor	Total	Price O&P
6" diameter	Inst	EA	U-2	30	39.00	---	12.68	51.68	57.64	U-2	24	44.85	---	15.85	60.70	68.15
7" diameter	Inst	EA	U-2	28	43.00	---	13.59	56.59	62.97	U-2	22	49.45	---	16.98	66.43	74.42
8" diameter	Inst	EA	U-2	26	49.75	---	14.63	64.38	71.26	U-2	21	57.21	---	18.29	75.50	84.10
10" diameter	Inst	EA	U-2	24	65.00	---	15.85	80.85	88.30	U-2	19	74.75	---	19.81	94.56	103.88
12" diameter	Inst	EA	U-2	22	78.00	---	17.29	95.29	103.42	U-2	18	89.70	---	21.61	111.31	121.47
14" diameter	Inst	EA	U-2	20	92.00	---	19.02	111.02	119.96	U-2	16	105.80	---	23.78	129.58	140.75

Insulated tee with insulated tee cap

Description	Oper	Unit	Crew Size	Day Prod	Mat'l	Equip	Labor	Total	Price O&P	Crew Size	Day Prod	Mat'l	Equip	Labor	Total	Price O&P
6" diameter	Inst	EA	U-2	30	73.00	---	12.68	85.68	91.64	U-2	24	83.95	---	15.85	99.80	107.25
7" diameter	Inst	EA	U-2	28	95.00	---	13.59	108.59	114.97	U-2	22	109.25	---	16.98	126.23	134.22
8" diameter	Inst	EA	U-2	26	107.00	---	14.63	121.63	128.51	U-2	21	123.05	---	18.29	141.34	149.94
10" diameter	Inst	EA	U-2	24	150.00	---	15.85	165.85	173.30	U-2	19	172.50	---	19.81	192.31	201.63
12" diameter	Inst	EA	U-2	22	210.00	---	17.29	227.29	235.42	U-2	18	241.50	---	21.61	263.11	273.27
14" diameter	Inst	EA	U-2	20	275.00	---	19.02	294.02	302.96	U-2	16	316.25	---	23.78	340.03	351.20

Joist shield

Description	Oper	Unit	Crew Size	Day Prod	Mat'l	Equip	Labor	Total	Price O&P	Crew Size	Day Prod	Mat'l	Equip	Labor	Total	Price O&P
6" diameter	Inst	EA	U-2	30	28.50	---	12.68	41.18	47.14	U-2	24	32.78	---	15.85	48.63	56.08
7" diameter	Inst	EA	U-2	28	31.00	---	13.59	44.59	50.97	U-2	22	35.65	---	16.98	52.63	60.62
8" diameter	Inst	EA	U-2	26	32.00	---	14.63	46.63	53.51	U-2	21	36.80	---	18.29	55.09	63.69
10" diameter	Inst	EA	U-2	24	40.00	---	15.85	55.85	63.30	U-2	19	46.00	---	19.81	65.81	75.13
12" diameter	Inst	EA	U-2	22	51.00	---	17.29	68.29	76.42	U-2	18	58.65	---	21.61	80.26	90.42
14" diameter	Inst	EA	U-2	20	63.00	---	19.02	82.02	90.96	U-2	16	72.45	---	23.78	96.23	107.40

Round top

Description	Oper	Unit	Crew Size	Day Prod	Mat'l	Equip	Labor	Total	Price O&P	Crew Size	Day Prod	Mat'l	Equip	Labor	Total	Price O&P
6" diameter	Inst	EA	U-2	30	26.00	---	12.68	38.68	44.64	U-2	24	29.90	---	15.85	45.75	53.20
7" diameter	Inst	EA	U-2	28	35.00	---	13.59	48.59	54.97	U-2	22	40.25	---	16.98	57.23	65.22
8" diameter	Inst	EA	U-2	26	45.00	---	14.63	59.63	66.51	U-2	21	51.75	---	18.29	70.04	78.64
10" diameter	Inst	EA	U-2	24	80.00	---	15.85	95.85	103.30	U-2	19	92.00	---	19.81	111.81	121.13
12" diameter	Inst	EA	U-2	22	115.00	---	17.29	132.29	140.42	U-2	18	132.25	---	21.61	153.86	164.02
14" diameter	Inst	EA	U-2	20	155.00	---	19.02	174.02	182.96	U-2	16	178.25	---	23.78	202.03	213.20

Adjustable roof flashing

Description	Oper	Unit	Crew Size	Day Prod	Mat'l	Equip	Labor	Total	Price O&P	Crew Size	Day Prod	Mat'l	Equip	Labor	Total	Price O&P
6" diameter	Inst	EA	U-2	30	30.00	---	12.68	42.68	48.64	U-2	24	34.50	---	15.85	50.35	57.80
7" diameter	Inst	EA	U-2	28	33.00	---	13.59	46.59	52.97	U-2	22	37.95	---	16.98	54.93	62.92
8" diameter	Inst	EA	U-2	26	36.00	---	14.63	50.63	57.51	U-2	21	41.40	---	18.29	59.69	68.29

			Costs Based On Large Volume							Costs Based On Small Volume						
Description	Oper	Unit	Crew Size	Avg Day Prod	Avg Mat'l Unit Cost	Avg Equip Unit Cost	Avg Labor Unit Cost	Avg Total Unit Cost	Avg Price Incl O&P	Crew Size	Avg Day Prod	Avg Mat'l Unit Cost	Avg Equip Unit Cost	Avg Labor Unit Cost	Avg Total Unit Cost	Avg Price Incl O&P

HVAC (continued)

10" diameter	Inst	EA	U-2	24	47.00	---	15.85	62.85	70.30	U-2	19	54.05	---	19.81	73.86	83.18
12" diameter	Inst	EA	U-2	22	60.00	---	17.29	77.29	85.42	U-2	18	69.00	---	21.61	90.61	100.77
14" diameter	Inst	EA	U-2	20	75.00	---	19.02	94.02	102.96	U-2	16	86.25	---	23.78	110.03	121.20
Minimum Job Charge	Inst	Job	U-2		---	---	126.81	126.81	186.41	U-2		---	---	158.51	158.51	233.01

Air conditioning and ventilating

System componets

Condensing units

Air cooled, compressor, standard controls

1.0 ton	Inst	EA	S-2	2.00	500.00	---	164	664	740	S-2	1.60	575.00	---	206	781	875
1.5 ton	Inst	EA	S-2	1.75	600.00	---	188	788	874	S-2	1.40	690.00	---	235	925	1033
2.0 ton	Inst	EA	S-2	1.50	700.00	---	219	919	1020	S-2	1.20	805.00	---	274	1079	1205
2.5 ton	Inst	EA	S-2	1.25	850.00	---	263	1113	1234	S-2	1.00	977.50	---	329	1306	1458
3.0 ton	Inst	EA	S-2	1.00	900.00	---	329	1229	1380	S-2	0.80	1035.00	---	411	1446	1635
4.0 ton	Inst	EA	S-2	0.75	1300.00	---	439	1739	1940	S-2	0.60	1495.00	---	548	2043	2296
5.0 ton	Inst	EA	S-2	0.50	1600.00	---	658	2258	2561	S-2	0.40	1840.00	---	822	2662	3041
Minimum Job Charge	Inst	Job	S-2		---	---	110	110	160	S-2		---	---	137	137	200

Diffusers

Aluminum, opposed blade damper unless noted otherwise

Ceiling or sidewall, linear

2" wide	Inst	LF	U-1	30	19.00	---	6.34	25.34	28.32	U-1	24	21.85	---	7.93	29.78	33.50
4" wide	Inst	LF	U-1	27	25.00	---	7.04	32.04	35.36	U-1	22	28.75	---	8.81	37.56	41.69
6" wide	Inst	LF	U-1	24	31.00	---	7.93	38.93	42.65	U-1	19	35.65	---	9.91	45.56	50.21
8" wide	Inst	LF	U-1	20	36.00	---	9.51	45.51	49.98	U-1	16	41.40	---	11.89	53.29	58.88

Perforated, 24" x 24" panel size

6" x 6"	Inst	EA	U-1	16	52.00	---	11.89	63.89	69.48	U-1	13	59.80	---	14.86	74.66	81.64
8" x 8"	Inst	EA	U-1	14	54.00	---	13.59	67.59	73.97	U-1	11	62.10	---	16.98	79.08	87.07
10" x 10"	Inst	EA	U-1	12	55.00	---	15.85	70.85	78.30	U-1	10	63.25	---	19.81	83.06	92.38
12" x 12"	Inst	EA	U-1	10	57.00	---	19.02	76.02	84.96	U-1	8	65.55	---	23.78	89.33	100.50
18" x 18"	Inst	EA	U-1	8	72.00	---	23.78	95.78	106.95	U-1	6	82.80	---	29.72	112.52	126.49

			Costs Based On Large Volume							Costs Based On Small Volume					
Description	Oper	Unit	Crew Size	Avg Day Prod	Avg Mat'l Unit Cost	Avg Equip Unit Cost	Avg Labor Unit Cost	Avg Total Unit Cost	Avg Price Incl O&P	Crew Size	Avg Day Prod	Avg Mat'l Unit Cost	Avg Equip Unit Cost	Avg Labor Unit Cost	Avg Total Unit Cost	Avg Price Incl O&P

HVAC (continued)

Rectangular, 1 to 4 way blow

Description	Oper	Unit	Crew Size	Avg Day Prod	Avg Mat'l Unit Cost	Avg Equip Unit Cost	Avg Labor Unit Cost	Avg Total Unit Cost	Avg Price Incl O&P	Crew Size	Avg Day Prod	Avg Mat'l Unit Cost	Avg Equip Unit Cost	Avg Labor Unit Cost	Avg Total Unit Cost	Avg Price Incl O&P
6" x 6"	Inst	EA	U-1	16	31.00	---	11.89	42.89	48.48	U-1	13	35.65	---	14.86	50.51	57.49
12" x 6"	Inst	EA	U-1	14	40.00	---	13.59	53.59	59.97	U-1	11	46.00	---	16.98	62.98	70.97
12" x 9"	Inst	EA	U-1	12	47.00	---	15.85	62.85	70.30	U-1	10	54.05	---	19.81	73.86	83.18
12" x 12"	Inst	EA	U-1	10	55.00	---	19.02	74.02	82.96	U-1	8	63.25	---	23.78	87.03	98.20
24" x 12"	Inst	EA	U-1	8	91.00	---	23.78	114.78	125.95	U-1	6	104.65	---	29.72	134.37	148.34

1 bar mounting, 24" x 24" lay in frame

Description	Oper	Unit	Crew Size	Avg Day Prod	Avg Mat'l Unit Cost	Avg Equip Unit Cost	Avg Labor Unit Cost	Avg Total Unit Cost	Avg Price Incl O&P	Crew Size	Avg Day Prod	Avg Mat'l Unit Cost	Avg Equip Unit Cost	Avg Labor Unit Cost	Avg Total Unit Cost	Avg Price Incl O&P
6" x 6"	Inst	EA	U-1	16	50.00	---	11.89	61.89	67.48	U-1	13	57.50	---	14.86	72.36	79.34
9" x 9"	Inst	EA	U-1	14	58.00	---	13.59	71.59	77.97	U-1	11	66.70	---	16.98	83.68	91.67
12" x 12"	Inst	EA	U-1	12	74.00	---	15.85	89.85	97.30	U-1	10	85.10	---	19.81	104.91	114.23
15" x 15"	Inst	EA	U-1	10	95.00	---	19.02	114.02	122.96	U-1	8	109.25	---	23.78	133.03	144.20
18" x 18"	Inst	EA	U-1	8	97.00	---	23.78	120.78	131.95	U-1	6	111.55	---	29.72	141.27	155.24
Minimum Job Charge	Inst	Job	U-1		---	---	54.35	54.35	79.89	U-1		---	---	67.93	67.93	99.86

Ductwork

Fabricated rectangular, includes fittings, joints, supports

Aluminum alloy

Description	Oper	Unit	Crew Size	Avg Day Prod	Avg Mat'l Unit Cost	Avg Equip Unit Cost	Avg Labor Unit Cost	Avg Total Unit Cost	Avg Price Incl O&P	Crew Size	Avg Day Prod	Avg Mat'l Unit Cost	Avg Equip Unit Cost	Avg Labor Unit Cost	Avg Total Unit Cost	Avg Price Incl O&P
Under 300 lbs	Inst	lb	U-6	70	2.55	---	7.29	9.84	13.27	U-6	56	2.93	---	9.12	12.05	16.33
500 to 1000 lbs	Inst	lb	U-6	90	1.95	---	5.67	7.62	10.29	U-6	72	2.24	---	7.09	9.33	12.66
1000 to 2000 lbs	Inst	lb	U-6	110	1.85	---	4.64	6.49	8.67	U-6	88	2.13	---	5.80	7.93	10.65
Over 10,000 lbs	Inst	lb	U-6	130	1.70	---	3.93	5.63	7.47	U-6	104	1.96	---	4.91	6.86	9.17

Galvanized steel

Description	Oper	Unit	Crew Size	Avg Day Prod	Avg Mat'l Unit Cost	Avg Equip Unit Cost	Avg Labor Unit Cost	Avg Total Unit Cost	Avg Price Incl O&P	Crew Size	Avg Day Prod	Avg Mat'l Unit Cost	Avg Equip Unit Cost	Avg Labor Unit Cost	Avg Total Unit Cost	Avg Price Incl O&P
Under 400 lbs	Inst	lb	U-6	200	1.20	---	2.55	3.75	4.95	U-6	160	1.38	---	3.19	4.57	6.07
400 to 1000 lbs	Inst	lb	U-6	215	0.90	---	2.37	3.27	4.39	U-6	172	1.04	---	2.97	4.00	5.40
1000 to 2000 lbs	Inst	lb	U-6	230	0.70	---	2.22	2.92	3.96	U-6	184	0.81	---	2.77	3.58	4.88
2000 to 5000 lbs	Inst	lb	U-6	240	0.65	---	2.13	2.78	3.78	U-6	192	0.75	---	2.66	3.41	4.66
Over 10,000 lbs	Inst	lb	U-6	250	0.60	---	2.04	2.64	3.60	U-6	200	0.69	---	2.55	3.24	4.44

Flexible, coated fabric on spring steel, aluminum, or corrosion resistant metal.

Non-insulated

Description	Oper	Unit	Crew Size	Avg Day Prod	Avg Mat'l Unit Cost	Avg Equip Unit Cost	Avg Labor Unit Cost	Avg Total Unit Cost	Avg Price Incl O&P	Crew Size	Avg Day Prod	Avg Mat'l Unit Cost	Avg Equip Unit Cost	Avg Labor Unit Cost	Avg Total Unit Cost	Avg Price Incl O&P
3" diameter	Inst	LF	U-4	275	0.75	---	1.16	1.91	2.46	U-4	220	0.86	---	1.46	2.32	3.00
5" diameter	Inst	LF	U-4	225	1.05	---	1.42	2.47	3.14	U-4	180	1.21	---	1.78	2.99	3.82
6" diameter	Inst	LF	U-4	200	1.20	---	1.60	2.80	3.55	U-4	160	1.38	---	2.00	3.38	4.32
7" diameter	Inst	LF	U-4	175	1.35	---	1.83	3.18	4.04	U-4	140	1.55	---	2.29	3.84	4.92

Description	Oper	Unit	Costs Based On Large Volume							Costs Based On Small Volume						
			Crew Size	Avg Day Prod	Avg Mat'l Unit Cost	Avg Equip Unit Cost	Avg Labor Unit Cost	Avg Total Unit Cost	Avg Price Incl O&P	Crew Size	Avg Day Prod	Avg Mat'l Unit Cost	Avg Equip Unit Cost	Avg Labor Unit Cost	Avg Total Unit Cost	Avg Price Incl O&P

HVAC (continued)

Description	Oper	Unit	Crew Size	Avg Day Prod	Avg Mat'l Unit Cost	Avg Equip Unit Cost	Avg Labor Unit Cost	Avg Total Unit Cost	Avg Price Incl O&P	Crew Size	Avg Day Prod	Avg Mat'l Unit Cost	Avg Equip Unit Cost	Avg Labor Unit Cost	Avg Total Unit Cost	Avg Price Incl O&P
8" diameter	Inst	LF	U-4	150	1.50	---	2.14	3.64	4.64	U-4	120	1.73	---	2.67	4.39	5.65
10" diameter	Inst	LF	U-4	125	1.80	---	2.56	4.36	5.57	U-4	100	2.07	---	3.20	5.27	6.78
12" diameter	Inst	LF	U-4	100	2.00	---	3.20	5.20	6.71	U-4	80	2.30	---	4.00	6.30	8.18
Insulated																
3" diameter	Inst	LF	U-4	250	1.20	---	1.28	2.48	3.08	U-4	200	1.38	---	1.60	2.98	3.73
4" diameter	Inst	LF	U-4	225	1.25	---	1.42	2.67	3.34	U-4	180	1.44	---	1.78	3.22	4.05
5" diameter	Inst	LF	U-4	200	1.50	---	1.60	3.10	3.85	U-4	160	1.73	---	2.00	3.73	4.67
6" diameter	Inst	LF	U-4	175	1.70	---	1.83	3.53	4.39	U-4	140	1.96	---	2.29	4.24	5.32
7" diameter	Inst	LF	U-4	150	1.90	---	2.14	4.04	5.04	U-4	120	2.19	---	2.67	4.85	6.11
8" diameter	Inst	LF	U-4	125	2.20	---	2.56	4.76	5.97	U-4	100	2.53	---	3.20	5.73	7.24
10" diameter	Inst	LF	U-4	100	2.60	---	3.20	5.80	7.31	U-4	80	2.99	---	4.00	6.99	8.87
12" diameter	Inst	LF	U-4	75	3.00	---	4.27	7.27	9.28	U-4	60	3.45	---	5.34	8.79	11.30
14" diameter	Inst	LF	U-4	50	3.60	---	6.41	10.01	13.02	U-4	40	4.14	---	8.01	12.15	15.91
Duct accessories																
Air extractors																
12" x 4"	Inst	EA	U-1	24	8.25	---	7.93	16.18	19.90	U-1	19	9.49	---	9.91	19.39	24.05
8" x 6"	Inst	EA	U-1	20	8.00	---	9.51	17.51	21.98	U-1	16	9.20	---	11.89	21.09	26.68
20" x 8"	Inst	EA	U-1	16	16.00	---	11.89	27.89	33.48	U-1	13	18.40	---	14.86	33.26	40.24
18" x 10"	Inst	EA	U-1	14	20.00	---	13.59	33.59	39.97	U-1	11	23.00	---	16.98	39.98	47.97
24" x 12"	Inst	EA	U-1	10	25.00	---	19.02	44.02	52.96	U-1	8	28.75	---	23.78	52.53	63.70
Dampers, fire, curtain type, vertical																
8" x 4"	Inst	EA	U-1	24	16.00	---	7.93	23.93	27.65	U-1	19	18.40	---	9.91	28.31	32.96
12" x 4"	Inst	EA	U-1	22	18.00	---	8.65	26.65	30.71	U-1	18	20.70	---	10.81	31.51	36.59
16" x 14"	Inst	EA	U-1	12	26.00	---	15.85	41.85	49.30	U-1	10	29.90	---	19.81	49.71	59.03
24" x 20"	Inst	EA	U-1	8	33.00	---	23.78	56.78	67.95	U-1	6	37.95	---	29.72	67.67	81.64
Dampers, multi-blade, opposed blade																
12" x 12"	Inst	EA	U-1	16	10.00	---	11.89	21.89	27.48	U-1	13	11.50	---	14.86	26.36	33.34
12" x 18"	Inst	EA	U-1	12	14.00	---	15.85	29.85	37.30	U-1	10	16.10	---	19.81	35.91	45.23
24" x 24"	Inst	EA	U-1	8	30.00	---	23.78	53.78	64.95	U-1	6	34.50	---	29.72	64.22	78.19
48" x 36"	Inst	EA	U-4	6	90.00	---	53.38	143.38	168.46	U-4	5	103.50	---	66.72	170.22	201.58

			Costs Based On Large Volume							Costs Based On Small Volume						
Description	Oper	Unit	Crew Size	Avg Day Prod	Avg Mat'l Unit Cost	Avg Equip Unit Cost	Avg Labor Unit Cost	Avg Total Unit Cost	Avg Price Incl O&P	Crew Size	Avg Day Prod	Avg Mat'l Unit Cost	Avg Equip Unit Cost	Avg Labor Unit Cost	Avg Total Unit Cost	Avg Price Incl O&P

HVAC (continued)

Dampers, variable volume modulating, motorized

Description	Oper	Unit	Crew Size	Avg Day Prod	Avg Mat'l Unit Cost	Avg Equip Unit Cost	Avg Labor Unit Cost	Avg Total Unit Cost	Avg Price Incl O&P	Crew Size	Avg Day Prod	Avg Mat'l Unit Cost	Avg Equip Unit Cost	Avg Labor Unit Cost	Avg Total Unit Cost	Avg Price Incl O&P
12" x 12"	Inst	EA	U-1	10	110	---	32	142	157	U-1	8	127	---	40	167	185
24" x 12"	Inst	EA	U-1	6	130	---	53	183	208	U-1	5	150	---	67	216	248
30" x 18"	Inst	EA	U-1	4	150	---	80	230	268	U-1	3	173	---	100	273	320
Thermostat, ADD	Inst	EA	U-1	6	50	---	53	103	128	U-1	5	58	---	67	124	156

Dampers, volume control

Description	Oper	Unit	Crew Size	Avg Day Prod	Avg Mat'l Unit Cost	Avg Equip Unit Cost	Avg Labor Unit Cost	Avg Total Unit Cost	Avg Price Incl O&P	Crew Size	Avg Day Prod	Avg Mat'l Unit Cost	Avg Equip Unit Cost	Avg Labor Unit Cost	Avg Total Unit Cost	Avg Price Incl O&P
8" x 8"	Inst	EA	U-1	20	18.00	---	16.01	34.01	41.54	U-1	16	20.70	---	20.02	40.72	50.12
16" x 10"	Inst	EA	U-1	16	25.00	---	20.02	45.02	54.42	U-1	13	28.75	---	25.02	53.77	65.53
18" x 12"	Inst	EA	U-1	12	30.00	---	26.69	56.69	69.23	U-1	10	34.50	---	33.36	67.86	83.54
28" x 16"	Inst	EA	U-1	8	55.00	---	40.03	95.03	113.85	U-1	6	63.25	---	50.04	113.29	136.81

Mixing box, with electric or pneumatic motor, with silencer

Description	Oper	Unit	Crew Size	Avg Day Prod	Avg Mat'l Unit Cost	Avg Equip Unit Cost	Avg Labor Unit Cost	Avg Total Unit Cost	Avg Price Incl O&P	Crew Size	Avg Day Prod	Avg Mat'l Unit Cost	Avg Equip Unit Cost	Avg Labor Unit Cost	Avg Total Unit Cost	Avg Price Incl O&P
150 to 250 CFM	Inst	EA	U-4	10	475	---	32	507	522	U-4	8	546	---	40	586	605
270 to 600 CFM	Inst	EA	U-4	8	540	---	40	580	599	U-4	6	621	---	50	671	695

Fans

Air Conditioning or processed air handling
Axial flow, compact, low sound
2.5" self-propelled

Description	Oper	Unit	Crew Size	Avg Day Prod	Avg Mat'l Unit Cost	Avg Equip Unit Cost	Avg Labor Unit Cost	Avg Total Unit Cost	Avg Price Incl O&P	Crew Size	Avg Day Prod	Avg Mat'l Unit Cost	Avg Equip Unit Cost	Avg Labor Unit Cost	Avg Total Unit Cost	Avg Price Incl O&P
3800 CFM, 5 HP	Inst	EA	U-5	3.0	2750	---	139	2889	2955	U-5	2.4	3163	---	174	3337	3419
15,600 CFM, 10 HP	Inst	EA	U-5	1.5	4800	---	279	5079	5210	U-5	1.2	5520	---	349	5869	6032

In-line centrifugal, supply/exhaust booster, aluminum wheel/hub, disconnect switch, 1/4"
self-propelled

Description	Oper	Unit	Crew Size	Avg Day Prod	Avg Mat'l Unit Cost	Avg Equip Unit Cost	Avg Labor Unit Cost	Avg Total Unit Cost	Avg Price Incl O&P	Crew Size	Avg Day Prod	Avg Mat'l Unit Cost	Avg Equip Unit Cost	Avg Labor Unit Cost	Avg Total Unit Cost	Avg Price Incl O&P
500 CFM, 10" dia. connect	Inst	EA	U-5	3.0	475	---	139	614	680	U-5	2.4	546	---	174	721	802
1520 CFM, 16" dia. connect	Inst	EA	U-5	2.0	800	---	209	1009	1107	U-5	1.6	920	---	261	1181	1304
3480 CFM, 20" dia. connect	Inst	EA	U-5	1.0	1200	---	418	1618	1815	U-5	0.8	1380	---	523	1903	2148

Ceiling fan, right angle, extra quiet 0.10" self-propelled

Description	Oper	Unit	Crew Size	Avg Day Prod	Avg Mat'l Unit Cost	Avg Equip Unit Cost	Avg Labor Unit Cost	Avg Total Unit Cost	Avg Price Incl O&P	Crew Size	Avg Day Prod	Avg Mat'l Unit Cost	Avg Equip Unit Cost	Avg Labor Unit Cost	Avg Total Unit Cost	Avg Price Incl O&P
200 CFM	Inst	EA	U-5	16	175	---	26	201	213	U-5	13	201	---	33	234	249
900 CFM	Inst	EA	U-5	12	400	---	35	435	451	U-5	10	460	---	44	504	524
3000 CFM	Inst	EA	U-5	8	700	---	52	752	777	U-5	6	805	---	65	870	901
Exterior wall or roof cap	Inst	EA	U-1	12	95	---	27	122	134	U-1	10	109	---	33	143	158

			Costs Based On Large Volume							Costs Based On Small Volume					
Description	Oper	Unit	Crew Size	Avg Day Prod	Avg Mat'l Unit Cost	Avg Equip Unit Cost	Avg Labor Unit Cost	Avg Total Unit Cost	Avg Price Incl O&P	Crew Size	Avg Day Prod	Avg Mat'l Unit Cost	Avg Equip Unit Cost	Avg Labor Unit Cost	Avg Total Unit Cost	Avg Price Incl O&P

HVAC (continued)

Corrosive fume resistant, plastic blades roof ventilators,
  Roof ventilator, centrifugal V belt drive motor, 1/4" self-propelled

Description	Oper	Unit	Crew Size	Avg Day Prod	Avg Mat'l Unit Cost	Avg Equip	Avg Labor Unit Cost	Avg Total Unit Cost	Avg Price Incl O&P	Crew Size	Avg Day Prod	Avg Mat'l Unit Cost	Avg Equip	Avg Labor Unit Cost	Avg Total Unit Cost	Avg Price Incl O&P
250 CFM, 1/4 HP	Inst	EA	U-5	4.0	1750.00	---	105	1855	1904	U-5	3.2	2012.50	---	131	2143	2205
900 CFM, 1/3 HP	Inst	EA	U-5	3.0	1900.00	---	139	2039	2105	U-5	2.4	2185.00	---	174	2359	2441
1650 CFM, 1/2 HP	Inst	EA	U-5	2.5	2250.00	---	167	2417	2496	U-5	2.0	2587.50	---	209	2797	2895
2250 CFM, 1 HP	Inst	EA	U-5	2.0	2350.00	---	209	2559	2657	U-5	1.6	2702.50	---	261	2964	3087
Utility set centrifugal V belt drive motor, 1/4" self-propelled																
1900 CFM, 1/4 hp	Inst	EA	U-5	4.0	3000.00	---	105	3105	3154	U-5	3.2	3450.00	---	131	3581	3642
2175 CFM, 1/3 hp	Inst	EA	U-5	3.0	3050.00	---	139	3189	3255	U-5	2.4	3507.50	---	174	3682	3764
2650 CFM, 1/2 hp	Inst	EA	U-5	2.5	3100.00	---	167	3267	3346	U-5	2.0	3565.00	---	209	3774	3872
3000 CFM, 3/4 hp	Inst	EA	U-5	2.0	3150.00	---	209	3359	3457	U-5	1.6	3622.50	---	261	3884	4007
Direct drive																
400 CFM, 8" x 8" damper	Inst	EA	U-5	4.0	425.00	---	105	530	579	U-5	3.2	488.75	---	131	619	681
650 CFM, 12" x 12" damper	Inst	EA	U-5	4.0	600.00	---	105	705	754	U-5	3.2	690.00	---	131	821	882
Ventilation, residential																
Attic																
Roof, type ventilators																
Aluminum dome, damper and curb																
6" dia., 300 CFM	Inst	EA	E-1	12.0	147	---	16.33	163.33	170.68	E-1	9.6	169	---	20.41	189.46	198.64
7" dia., 450 CFM	Inst	EA	E-1	11.0	165	---	17.81	182.81	190.83	E-1	8.8	190	---	22.26	212.01	222.03
9" dia., 900 CFM	Inst	EA	E-1	10.0	295	---	19.59	314.59	323.41	E-1	8.0	339	---	24.49	363.74	374.76
12" dia., 1000 CFM	Inst	EA	E-1	8.0	210	---	24.49	234.49	245.51	E-1	6.4	242	---	30.61	272.11	285.89
16" dia., 1500 CFM	Inst	EA	E-1	7.0	250	---	27.99	277.99	290.59	E-1	5.6	288	---	34.99	322.49	338.23
20" dia., 2500 CFM	Inst	EA	E-1	6.0	310	---	32.66	342.66	357.35	E-1	4.8	357	---	40.82	397.32	415.69
26" dia., 4000 CFM	Inst	EA	E-1	5.0	390	---	39.19	429.19	446.82	E-1	4.0	449	---	48.98	497.48	519.53
32" dia., 6500 CFM	Inst	EA	E-1	4.0	525	---	48.98	573.98	596.03	E-1	3.2	604	---	61.23	664.98	692.53
38" dia., 8000 CFM	Inst	EA	E-1	3.0	730	---	65.31	795.31	824.70	E-1	2.4	840	---	81.64	921.14	957.88
50" dia., 13,000 CFM	Inst	EA	E-1	2.0	1200	---	97.97	1297.97	1342.05	E-1	1.6	1380	---	122.46	1502.46	1557.56
Plastic ABS dome																
900 CFM	Inst	EA	E-1	12.0	75	---	16.33	91.33	98.68	E-1	9.6	86	---	20.41	106.66	115.84
1600 CFM	Inst	EA	E-1	10.0	110	---	19.59	129.59	138.41	E-1	8.0	127	---	24.49	150.99	162.01

			Costs Based On Large Volume							Costs Based On Small Volume						
Description	Oper	Unit	Crew Size	Avg Day Prod	Avg Mat'l Unit Cost	Avg Equip Unit Cost	Avg Labor Unit Cost	Avg Total Unit Cost	Avg Price Incl O&P	Crew Size	Avg Day Prod	Avg Mat'l Unit Cost	Avg Equip Unit Cost	Avg Labor Unit Cost	Avg Total Unit Cost	Avg Price Incl O&P

HVAC (continued)

Wall type ventilators, one speed, with shutter

Description	Oper	Unit	Crew Size	Avg Day Prod	Avg Mat'l Unit Cost	Avg Equip Unit Cost	Avg Labor Unit Cost	Avg Total Unit Cost	Avg Price Incl O&P	Crew Size	Avg Day Prod	Avg Mat'l Unit Cost	Avg Equip Unit Cost	Avg Labor Unit Cost	Avg Total Unit Cost	Avg Price Incl O&P
12" dia. 1000 CFM	Inst	EA	E-1	12.0	125	---	16.33	141.33	**148.68**	E-1	9.6	144	---	20.41	164.16	**173.34**
14" dia., 1500 CFM	Inst	EA	E-1	10.0	150	---	19.59	169.59	**178.41**	E-1	8.0	173	---	24.49	196.99	**208.01**
16" dia., 2000 CFM	Inst	EA	E-1	8.0	175	---	24.49	199.49	**210.51**	E-1	6.4	201	---	30.61	231.86	**245.64**
Entire structure, wall type, one speed, with shutter																
30" dia., 4800 CFM	Inst	EA	E-1	5.0	315	---	39.19	354.19	**371.82**	E-1	4.0	362	---	48.98	411.23	**433.28**
36" dia., 7000 CFM	Inst	EA	E-1	4.0	375	---	48.98	423.98	**446.03**	E-1	3.2	431	---	61.23	492.48	**520.03**
42" dia., 10,000 CFM	Inst	EA	E-1	3.0	450	---	65.31	515.31	**544.70**	E-1	2.4	518	---	81.64	599.14	**635.88**
48" dia., 16,000 CFM	Inst	EA	E-1	2.0	525	---	97.97	622.97	**667.05**	E-1	1.6	604	---	122.46	726.21	**781.31**
Two speeds, ADD	Inst	EA			15			15.00	**15.00**			17			17.25	**17.25**
Entire structure, lay-down type one speed, with shutter																
30" dia., 4500 CFM	Inst	EA	E-1	6.0	325	---	32.66	357.66	**372.35**	E-1	4.8	374	---	40.82	414.57	**432.94**
36" dia., 6500 CFM	Inst	EA	E-1	5.0	395	---	39.19	434.19	**451.82**	E-1	4.0	454	---	48.98	503.23	**525.28**
42" dia., 9000 CFM	Inst	EA	E-1	4.0	475	---	48.98	523.98	**546.03**	E-1	3.2	546	---	61.23	607.48	**635.03**
48" dia., 12,000 CFM	Inst	EA	E-1	3.0	550	---	65.31	615.31	**644.70**	E-1	2.4	633	---	81.64	714.14	**750.88**
Two speeds, ADD	Inst	EA	---	---	15	---		15.00	**15.00**	---	---	17	---		17.25	**17.25**
12 hr. timer, ADD	Inst	EA	E-1	16	20	---	12.25	32.25	**37.76**	E-1	13	23	---	15.31	38.31	**45.20**
Minimum Job Charge	Inst	Job	E-1		---	---	55.98	55.98	**81.17**	E-1			---	---	69.98	69.98	**101.46**

Grilles
Aluminum
Air return

Description	Oper	Unit	Crew Size	Avg Day Prod	Avg Mat'l Unit Cost	Avg Equip Unit Cost	Avg Labor Unit Cost	Avg Total Unit Cost	Avg Price Incl O&P	Crew Size	Avg Day Prod	Avg Mat'l Unit Cost	Avg Equip Unit Cost	Avg Labor Unit Cost	Avg Total Unit Cost	Avg Price Incl O&P
6" x 6"	Inst	EA	U-1	24	8.00	---	7.93	15.93	**19.65**	U-1	19	9.20	---	9.91	19.11	**23.76**
10" x 6"	Inst	EA	U-1	22	9.00	---	8.65	17.65	**21.71**	U-1	18	10.35	---	10.81	21.16	**26.24**
16" x 8"	Inst	EA	U-1	20	14.00	---	9.51	23.51	**27.98**	U-1	16	16.10	---	11.89	27.99	**33.58**
12" x 12"	Inst	EA	U-1	16	15.00	---	11.89	26.89	**32.48**	U-1	13	17.25	---	14.86	32.11	**39.09**
24" x 12"	Inst	EA	U-1	14	25.00	---	13.59	38.59	**44.97**	U-1	11	28.75	---	16.98	45.73	**53.72**
48" x 24"	Inst	EA	U-1	10	85.00	---	19.02	104.02	**112.96**	U-1	8	97.75	---	23.78	121.53	**132.70**
Minimum Job Charge	Inst	Job	U-1	4	---	---	54.35	54.35	**79.89**	U-1			---	---	67.93	67.93	**99.86**

Registers
Air supply
Ceiling/wall, O.B. damper, anodized aluminum
One or two way deflection, adjustable curved face bars

Description	Oper	Unit	Crew Size	Avg Day Prod	Avg Mat'l Unit Cost	Avg Equip Unit Cost	Avg Labor Unit Cost	Avg Total Unit Cost	Avg Price Incl O&P	Crew Size	Avg Day Prod	Avg Mat'l Unit Cost	Avg Equip Unit Cost	Avg Labor Unit Cost	Avg Total Unit Cost	Avg Price Incl O&P
14" x 8"	Inst	EA	U-1	16	22.00	---	11.89	33.89	**39.48**	U-1	13	25.30	---	14.86	40.16	**47.14**

Description	Oper	Unit	Costs Based On Large Volume							Costs Based On Small Volume						
			Crew Size	Avg Day Prod	Avg Mat'l Unit Cost	Avg Equip Unit Cost	Avg Labor Unit Cost	Avg Total Unit Cost	Avg Price incl O&P	Crew Size	Avg Day Prod	Avg Mat'l Unit Cost	Avg Equip Unit Cost	Avg Labor Unit Cost	Avg Total Unit Cost	Avg Price incl O&P

HVAC (continued)

Baseboard, adjustable damper, enameled steel

Description	Oper	Unit	Crew Size	Avg Day Prod	Avg Mat'l Unit Cost	Avg Equip Unit Cost	Avg Labor Unit Cost	Avg Total Unit Cost	Avg Price incl O&P	Crew Size	Avg Day Prod	Avg Mat'l Unit Cost	Avg Equip Unit Cost	Avg Labor Unit Cost	Avg Total Unit Cost	Avg Price incl O&P
10" x 6"	Inst	EA	U-1	24	6.50	---	7.93	14.43	18.15	U-1	19	7.48	---	9.91	17.38	22.04
12" x 5"	Inst	EA	U-1	22	7.75	---	8.65	16.40	20.46	U-1	18	8.91	---	10.81	19.72	24.80
12" x 6"	Inst	EA	U-1	22	7.25	---	8.65	15.90	19.96	U-1	18	8.34	---	10.81	19.14	24.22
12" x 8"	Inst	EA	U-1	20	10.50	---	9.51	20.01	24.48	U-1	16	12.08	---	11.89	23.96	29.55
14" x 6"	Inst	EA	U-1	18	7.90	---	10.57	18.47	23.43	U-1	14	9.09	---	13.21	22.29	28.50
Minimum Job Charge	Inst	Job	U-1		---	---	54.35	54.35	79.89	U-1		---	---	67.93	67.93	99.86

Ventilators

Base, damper, screen; 8" neck dia.

Description	Oper	Unit	Crew Size	Avg Day Prod	Avg Mat'l Unit Cost	Avg Equip Unit Cost	Avg Labor Unit Cost	Avg Total Unit Cost	Avg Price incl O&P	Crew Size	Avg Day Prod	Avg Mat'l Unit Cost	Avg Equip Unit Cost	Avg Labor Unit Cost	Avg Total Unit Cost	Avg Price incl O&P
215 CFM @ 5 MPH wind	Inst	EA	U-4	6	50.00	---	53.38	103.38	128.46	U-1	4	57.50	---	80.06	137.56	175.19

System units, complete

Fan coil air conditioning

Cabinet mounted, with filters

Chilled water

Description	Oper	Unit	Crew Size	Avg Day Prod	Avg Mat'l Unit Cost	Avg Equip Unit Cost	Avg Labor Unit Cost	Avg Total Unit Cost	Avg Price incl O&P	Crew Size	Avg Day Prod	Avg Mat'l Unit Cost	Avg Equip Unit Cost	Avg Labor Unit Cost	Avg Total Unit Cost	Avg Price incl O&P
0.5 ton cooling	Inst	EA	S-2	4.00	550	---	82	632	670	S-2	3.20	633	---	103	735	783
1 ton cooling	Inst	EA	S-2	3.00	625	---	110	735	785	S-2	2.40	719	---	137	856	919
2.5 ton cooling	Inst	EA	S-2	2.00	1100	---	164	1264	1340	S-2	1.60	1265	---	206	1471	1565
3 ton cooling	Inst	EA	S-2	1.00	1200	---	329	1529	1680	S-2	0.80	1380	---	411	1791	1980
10 ton cooling	Inst	EA	S-6	2.00	1725	---	264	1989	2110	S-6	1.60	1984	---	330	2314	2465
15 ton cooling	Inst	EA	S-6	1.50	2400	---	352	2752	2914	S-6	1.20	2760	---	440	3200	3402
20 ton cooling	Inst	EA	S-6	0.75	3050	---	704	3754	4078	S-6	0.60	3508	---	880	4387	4792
30 ton cooling	Inst	EA	S-6	0.50	4450	---	1056	5506	5992	S-6	0.40	5118	---	1320	6437	7044
Minimum Job Charge	Inst	Job	S-2		---	---	110	110	160	S-2		---	---	137	137	200

Heat pumps

Air to air, split system, not including curbs or pads

Description	Oper	Unit	Crew Size	Avg Day Prod	Avg Mat'l Unit Cost	Avg Equip Unit Cost	Avg Labor Unit Cost	Avg Total Unit Cost	Avg Price incl O&P	Crew Size	Avg Day Prod	Avg Mat'l Unit Cost	Avg Equip Unit Cost	Avg Labor Unit Cost	Avg Total Unit Cost	Avg Price incl O&P
2 ton cooling, 8.5 MBH heat	Inst	EA	S-2	1.00	1500	---	329	1829	1980	S-2	0.80	1725	---	411	2136	2325
3 ton cooling, 13 MBH heat	Inst	EA	S-2	0.50	1950	---	658	2608	2911	S-2	0.40	2243	---	822	3065	3443
7.5 ton cooling, 33 MBH heat	Inst	EA	S-2	0.33	4700	---	997	5697	6155	S-2	0.26	5405	---	1246	6651	7224
15 ton cooling, 64 MBH heat	Inst	EA	S-2	0.25	8300	---	1316	9616	10221	S-2	0.20	9545	---	1645	11190	11947

			Costs Based On Large Volume							Costs Based On Small Volume						
Description	Oper	Unit	Crew Size	Avg Day Prod	Avg Mat'l Unit Cost	Avg Equip Unit Cost	Avg Labor Unit Cost	Avg Total Unit Cost	Avg Price Incl O&P	Crew Size	Avg Day Prod	Avg Mat'l Unit Cost	Avg Equip Unit Cost	Avg Labor Unit Cost	Avg Total Unit Cost	Avg Price Incl O&P

HVAC (continued)

Air to air, single package, not including curbs, pads, or plenums

Description	Oper	Unit	Crew Size	Avg Day Prod	Avg Mat'l Unit Cost	Avg Equip Unit Cost	Avg Labor Unit Cost	Avg Total Unit Cost	Avg Price Incl O&P	Crew Size	Avg Day Prod	Avg Mat'l Unit Cost	Avg Equip Unit Cost	Avg Labor Unit Cost	Avg Total Unit Cost	Avg Price Incl O&P
2 ton cooling, 6.5 MBH heat	Inst	EA	S-2	1.25	1800	---	263	2063	2184	S-2	1.00	2070	---	329	2399	2550
3 ton cooling, 10 MBH heat	Inst	EA	S-2	0.75	2250	---	439	2689	2890	S-2	0.60	2588	---	548	3136	3388

Water source to air, single package

Description	Oper	Unit	Crew Size	Avg Day Prod	Avg Mat'l Unit Cost	Avg Equip Unit Cost	Avg Labor Unit Cost	Avg Total Unit Cost	Avg Price Incl O&P	Crew Size	Avg Day Prod	Avg Mat'l Unit Cost	Avg Equip Unit Cost	Avg Labor Unit Cost	Avg Total Unit Cost	Avg Price Incl O&P
1 ton cooling, 13 MBH heat	Inst	EA	S-2	1.75	800	---	188	988	1074	S-2	1.40	920	---	235	1155	1263
2 ton cooling, 19 MBH heat	Inst	EA	S-2	1.50	1000	---	219	1219	1320	S-2	1.20	1150	---	274	1424	1550
5 ton cooling, 29 MBH heat	Inst	EA	S-2	0.75	1725	---	439	2164	2365	S-2	0.60	1984	---	548	2532	2784
Minimum Job Charge	Inst	Job	S-2		---	---	164	164	240	S-2		---	---	206	206	300

Roof top air conditioners

Standard controls, curb, energy economizer

Single zone, electric fired cooling, gas fired heating

Description	Oper	Unit	Crew Size	Avg Day Prod	Avg Mat'l Unit Cost	Avg Equip Unit Cost	Avg Labor Unit Cost	Avg Total Unit Cost	Avg Price Incl O&P	Crew Size	Avg Day Prod	Avg Mat'l Unit Cost	Avg Equip Unit Cost	Avg Labor Unit Cost	Avg Total Unit Cost	Avg Price Incl O&P
3 ton cooling, 60 MBH heat	Inst	EA	S-2	1.20	3100	---	274	3374	3500	S-2	0.96	3565	---	343	3908	4065
4 ton cooling, 95 MBH heat	Inst	EA	S-2	1.00	3300	---	329	3629	3780	S-2	0.80	3795	---	411	4206	4395
10 ton cooling, 200 MBH heat	Inst	EA	S-6	0.40	6800	---	1320	8120	8727	S-6	0.32	7820	---	1650	9470	10229
30 ton cooling, 540 MBH heat	Inst	EA	S-7	0.20	20500	---	3634	24134	25806	S-7	0.16	23575	---	4543	28118	30207
40 ton cooling, 675 MBH heat	Inst	EA	S-7	0.16	28000	---	4543	32543	34632	S-7	0.13	32200	---	5679	37879	40491
50 ton cooling, 810 MBH heat	Inst	EA	S-7	0.12	33500	---	6057	39557	42343	S-7	0.10	38525	---	7571	46096	49579

Single zone, gas fired cooling and heating

Description	Oper	Unit	Crew Size	Avg Day Prod	Avg Mat'l Unit Cost	Avg Equip Unit Cost	Avg Labor Unit Cost	Avg Total Unit Cost	Avg Price Incl O&P	Crew Size	Avg Day Prod	Avg Mat'l Unit Cost	Avg Equip Unit Cost	Avg Labor Unit Cost	Avg Total Unit Cost	Avg Price Incl O&P
3 ton cooling, 90 MBH heat	Inst	EA	S-2	1.00	3150	---	329	3479	3630	S-2	0.80	3623	---	411	4034	4223

Multizone, electric fired cooling, gas fired cooling

Description	Oper	Unit	Crew Size	Avg Day Prod	Avg Mat'l Unit Cost	Avg Equip Unit Cost	Avg Labor Unit Cost	Avg Total Unit Cost	Avg Price Incl O&P	Crew Size	Avg Day Prod	Avg Mat'l Unit Cost	Avg Equip Unit Cost	Avg Labor Unit Cost	Avg Total Unit Cost	Avg Price Incl O&P
20 ton cooling, 360 MBH heat	Inst	EA	S-7	0.20	30500	---	3634	34134	35806	S-7	0.16	35075	---	4543	39618	41707
30 ton cooling, 540 MBH heat	Inst	EA	S-7	0.14	40000	---	5192	45192	47580	S-7	0.11	46000	---	6490	52490	55475
70 ton cooling, 1500 MBH heat	Inst	EA	S-7	0.08	58000	---	9086	67086	71265	S-7	0.06	66700	---	11357	78057	83281
80 ton cooling, 1500 MBH heat	Inst	EA	S-7	0.07	66000	---	10384	76384	81160	S-7	0.06	75900	---	12979	88879	94850
105 ton cooling,1500 MBH heat	Inst	EA	S-7	0.05	86000	---	14537	100537	107224	S-7	0.04	98900	---	18171	117071	125430
Minimum Job Charge	Inst	Job	S-2		---	---	103	103	150	S-2		---	---	129	129	188

Window unit air conditioners

Semi-permanent installation, 3 speed fan, 125 volt GFI receptacle, energy efficient models

Description	Oper	Unit	Crew Size	Avg Day Prod	Avg Mat'l Unit Cost	Avg Equip Unit Cost	Avg Labor Unit Cost	Avg Total Unit Cost	Avg Price Incl O&P	Crew Size	Avg Day Prod	Avg Mat'l Unit Cost	Avg Equip Unit Cost	Avg Labor Unit Cost	Avg Total Unit Cost	Avg Price Incl O&P
6,000 BTUH (0.5 ton cooling)	Inst	EA	E-2	4.0	500	---	98	598	642	E-2	3.2	575	---	122	697	753
9,000 BTUH (0.75 ton cooling)	Inst	EA	E-2	4.0	550	---	98	648	692	E-2	3.2	633	---	122	755	810
12,000 BTUH (1 ton cooling)	Inst	EA	E-2	4.0	600	---	98	698	742	E-2	3.2	690	---	122	812	868

			Costs Based On Large Volume							Costs Based On Small Volume						
Description	Oper	Unit	Crew Size	Avg Day Prod	Avg Mat'l Unit Cost	Avg Equip Unit Cost	Avg Labor Unit Cost	Avg Total Unit Cost	Avg Price Incl O&P	Crew Size	Avg Day Prod	Avg Mat'l Unit Cost	Avg Equip Unit Cost	Avg Labor Unit Cost	Avg Total Unit Cost	Avg Price Incl O&P

HVAC (continued)

Description	Oper	Unit	Crew Size	Avg Day Prod	Avg Mat'l Unit Cost	Avg Equip Unit Cost	Avg Labor Unit Cost	Avg Total Unit Cost	Avg Price Incl O&P	Crew Size	Avg Day Prod	Avg Mat'l Unit Cost	Avg Equip Unit Cost	Avg Labor Unit Cost	Avg Total Unit Cost	Avg Price Incl O&P
18,000 BTUH (1.5 ton cooling)	Inst	EA	E-2	3.0	700	---	131	831	889	E-2	2.4	805	---	163	968	1042
24,000 BTUH (2.0 ton cooling)	Inst	EA	E-2	3.0	800	---	131	931	989	E-2	2.4	920	---	163	1083	1157
36,000 BTUH (3.0 ton cooling)	Inst	EA	E-2	3.0	1000	---	131	1131	1189	E-2	2.4	1150	---	163	1313	1387
Minimum Job Charge	Inst	Job	E-1		---	---	65	65	95	E-1		---	---	82	82	118

Insulation

Batt or Roll, with wall or ceiling finish already removed

Description	Oper	Unit	Crew Size	Avg Day Prod	Avg Mat'l Unit Cost	Avg Equip Unit Cost	Avg Labor Unit Cost	Avg Total Unit Cost	Avg Price Incl O&P	Crew Size	Avg Day Prod	Avg Mat'l Unit Cost	Avg Equip Unit Cost	Avg Labor Unit Cost	Avg Total Unit Cost	Avg Price Incl O&P
Joists, 16" or 24" o.c.	Demo	SF	L-2	2935	---	---	0.09	0.09	0.13	L-2	1908	---	---	0.14	0.14	0.20
Rafters, 16" or 24" o.c.	Demo	SF	L-2	2560	---	---	0.10	0.10	0.15	L-2	1664	---	---	0.16	0.16	0.23
Studs, 16" or 24" o.c.	Demo	SF	L-2	3285	---	---	0.08	0.08	0.12	L-2	2135	---	---	0.12	0.12	0.18

Batt or Roll, place and/or staple, Johns-Manville fiberglass; allowance made for joists, rafters, studs

Joists
Un-faced

Description	Oper	Unit	Crew Size	Avg Day Prod	Avg Mat'l Unit Cost	Avg Equip Unit Cost	Avg Labor Unit Cost	Avg Total Unit Cost	Avg Price Incl O&P	Crew Size	Avg Day Prod	Avg Mat'l Unit Cost	Avg Equip Unit Cost	Avg Labor Unit Cost	Avg Total Unit Cost	Avg Price Incl O&P
3-1/2" T (R-13)																
16" o.c.	Inst	SF	C-2	975	0.35	---	0.19	0.54	0.63	C-2	634	0.41	---	0.29	0.70	0.84
6-1/2" T (R-19)																
16" o.c.	Inst	SF	C-2	975	0.42	---	0.19	0.61	0.70	C-2	634	0.49	---	0.29	0.78	0.92
24" o.c.	Inst	SF	C-2	1460	0.42	---	0.13	0.55	0.61	C-2	949	0.49	---	0.19	0.68	0.78
7" T (R-22)																
16" o.c.	Inst	SF	C-2	975	0.55	---	0.19	0.74	0.83	C-2	634	0.64	---	0.29	0.94	1.08
24" o.c.	Inst	SF	C-2	1460	0.55	---	0.13	0.68	0.74	C-2	949	0.64	---	0.19	0.84	0.93
9-1/4" T (R-30)																
16" o.c.	Inst	SF	C-2	975	0.68	---	0.19	0.87	0.97	C-2	634	0.80	---	0.29	1.09	1.23
24" o.c.	Inst	SF	C-2	1460	0.68	---	0.13	0.81	0.87	C-2	949	0.80	---	0.19	0.99	1.09
Kraft-faced																
3-1/2" T (R-11)																
16" o.c.	Inst	SF	C-2	975	0.26	---	0.19	0.45	0.55	C-2	634	0.31	---	0.29	0.60	0.74
24" o.c.	Inst	SF	C-2	1460	0.26	---	0.13	0.39	0.45	C-2	949	0.31	---	0.19	0.50	0.60
3-1/2" T (R-13)																
16" o.c.	Inst	SF	C-2	975	0.37	---	0.19	0.56	0.65	C-2	634	0.43	---	0.29	0.73	0.87
6-1/2" T (R-19)																
16" o.c.	Inst	SF	C-2	975	0.44	---	0.19	0.63	0.73	C-2	634	0.52	---	0.29	0.81	0.95
24" o.c.	Inst	SF	C-2	1460	0.44	---	0.13	0.57	0.63	C-2	949	0.52	---	0.19	0.71	0.81

Description	Oper	Unit	Costs Based On Large Volume							Costs Based On Small Volume						
			Crew Size	Avg Day Prod	Avg Mat'l Unit Cost	Avg Equip Unit Cost	Avg Labor Unit Cost	Avg Total Unit Cost	Avg Price Incl O&P	Crew Size	Avg Day Prod	Avg Mat'l Unit Cost	Avg Equip Unit Cost	Avg Labor Unit Cost	Avg Total Unit Cost	Avg Price Incl O&P

Insulation (continued)

Description	Oper	Unit	Crew	Prod	Mat'l	Equip	Labor	Total	O&P	Crew	Prod	Mat'l	Equip	Labor	Total	O&P
7" T (R-22)																
16" o.c.	Inst	SF	C-2	975	0.60	---	0.19	0.79	0.88	C-2	634	0.70	---	0.29	0.99	1.13
24" o.c.	Inst	SF	C-2	1460	0.60	---	0.13	0.73	0.79	C-2	949	0.70	---	0.19	0.89	0.99
9-1/4" T (R-30)																
16" o.c.	Inst	SF	C-2	975	0.72	---	0.19	0.91	1.00	C-2	634	0.84	---	0.29	1.13	1.27
24" o.c.	Inst	SF	C-2	1460	0.72	---	0.13	0.85	0.91	C-2	949	0.84	---	0.19	1.03	1.13
Foil-faced																
4" T (R-11)																
16" o.c.	Inst	SF	C-2	975	0.30	---	0.19	0.49	0.58	C-2	634	0.35	---	0.29	0.64	0.78
24" o.c.	Inst	SF	C-2	1460	0.30	---	0.13	0.43	0.49	C-2	949	0.35	---	0.19	0.54	0.64
6-1/2" T (R-19)																
16" o.c.	Inst	SF	C-2	975	0.48	---	0.19	0.67	0.76	C-2	634	0.56	---	0.29	0.85	0.99
24" o.c.	Inst	SF	C-2	1460	0.48	---	0.13	0.61	0.67	C-2	949	0.56	---	0.19	0.75	0.85
Rafters																
Un-faced																
3-1/2" T (R-13)																
16" o.c.	Inst	SF	C-2	650	0.35	---	0.28	0.63	0.77	C-2	423	0.41	---	0.44	0.84	1.06
6-1/2" T (R-19)																
16" o.c.	Inst	SF	C-2	650	0.42	---	0.28	0.70	0.84	C-2	423	0.49	---	0.44	0.93	1.14
24" o.c.	Inst	SF	C-2	975	0.42	---	0.19	0.61	0.70	C-2	634	0.49	---	0.29	0.78	0.92
7" T (R-22)																
16" o.c.	Inst	SF	C-2	650	0.55	---	0.28	0.84	0.98	C-2	423	0.64	---	0.44	1.08	1.30
24" o.c.	Inst	SF	C-2	975	0.55	---	0.19	0.74	0.83	C-2	634	0.64	---	0.29	0.94	1.08
9-1/4" T (R-30)																
16" o.c.	Inst	SF	C-2	650	0.68	---	0.28	0.97	1.11	C-2	423	0.80	---	0.44	1.24	1.45
24" o.c.	Inst	SF	C-2	975	0.68	---	0.19	0.87	0.97	C-2	634	0.80	---	0.29	1.09	1.23
Kraft-faced																
3-1/2" T (R-11)																
16" o.c.	Inst	SF	C-2	650	0.26	---	0.28	0.55	0.69	C-2	423	0.31	---	0.44	0.75	0.96
24" o.c.	Inst	SF	C-2	975	0.26	---	0.19	0.45	0.55	C-2	634	0.31	---	0.29	0.60	0.74
3-1/2" T (R-13)																
16" o.c.	Inst	SF	C-2	650	0.37	---	0.28	0.66	0.80	C-2	423	0.43	---	0.44	0.87	1.09

Description	Oper	Unit	Costs Based On Large Volume							Costs Based On Small Volume						
			Crew Size	Avg Day Prod	Avg Mat'l Unit Cost	Avg Equip Unit Cost	Avg Labor Unit Cost	Avg Total Unit Cost	Avg Price Incl O&P	Crew Size	Avg Day Prod	Avg Mat'l Unit Cost	Avg Equip Unit Cost	Avg Labor Unit Cost	Avg Total Unit Cost	Avg Price Incl O&P

Insulation (continued)

Description	Oper	Unit	Crew Size	Avg Day Prod	Avg Mat'l Unit Cost	Avg Equip Unit Cost	Avg Labor Unit Cost	Avg Total Unit Cost	Avg Price Incl O&P	Crew Size	Avg Day Prod	Avg Mat'l Unit Cost	Avg Equip Unit Cost	Avg Labor Unit Cost	Avg Total Unit Cost	Avg Price Incl O&P
6-1/2" T (R-19)																
16" o.c.	Inst	SF	C-2	650	0.44	---	0.28	0.73	0.87	C-2	423	0.52	---	0.44	0.96	1.17
24" o.c.	Inst	SF	C-2	975	0.44	---	0.19	0.63	0.73	C-2	634	0.52	---	0.29	0.81	0.95
7" T (R-22)																
16" o.c.	Inst	SF	C-2	650	0.60	---	0.28	0.88	1.02	C-2	423	0.70	---	0.44	1.14	1.35
24" o.c.	Inst	SF	C-2	975	0.60	---	0.19	0.79	0.88	C-2	634	0.70	---	0.29	0.99	1.13
9-1/4" T (R-30)																
16" o.c.	Inst	SF	C-2	650	0.72	---	0.28	1.00	1.14	C-2	423	0.84	---	0.44	1.28	1.49
24" o.c.	Inst	SF	C-2	975	0.72	---	0.19	0.91	1.00	C-2	634	0.84	---	0.29	1.13	1.27
Foil-faced																
4" T (R-11)																
16" o.c.	Inst	SF	C-2	650	0.30	---	0.28	0.58	0.72	C-2	423	0.35	---	0.44	0.79	1.00
24" o.c.	Inst	SF	C-2	975	0.30	---	0.19	0.49	0.58	C-2	634	0.35	---	0.29	0.64	0.78
6-1/2" T (R-19)																
16" o.c.	Inst	SF	C-2	650	0.48	---	0.28	0.76	0.90	C-2	423	0.56	---	0.44	1.00	1.21
24" o.c.	Inst	SF	C-2	975	0.48	---	0.19	0.67	0.76	C-2	634	0.56	---	0.29	0.85	0.99
Studs																
Un-faced																
3-1/2" T (R-13)																
16" o.c.	Inst	SF	C-2	815	0.35	---	0.23	0.57	0.69	C-2	530	0.41	---	0.35	0.75	0.93
6-1/2" T (R-19)																
16" o.c.	Inst	SF	C-2	815	0.42	---	0.23	0.65	0.76	C-2	530	0.49	---	0.35	0.84	1.01
24" o.c.	Inst	SF	C-2	1220	0.42	---	0.15	0.57	0.65	C-2	793	0.49	---	0.23	0.72	0.84
7" T (R-22)																
16" o.c.	Inst	SF	C-2	815	0.55	---	0.23	0.78	0.89	C-2	530	0.64	---	0.35	0.99	1.16
24" o.c.	Inst	SF	C-2	1220	0.55	---	0.15	0.70	0.78	C-2	793	0.64	---	0.23	0.88	0.99
9-1/4" T (R-30)																
16" o.c.	Inst	SF	C-2	815	0.68	---	0.23	0.91	1.02	C-2	530	0.80	---	0.35	1.15	1.32
24" o.c.	Inst	SF	C-2	1220	0.68	---	0.15	0.84	0.91	C-2	793	0.80	---	0.23	1.03	1.15
Kraft-faced																
3-1/2" T (R-11)																
16" o.c.	Inst	SF	C-2	815	0.26	---	0.23	0.49	0.60	C-2	530	0.31	---	0.35	0.66	0.83
24" o.c.	Inst	SF	C-2	1220	0.26	---	0.15	0.42	0.49	C-2	793	0.31	---	0.23	0.54	0.66

| | | | Costs Based On Large Volume | | | | | | | | Costs Based On Small Volume | | | | | |
Description	Oper	Unit	Crew Size	Avg Day Prod	Avg Mat'l Unit Cost	Avg Equip Unit Cost	Avg Labor Unit Cost	Avg Total Unit Cost	Avg Price Incl O&P	Crew Size	Avg Day Prod	Avg Mat'l Unit Cost	Avg Equip Unit Cost	Avg Labor Unit Cost	Avg Total Unit Cost	Avg Price Incl O&P
Insulation (continued)																
3-1/2" T (R-13)																
16" o.c.	Inst	SF	C-2	815	0.37	---	0.23	0.60	0.71	C-2	530	0.43	---	0.35	0.78	0.95
6-1/2" T (R-19)																
16" o.c.	Inst	SF	C-2	815	0.44	---	0.23	0.67	0.78	C-2	530	0.52	---	0.35	0.87	1.04
24" o.c.	Inst	SF	C-2	1220	0.44	---	0.15	0.60	0.67	C-2	793	0.52	---	0.23	0.75	0.87
7" T (R-22)																
16" o.c.	Inst	SF	C-2	815	0.60	---	0.23	0.83	0.94	C-2	530	0.70	---	0.35	1.05	1.22
24" o.c.	Inst	SF	C-2	1220	0.60	---	0.15	0.75	0.83	C-2	793	0.70	---	0.23	0.93	1.05
9-1/4" T (R-30)																
16" o.c.	Inst	SF	C-2	815	0.72	---	0.23	0.95	1.06	C-2	530	0.84	---	0.35	1.19	1.36
24" o.c.	Inst	SF	C-2	1220	0.72	---	0.15	0.87	0.95	C-2	793	0.84	---	0.23	1.07	1.19
Foil-faced																
4" T (R-11)																
16" o.c.	Inst	SF	C-2	815	0.30	---	0.23	0.53	0.64	C-2	530	0.35	---	0.35	0.70	0.87
24" o.c.	Inst	SF	C-2	1220	0.30	---	0.15	0.45	0.53	C-2	793	0.35	---	0.23	0.58	0.70
6-1/2" T (R-19)																
16" o.c.	Inst	SF	C-2	815	0.48	---	0.23	0.71	0.82	C-2	530	0.56	---	0.35	0.91	1.08
24" o.c.	Inst	SF	C-2	1220	0.48	---	0.15	0.63	0.71	C-2	793	0.56	---	0.23	0.79	0.91
Loose fill with ceiling finish already removed																
Joists, 16" or 24" o.c.																
4" T	Demo	SF	L-2	3900	---	---	0.07	0.07	0.10	L-2	2535		---	0.10	0.10	0.15
6" T	Demo	SF	L-2	2340	---	---	0.11	0.11	0.16	L-2	1521		---	0.17	0.17	0.25
Loose fill, allowance made for joists, cavities, and cores																
Insulating wood, granule or pellet (40 lbs/bag, 4 CF/bag)																
Joists, @ 7 lbs/CF density																
16" o.c.																
4" T	Inst	SF	C-4	960	0.67	---	0.23	0.90	1.01	C-4	624	0.78	---	0.35	1.13	1.30
6" T	Inst	SF	C-6	720	1.02	---	0.35	1.37	1.54	C-6	468	1.19	---	0.53	1.72	1.99
24" o.c.																
4" T	Inst	SF	C-4	930	0.70	---	0.23	0.93	1.04	C-4	605	0.81	---	0.36	1.17	1.35
6" T	Inst	SF	C-6	690	1.04	---	0.36	1.41	1.58	C-6	449	1.22	---	0.56	1.78	2.05

Description	Oper	Unit	Costs Based On Large Volume							Costs Based On Small Volume						
			Crew Size	Avg Day Prod	Avg Mat'l Unit Cost	Avg Equip Unit Cost	Avg Labor Unit Cost	Avg Total Unit Cost	Avg Price Incl O&P	Crew Size	Avg Day Prod	Avg Mat'l Unit Cost	Avg Equip Unit Cost	Avg Labor Unit Cost	Avg Total Unit Cost	Avg Price Incl O&P

Insulation (continued)

Vermiculite/Perlite (4 CF/bag, approximately 10 lbs/bag)

Joists
16" o.c.

Description	Oper	Unit	Crew Size	Avg Day Prod	Avg Mat'l Unit Cost	Avg Equip Unit Cost	Avg Labor Unit Cost	Avg Total Unit Cost	Avg Price Incl O&P	Crew Size	Avg Day Prod	Avg Mat'l Unit Cost	Avg Equip Unit Cost	Avg Labor Unit Cost	Avg Total Unit Cost	Avg Price Incl O&P
4" T	Inst	SF	C-6	1340	0.49	---	0.19	0.68	0.77	C-6	871	0.57	---	0.29	0.86	1.00
6" T	Inst	SF	C-6	1000	0.73	---	0.25	0.98	1.10	C-6	650	0.85	---	0.38	1.24	1.43
24" o.c.																
4" T	Inst	SF	C-6	1300	0.50	---	0.19	0.70	0.79	C-6	845	0.59	---	0.30	0.88	1.03
6" T	Inst	SF	C-6	960	0.76	---	0.26	1.02	1.14	C-6	624	0.88	---	0.40	1.28	1.48
Cavity walls																
1" T	Inst	SF	C-4	3070	0.12	---	0.07	0.19	0.23	C-4	1996	0.14	---	0.11	0.25	0.30
2" T	Inst	SF	C-4	1690	0.24	---	0.13	0.37	0.43	C-4	1099	0.28	---	0.20	0.48	0.57
Block walls (2 cores/block)																
8" T block	Inst	SF	C-4	815	0.41	---	0.27	0.67	0.81	C-4	530	0.48	---	0.41	0.89	1.09
12" T block	Inst	SF	C-4	610	0.77	---	0.36	1.12	1.30	C-4	397	0.90	---	0.55	1.44	1.71

Rigid

Roofs

Description	Oper	Unit	Crew Size	Avg Day Prod	Avg Mat'l Unit Cost	Avg Equip Unit Cost	Avg Labor Unit Cost	Avg Total Unit Cost	Avg Price Incl O&P	Crew Size	Avg Day Prod	Avg Mat'l Unit Cost	Avg Equip Unit Cost	Avg Labor Unit Cost	Avg Total Unit Cost	Avg Price Incl O&P
1/2" T	Demo	SQ	L-2	17	---	---	15.30	15.30	22.64	L-2	11	---	---	23.54	23.54	34.83
1" T	Demo	SQ	L-2	15	---	---	17.34	15.30	25.66	L-2	10	---	---	26.68	26.68	39.48
Walls, 1/2" T	Demo	SF	L-2	2140	---	---	0.12	0.12	0.18	L-2	1391	---	---	0.19	0.19	0.28

Rigid insulating board

Roofs

Over wood decks, 5% waste included
Normal (dry) moisture conditions within building
Nail 1-ply 15 lb felt, set and nail:
2' x 8' x 1/2" T&G asphalt

Description	Oper	Unit	Crew Size	Avg Day Prod	Avg Mat'l Unit Cost	Avg Equip Unit Cost	Avg Labor Unit Cost	Avg Total Unit Cost	Avg Price Incl O&P	Crew Size	Avg Day Prod	Avg Mat'l Unit Cost	Avg Equip Unit Cost	Avg Labor Unit Cost	Avg Total Unit Cost	Avg Price Incl O&P
sheathing	Inst	SF	C-13	1020	0.29	---	0.39	0.68	0.88	C-13	663	0.34	---	0.61	0.94	1.24
4' x 8' x 1/2" asphalt sheathing	Inst	SF	C-11	1120	0.29	---	0.33	0.62	0.78	C-11	728	0.34	---	0.51	0.84	1.09
4' x 8' x 1/2" building block	Inst	SF	C-11	1120	0.32	---	0.33	0.65	0.82	C-11	728	0.38	---	0.51	0.89	1.13

			Costs Based On Large Volume							Costs Based On Small Volume					
Description	Oper	Unit	Crew Size	Avg Day Prod	Avg Mat'l Unit Cost	Avg Equip Unit Cost	Avg Labor Unit Cost	Avg Total Unit Cost	Avg Price Incl O&P	Crew Size	Avg Day Prod	Avg Mat'l Unit Cost	Avg Equip Unit Cost	Avg Labor Unit Cost	Avg Total Unit Cost	Avg Price Incl O&P

Insulation (continued)

Excessive (humid) moisture conditions within building
Nail and overlap 3-ply 15 lb felt, mop laps and surface one coat and embed:

Description	Oper	Unit	Crew Size	Avg Day Prod	Avg Mat'l Unit Cost	Avg Equip Unit Cost	Avg Labor Unit Cost	Avg Total Unit Cost	Avg Price Incl O&P	Crew Size	Avg Day Prod	Avg Mat'l Unit Cost	Avg Equip Unit Cost	Avg Labor Unit Cost	Avg Total Unit Cost	Avg Price Incl O&P
2' x 8' x 1/2" T&G asphalt sheathing	Inst	SF	R-18	780	0.50	---	0.58	1.08	1.40	R-18	507	0.59	---	0.89	1.48	1.96
4' x 8' x 1/2" asphalt sheathing	Inst	SF	R-18	810	0.50	---	0.56	1.06	1.36	R-18	527	0.59	---	0.86	1.45	1.91
4' x 8' x 1/2" building block	Inst	SF	R-18	810	0.53	---	0.56	1.09	1.39	R-18	527	0.62	---	0.86	1.47	1.94

Over non-combustible decks, 5% waste included
Normal (dry) moisture conditions within building
Mop one coat and embed

Description	Oper	Unit	Crew Size	Avg Day Prod	Avg Mat'l Unit Cost	Avg Equip Unit Cost	Avg Labor Unit Cost	Avg Total Unit Cost	Avg Price Incl O&P	Crew Size	Avg Day Prod	Avg Mat'l Unit Cost	Avg Equip Unit Cost	Avg Labor Unit Cost	Avg Total Unit Cost	Avg Price Incl O&P
2' x 8' x 1/2" T&G asphalt sheathing	Inst	SF	R-18	1100	0.36	---	0.41	0.77	0.99	R-18	715	0.42	---	0.63	1.05	1.39
4' x 8' x 1/2" asphalt sheathing	Inst	SF	R-18	1160	0.36	---	0.39	0.75	0.96	R-18	754	0.42	---	0.60	1.02	1.34
4' x 8' x 1/2" building block	Inst	SF	R-18	1160	0.38	---	0.39	0.77	0.98	R-18	754	0.45	---	0.60	1.05	1.37

Excessive (humid) moisture conditions within building
Mop one coat, embedded 2-ply 15 lb. felt, mop and embed:

Description	Oper	Unit	Crew Size	Avg Day Prod	Avg Mat'l Unit Cost	Avg Equip Unit Cost	Avg Labor Unit Cost	Avg Total Unit Cost	Avg Price Incl O&P	Crew Size	Avg Day Prod	Avg Mat'l Unit Cost	Avg Equip Unit Cost	Avg Labor Unit Cost	Avg Total Unit Cost	Avg Price Incl O&P
2' x 8' x 1/2" T&G asphalt sheathing	Inst	SF	R-18	630	0.50	---	0.72	1.22	1.61	R-18	410	0.59	---	1.10	1.69	2.29
4' x 8' x 1/2" asphalt sheathing	Inst	SF	R-18	650	0.50	---	0.70	1.20	1.58	R-18	423	0.59	---	1.07	1.66	2.24
4' x 8' x 1/2" building block	Inst	SF	R-18	650	0.53	---	0.70	1.22	1.60	R-18	423	0.62	---	1.07	1.69	2.26

Walls, nailed, 5% waste included
4' x 8' x 1/2" asphalt sheathing

Description	Oper	Unit	Crew Size	Avg Day Prod	Avg Mat'l Unit Cost	Avg Equip Unit Cost	Avg Labor Unit Cost	Avg Total Unit Cost	Avg Price Incl O&P	Crew Size	Avg Day Prod	Avg Mat'l Unit Cost	Avg Equip Unit Cost	Avg Labor Unit Cost	Avg Total Unit Cost	Avg Price Incl O&P
Straight wall	Inst	SF	C-11	1440	0.26	---	0.26	0.52	0.65	C-11	936	0.31	---	0.39	0.70	0.90
Cut-up wall	Inst	SF	C-10	1200	0.26	---	0.29	0.56	0.70	C-10	780	0.31	---	0.45	0.76	0.98
4' x 8' x 1/2" building board																
Straight wall	Inst	SF	C-11	1440	0.29	---	0.26	0.54	0.67	C-11	936	0.34	---	0.39	0.73	0.92
Cut-up wall	Inst	SF	C-10	1200	0.29	---	0.29	0.58	0.73	C-10	780	0.34	---	0.45	0.79	1.01

			Costs Based On Large Volume							Costs Based On Small Volume					
Description	Oper	Unit	Crew Size	Avg Day Prod	Avg Mat'l Unit Cost	Avg Equip Unit Cost	Avg Labor Unit Cost	Avg Total Unit Cost	Avg Price Incl O&P	Crew Size	Avg Day Prod	Avg Mat'l Unit Cost	Avg Equip Unit Cost	Avg Labor Unit Cost	Avg Total Unit Cost	Avg Price Incl O&P

Intercom systems. See Electrical.

Jacuzzi whirlpools. See Plumbing.

Lath & Plaster. See Plaster & Stucco.

Lighting fixtures. Labor includes hanging and connecting fixtures

Indoor lighting

Fluorescent

Description	Oper	Unit	Crew Size	Avg Day Prod	Avg Mat'l Unit Cost	Avg Equip Unit Cost	Avg Labor Unit Cost	Avg Total Unit Cost	Avg Price Incl O&P	Crew Size	Avg Day Prod	Avg Mat'l Unit Cost	Avg Equip Unit Cost	Avg Labor Unit Cost	Avg Total Unit Cost	Avg Price Incl O&P
Pendant mounted worklights																
4' L, two 40 watt RS	Inst	EA	E-1	10.0	25.00	---	19.59	44.59	53.41	E-1	7.50	30.00	---	26.12	56.12	67.88
4' L, two 60 watt RS	Inst	EA	E-1	10.0	31.25	---	19.59	50.84	59.66	E-1	7.50	37.50	---	26.12	63.62	75.38
8' L, two 75 watt RS	Inst	EA	E-1	8.0	37.50	---	24.49	61.99	73.01	E-1	6.00	45.00	---	32.66	77.66	92.35
Recessed mounted light fixture with acrylic diffuser																
1' W x 4' L, two 40 WRS	Inst	EA	E-1	10.0	93.75	---	19.59	113.34	122.16	E-1	7.50	112.50	---	26.12	138.62	150.38
2' W x 2' L, two U 40 WRS	Inst	EA	E-1	10.0	93.75	---	19.59	113.34	122.16	E-1	7.50	112.50	---	26.12	138.62	150.38
2' W x 4' L, four 40 WRS	Inst	EA	E-1	8.0	131.25	---	24.49	155.74	166.76	E-1	6.00	157.50	---	32.66	190.16	204.85
Strip lighting fixtures																
2' L, one 40 watt RS	Inst	EA	E-1	10.0	81.25	---	19.59	100.84	109.66	E-1	7.50	97.50	---	26.12	123.62	135.38
2' L, two 40 watt RS	Inst	EA	E-1	10.0	100.00	---	19.59	119.59	128.41	E-1	7.50	120.00	---	26.12	146.12	157.88
4' L, two 40 watt SL	Inst	EA	E-1	8.0	112.50	---	24.49	136.99	148.01	E-1	6.00	135.00	---	32.66	167.66	182.35
4' L, four 40 watt SL	Inst	EA	E-1	8.0	125.00	---	24.49	149.49	160.51	E-1	6.00	150.00	---	32.66	182.66	197.35
Surface mounted acrylic diffuser																
1' W x 4' L, two 40 watt RS	Inst	EA	E-1	10.0	81.25	---	19.59	100.84	109.66	E-1	7.50	97.50	---	26.12	123.62	135.38
2' W x 2' L, two U 40 watt RS	Inst	EA	E-1	10.0	100.00	---	19.59	119.59	128.41	E-1	7.50	120.00	---	26.12	146.12	157.88
2' W x 4' L, four 40 watt RS	Inst	EA	E-1	8.0	125.00	---	24.49	149.49	160.51	E-1	6.00	150.00	---	32.66	182.66	197.35
White enameled circline steel ceiling fixtures																
8" W, 22 watt RS	Inst	EA	E-1	12.0	25.00	---	16.33	41.33	48.68	E-1	9.00	30.00	---	21.77	51.77	61.57
12" W, 22 to 32 watt RS	Inst	EA	E-1	12.0	37.50	---	16.33	53.83	61.18	E-1	9.00	45.00	---	21.77	66.77	76.57
16" W, 22 to 40 watt RS	Inst	EA	E-1	12.0	50.00	---	16.33	66.33	73.68	E-1	9.00	60.00	---	21.77	81.77	91.57
Decorative circline fixtures																
12" W, 22 to 32 watt RS	Inst	EA	E-1	12.0	87.50	---	16.33	103.83	111.18	E-1	9.00	105.00	---	21.77	126.77	136.57
16" W, 22 to 40 watt RS	Inst	EA	E-1	12.0	150.00	---	16.33	166.33	173.68	E-1	9.00	180.00	---	21.77	201.77	211.57

			Costs Based On Large Volume							Costs Based On Small Volume					
Description	Oper	Unit	Crew Size	Avg Day Prod	Avg Mat'l Unit Cost	Avg Equip Unit Cost	Avg Labor Unit Cost	Avg Total Unit Cost	Avg Price incl O&P	Crew Size	Avg Day Prod	Avg Mat'l Unit Cost	Avg Equip Unit Cost	Avg Labor Unit Cost	Avg Total Unit Cost	Avg Price incl O&P

Lighting fixtures (continued)

Incandescent

Ceiling fixture surface mounted

Description	Oper	Unit	Crew Size	Avg Day Prod	Avg Mat'l Unit Cost	Avg Equip Unit Cost	Avg Labor Unit Cost	Avg Total Unit Cost	Avg Price incl O&P	Crew Size	Avg Day Prod	Avg Mat'l Unit Cost	Avg Equip Unit Cost	Avg Labor Unit Cost	Avg Total Unit Cost	Avg Price incl O&P
15" x 5" white bent glass	Inst	EA	E-1	12.0	23.75	---	16.33	40.08	47.43	E-1	9.00	28.50	---	21.77	50.27	60.07
10" x 7" two light circular	Inst	EA	E-1	12.0	23.75	---	16.33	40.08	47.43	E-1	9.00	28.50	---	21.77	50.27	60.07
8" x 8" one light screw-in	Inst	EA	E-1	12.0	18.75	---	16.33	35.08	42.43	E-1	9.00	22.50	---	21.77	44.27	54.07

Ceiling fixture recessed

Square fixture drop or flat lens

Description	Oper	Unit	Crew Size	Avg Day Prod	Avg Mat'l Unit Cost	Avg Equip Unit Cost	Avg Labor Unit Cost	Avg Total Unit Cost	Avg Price incl O&P	Crew Size	Avg Day Prod	Avg Mat'l Unit Cost	Avg Equip Unit Cost	Avg Labor Unit Cost	Avg Total Unit Cost	Avg Price incl O&P
8" frame	Inst	EA	E-1	8.0	25.00	---	24.49	49.49	60.51	E-1	6.00	30.00	---	32.66	62.66	77.35
10 " frame	Inst	EA	E-1	8.0	27.50	---	24.49	51.99	63.01	E-1	6.00	33.00	---	32.66	65.66	80.35
12" frame	Inst	EA	E-1	8.0	31.25	---	24.49	55.74	66.76	E-1	6.00	37.50	---	32.66	70.16	84.85

Round fixture for concentrated light over small areas

Description	Oper	Unit	Crew Size	Avg Day Prod	Avg Mat'l Unit Cost	Avg Equip Unit Cost	Avg Labor Unit Cost	Avg Total Unit Cost	Avg Price incl O&P	Crew Size	Avg Day Prod	Avg Mat'l Unit Cost	Avg Equip Unit Cost	Avg Labor Unit Cost	Avg Total Unit Cost	Avg Price incl O&P
7" shower fixture frame	Inst	EA	E-1	8.0	32.50	---	24.49	56.99	68.01	E-1	6.00	39.00	---	32.66	71.66	86.35
8" spotlight fixture	Inst	EA	E-1	8.0	31.25	---	24.49	55.74	66.76	E-1	6.00	37.50	---	32.66	70.16	84.85

8" flat lens or stepped

Description	Oper	Unit	Crew Size	Avg Day Prod	Avg Mat'l Unit Cost	Avg Equip Unit Cost	Avg Labor Unit Cost	Avg Total Unit Cost	Avg Price incl O&P	Crew Size	Avg Day Prod	Avg Mat'l Unit Cost	Avg Equip Unit Cost	Avg Labor Unit Cost	Avg Total Unit Cost	Avg Price incl O&P
baffle frame	Inst	EA	E-1	8.0	32.50	---	24.49	56.99	68.01	E-1	6.00	39.00	---	32.66	71.66	86.35

Track lighting for highlighting effects from a ceiling or wall. Swivel track heads with the ability to slide head along track to a new position

Track heads

Description	Oper	Unit	Crew Size	Avg Day Prod	Avg Mat'l Unit Cost	Avg Equip Unit Cost	Avg Labor Unit Cost	Avg Total Unit Cost	Avg Price incl O&P	Crew Size	Avg Day Prod	Avg Mat'l Unit Cost	Avg Equip Unit Cost	Avg Labor Unit Cost	Avg Total Unit Cost	Avg Price incl O&P
Large cylinder	Inst	EA	E-1	---	37.50	---	---	37.50	37.50	E-1	---	45.00	---	---	45.00	45.00
Small cylinder	Inst	EA	E-1	---	31.25	---	---	31.25	31.25	E-1	---	37.50	---	---	37.50	37.50
Sphere cylinder	Inst	EA	E-1	---	37.50	---	---	37.50	37.50	E-1	---	45.00	---	---	45.00	45.00

Track; 1-7/16" W x 3/4" D

Description	Oper	Unit	Crew Size	Avg Day Prod	Avg Mat'l Unit Cost	Avg Equip Unit Cost	Avg Labor Unit Cost	Avg Total Unit Cost	Avg Price incl O&P	Crew Size	Avg Day Prod	Avg Mat'l Unit Cost	Avg Equip Unit Cost	Avg Labor Unit Cost	Avg Total Unit Cost	Avg Price incl O&P
2' track	Inst	EA	E-1	24.0	15.00	---	8.16	23.16	26.84	E-1	18.00	18.00	---	10.89	28.89	33.78
4' track	Inst	EA	E-1	16.0	25.00	---	12.25	37.25	42.76	E-1	12.00	30.00	---	16.33	46.33	53.68
8' track	Inst	EA	E-1	12.0	50.00	---	16.33	66.33	73.68	E-1	9.00	60.00	---	21.77	81.77	91.57

Straight connector; joins two track sections end to

Description	Oper	Unit	Crew Size	Avg Day Prod	Avg Mat'l Unit Cost	Avg Equip Unit Cost	Avg Labor Unit Cost	Avg Total Unit Cost	Avg Price incl O&P	Crew Size	Avg Day Prod	Avg Mat'l Unit Cost	Avg Equip Unit Cost	Avg Labor Unit Cost	Avg Total Unit Cost	Avg Price incl O&P
end; 7-5/8" L	Inst	EA	E-1	---	11.25	---	---	11.25	11.25	E-1	---	13.50	---	---	13.50	13.50

L - connector, joins two track sections for 90 degree angle turns;

Description	Oper	Unit	Crew Size	Avg Day Prod	Avg Mat'l Unit Cost	Avg Equip Unit Cost	Avg Labor Unit Cost	Avg Total Unit Cost	Avg Price incl O&P	Crew Size	Avg Day Prod	Avg Mat'l Unit Cost	Avg Equip Unit Cost	Avg Labor Unit Cost	Avg Total Unit Cost	Avg Price incl O&P
7-5/8" L	Inst	EA	E-1	---	11.25	---	---	11.25	11.25	E-1	---	13.50	---	---	13.50	13.50

Feed in unit, attaches to ceiling or wall outlet box. Supplies electrical current

Description	Oper	Unit	Crew Size	Avg Day Prod	Avg Mat'l Unit Cost	Avg Equip Unit Cost	Avg Labor Unit Cost	Avg Total Unit Cost	Avg Price incl O&P	Crew Size	Avg Day Prod	Avg Mat'l Unit Cost	Avg Equip Unit Cost	Avg Labor Unit Cost	Avg Total Unit Cost	Avg Price incl O&P
to all heads on track(s)	Inst	EA	E-1	12.0	12.50	---	16.33	28.83	36.18	E-1	9.00	15.00	---	21.77	36.77	46.57

			Costs Based On Large Volume							Costs Based On Small Volume						
Description	Oper	Unit	Crew Size	Avg Day Prod	Avg Mat'l Unit Cost	Avg Equip Unit Cost	Avg Labor Unit Cost	Avg Total Unit Cost	Avg Price Incl O&P	Crew Size	Avg Day Prod	Avg Mat'l Unit Cost	Avg Equip Unit Cost	Avg Labor Unit Cost	Avg Total Unit Cost	Avg Price Incl O&P

Lighting fixtures (continued)

Wall fixtures
White glass with on/off switch

Description	Oper	Unit	Crew Size	Avg Day Prod	Avg Mat'l Unit Cost	Avg Equip Unit Cost	Avg Labor Unit Cost	Avg Total Unit Cost	Avg Price Incl O&P	Crew Size	Avg Day Prod	Avg Mat'l Unit Cost	Avg Equip Unit Cost	Avg Labor Unit Cost	Avg Total Unit Cost	Avg Price Incl O&P
1 light, 5" W x 5" H	Inst	EA	E-1	12.0	12.50	---	16.33	28.83	36.18	E-1	9.00	15.00	---	21.77	36.77	46.57
2 light 14" W x 5" H	Inst	EA	E-1	12.0	18.75	---	16.33	35.08	42.43	E-1	9.00	22.50	---	21.77	44.27	54.07
4 light, 24" W x 4" H	Inst	EA	E-1	12.0	27.50	---	16.33	43.83	51.18	E-1	9.00	33.00	---	21.77	54.77	64.57
Swivel wall fixture																
1 light, 4" W x 8" H	Inst	EA	E-1	12.0	22.50	---	16.33	38.83	46.18	E-1	9.00	27.00	---	21.77	48.77	58.57
2 light, 9" W x 9" H	Inst	EA	E-1	12.0	33.75	---	16.33	50.08	57.43	E-1	9.00	40.50	---	21.77	62.27	72.07

Outdoor lighting

Ceiling fixture for porch

Description	Oper	Unit	Crew Size	Avg Day Prod	Avg Mat'l Unit Cost	Avg Equip Unit Cost	Avg Labor Unit Cost	Avg Total Unit Cost	Avg Price Incl O&P	Crew Size	Avg Day Prod	Avg Mat'l Unit Cost	Avg Equip Unit Cost	Avg Labor Unit Cost	Avg Total Unit Cost	Avg Price Incl O&P
7" W x 13" L, one 75 watt RS	Inst	EA	E-1	12.0	40.00	---	16.33	56.33	63.68	E-1	9.00	48.00	---	21.77	69.77	79.57
11" W x 4" L, two 60 watt RS	Inst	EA	E-1	12.0	48.75	---	16.33	65.08	72.43	E-1	9.00	58.50	---	21.77	80.27	90.07
15" W x 4" L, three 60 watt RS	Inst	EA	E-1	12.0	52.50	---	16.33	68.83	76.18	E-1	9.00	63.00	---	21.77	84.77	94.57
Wall fixture for porch																
6" W x 10" L, one 75 watt RS	Inst	EA	E-1	12.0	35.00	---	16.33	51.33	58.68	E-1	9.00	42.00	---	21.77	63.77	73.57
6" W x 16" L, one 75 watt RS	Inst	EA	E-1	12.0	43.75	---	16.33	60.08	67.43	E-1	9.00	52.50	---	21.77	74.27	84.07

Post lantern fixture
Aluminum cast posts
84" H post with lantern, set in concrete, includes 50' of conduit, circuit, and

Description	Oper	Unit	Crew Size	Avg Day Prod	Avg Mat'l Unit Cost	Avg Equip Unit Cost	Avg Labor Unit Cost	Avg Total Unit Cost	Avg Price Incl O&P	Crew Size	Avg Day Prod	Avg Mat'l Unit Cost	Avg Equip Unit Cost	Avg Labor Unit Cost	Avg Total Unit Cost	Avg Price Incl O&P
cement	Inst	EA	E-1	1.0	312.50	---	195.93	508.43	596.60	E-1	0.75	375.00	---	261.24	636.24	753.80

Urethane (in form of simulated redwood) over steel post
Post with matching lantern, set in concrete, includes 50' of conduit, circuit,

Description	Oper	Unit	Crew Size	Avg Day Prod	Avg Mat'l Unit Cost	Avg Equip Unit Cost	Avg Labor Unit Cost	Avg Total Unit Cost	Avg Price Incl O&P	Crew Size	Avg Day Prod	Avg Mat'l Unit Cost	Avg Equip Unit Cost	Avg Labor Unit Cost	Avg Total Unit Cost	Avg Price Incl O&P
and cement	Inst	EA	E-1	1.0	375.00	---	195.93	570.93	659.10	E-1	0.75	450.00	---	261.24	711.24	828.80

Linoleum. See Resilient Flooring.

Lumber. See Framing.

Mantels, Fireplace

Ponderosa pine, kiln-dried, unfinished, assembled
Versailles, ornate, French design

Description	Oper	Unit	Crew Size	Avg Day Prod	Avg Mat'l Unit Cost	Avg Equip Unit Cost	Avg Labor Unit Cost	Avg Total Unit Cost	Avg Price Incl O&P	Crew Size	Avg Day Prod	Avg Mat'l Unit Cost	Avg Equip Unit Cost	Avg Labor Unit Cost	Avg Total Unit Cost	Avg Price Incl O&P
59" W x 46" H	Inst	EA	C-9	4.0	1056.25	---	74.65	1130.90	1167.5	C-9	3.00	1183.00	---	99.53	1282.53	1331.3

			Costs Based On Large Volume							Costs Based On Small Volume						
Description	Oper	Unit	Crew Size	Avg Day Prod	Avg Mat'l Unit Cost	Avg Equip Unit Cost	Avg Labor Unit Cost	Avg Total Unit Cost	Avg Price Incl O&P	Crew Size	Avg Day Prod	Avg Mat'l Unit Cost	Avg Equip Unit Cost	Avg Labor Unit Cost	Avg Total Unit Cost	Avg Price Incl O&P

Mantels (continued)

Victorian, ornate, English design

| 63" W x 52" H | Inst | EA | C-9 | 4.0 | 818.75 | --- | 74.65 | 893.40 | 929.98 | C-9 | 3.00 | 917.00 | --- | 99.53 | 1016.53 | 1065.3 |

Chelsea, plain, English design

| 70" W x 52" H | Inst | EA | C-9 | 4.0 | 740.00 | --- | 74.65 | 814.65 | 851.23 | C-9 | 3.00 | 828.80 | --- | 99.53 | 928.33 | 977.10 |

Jamestown, plain, Early American

| 68" W x 53" H | Inst | EA | C-9 | 4.0 | 375.00 | --- | 74.65 | 449.65 | 486.23 | C-9 | 3.00 | 420.00 | --- | 99.53 | 519.53 | 568.30 |

Marlite

| Panels, 4' x 8', adhesive set | Demo | SF | L-2 | 1850 | --- | --- | 0.14 | 0.14 | 0.21 | L-2 | 1295 | --- | --- | 0.20 | 0.20 | 0.30 |

Plastic coated masonite panels; 4' x 8', 5' x 5'; screw applied;

channel molding around perimeter; 1/8" T

| Solid colors | Inst | SF | C-9 | 340 | 1.64 | --- | 0.88 | 2.52 | 2.95 | C-9 | 238 | 2.46 | --- | 1.25 | 3.71 | 4.33 |
| Patterned panels | Inst | SF | C-9 | 340 | 1.97 | --- | 0.88 | 2.84 | 3.27 | C-9 | 238 | 2.95 | --- | 1.25 | 4.20 | 4.82 |

Molding, 1/8" panels; corners, divisions, or edging; nailed to framing or sheathing

Bright anodized	Inst	LF	C-9	255	0.18	---	1.17	1.35	1.93	C-9	179	0.28	---	1.67	1.95	2.77
Gold anodized	Inst	LF	C-9	255	0.21	---	1.17	1.38	1.95	C-9	179	0.32	---	1.67	1.99	2.81
Colors	Inst	LF	C-9	255	0.26	---	1.17	1.43	2.01	C-9	179	0.39	---	1.67	2.07	2.89

Masonry

Brick, Standard

Running bond, 3/8" mortar joints

Chimneys; flue lining included; no scaffolding included

4" T wall with standard brick

16" x 16" with one 8" x 8" flue	Demo	VLF	A-2	15.4	---	3.22	19.41	22.63	31.76	L-2	13.1	---	---	19.87	19.87	29.41
20" x 16" with one 12" x 8" flue	Demo	VLF	A-2	13.3	---	3.73	22.48	26.21	36.77	L-2	11.3	---	---	23.01	23.01	34.05
20" x 20" with one 12" x 12" flue	Demo	VLF	A-2	12.2	---	4.07	24.50	28.57	40.08	L-2	10.4	---	---	25.08	25.08	37.12
28" x 16" with two 8" x 8" flues	Demo	VLF	A-2	10.0	---	4.96	29.89	34.85	48.90	L-2	8.5	---	---	30.60	30.60	45.29
32" x 16" with one 8" x 8" & one 12" x 8" flue	Demo	VLF	A-2	9.4	---	5.28	31.80	37.08	52.03	L-2	8.0	---	---	32.55	32.55	48.18
36" x 16" with two 12" x 8" flues	Demo	VLF	A-2	8.8	---	5.64	33.97	39.61	55.57	L-2	7.5	---	---	34.77	34.77	51.46
36" x 20" with two 12" x 12" flues	Demo	VLF	A-2	7.9	---	6.28	37.84	44.12	61.90	L-2	6.7	---	---	38.73	38.73	57.32

Masonry

Brick. All material costs include mortar and waste, 5% on brick and 30% on mortar.

1. **Dimensions:**

   a. Standard or regular: 8" L x 2-1/4" H x 3-3/4" W.

   b. Modular: 7-5/8" L x 2-1/4" H x 3-5/8" W.

   c. Norman: 11-5/8" L x 2-2/3" H x 3-5/8" W.

   d. Roman: 11-5/8" L x 1-5/8" H x 3-5/8" W.

2. **Installation:**

   a. Mortar joints are 3/8" T both horizontally and vertically.

   b. Mortar mix is 1:3, 1 part masonry cement and 3 parts sand.

   c. Galvanized, corrugated wall ties are used on veneers at 1 tie per SF wall area. The tie is 7/8" x 7" x 16 ga.

   d. Running bond used on veneers and walls.

3. **Notes — Labor**: Output is based on a crew composed of bricklayers and bricktenders at a 1:1 ratio.

4. **Estimating technique**: Chimneys and columns figured per vertical linear foot with allowances already made for brick waste and mortar waste. Veneers and walls are computed per square foot of wall area.

Concrete (masonry) block. All material costs include 3% block waste and 30% mortar waste.

1. **Dimension:** All blocks are two core.

   a. Heavyweight: 8" T blocks weigh approximately 46 lbs/block; 12" T blocks weigh approximately 65 lbs/block.

   b. Lightweight: Also known as haydite blocks. 8" T blocks weigh approximately 30 lbs/block; 12" T blocks weigh approximately 41 lbs/block.

2. **Installation:**

   a. Mortar joints are 3/8" T both horizontally and vertically.

   b. Mortar mix is 1:3, 1 part masonry cement and 3 parts sand.

   c. Reinforcing: Lateral metal is regular truss with 9 gauge sides and ties. Vertical steel is #4 (1/2" dia.) rods, at 0.668 lbs/LF.

3. **Notes — Labor**: Output is based on a crew composed of bricklayers and bricktenders.

4. **Estimating technique:** Chimneys and columns figured per vertical lineal foot with allowances already made for block waste and mortar waste. Veneers and walls are computed per square foot of wall area.

Quarry Tile (on floor). Includes 5% tile waste.

1. Dimensions: 6" square tile is 1/2" T and 9" square tile is 3/4" T.

2. **Installation:**

   a. Conventional mortar set utilizes portland cement, mortar mix, sand, and water. The mortar dry cures and bonds to tile.

   b. Dry-set mortar utilizes dry-set portland cement, mortar mix, sand, and water. The mortar dry cures and bonds to tile.

3. **Notes — Labor:** Output is based on a crew composed of bricklayers and bricktenders.

4. **Estimating technique:** Compute square foot of floor area.

4" Wall thickness

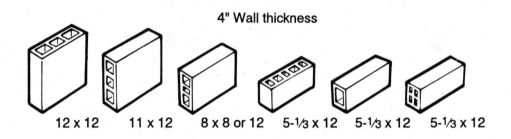

12 x 12   11 x 12   8 x 8 or 12   5-1/3 x 12   5-1/3 x 12   5-1/3 x 12

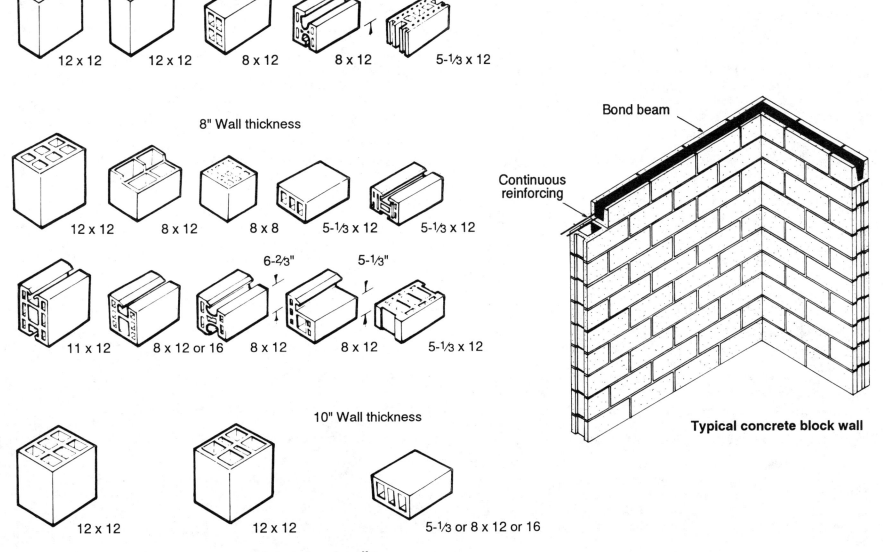

6" Wall thickness

12 x 12   12 x 12   8 x 12   6-2/3" 8 x 12   5-1/3 x 12

8" Wall thickness

12 x 12   8 x 12   8 x 8   5-1/3 x 12   5-1/3 x 12

6-2/3"   5-1/3"

11 x 12   8 x 12 or 16   8 x 12   8 x 12   5-1/3 x 12

10" Wall thickness

12 x 12   12 x 12   5-1/3 or 8 x 12 or 16

Common sizes and shapes for clay tile

Bond beam

Continuous reinforcing

Typical concrete block wall

Description	Oper	Unit	Costs Based On Large Volume							Costs Based On Small Volume						
			Crew Size	Avg Day Prod	Avg Mat'l Unit Cost	Avg Equip Unit Cost	Avg Labor Unit Cost	Avg Total Unit Cost	Avg Price Incl O&P	Crew Size	Avg Day Prod	Avg Mat'l Unit Cost	Avg Equip Unit Cost	Avg Labor Unit Cost	Avg Total Unit Cost	Avg Price Incl O&P

Masonry (continued)

Description	Oper	Unit	Crew Size	Avg Day Prod	Avg Mat'l Unit Cost	Avg Equip Unit Cost	Avg Labor Unit Cost	Avg Total Unit Cost	Avg Price Incl O&P	Crew Size	Avg Day Prod	Avg Mat'l Unit Cost	Avg Equip Unit Cost	Avg Labor Unit Cost	Avg Total Unit Cost	Avg Price Incl O&P
16" x 16" with one 8" x 8" flue	Inst	VLF	B-2	14.3	12.54	---	21.38	33.92	44.18	B-2	12.6	14.42	---	24.30	38.71	50.37
20" x 16" with one 12" x 8" flue	Inst	VLF	B-2	12.6	15.20	---	24.27	39.46	51.11	B-2	11.1	17.48	---	27.57	45.05	58.29
20" x 20" with one 12" x 12" flue	Inst	VLF	B-2	11.3	17.69	---	27.06	44.74	57.73	B-2	9.9	20.34	---	30.75	51.09	65.84
28" x 16" with two 8" x 8" flues	Inst	VLF	B-2	9.4	22.07	---	32.53	54.60	70.21	B-2	8.3	25.38	---	36.96	62.34	80.08
32" x 16" with one 8" x 8" & one 12" x 8" flue	Inst	VLF	B-2	8.8	24.40	---	34.74	59.14	75.82	B-2	7.7	28.06	---	39.48	67.54	86.49
36" x 16" with two 12" x 8" flues	Inst	VLF	B-2	8.2	27.07	---	37.29	64.35	82.25	B-2	7.2	31.12	---	42.37	73.49	93.83
36" x 20" with two 12" x 12" flues	Inst	VLF	B-2	7.4	32.37	---	41.32	73.69	93.52	B-2	6.5	37.23	---	46.95	84.18	106.72
8" T wall with standard brick																
24" x 24" with one 8" x 8" flue	Demo	VLF	A-2	7.2	---	6.89	41.52	48.41	67.92	L-2	6.1	---	---	42.50	42.50	62.90
28" x 24" with one 12" x 8" flue	Demo	VLF	A-2	6.7	---	7.40	44.62	52.02	72.99	L-2	5.7	---	---	45.67	45.67	67.59
28" x 28" with one 12" x 12" flue	Demo	VLF	A-2	6.1	---	8.13	49.01	57.14	80.17	L-2	5.2	---	---	50.16	50.16	74.24
36" x 24" with two 8" x 8" flues	Demo	VLF	A-2	5.5	---	9.02	54.35	63.37	88.92	L-2	4.7	---	---	55.63	55.63	82.34
40" x 24" with one 8" x 8" & one 12" x 8" flue	Demo	VLF	A-2	5.0	---	9.92	59.79	69.71	97.81	L-2	4.3	---	---	61.20	61.20	90.57
44" x 24" with two 12" x 8" flues	Demo	VLF	A-2	4.8	---	10.33	62.28	72.61	101.88	L-2	4.1	---	---	63.75	63.75	94.34
44" x 28" with two 12" x 12" flues	Demo	VLF	A-2	4.5	---	11.02	66.43	77.45	108.67	L-2	3.8	---	---	68.00	68.00	100.63
24" x 24" with one 8" x 8" flue	Inst	VLF	B-2	6.4	28.17	---	47.77	75.94	98.87	B-2	5.6	32.40	---	54.29	86.68	112.74
28" x 24" with one 12" x 8" flue	Inst	VLF	B-2	5.8	32.16	---	52.71	84.88	110.18	B-2	5.1	36.99	---	59.90	96.89	125.64
28" x 28" with one 12" x 12" flue	Inst	VLF	B-2	5.4	36.32	---	56.62	92.93	120.11	B-2	4.8	41.76	---	64.34	106.10	136.99
36" x 24" with two 8" x 8" flues	Inst	VLF	B-2	4.8	42.03	---	63.70	105.72	136.30	B-2	4.2	48.33	---	72.38	120.71	155.46
40" x 24" with one 8" x 8" & one 12" x 8" flue	Inst	VLF	B-2	4.3	47.68	---	71.10	118.79	152.92	B-2	3.8	54.84	---	80.80	135.64	174.42
44" x 24" with two 12" x 8" flues	Inst	VLF	B-2	4.2	50.35	---	72.80	123.14	158.08	B-2	3.7	57.90	---	82.72	140.62	180.33
44" x 28" with two 12" x 12" flues	Inst	VLF	B-2	3.9	57.32	---	78.40	135.71	173.34	B-2	3.4	65.92	---	89.09	155.00	197.76
Columns, outside dimension; no shoring; solid centers; no scaffolding included																
8" x 8"	Demo	VLF	A-2	48.3	---	1.03	6.19	7.22	10.12	L-2	38.6	---	---	6.73	6.73	9.96
12" x 8"	Demo	VLF	A-2	29.3	---	1.69	10.20	11.90	16.69	L-2	23.4	---	---	11.10	11.10	16.42
16" x 8"	Demo	VLF	A-2	22.5	---	2.20	13.29	15.49	21.73	L-2	18.0	---	---	14.45	14.45	21.38
20" x 8"	Demo	VLF	A-2	18.4	---	2.70	16.25	18.94	26.58	L-2	14.7	---	---	17.67	17.67	26.15
24" x 8"	Demo	VLF	A-2	15.7	---	3.16	19.04	22.20	31.15	L-2	12.6	---	---	20.71	20.71	30.65

			Costs Based On Large Volume							Costs Based On Small Volume						
Description	Oper	Unit	Crew Size	Avg Day Prod	Avg Mat'l Unit Cost	Avg Equip Unit Cost	Avg Labor Unit Cost	Avg Total Unit Cost	Avg Price Incl O&P	Crew Size	Avg Day Prod	Avg Mat'l Unit Cost	Avg Equip Unit Cost	Avg Labor Unit Cost	Avg Total Unit Cost	Avg Price Incl O&P

Masonry (continued)

Description	Oper	Unit	Crew Size	Avg Day Prod	Avg Mat'l Unit Cost	Avg Equip Unit Cost	Avg Labor Unit Cost	Avg Total Unit Cost	Avg Price Incl O&P	Crew Size	Avg Day Prod	Avg Mat'l Unit Cost	Avg Equip Unit Cost	Avg Labor Unit Cost	Avg Total Unit Cost	Avg Price Incl O&P
12" x 12"	Demo	VLF	A-2	20.2	---	2.46	14.80	17.25	24.21	L-2	16.2	---	---	16.09	16.09	23.82
16" x 12"	Demo	VLF	A-2	15.7	---	3.16	19.04	22.20	31.15	L-2	12.6	---	---	20.71	20.71	30.65
20" x 12"	Demo	VLF	A-2	13.0	---	3.82	23.00	26.81	37.62	L-2	10.4	---	---	25.01	25.01	37.01
24" x 12"	Demo	VLF	A-2	11.2	---	4.43	26.69	31.12	43.66	L-2	9.0	---	---	29.03	29.03	42.96
28" x 12"	Demo	VLF	A-2	9.9	---	5.01	30.20	35.21	49.40	L-2	7.9	---	---	32.84	32.84	48.60
32" x 12"	Demo	VLF	A-2	8.9	---	5.57	33.59	39.16	54.95	L-2	7.1	---	---	36.53	36.53	54.06
16" x 16"	Demo	VLF	A-2	12.3	---	4.03	24.30	28.34	39.76	A-2	11.1	---	4.48	27.00	31.48	44.18
20" x 16"	Demo	VLF	A-2	10.3	---	4.82	29.02	33.84	47.48	A-2	9.3	---	5.35	32.25	37.60	52.75
24" x 16"	Demo	VLF	A-2	8.9	---	5.57	33.59	39.16	54.95	A-2	8.0	---	6.19	37.32	43.51	61.05
28" x 16"	Demo	VLF	A-2	8.0	---	6.20	37.37	43.57	61.13	A-2	7.2	---	6.89	41.52	48.41	67.92
32" x 16"	Demo	VLF	A-2	7.2	---	6.89	41.52	48.41	67.92	A-2	6.5	---	7.65	46.13	53.79	75.47
36" x 16"	Demo	VLF	A-2	6.7	---	7.40	44.62	52.02	72.99	A-2	6.0	---	8.23	49.57	57.80	81.10
20" x 20"	Demo	VLF	A-2	8.7	---	5.70	34.36	40.06	56.21	A-2	7.8	---	6.33	38.18	44.51	62.46
24" x 20"	Demo	VLF	A-2	7.4	---	6.70	40.40	47.10	66.09	A-2	6.7	---	7.45	44.89	52.33	73.43
28" x 20"	Demo	VLF	A-2	6.7	---	7.40	44.62	52.02	72.99	A-2	6.0	---	8.23	49.57	57.80	81.10
32" x 20"	Demo	VLF	A-2	6.1	---	8.13	49.01	57.14	80.17	A-2	5.5	---	9.03	54.45	63.49	89.08
36" x 20"	Demo	VLF	A-2	5.7	---	8.70	52.44	61.15	85.80	A-2	5.1	---	9.67	58.27	67.94	95.33
24" x 24"	Demo	VLF	A-2	6.7	---	7.40	44.62	52.02	72.99	A-2	6.0	---	8.23	49.57	57.80	81.10
28" x 24"	Demo	VLF	A-2	6.0	---	8.27	49.82	58.09	81.51	A-2	5.4	---	9.19	55.36	64.54	90.56
32" x 24"	Demo	VLF	A-2	5.5	---	9.02	54.35	63.37	88.92	A-2	5.0	---	10.02	60.39	70.41	98.80
36" x 24"	Demo	VLF	A-2	5.1	---	9.73	58.61	68.34	95.89	A-2	4.6	---	10.81	65.13	75.93	106.54
28" x 28"	Demo	VLF	A-2	5.4	---	9.19	55.36	64.54	90.56	A-2	4.9	---	10.21	61.51	71.72	100.62
32" x 28"	Demo	VLF	A-2	4.9	---	10.12	61.01	71.13	99.80	A-2	4.4	---	11.25	67.79	79.03	110.89
36" x 28"	Demo	VLF	A-2	4.5	---	11.02	66.43	77.45	108.67	A-2	4.1	---	12.25	73.81	86.06	120.75
32" x 32"	Demo	VLF	A-2	4.5	---	11.02	66.43	77.45	108.67	A-2	4.1	---	12.25	73.81	86.06	120.75
36" x 32"	Demo	VLF	A-2	4.1	---	12.10	72.91	85.01	119.28	A-2	3.7	---	13.44	81.01	94.45	132.53
36" x 36"	Demo	VLF	A-2	3.8	---	13.05	78.67	91.72	128.69	A-2	3.4	---	14.50	87.41	101.91	142.99
8" x 8" (9.33 Brick/VLF)	Inst	VLF	B-2	38.6	3.09	---	7.92	11.02	14.82	B-2	34.7	3.56	---	8.80	12.36	16.58
12" x 8" (14.00 Brick/VLF)	Inst	VLF	B-2	26.4	4.64	---	11.58	16.23	21.78	B-2	23.8	5.34	---	12.87	18.21	24.39
16" x 8" (18.67 Brick/VLF)	Inst	VLF	B-2	20.4	6.19	---	14.99	21.18	28.37	B-2	18.4	7.12	---	16.65	23.77	31.77
20" x 8" (23.33 Brick/VLF)	Inst	VLF	B-2	16.7	7.74	---	18.31	26.05	34.84	B-2	15.0	8.90	---	20.34	29.24	39.01
24" x 8" (28.00 Brick/VLF)	Inst	VLF	B-2	14.3	9.29	---	21.38	30.67	40.93	B-2	12.9	10.68	---	23.76	34.44	45.84

			Costs Based On Large Volume							Costs Based On Small Volume						
Description	Oper	Unit	Crew Size	Avg Day Prod	Avg Mat'l Unit Cost	Avg Equip Unit Cost	Avg Labor Unit Cost	Avg Total Unit Cost	Avg Price Incl O&P	Crew Size	Avg Day Prod	Avg Mat'l Unit Cost	Avg Equip Unit Cost	Avg Labor Unit Cost	Avg Total Unit Cost	Avg Price Incl O&P

Masonry (continued)

Description	Oper	Unit	Crew Size	Avg Day Prod	Avg Mat'l Unit Cost	Avg Equip Unit Cost	Avg Labor Unit Cost	Avg Total Unit Cost	Avg Price Incl O&P	Crew Size	Avg Day Prod	Avg Mat'l Unit Cost	Avg Equip Unit Cost	Avg Labor Unit Cost	Avg Total Unit Cost	Avg Price Incl O&P
12" x 12" (21.00 Brick/VLF)	Inst	VLF	B-2	18.3	6.97	---	16.71	23.67	31.69	B-2	16.5	8.01	---	18.56	26.57	35.48
16" x 12" (28.00 Brick/VLF)	Inst	VLF	B-2	14.3	9.29	---	21.38	30.67	40.93	B-2	12.9	10.68	---	23.76	34.44	45.84
20" x 12" (35.00 Brick/VLF)	Inst	VLF	B-2	11.9	11.61	---	25.69	37.30	49.64	B-2	10.7	13.35	---	28.55	41.90	55.60
24" x 12" (42.00 Brick/VLF)	Inst	VLF	B-2	10.2	13.93	---	29.97	43.91	58.29	B-2	9.2	16.02	---	33.31	49.33	65.31
28" x 12" (49.00 Brick/VLF)	Inst	VLF	B-2	9.1	16.25	---	33.60	49.85	65.98	B-2	8.2	18.69	---	37.33	56.02	73.94
32" x 12" (56.00 Brick/VLF)	Inst	VLF	B-2	8.2	18.58	---	37.29	55.86	73.76	B-2	7.4	21.36	---	41.43	62.79	82.68
16" x 16" (37.33 Brick/VLF)	Inst	VLF	B-2	11.3	12.38	---	27.06	39.44	52.43	B-2	10.2	14.24	---	30.06	44.30	58.73
20" x 16" (46.67 Brick/VLF)	Inst	VLF	B-2	9.4	15.48	---	32.53	48.01	63.62	B-2	8.5	17.80	---	36.14	53.94	71.29
24" x 16" (56.00 Brick/VLF)	Inst	VLF	B-2	8.2	18.58	---	37.29	55.86	73.76	B-2	7.4	21.36	---	41.43	62.79	82.68
28" x 16" (65.33 Brick/VLF)	Inst	VLF	B-2	7.4	21.67	---	41.32	62.99	82.82	B-2	6.7	24.92	---	45.91	70.83	92.86
32" x 16" (74.67 Brick/VLF)	Inst	VLF	B-2	6.7	24.77	---	45.63	70.40	92.31	B-2	6.0	28.49	---	50.70	79.19	103.53
36" x 16" (84.00 Brick/VLF)	Inst	VLF	B-2	6.2	27.86	---	49.31	77.18	100.85	B-2	5.6	32.04	---	54.79	86.84	113.14
20" x 20" (58.33 Brick/VLF)	Inst	VLF	B-2	8.0	19.35	---	38.22	57.57	75.91	B-2	7.2	22.25	---	42.46	64.72	85.10
24" x 20" (70.00 Brick/VLF)	Inst	VLF	B-2	6.9	23.22	---	44.31	67.53	88.80	B-2	6.2	26.70	---	49.23	75.94	99.57
28" x 20" (81.67 Brick/VLF)	Inst	VLF	B-2	6.2	27.09	---	49.31	76.41	100.08	B-2	5.6	31.16	---	54.79	85.95	112.25
32" x 20" (93.33 Brick/VLF)	Inst	VLF	B-2	5.7	30.96	---	53.64	84.60	110.35	B-2	5.1	35.60	---	59.60	95.20	123.81
36" x 20" (105.00 Brick/VLF)	Inst	VLF	B-2	5.3	34.83	---	57.69	92.52	120.21	B-2	4.8	40.06	---	64.10	104.15	134.92
24" x 24" (84.00 Brick/VLF)	Inst	VLF	B-2	6.2	27.86	---	49.31	77.18	100.85	B-2	5.6	32.04	---	54.79	86.84	113.14
28" x 24" (98.00 Brick/VLF)	Inst	VLF	B-2	5.6	32.51	---	54.60	87.11	113.31	B-2	5.0	37.38	---	60.66	98.05	127.17
32" x 24" (112.00 Brick/VLF)	Inst	VLF	B-2	5.2	37.15	---	58.80	95.95	124.17	B-2	4.7	42.73	---	65.33	108.06	139.41
36" x 24" (126.00 Brick/VLF)	Inst	VLF	B-2	4.8	41.80	---	63.70	105.49	136.07	B-2	4.3	48.07	---	70.77	118.84	152.81
28" x 28" (114.33 Brick/VLF)	Inst	VLF	B-2	5.0	37.93	---	61.15	99.07	128.43	B-2	4.5	43.61	---	67.94	111.56	144.17
32" x 28" (130.67 Brick/VLF)	Inst	VLF	B-2	4.6	43.35	---	66.47	109.81	141.72	B-2	4.1	49.85	---	73.85	123.70	159.15
36" x 28" (147.00 Brick/VLF)	Inst	VLF	B-2	4.1	48.76	---	74.57	123.33	159.13	B-2	3.7	56.08	---	82.86	138.93	178.71
32" x 32" (149.33 Brick/VLF)	Inst	VLF	B-2	4.0	49.54	---	76.44	125.97	162.66	B-2	3.6	56.97	---	84.93	141.89	182.66
36" x 32" (168.00 Brick/VLF)	Inst	VLF	B-2	3.6	55.73	---	84.93	140.66	181.42	B-2	3.2	64.09	---	94.36	158.45	203.75
36" x 36" (189.00 Brick/VLF)	Inst	VLF	B-2	3.2	62.70	---	95.54	158.24	204.10	B-2	2.9	72.10	---	106.16	178.26	229.22
Veneer, 4" T, with air tools	Demo	SF	A-2	325	---	0.15	0.92	1.07	1.50	A-2	286	---	0.17	1.05	1.22	1.71
Veneers, 4" T, with wall ties; no scaffolding included																
Common, 8" x 2-2/3" x 4"	Inst	SF	B-4	235	2.54	0.15	3.35	6.04	7.65	B-3	175	2.92	0.20	3.49	6.62	8.29
Standard Face, 8" x 2-2/3" x 4"	Inst	SF	B-4	220	2.74	0.16	3.58	6.48	8.20	B-3	165	3.15	0.21	3.71	7.07	8.85
Glazed, 8" x 2-2/3" x 4"	Inst	SF	B-4	210	6.86	0.17	3.75	10.78	12.58	B-3	155	7.89	0.23	3.95	12.06	13.96

			Costs Based On Large Volume							Costs Based On Small Volume					
Description	Oper	Unit	Crew Size	Avg Day Prod	Avg Mat'l Unit Cost	Avg Equip Unit Cost	Avg Labor Unit Cost	Avg Total Unit Cost	Avg Price Incl O&P	Crew Size	Avg Day Prod	Avg Mat'l Unit Cost	Avg Equip Unit Cost	Avg Labor Unit Cost	Avg Total Unit Cost	Avg Price Incl O&P

Masonry (continued)

Other brick types/sizes

Description	Oper	Unit	Crew Size	Avg Day Prod	Avg Mat'l Unit Cost	Avg Equip Unit Cost	Avg Labor Unit Cost	Avg Total Unit Cost	Avg Price Incl O&P	Crew Size	Avg Day Prod	Avg Mat'l Unit Cost	Avg Equip Unit Cost	Avg Labor Unit Cost	Avg Total Unit Cost	Avg Price Incl O&P
8" x 4" x 4"	Inst	SF	B-4	315	2.67	0.11	2.50	5.28	6.48	B-3	235	3.07	0.15	2.60	5.82	7.07
8" x 3-1/5" x 4"	Inst	SF	B-4	265	3.72	0.13	2.97	6.83	8.25	B-3	195	4.28	0.18	3.14	7.60	9.10
12" x 4" x 6"	Inst	SF	B-4	440	3.90	0.08	1.79	5.77	6.63	B-3	325	4.48	0.11	1.88	6.47	7.37
12" x 2-2/3" x 4"	Inst	SF	B-4	325	2.85	0.11	2.42	5.38	6.54	B-3	240	3.28	0.15	2.55	5.97	7.20
12" x 3-1/5" x 4"	Inst	SF	B-4	375	2.65	0.09	2.10	4.84	5.85	B-3	280	3.04	0.13	2.18	5.35	6.40
12" x 2" x 4"	Inst	SF	B-4	250	4.27	0.14	3.15	7.56	9.07	B-3	185	4.91	0.19	3.31	8.41	9.99
12" x 2-2/3" x 6"	Inst	SF	B-4	315	3.75	0.11	2.50	6.36	7.56	B-3	235	4.31	0.15	2.60	7.06	8.31
12" x 4" x 4"	Inst	SF	B-4	450	3.03	0.08	1.75	4.86	5.70	B-3	335	3.48	0.11	1.83	5.41	6.29

Walls, with air tools

Description	Oper	Unit	Crew Size	Avg Day Prod	Avg Mat'l Unit Cost	Avg Equip Unit Cost	Avg Labor Unit Cost	Avg Total Unit Cost	Avg Price Incl O&P	Crew Size	Avg Day Prod	Avg Mat'l Unit Cost	Avg Equip Unit Cost	Avg Labor Unit Cost	Avg Total Unit Cost	Avg Price Incl O&P
8" T	Demo	SF	A-2	185	---	0.27	1.62	1.88	2.64	A-2	165	---	0.30	1.81	2.11	2.96
12" T	Demo	SF	A-2	130	---	0.38	2.30	2.68	3.76	A-2	115	---	0.43	2.60	3.03	4.25
16" T	Demo	SF	A-2	100	---	0.50	2.99	3.49	4.89	A-2	90	---	0.55	3.32	3.87	5.43
24" T	Demo	SF	A-2	85	---	0.58	3.52	4.10	5.75	A-2	75	---	0.66	3.99	4.65	6.52

Walls with common brick; no scaffolding included

Description	Oper	Unit	Crew Size	Avg Day Prod	Avg Mat'l Unit Cost	Avg Equip Unit Cost	Avg Labor Unit Cost	Avg Total Unit Cost	Avg Price Incl O&P	Crew Size	Avg Day Prod	Avg Mat'l Unit Cost	Avg Equip Unit Cost	Avg Labor Unit Cost	Avg Total Unit Cost	Avg Price Incl O&P
8" T (13.50 Brick/SF)	Inst	SF	B-3	95	4.08	0.37	6.44	10.89	13.98	B-5	60	4.69	0.59	8.03	13.31	17.16
12" T (20.25 Brick/SF)	Inst	SF	B-3	70	6.12	0.50	6.88	13.50	16.81	B-5	45	7.04	0.78	10.70	18.53	23.66
16" T (27.00 Brick/SF)	Inst	SF	B-3	55	8.16	0.64	8.76	17.56	21.76	B-5	35	9.38	1.01	13.76	24.15	30.76
24" T (40.50 Brick/SF)	Inst	SF	B-3	40	12.24	0.88	12.04	25.16	30.94	B-5	25	14.08	1.41	19.27	34.75	44.00

Brick, Adobe

Running bond, 3/8" mortar joints

Walls, with air tools

Description	Oper	Unit	Crew Size	Avg Day Prod	Avg Mat'l Unit Cost	Avg Equip Unit Cost	Avg Labor Unit Cost	Avg Total Unit Cost	Avg Price Incl O&P	Crew Size	Avg Day Prod	Avg Mat'l Unit Cost	Avg Equip Unit Cost	Avg Labor Unit Cost	Avg Total Unit Cost	Avg Price Incl O&P
4" T	Demo	SF	A-2	390	---	0.13	0.77	0.89	1.25	A-2	330	---	0.15	0.91	1.06	1.48
6" T	Demo	SF	A-2	380	---	0.13	0.79	0.92	1.29	A-2	325	---	0.15	0.92	1.07	1.50
8" T	Demo	SF	A-2	370	---	0.13	0.81	0.94	1.32	A-2	315	---	0.16	0.95	1.11	1.55
12" T	Demo	SF	A-2	350	---	0.14	0.85	1.00	1.40	A-2	295	---	0.17	1.01	1.18	1.66

Walls; no scaffolding included

Description	Oper	Unit	Crew Size	Avg Day Prod	Avg Mat'l Unit Cost	Avg Equip Unit Cost	Avg Labor Unit Cost	Avg Total Unit Cost	Avg Price Incl O&P	Crew Size	Avg Day Prod	Avg Mat'l Unit Cost	Avg Equip Unit Cost	Avg Labor Unit Cost	Avg Total Unit Cost	Avg Price Incl O&P
4" x 4" x 16"	Inst	SF	B-3	230	3.98	0.15	2.66	6.79	8.07	B-5	150	4.58	0.24	3.21	8.02	9.56
6" x 4" x 16"	Inst	SF	B-3	220	5.12	0.16	2.19	7.47	8.52	B-5	145	5.89	0.24	3.32	9.45	11.05
8" x 4" x 16"	Inst	SF	B-3	210	6.15	0.17	2.29	8.61	9.71	B-5	135	7.07	0.26	3.57	10.90	12.61
12" x 4" x 16"	Inst	SF	B-3	190	7.97	0.19	2.53	10.69	11.91	B-5	125	9.16	0.28	3.85	13.30	15.15

			Costs Based On Large Volume							Costs Based On Small Volume						
Description	Oper	Unit	Crew Size	Avg Day Prod	Avg Mat'l Unit Cost	Avg Equip Unit Cost	Avg Labor Unit Cost	Avg Total Unit Cost	Avg Price Incl O&P	Crew Size	Avg Day Prod	Avg Mat'l Unit Cost	Avg Equip Unit Cost	Avg Labor Unit Cost	Avg Total Unit Cost	Avg Price Incl O&P

Masonry (continued)

Concrete block

Lightweight (haydite) or heavyweight blocks;
2 cores/block, solid face; includes allowances for lintels, bond beams.
Foundations and retaining walls; no excavation included
Without reinforcing or with lateral reinforcing only
With air tools

Description	Oper	Unit	Crew Size	Avg Day Prod	Avg Mat'l Unit Cost	Avg Equip Unit Cost	Avg Labor Unit Cost	Avg Total Unit Cost	Avg Price Incl O&P	Crew Size	Avg Day Prod	Avg Mat'l Unit Cost	Avg Equip Unit Cost	Avg Labor Unit Cost	Avg Total Unit Cost	Avg Price Incl O&P
8" W x 8" H x 16" L	Demo	SF	A-2	335	---	0.15	0.89	1.04	1.46	A-2	295	---	0.17	1.01	1.18	1.66
10" W x 8" H x 16" L	Demo	SF	A-2	310	---	0.16	0.96	1.12	1.58	A-2	275	---	0.18	1.09	1.27	1.78
12" W x 8" H x 16" L	Demo	SF	A-2	290	---	0.17	1.03	1.20	1.69	A-2	255	---	0.19	1.17	1.37	1.92

Without air tools

Description	Oper	Unit	Crew Size	Avg Day Prod	Avg Mat'l Unit Cost	Avg Equip Unit Cost	Avg Labor Unit Cost	Avg Total Unit Cost	Avg Price Incl O&P	Crew Size	Avg Day Prod	Avg Mat'l Unit Cost	Avg Equip Unit Cost	Avg Labor Unit Cost	Avg Total Unit Cost	Avg Price Incl O&P
8" W x 8" H x 16" L	Demo	SF	L-2	270	---	---	0.96	0.96	1.43	L-2	235	---	---	1.11	1.11	1.64
10" W x 8" H x 16" L	Demo	SF	L-2	250	---	---	1.04	1.04	1.54	L-2	215	---	---	1.21	1.21	1.79
12" W x 8" H x 16" L	Demo	SF	L-2	230	---	---	1.13	1.13	1.67	L-2	200	---	---	1.30	1.30	1.92

With vertical reinforcing in every other core (2 core blocks) with core concrete filled
With air tools

Description	Oper	Unit	Crew Size	Avg Day Prod	Avg Mat'l Unit Cost	Avg Equip Unit Cost	Avg Labor Unit Cost	Avg Total Unit Cost	Avg Price Incl O&P	Crew Size	Avg Day Prod	Avg Mat'l Unit Cost	Avg Equip Unit Cost	Avg Labor Unit Cost	Avg Total Unit Cost	Avg Price Incl O&P
8" W x 8" H x 16" L	Demo	SF	A-2	205	---	0.24	1.46	1.70	2.39	A-2	180	---	0.28	1.66	1.94	2.72
10" W x 8" H x 16" L	Demo	SF	A-2	190	---	0.26	1.57	1.83	2.57	A-2	165	---	0.30	1.81	2.11	2.96
12" W x 8" H x 16" L	Demo	SF	A-2	175	---	0.28	1.71	1.99	2.79	A-2	155	---	0.32	1.93	2.25	3.16

Foundations and retaining walls
No reinforcing
Heavyweight blocks

Description	Oper	Unit	Crew Size	Avg Day Prod	Avg Mat'l Unit Cost	Avg Equip Unit Cost	Avg Labor Unit Cost	Avg Total Unit Cost	Avg Price Incl O&P	Crew Size	Avg Day Prod	Avg Mat'l Unit Cost	Avg Equip Unit Cost	Avg Labor Unit Cost	Avg Total Unit Cost	Avg Price Incl O&P
8" x 8" x 16"	Inst	SF	B-4	315	1.77	0.11	2.50	4.38	5.58	B-3	225	2.04	0.16	2.72	4.91	6.22
10" x 8" x 16"	Inst	SF	B-4	290	2.31	0.12	2.72	5.14	6.45	B-3	205	2.65	0.17	2.98	5.81	7.24
12" x 8" x 16"	Inst	SF	B-4	265	2.61	0.13	2.97	5.71	7.14	B-3	190	3.00	0.19	3.22	6.41	7.95

Lightweight blocks

Description	Oper	Unit	Crew Size	Avg Day Prod	Avg Mat'l Unit Cost	Avg Equip Unit Cost	Avg Labor Unit Cost	Avg Total Unit Cost	Avg Price Incl O&P	Crew Size	Avg Day Prod	Avg Mat'l Unit Cost	Avg Equip Unit Cost	Avg Labor Unit Cost	Avg Total Unit Cost	Avg Price Incl O&P
8" x 8" x 16"	Inst	SF	B-4	350	2.06	0.10	2.25	4.41	5.49	B-3	250	2.37	0.14	2.45	4.96	6.13
10" x 8" x 16"	Inst	SF	B-4	320	2.60	0.11	2.46	5.17	6.35	B-3	230	2.99	0.15	2.66	5.80	7.07
12" x 8" x 16"	Inst	SF	B-4	295	2.90	0.12	2.67	5.69	6.97	B-3	210	3.33	0.17	2.91	6.41	7.81

Lateral reinforcing every second course
Heavyweight blocks

Description	Oper	Unit	Crew Size	Avg Day Prod	Avg Mat'l Unit Cost	Avg Equip Unit Cost	Avg Labor Unit Cost	Avg Total Unit Cost	Avg Price Incl O&P	Crew Size	Avg Day Prod	Avg Mat'l Unit Cost	Avg Equip Unit Cost	Avg Labor Unit Cost	Avg Total Unit Cost	Avg Price Incl O&P
8" x 8" x 16"	Inst	SF	B-4	300	2.07	0.12	2.62	4.81	6.07	B-3	215	2.38	0.16	2.84	5.39	6.76
10" x 8" x 16"	Inst	SF	B-4	275	2.61	0.13	2.86	5.60	6.97	B-3	195	3.00	0.18	3.14	6.31	7.82
12" x 8" x 16"	Inst	SF	B-4	250	2.91	0.14	3.15	6.20	7.71	B-3	180	3.35	0.20	3.40	6.94	8.57

Masonry (continued)

			Costs Based On Large Volume							Costs Based On Small Volume						
Description	Oper	Unit	Crew Size	Avg Day Prod	Avg Mat'l Unit Cost	Avg Equip Unit Cost	Avg Labor Unit Cost	Avg Total Unit Cost	Avg Price Incl O&P	Crew Size	Avg Day Prod	Avg Mat'l Unit Cost	Avg Equip Unit Cost	Avg Labor Unit Cost	Avg Total Unit Cost	Avg Price Incl O&P
Lightweight blocks																
8" x 8" x 16"	Inst	SF	B-4	335	2.36	0.11	2.35	4.82	5.95	B-3	240	2.72	0.15	2.55	5.41	6.63
10" x 8" x 16"	Inst	SF	B-4	305	2.90	0.12	2.58	5.59	6.83	B-3	220	3.33	0.16	2.78	6.27	7.61
12" x 8" x 16"	Inst	SF	B-4	280	3.20	0.13	2.81	6.14	7.49	B-3	200	3.68	0.18	3.06	6.91	8.38
Vertical reinforcing (No. 4 rod) every second core with core concrete filled																
Heavyweight blocks																
8" x 8" x 16"	Inst	SF	B-4	220	2.30	0.16	3.58	6.04	7.76	B-4	195	2.65	0.18	4.04	6.87	8.80
10" x 8" x 16"	Inst	SF	B-4	200	2.94	0.18	3.94	7.05	8.94	B-4	180	3.38	0.20	4.37	7.95	10.05
12" x 8" x 16"	Inst	SF	B-4	185	3.36	0.19	4.26	7.81	9.85	B-4	165	3.87	0.21	4.77	8.85	11.14
Lightweight blocks																
8" x 8" x 16"	Inst	SF	B-4	240	2.59	0.15	3.28	6.02	7.59	B-4	215	2.98	0.16	3.66	6.81	8.56
10" x 8" x 16"	Inst	SF	B-4	220	3.23	0.16	3.58	6.97	8.69	B-4	195	3.71	0.18	4.04	7.93	9.87
12" x 8" x 16"	Inst	SF	B-4	200	3.65	0.18	3.94	7.77	9.66	B-4	180	4.20	0.20	4.37	8.77	10.87
Exterior walls (above grade); no bracing or shoring included																
Without reinforcing or with lateral reinforcing only																
With air tools																
8" W x 8" H x 16" L	Demo	SF	A-2	335	---	0.15	0.89	1.04	1.46	A-2	295	---	0.17	1.01	1.18	1.66
10" W x 8" H x 16" L	Demo	SF	A-2	305	---	0.16	0.98	1.14	1.60	A-2	265	---	0.19	1.13	1.32	1.85
12" W x 8" H x 16" L	Demo	SF	A-2	275	---	0.18	1.09	1.27	1.78	A-2	240	---	0.21	1.25	1.45	2.04
Without air tools																
8" W x 8" H x 16" L	Demo	SF	L-2	265	---	---	0.98	0.98	1.45	L-2	230	---	---	1.13	1.13	1.67
10" W x 8" H x 16" L	Demo	SF	L-2	240	---	---	1.08	1.08	1.60	L-2	210	---	---	1.24	1.24	1.83
12" W x 8" H x 16" L	Demo	SF	L-2	220	---	---	1.18	1.18	1.75	L-2	190	---	---	1.37	1.37	2.03
Exterior walls																
No reinforcing																
Heavyweight blocks																
8" x 8" x 16"	Inst	SF	B-4	295	1.77	0.12	2.67	4.56	5.84	B-3	210	2.04	0.17	2.91	5.12	6.52
10" x 8" x 16"	Inst	SF	B-4	270	2.31	0.13	2.92	5.35	6.75	B-3	195	2.65	0.18	3.14	5.97	7.47
12" x 8" x 16"	Inst	SF	B-4	250	2.61	0.14	3.15	5.90	7.41	B-3	180	3.00	0.20	3.40	6.59	8.22
Lightweight blocks																
8" x 8" x 16"	Inst	SF	B-4	330	2.06	0.11	2.39	4.55	5.70	B-3	235	2.37	0.15	2.60	5.12	6.37
10" x 8" x 16"	Inst	SF	B-4	305	2.60	0.12	2.58	5.29	6.53	B-3	215	2.99	0.16	2.84	5.99	7.36
12" x 8" x 16"	Inst	SF	B-4	280	2.90	0.13	2.81	5.84	7.19	B-3	200	3.33	0.18	3.06	6.57	8.04

			Costs Based On Large Volume								Costs Based On Small Volume					
Description	Oper	Unit	Crew Size	Avg Day Prod	Avg Mat'l Unit Cost	Avg Equip Unit Cost	Avg Labor Unit Cost	Avg Total Unit Cost	Avg Price Incl O&P	Crew Size	Avg Day Prod	Avg Mat'l Unit Cost	Avg Equip Unit Cost	Avg Labor Unit Cost	Avg Total Unit Cost	Avg Price Incl O&P

Masonry (continued)

Lateral reinforcing every second course
Heavyweight blocks

Description	Oper	Unit	Crew Size	Avg Day Prod	Avg Mat'l Unit Cost	Avg Equip Unit Cost	Avg Labor Unit Cost	Avg Total Unit Cost	Avg Price Incl O&P	Crew Size	Avg Day Prod	Avg Mat'l Unit Cost	Avg Equip Unit Cost	Avg Labor Unit Cost	Avg Total Unit Cost	Avg Price Incl O&P
8" x 8" x 16"	Inst	SF	B-4	280	1.97	0.13	2.81	4.91	6.26	B-3	200	2.27	0.18	3.06	5.50	6.97
10" x 8" x 16"	Inst	SF	B-4	260	2.51	0.14	3.03	5.67	7.12	B-3	185	2.88	0.19	3.31	6.38	7.97
12" x 8" x 16"	Inst	SF	B-4	240	2.81	0.15	3.28	6.24	7.81	B-3	170	3.23	0.21	3.60	7.04	8.76

Lightweight blocks

Description	Oper	Unit	Crew Size	Avg Day Prod	Avg Mat'l Unit Cost	Avg Equip Unit Cost	Avg Labor Unit Cost	Avg Total Unit Cost	Avg Price Incl O&P	Crew Size	Avg Day Prod	Avg Mat'l Unit Cost	Avg Equip Unit Cost	Avg Labor Unit Cost	Avg Total Unit Cost	Avg Price Incl O&P
8" x 8" x 16"	Inst	SF	B-4	315	2.26	0.11	2.50	4.87	6.07	B-3	225	2.60	0.16	2.72	5.48	6.78
10" x 8" x 16"	Inst	SF	B-4	290	2.80	0.12	2.72	5.63	6.94	B-3	205	3.22	0.17	2.98	6.37	7.80
12" x 8" x 16"	Inst	SF	B-4	265	3.10	0.13	2.97	6.20	7.63	B-3	190	3.56	0.19	3.22	6.97	8.51

Partitions (above grade); no bracing or shoring included
Without reinforcing or with lateral reinforcing only
With air tools

Description	Oper	Unit	Crew Size	Avg Day Prod	Avg Mat'l Unit Cost	Avg Equip Unit Cost	Avg Labor Unit Cost	Avg Total Unit Cost	Avg Price Incl O&P	Crew Size	Avg Day Prod	Avg Mat'l Unit Cost	Avg Equip Unit Cost	Avg Labor Unit Cost	Avg Total Unit Cost	Avg Price Incl O&P
4" W x 8" H x 16" L	Demo	SF	A-2	330	---	0.15	0.91	1.06	1.48	A-2	290	---	0.17	1.03	1.20	1.69
6" W x 8" H x 16" L	Demo	SF	A-2	310	---	0.16	0.96	1.12	1.58	A-2	275	---	0.18	1.09	1.27	1.78
8" W x 8" H x 16" L	Demo	SF	A-2	295	---	0.17	1.01	1.18	1.66	A-2	260	---	0.19	1.15	1.34	1.88
10" W x 8" H x 16" L	Demo	SF	A-2	265	---	0.19	1.13	1.32	1.85	A-2	230	---	0.22	1.30	1.52	2.13
12" W x 8" H x 16" L	Demo	SF	A-2	235	---	0.21	1.27	1.48	2.08	A-2	205	---	0.24	1.46	1.70	2.39

Partitions
No reinforcing
Heavyweight blocks

Description	Oper	Unit	Crew Size	Avg Day Prod	Avg Mat'l Unit Cost	Avg Equip Unit Cost	Avg Labor Unit Cost	Avg Total Unit Cost	Avg Price Incl O&P	Crew Size	Avg Day Prod	Avg Mat'l Unit Cost	Avg Equip Unit Cost	Avg Labor Unit Cost	Avg Total Unit Cost	Avg Price Incl O&P
4" W x 8" H x 16" L	Inst	SF	B-4	310	1.12	0.11	2.54	3.77	4.99	B-3	220	1.28	0.16	2.78	4.22	5.56
6" W x 8" H x 16" L	Inst	SF	B-4	290	1.38	0.12	2.72	4.22	5.52	B-3	205	1.59	0.17	2.98	4.74	6.18
8" W x 8" H x 16" L	Inst	SF	B-4	265	1.77	0.13	2.97	4.88	6.30	B-3	190	2.04	0.19	3.22	5.44	6.99
10" W x 8" H x 16" L	Inst	SF	B-4	245	2.31	0.14	3.21	5.66	7.21	B-3	175	2.65	0.20	3.49	6.35	8.03
12" W x 8" H x 16" L	Inst	SF	B-4	225	2.61	0.16	3.50	6.26	7.94	B-3	160	3.00	0.22	3.82	7.04	8.88

Lightweight blocks

Description	Oper	Unit	Crew Size	Avg Day Prod	Avg Mat'l Unit Cost	Avg Equip Unit Cost	Avg Labor Unit Cost	Avg Total Unit Cost	Avg Price Incl O&P	Crew Size	Avg Day Prod	Avg Mat'l Unit Cost	Avg Equip Unit Cost	Avg Labor Unit Cost	Avg Total Unit Cost	Avg Price Incl O&P
4" W x 8" H x 16" L	Inst	SF	B-4	335	1.41	0.11	2.35	3.86	4.99	B-3	240	1.62	0.15	2.55	4.31	5.54
6" W x 8" H x 16" L	Inst	SF	B-4	315	1.67	0.11	2.50	4.28	5.48	B-3	225	1.92	0.16	2.72	4.80	6.10
8" W x 8" H x 16" L	Inst	SF	B-4	295	2.06	0.12	2.67	4.85	6.13	B-3	210	2.37	0.17	2.91	5.45	6.85
10" W x 8" H x 16" L	Inst	SF	B-4	270	2.60	0.13	2.92	5.64	7.04	B-3	195	2.99	0.18	3.14	6.30	7.81
12" W x 8" H x 16" L	Inst	SF	B-4	250	2.90	0.14	3.15	6.19	7.70	B-3	180	3.33	0.20	3.40	6.93	8.56

			Costs Based On Large Volume							Costs Based On Small Volume					
Description	Oper	Unit	Crew Size	Avg Day Prod	Avg Mat'l Unit Cost	Avg Equip Unit Cost	Avg Labor Unit Cost	Avg Total Unit Cost	Avg Price Incl O&P	Crew Size	Avg Day Prod	Avg Mat'l Unit Cost	Avg Equip Unit Cost	Avg Labor Unit Cost	Avg Total Unit Cost	Avg Price Incl O&P

Masonry (continued)

Lateral reinforcing every second course
Heavyweight blocks

Description	Oper	Unit	Crew Size	Avg Day Prod	Avg Mat'l Unit Cost	Avg Equip Unit Cost	Avg Labor Unit Cost	Avg Total Unit Cost	Avg Price Incl O&P	Crew Size	Avg Day Prod	Avg Mat'l Unit Cost	Avg Equip Unit Cost	Avg Labor Unit Cost	Avg Total Unit Cost	Avg Price Incl O&P
4" W x 8" H x 16" L	Inst	SF	B-4	295	1.32	0.12	2.67	4.11	5.39	B-3	210	1.51	0.17	2.91	4.59	5.99
6" W x 8" H x 16" L	Inst	SF	B-4	275	1.58	0.13	2.86	4.57	5.95	B-3	195	1.82	0.18	3.14	5.14	6.64
8" W x 8" H x 16" L	Inst	SF	B-4	250	1.97	0.14	3.15	5.26	6.78	B-3	180	2.27	0.20	3.40	5.86	7.49
10" W x 8" H x 16" L	Inst	SF	B-4	230	2.51	0.15	3.42	6.08	7.73	B-3	165	2.88	0.21	3.71	6.80	8.58
12" W x 8" H x 16" L	Inst	SF	B-4	210	2.81	0.17	3.75	6.73	8.53	B-3	150	3.23	0.24	4.08	7.54	9.50
Lightweight blocks																
4" W x 8" H x 16" L	Inst	SF	B-4	320	1.61	0.11	2.46	4.18	5.36	B-3	230	1.85	0.15	2.66	4.66	5.94
6" W x 8" H x 16" L	Inst	SF	B-4	300	1.87	0.12	2.62	4.61	5.87	B-3	215	2.15	0.16	2.84	5.16	6.53
8" W x 8" H x 16" L	Inst	SF	B-4	280	2.26	0.13	2.81	5.20	6.55	B-3	200	2.60	0.18	3.06	5.84	7.30
10" W x 8" H x 16" L	Inst	SF	B-4	260	2.80	0.14	3.03	5.96	7.41	B-3	185	3.22	0.19	3.31	6.71	8.30
12" W x 8" H x 16" L	Inst	SF	B-4	240	3.10	0.15	3.28	6.53	8.10	B-3	170	3.56	0.21	3.60	7.37	9.09

Fences, without reinforcing or with lateral reinforcing only
With air tool

Description	Oper	Unit	Crew Size	Avg Day Prod	Avg Mat'l Unit Cost	Avg Equip Unit Cost	Avg Labor Unit Cost	Avg Total Unit Cost	Avg Price Incl O&P	Crew Size	Avg Day Prod	Avg Mat'l Unit Cost	Avg Equip Unit Cost	Avg Labor Unit Cost	Avg Total Unit Cost	Avg Price Incl O&P
6" W x 4" H x 16" L	Demo	SF	A-2	390	---	0.13	0.77	0.89	1.25	A-2	365	---	0.14	0.82	0.95	1.34
6" W x 6" H x 16" L	Demo	SF	A-2	370	---	0.13	0.81	0.94	1.32	A-2	345	---	0.14	0.87	1.01	1.42
8" W x 8" H x 16" L	Demo	SF	A-2	350	---	0.14	0.85	1.00	1.40	A-2	325	---	0.15	0.92	1.07	1.50
10" W x 8" H x 16" L	Demo	SF	A-2	325	---	0.15	0.92	1.07	1.50	A-2	300	---	0.17	1.00	1.16	1.63
12" W x 8" H x 16" L	Demo	SF	A-2	300	---	0.17	1.00	1.16	1.63	A-2	275	---	0.18	1.09	1.27	1.78
Without air tools																
6" W x 4" H x 16" L	Demo	SF	L-2	310	---	---	0.84	0.84	1.24	L-2	295	---	---	0.88	0.88	1.30
6" W x 6" H x 16" L	Demo	SF	L-2	295	---	---	0.88	0.88	1.30	L-2	270	---	---	0.96	0.96	1.43
8" W x 8" H x 16" L	Demo	SF	L-2	280	---	---	0.93	0.93	1.37	L-2	265	---	---	0.98	0.98	1.45
10" W x 8" H x 16" L	Demo	SF	L-2	220	---	---	1.18	1.18	1.75	L-2	245	---	---	1.06	1.06	1.57
12" W x 8" H x 16" L	Demo	SF	L-2	240	---	---	1.08	1.08	1.60	L-2	225	---	---	1.16	1.16	1.71

Fences, lightweight blocks
No reinforcing

Description	Oper	Unit	Crew Size	Avg Day Prod	Avg Mat'l Unit Cost	Avg Equip Unit Cost	Avg Labor Unit Cost	Avg Total Unit Cost	Avg Price Incl O&P	Crew Size	Avg Day Prod	Avg Mat'l Unit Cost	Avg Equip Unit Cost	Avg Labor Unit Cost	Avg Total Unit Cost	Avg Price Incl O&P
4" W x 8" H x 16" L	Inst	SF	B-4	370	1.41	0.10	2.13	3.63	4.65	B-3	265	1.62	0.13	2.31	4.06	5.17
6" W x 4" H x 16" L	Inst	SF	B-4	180	3.00	0.20	4.37	7.57	9.67	B-3	130	3.45	0.27	4.70	8.42	10.68
6" W x 6" H x 16" L	Inst	SF	B-4	260	2.14	0.14	3.03	5.30	6.76	B-3	185	2.46	0.19	3.31	5.96	7.54

Description	Oper	Unit	Costs Based On Large Volume							Costs Based On Small Volume						
			Crew Size	Avg Day Prod	Avg Mat'l Unit Cost	Avg Equip Unit Cost	Avg Labor Unit Cost	Avg Total Unit Cost	Avg Price Incl O&P	Crew Size	Avg Day Prod	Avg Mat'l Unit Cost	Avg Equip Unit Cost	Avg Labor Unit Cost	Avg Total Unit Cost	Avg Price Incl O&P

Masonry (continued)

Description	Oper	Unit	Crew Size	Avg Day Prod	Avg Mat'l Unit Cost	Avg Equip Unit Cost	Avg Labor Unit Cost	Avg Total Unit Cost	Avg Price Incl O&P	Crew Size	Avg Day Prod	Avg Mat'l Unit Cost	Avg Equip Unit Cost	Avg Labor Unit Cost	Avg Total Unit Cost	Avg Price Incl O&P
6" W x 8" H x 16" L	Inst	SF	B-4	350	1.67	0.10	2.25	4.02	5.10	B-3	250	1.92	0.14	2.45	4.51	5.68
8" W x 8" H x 16" L	Inst	SF	B-4	330	2.06	0.11	2.39	4.56	5.70	B-3	235	2.37	0.15	2.60	5.13	6.37
10" W x 8" H x 16" L	Inst	SF	B-4	305	2.60	0.12	2.58	5.29	6.53	B-3	215	2.99	0.16	2.84	5.99	7.36
12" W x 8" H x 16" L	Inst	SF	B-4	280	2.90	0.13	2.81	5.84	7.19	B-3	200	3.33	0.18	3.06	6.57	8.03
Lateral reinforcing every third course																
4" W x 8" H x 16" L	Inst	SF	B-4	355	1.61	0.10	2.22	3.92	4.99	B-3	255	1.85	0.14	2.40	4.38	5.53
6" W x 4" H x 16" L	Inst	SF	B-4	175	3.20	0.20	4.50	7.90	10.06	B-3	125	3.68	0.28	4.89	8.85	11.20
6" W x 6" H x 16" L	Inst	SF	B-4	245	2.34	0.14	3.21	5.70	7.24	B-3	175	2.69	0.20	3.49	6.39	8.06
6" W x 8" H x 16" L	Inst	SF	B-4	335	1.87	0.11	2.35	4.33	5.46	B-3	240	2.15	0.15	2.55	4.85	6.07
8" W x 8" H x 16" L	Inst	SF	B-4	315	2.26	0.11	2.50	4.88	6.08	B-3	225	2.60	0.16	2.72	5.48	6.78
10" W x 8" H x 16" L	Inst	SF	B-4	290	2.80	0.12	2.72	5.63	6.94	B-3	205	3.22	0.17	2.98	6.37	7.80
12" W x 8" H x 16" L	Inst	SF	B-4	265	3.10	0.13	2.97	6.20	7.63	B-3	190	3.56	0.19	3.22	6.97	8.51

Concrete slump block
Running bond, 3/8" mortar joints

Walls, with air tools

Description	Oper	Unit	Crew Size	Avg Day Prod	Avg Mat'l Unit Cost	Avg Equip Unit Cost	Avg Labor Unit Cost	Avg Total Unit Cost	Avg Price Incl O&P	Crew Size	Avg Day Prod	Avg Mat'l Unit Cost	Avg Equip Unit Cost	Avg Labor Unit Cost	Avg Total Unit Cost	Avg Price Incl O&P
4" T	Demo	SF	A-2	370	---	0.13	0.81	0.94	1.32	A-2	315	---	0.16	0.95	1.11	1.55
6" T	Demo	SF	A-2	350	---	0.14	0.85	1.00	1.40	A-2	300	---	0.17	1.00	1.16	1.63
8" T	Demo	SF	A-2	335	---	0.15	0.89	1.04	1.46	A-2	285	---	0.17	1.05	1.22	1.72
12" T	Demo	SF	A-2	275	---	0.18	1.09	1.27	1.78	A-2	235	---	0.21	1.27	1.48	2.08

Walls; no scaffolding included

Description	Oper	Unit	Crew Size	Avg Day Prod	Avg Mat'l Unit Cost	Avg Equip Unit Cost	Avg Labor Unit Cost	Avg Total Unit Cost	Avg Price Incl O&P	Crew Size	Avg Day Prod	Avg Mat'l Unit Cost	Avg Equip Unit Cost	Avg Labor Unit Cost	Avg Total Unit Cost	Avg Price Incl O&P
4" x 4" x 16"	Inst	SF	B-3	250	3.48	0.14	2.45	6.07	7.24	B-5	165	4.00	0.21	2.92	7.13	8.54
6" x 4" x 16"	Inst	SF	B-3	240	4.31	0.15	2.01	6.47	7.43	B-5	155	4.96	0.23	3.11	8.29	9.79
6" x 6" x 16"	Inst	SF	B-3	270	2.27	0.13	1.78	4.19	5.04	B-5	175	2.61	0.20	2.75	5.57	6.89
8" x 4" x 16"	Inst	SF	B-3	230	5.37	0.15	2.09	7.62	8.62	B-5	150	6.18	0.24	3.21	9.62	11.16
8" x 6" x 16"	Inst	SF	B-3	260	2.86	0.14	1.85	4.85	5.73	B-5	170	3.29	0.21	2.83	6.33	7.69
12" x 4" x 16"	Inst	SF	B-3	210	9.02	0.17	2.29	11.48	12.58	B-5	135	10.37	0.26	3.57	14.20	15.91
12" x 6" x 16"	Inst	SF	B-3	240	4.58	0.15	2.01	6.74	7.70	B-5	155	5.27	0.23	3.11	8.60	10.10

Concrete slump brick
Running bond, 3/8" mortar joints

Walls, with air tools

Description	Oper	Unit	Crew Size	Avg Day Prod	Avg Mat'l Unit Cost	Avg Equip Unit Cost	Avg Labor Unit Cost	Avg Total Unit Cost	Avg Price Incl O&P	Crew Size	Avg Day Prod	Avg Mat'l Unit Cost	Avg Equip Unit Cost	Avg Labor Unit Cost	Avg Total Unit Cost	Avg Price Incl O&P
4" T	Demo	SF	A-2	370	---	0.13	0.81	0.94	1.32	A-2	315	---	0.16	0.95	1.11	1.55

| Description | Oper | Unit | Costs Based On Large Volume | | | | | | | Costs Based On Small Volume | | | | | | |
			Crew Size	Avg Day Prod	Avg Mat'l Unit Cost	Avg Equip Unit Cost	Avg Labor Unit Cost	Avg Total Unit Cost	Avg Price Incl O&P	Crew Size	Avg Day Prod	Avg Mat'l Unit Cost	Avg Equip Unit Cost	Avg Labor Unit Cost	Avg Total Unit Cost	Avg Price Incl O&P
Masonry (continued)																
Walls; no scaffolding included																
4" x 4" x 8"	Inst	SF	B-3	220	10.34	0.16	2.78	13.28	14.62	B-5	145	11.89	0.24	3.32	15.46	17.05
4" x 4" x 12"	Inst	SF	B-3	240	5.32	0.15	2.01	7.47	8.43	B-5	155	6.12	0.23	3.11	9.45	10.94
Concrete textured screen block																
Running bond, 3/8" mortar joints																
Walls, with air tools																
4" T	Demo	SF	A-2	480	---	0.10	0.62	0.73	1.02	A-2	410	---	0.12	0.73	0.85	1.19
Walls; no scaffolding included																
4" x 6" x 6"	Inst	SF	B-3	160	9.22	0.22	3.82	13.26	15.09	B-5	105	10.60	0.34	4.59	15.52	17.73
4" x 8" x 8"	Inst	SF	B-3	240	6.97	0.15	2.01	9.12	10.08	B-5	155	8.01	0.23	3.11	11.35	12.84
4" x 12" x 12"	Inst	SF	B-3	360	2.00	0.10	1.34	3.43	4.08	B-5	235	2.30	0.15	2.05	4.50	5.48
Glass block																
Plain, 4" Thick																
6" x 6"	Demo	SF	L-2	250	---	---	1.04	1.04	1.54	L-2	200	---	---	1.30	1.30	1.92
8" x 8"	Demo	SF	L-2	280	---	---	0.93	0.93	1.37	L-2	225	---	---	1.16	1.16	1.71
12" x 12"	Demo	SF	L-2	310	---	---	0.84	0.84	1.24	L-2	245	---	---	1.06	1.06	1.57
6" x 6"	Inst	SF	B-5	75	16.62	---	6.42	23.04	26.12	B-2	45	19.11	---	6.79	25.90	29.16
8" x 8"	Inst	SF	B-5	110	11.97	---	4.38	16.35	18.45	B-2	65	13.76	---	4.70	18.47	20.72
12" x 12"	Inst	SF	B-5	165	14.86	---	2.92	17.78	19.18	B-2	95	17.09	---	3.22	20.31	21.85
Glazed tile. 6-T Series																
Includes normal allowance for special shapes																
All Sizes	Demo	SF	L-2	310	---	---	0.84	0.84	1.24	L-2	245	---	---	1.06	1.06	1.57
Glazed 1 Side																
2" x 5-1/3" x 12"	Inst	SF	B-5	84	2.82	---	5.73	8.55	11.30	B-2	48	3.24	---	6.37	9.61	12.66
4" x 5-1/3" x 12"	Inst	SF	B-5	80	4.22	---	6.02	10.24	13.13	B-2	46	4.85	---	6.65	11.50	14.69
6" x 5-1/3" x 12"	Inst	SF	B-5	76	5.50	---	6.34	11.83	14.87	B-2	43	6.32	---	7.11	13.43	16.84
8" x 5-1/3" x 12"	Inst	SF	B-5	68	6.53	---	7.08	13.61	17.01	B-2	39	7.50	---	7.84	15.34	19.11

Description	Oper	Unit	Costs Based On Large Volume							Costs Based On Small Volume						
			Crew Size	Avg Day Prod	Avg Mat'l Unit Cost	Avg Equip Unit Cost	Avg Labor Unit Cost	Avg Total Unit Cost	Avg Price Incl O&P	Crew Size	Avg Day Prod	Avg Mat'l Unit Cost	Avg Equip Unit Cost	Avg Labor Unit Cost	Avg Total Unit Cost	Avg Price Incl O&P

Masonry (continued)

Glazed 2 Sides

Description	Oper	Unit	Crew	Prod	Mat'l	Equip	Labor	Total	O&P	Crew	Prod	Mat'l	Equip	Labor	Total	O&P
4" x 5-1/3" x 12"	Inst	SF	B-5	64	6.36	---	7.53	13.88	17.49	B-2	36	7.31	---	8.49	15.80	19.88
6" x 5-1/3" x 12"	Inst	SF	B-5	60	7.77	---	8.03	15.79	19.65	B-2	34	8.93	---	8.99	17.93	22.24
8" x 5-1/3" x 12"	Inst	SF	B-5	52	9.34	---	9.26	18.60	23.05	B-2	30	10.74	---	10.19	20.93	25.82

Quarry tile

Floors

Description	Oper	Unit	Crew	Prod	Mat'l	Equip	Labor	Total	O&P	Crew	Prod	Mat'l	Equip	Labor	Total	O&P
Conventional mortar set	Demo	SF	L-2	535	---	---	0.49	0.49	0.72	L-2	445	---	---	0.58	0.58	0.86
Dry-set mortar	Demo	SF	L-2	620	---	---	0.42	0.42	0.62	L-2	515	---	---	0.51	0.51	0.75

Floors

Conventional mortar set with unmounted tile

Description	Oper	Unit	Crew	Prod	Mat'l	Equip	Labor	Total	O&P	Crew	Prod	Mat'l	Equip	Labor	Total	O&P
6" x 6" x 1/2" T	Inst	SF	B-3	260	1.17	0.14	2.35	3.66	4.79	B-5	180	1.35	0.20	2.68	4.22	5.50
9" x 9" x 3/4" T	Inst	SF	B-3	310	1.51	0.11	1.97	3.60	4.54	B-5	215	1.74	0.16	2.24	4.14	5.22

Dry-set mortar with unmounted tile

Description	Oper	Unit	Crew	Prod	Mat'l	Equip	Labor	Total	O&P	Crew	Prod	Mat'l	Equip	Labor	Total	O&P
6" x 6" x 1/2" T	Inst	SF	B-3	315	3.63	0.11	1.94	5.68	6.62	B-5	220	4.18	0.16	2.19	6.52	7.58
9" x 9" x 3/4" T	Inst	SF	B-3	390	3.97	0.09	1.57	5.63	6.38	B-5	270	4.56	0.13	1.78	6.48	7.34

Medicine cabinets. See Plumbing.

Millwork. See specific items.

Mirrors. See Glass & Glazing.

Molding and trim

Removal

Description	Oper	Unit	Crew	Prod	Mat'l	Equip	Labor	Total	O&P	Crew	Prod	Mat'l	Equip	Labor	Total	O&P
At base (floor)	Demo	LF	L-2	1060	---	---	0.25	0.25	0.36	L-2	689	---	---	0.38	0.38	0.56
At ceiling	Demo	LF	L-2	1200	---	---	0.22	0.22	0.32	L-2	780	---	---	0.33	0.33	0.49
On wall or cabinets	Demo	LF	L-2	1600	---	---	0.16	0.16	0.24	L-2	1040	---	---	0.25	0.25	0.37

Hardwood

Base

1/2" x 1-3/4"

Description	Oper	Unit	Crew	Prod	Mat'l	Equip	Labor	Total	O&P	Crew	Prod	Mat'l	Equip	Labor	Total	O&P
Ash	Inst	LF	C-1	266	0.83	---	0.63	1.46	1.77	C-1	173	1.06	---	0.97	2.03	2.51
Birch	Inst	LF	C-1	266	0.80	---	0.63	1.43	1.74	C-1	173	1.02	---	0.97	2.00	2.48
Oak	Inst	LF	C-1	266	0.56	---	0.63	1.20	1.51	C-1	173	0.72	---	0.97	1.69	2.17
Philippine mahogany	Inst	LF	C-1	266	0.45	---	0.63	1.08	1.39	C-1	173	0.58	---	0.97	1.55	2.03

Molding

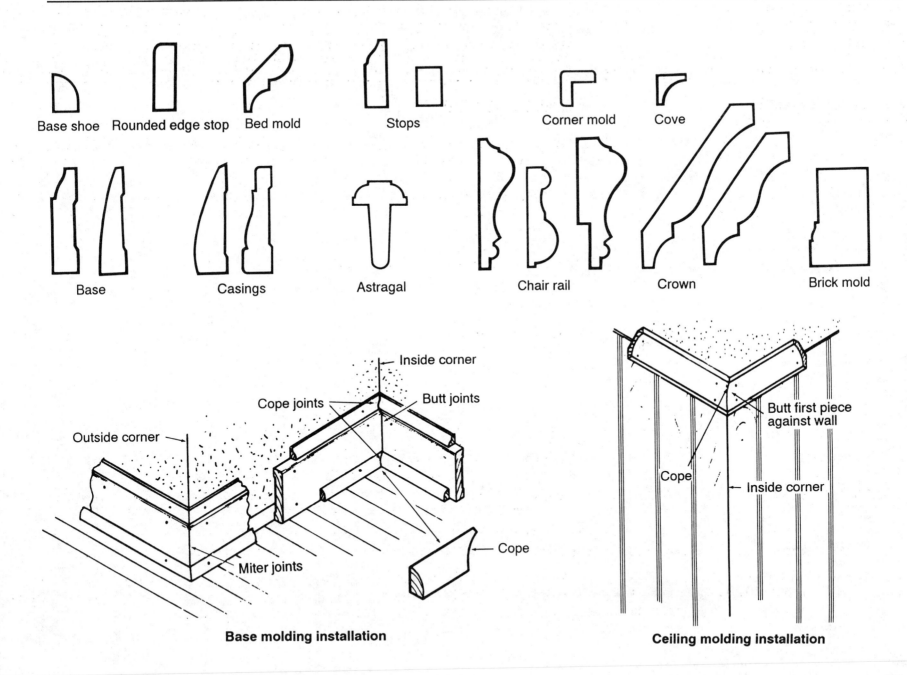

Base shoe

Rounded edge stop

Bed mold

Stops

Corner mold

Cove

Base

Casings

Astragal

Chair rail

Crown

Brick mold

Outside corner

Inside corner

Cope joints

Butt joints

Cope

Miter joints

Base molding installation

Butt first piece against wall

Cope

Inside corner

Ceiling molding installation

			Costs Based On Large Volume							Costs Based On Small Volume					
Description	Oper	Unit	Crew Size	Avg Day Prod	Avg Mat'l Unit Cost	Avg Equip Unit Cost	Avg Labor Unit Cost	Avg Total Unit Cost	Avg Price Incl O&P	Crew Size	Avg Day Prod	Avg Mat'l Unit Cost	Avg Equip Unit Cost	Avg Labor Unit Cost	Avg Total Unit Cost	Avg Price Incl O&P

Molding (continued)

5/8" x 1-5/8"
Ash	Inst	LF	C-1	266	0.88	---	0.63	1.51	1.82	C-1	173	1.12	---	0.97	2.09	2.57
Birch	Inst	LF	C-1	266	0.96	---	0.63	1.60	1.91	C-1	173	1.23	---	0.97	2.21	2.68
Oak	Inst	LF	C-1	266	0.65	---	0.63	1.28	1.59	C-1	173	0.83	---	0.97	1.81	2.28
Philippine mahogany	Inst	LF	C-1	266	0.50	---	0.63	1.13	1.44	C-1	173	0.64	---	0.97	1.61	2.09

1/2" x 2-1/2"
Ash	Inst	LF	C-1	266	1.19	---	0.63	1.82	2.13	C-1	173	1.52	---	0.97	2.49	2.97
Birch	Inst	LF	C-1	266	1.15	---	0.63	1.78	2.09	C-1	173	1.47	---	0.97	2.45	2.92
Oak	Inst	LF	C-1	266	0.81	---	0.63	1.45	1.76	C-1	173	1.04	---	0.97	2.01	2.49
Philippine mahogany	Inst	LF	C-1	266	0.65	---	0.63	1.28	1.59	C-1	173	0.83	---	0.97	1.81	2.28

Base shoe

3/8" x 3/4"
Ash	Inst	LF	C-1	266	0.34	---	0.63	0.97	1.28	C-1	173	0.43	---	0.97	1.41	1.88
Birch	Inst	LF	C-1	266	0.30	---	0.63	0.93	1.24	C-1	173	0.38	---	0.97	1.36	1.84
Oak	Inst	LF	C-1	266	0.25	---	0.63	0.88	1.19	C-1	173	0.32	---	0.97	1.29	1.77
Philippine mahogany	Inst	LF	C-1	266	0.20	---	0.63	0.83	1.14	C-1	173	0.26	---	0.97	1.23	1.71

Battens

1/4" x 3/4"
Ash	Inst	LF	C-1	266	0.34	---	0.63	0.97	1.28	C-1	173	0.43	---	0.97	1.41	1.88
Birch	Inst	LF	C-1	266	0.30	---	0.63	0.93	1.24	C-1	173	0.38	---	0.97	1.36	1.84
Oak	Inst	LF	C-1	266	0.25	---	0.63	0.88	1.19	C-1	173	0.32	---	0.97	1.29	1.77
Philippine mahogany	Inst	LF	C-1	266	0.20	---	0.63	0.83	1.14	C-1	173	0.26	---	0.97	1.23	1.71

Casing

1/2" x 1-1/2"
Ash	Inst	LF	C-1	266	0.70	---	0.63	1.33	1.64	C-1	173	0.90	---	0.97	1.87	2.35
Birch	Inst	LF	C-1	266	0.64	---	0.63	1.27	1.58	C-1	173	0.82	---	0.97	1.79	2.27
Oak	Inst	LF	C-1	266	0.59	---	0.63	1.22	1.53	C-1	173	0.75	---	0.97	1.73	2.20
Philippine mahogany	Inst	LF	C-1	266	0.38	---	0.63	1.01	1.32	C-1	173	0.48	---	0.97	1.45	1.93

5/8" x 1-5/8"
Ash	Inst	LF	C-1	266	0.83	---	0.63	1.46	1.77	C-1	173	1.06	---	0.97	2.03	2.51
Birch	Inst	LF	C-1	266	0.80	---	0.63	1.43	1.74	C-1	173	1.02	---	0.97	2.00	2.48
Oak	Inst	LF	C-1	266	0.56	---	0.63	1.20	1.51	C-1	173	0.72	---	0.97	1.69	2.17
Philippine mahogany	Inst	LF	C-1	266	0.45	---	0.63	1.08	1.39	C-1	173	0.58	---	0.97	1.55	2.03

Description	Oper	Unit	Costs Based On Large Volume							Costs Based On Small Volume						
			Crew Size	Avg Day Prod	Avg Mat'l Unit Cost	Avg Equip Unit Cost	Avg Labor Unit Cost	Avg Total Unit Cost	Avg Price Incl O&P	Crew Size	Avg Day Prod	Avg Mat'l Unit Cost	Avg Equip Unit Cost	Avg Labor Unit Cost	Avg Total Unit Cost	Avg Price Incl O&P

Molding (continued)

Chair rail
1/2" x 2"

Description	Oper	Unit	Crew Size	Avg Day Prod	Avg Mat'l Unit Cost	Avg Equip Unit Cost	Avg Labor Unit Cost	Avg Total Unit Cost	Avg Price Incl O&P	Crew Size	Avg Day Prod	Avg Mat'l Unit Cost	Avg Equip Unit Cost	Avg Labor Unit Cost	Avg Total Unit Cost	Avg Price Incl O&P
Ash	Inst	LF	C-1	200	0.99	---	0.84	1.83	2.24	C-1	130	1.26	---	1.30	2.56	3.20
Birch	Inst	LF	C-1	200	0.95	---	0.84	1.79	2.21	C-1	130	1.22	---	1.30	2.51	3.15
Oak	Inst	LF	C-1	200	0.69	---	0.84	1.53	1.94	C-1	130	0.88	---	1.30	2.18	2.81
Philippine mahogany	Inst	LF	C-1	200	0.53	---	0.84	1.37	1.78	C-1	130	0.67	---	1.30	1.97	2.60

Corner mold
3/4" x 3/4"

Ash	Inst	LF	C-1	266	0.80	---	0.63	1.43	1.74	C-1	173	1.02	---	0.97	2.00	2.48
Birch	Inst	LF	C-1	266	0.78	---	0.63	1.41	1.72	C-1	173	0.99	---	0.97	1.97	2.44
Oak	Inst	LF	C-1	266	0.50	---	0.63	1.13	1.44	C-1	173	0.64	---	0.97	1.61	2.09
Philippine mahogany	Inst	LF	C-1	266	0.34	---	0.63	0.97	1.28	C-1	173	0.43	---	0.97	1.41	1.88

1" x 1"

Ash	Inst	LF	C-1	266	0.91	---	0.63	1.55	1.86	C-1	173	1.17	---	0.97	2.14	2.62
Birch	Inst	LF	C-1	266	0.88	---	0.63	1.51	1.82	C-1	173	1.12	---	0.97	2.09	2.57
Oak	Inst	LF	C-1	266	0.58	---	0.63	1.21	1.52	C-1	173	0.74	---	0.97	1.71	2.19
Philippine mahogany	Inst	LF	C-1	266	0.49	---	0.63	1.12	1.43	C-1	173	0.62	---	0.97	1.60	2.08

Cove
1/2" x 1/2"

Ash	Inst	LF	C-1	200	0.40	---	0.84	1.24	1.66	C-1	130	0.51	---	1.30	1.81	2.44
Birch	Inst	LF	C-1	200	0.39	---	0.84	1.23	1.64	C-1	130	0.50	---	1.30	1.79	2.43
Oak	Inst	LF	C-1	200	0.28	---	0.84	1.12	1.53	C-1	130	0.35	---	1.30	1.65	2.28
Philippine mahogany	Inst	LF	C-1	200	0.21	---	0.84	1.06	1.47	C-1	130	0.27	---	1.30	1.57	2.20

3/4" x 3/4"

Ash	Inst	LF	C-1	200	0.63	---	0.84	1.47	1.88	C-1	130	0.80	---	1.30	2.10	2.73
Birch	Inst	LF	C-1	200	0.55	---	0.84	1.39	1.81	C-1	130	0.70	---	1.30	2.00	2.64
Oak	Inst	LF	C-1	200	0.46	---	0.84	1.31	1.72	C-1	130	0.59	---	1.30	1.89	2.52
Philippine mahogany	Inst	LF	C-1	200	0.36	---	0.84	1.21	1.62	C-1	130	0.46	---	1.30	1.76	2.40

Crown
1/2" x 1-5/8"

Ash	Inst	LF	C-1	200	0.83	---	0.84	1.67	2.08	C-1	130	1.06	---	1.30	2.35	2.99
Birch	Inst	LF	C-1	200	0.80	---	0.84	1.64	2.06	C-1	130	1.02	---	1.30	2.32	2.96
Oak	Inst	LF	C-1	200	0.56	---	0.84	1.41	1.82	C-1	130	0.72	---	1.30	2.02	2.65
Philippine mahogany	Inst	LF	C-1	200	0.45	---	0.84	1.29	1.71	C-1	130	0.58	---	1.30	1.87	2.51

			Costs Based On Large Volume							Costs Based On Small Volume					
Description	Oper	Unit	Crew Size	Avg Day Prod	Avg Mat'l Unit Cost	Avg Equip Unit Cost	Avg Labor Unit Cost	Avg Total Unit Cost	Avg Price Incl O&P	Crew Size	Avg Day Prod	Avg Mat'l Unit Cost	Avg Equip Unit Cost	Avg Labor Unit Cost	Avg Total Unit Cost	Avg Price Incl O&P

Molding (continued)

1/2" x 2-1/4"

Ash	Inst	LF	C-1	200	1.19	---	0.84	2.03	2.44	C-1	130	1.52	---	1.30	2.82	3.45
Birch	Inst	LF	C-1	200	1.15	---	0.84	1.99	2.41	C-1	130	1.47	---	1.30	2.77	3.40
Oak	Inst	LF	C-1	200	0.74	---	0.84	1.58	1.99	C-1	130	0.94	---	1.30	2.24	2.88
Philippine mahogany	Inst	LF	C-1	200	0.65	---	0.84	1.49	1.91	C-1	130	0.83	---	1.30	2.13	2.76

Stops, rounded edge

3/8" x 1-1/4"

Ash	Inst	LF	C-1	266	0.46	---	0.63	1.10	1.41	C-1	173	0.59	---	0.97	1.57	2.04
Birch	Inst	LF	C-1	266	0.41	---	0.63	1.05	1.36	C-1	173	0.53	---	0.97	1.50	1.98
Oak	Inst	LF	C-1	266	0.34	---	0.63	0.97	1.28	C-1	173	0.43	---	0.97	1.41	1.88
Philippine mahogany	Inst	LF	C-1	266	0.28	---	0.63	0.91	1.22	C-1	173	0.35	---	0.97	1.33	1.80

Pine

Astragal

3/4" x 1-5/8"

	Inst	LF	C-1	266	0.59	---	0.63	1.22	1.53	C-1	173	0.75	---	0.97	1.73	2.20

Base

3/8" x 2-1/4"	Inst	LF	C-1	266	0.43	---	0.63	1.06	1.37	C-1	173	0.54	---	0.97	1.52	2.00
1/2" x 1-3/4"	Inst	LF	C-1	266	0.44	---	0.63	1.07	1.38	C-1	173	0.56	---	0.97	1.53	2.01
1/2" x 2-1/4"	Inst	LF	C-1	266	0.60	---	0.63	1.23	1.54	C-1	173	0.77	---	0.97	1.74	2.22
1/2" x 2-1/2"	Inst	LF	C-1	266	0.68	---	0.63	1.31	1.62	C-1	173	0.86	---	0.97	1.84	2.32
1/2" x 3-1/2"	Inst	LF	C-1	266	0.98	---	0.63	1.61	1.92	C-1	173	1.25	---	0.97	2.22	2.70

Base (combination or cove)

7/16" x 1-5/8"	Inst	LF	C-1	200	0.40	---	0.84	1.24	1.66	C-1	130	0.51	---	1.30	1.81	2.44
7/16" x 3-1/4"	Inst	LF	C-1	200	0.78	---	0.84	1.62	2.03	C-1	130	0.99	---	1.30	2.29	2.92

Base shoe

3/8" x 3/4"	Inst	LF	C-1	266	0.18	---	0.63	0.81	1.12	C-1	173	0.22	---	0.97	1.20	1.68
1/2" x 3/4"	Inst	LF	C-1	266	0.20	---	0.63	0.83	1.14	C-1	173	0.26	---	0.97	1.23	1.71

Bed mold

3/4" x 1-5/8"	Inst	LF	C-1	200	0.53	---	0.84	1.37	1.78	C-1	130	0.67	---	1.30	1.97	2.60
3/4" x 2-1/4"	Inst	LF	C-1	200	0.71	---	0.84	1.56	1.97	C-1	130	0.91	---	1.30	2.21	2.84
3/4" x 3-1/2"	Inst	LF	C-1	200	1.05	---	0.84	1.89	2.31	C-1	130	1.34	---	1.30	2.64	3.28

Blind stop

3/4" x 1-1/4"	Inst	LF	C-1	266	0.50	---	0.63	1.13	1.44	C-1	173	0.64	---	0.97	1.61	2.09

Molding (continued)

Description	Oper	Unit	Costs Based On Large Volume							Costs Based On Small Volume						
			Crew Size	Avg Day Prod	Avg Mat'l Unit Cost	Avg Equip Unit Cost	Avg Labor Unit Cost	Avg Total Unit Cost	Avg Price Incl O&P	Crew Size	Avg Day Prod	Avg Mat'l Unit Cost	Avg Equip Unit Cost	Avg Labor Unit Cost	Avg Total Unit Cost	Avg Price Incl O&P
Casings																
1/2" x 1-5/8"	Inst	LF	C-1	266	0.45	---	0.63	1.08	1.39	C-1	173	0.58	---	0.97	1.55	2.03
5/8" x 1-5/8"	Inst	LF	C-1	266	0.46	---	0.63	1.10	1.41	C-1	173	0.59	---	0.97	1.57	2.04
5/8" x 2-1/2"	Inst	LF	C-1	266	0.64	---	0.63	1.27	1.58	C-1	173	0.82	---	0.97	1.79	2.27
Corner bead																
3/4" x 3/4"	Inst	LF	C-1	266	0.29	---	0.63	0.92	1.23	C-1	173	0.37	---	0.97	1.34	1.82
1" x 1"	Inst	LF	C-1	266	0.54	---	0.63	1.17	1.48	C-1	173	0.69	---	0.97	1.66	2.14
1-3/8" x 1-3/8"	Inst	LF	C-1	266	0.85	---	0.63	1.48	1.79	C-1	173	1.09	---	0.97	2.06	2.54
Cove																
3/8" x 3/8"	Inst	LF	C-1	200	0.15	---	0.84	0.99	1.41	C-1	130	0.19	---	1.30	1.49	2.12
1/2" x 1/2"	Inst	LF	C-1	200	0.16	---	0.84	1.01	1.42	C-1	130	0.21	---	1.30	1.50	2.14
5/8" x 5/8"	Inst	LF	C-1	200	0.25	---	0.84	1.09	1.51	C-1	130	0.32	---	1.30	1.62	2.25
3/4" x 3/4"	Inst	LF	C-1	200	0.29	---	0.84	1.13	1.54	C-1	130	0.37	---	1.30	1.66	2.30
1" x 1"	Inst	LF	C-1	200	0.54	---	0.84	1.38	1.79	C-1	130	0.69	---	1.30	1.98	2.62
Cove - sprung																
3/4" x 1-5/8"	Inst	LF	C-1	200	0.48	---	0.84	1.32	1.73	C-1	130	0.61	---	1.30	1.90	2.54
3/4" x 2-1/4"	Inst	LF	C-1	200	0.66	---	0.84	1.51	1.92	C-1	130	0.85	---	1.30	2.14	2.78
3/4" x 3-1/2"	Inst	LF	C-1	200	1.03	---	0.84	1.87	2.28	C-1	130	1.31	---	1.30	2.61	3.24
Chair rail																
1/2" x 1-1/2"	Inst	LF	C-1	200	0.48	---	0.84	1.32	1.73	C-1	130	0.61	---	1.30	1.90	2.54
5/8" x 2-1/2"	Inst	LF	C-1	200	0.81	---	0.84	1.66	2.07	C-1	130	1.04	---	1.30	2.34	2.97
Crown																
3/4" x 3/4"	Inst	LF	C-1	200	0.29	---	0.84	1.13	1.54	C-1	130	0.37	---	1.30	1.66	2.30
3/4" x 1-5/8"	Inst	LF	C-1	200	0.48	---	0.84	1.32	1.73	C-1	130	0.61	---	1.30	1.90	2.54
3/4" x 2-1/4"	Inst	LF	C-1	200	0.64	---	0.84	1.48	1.89	C-1	130	0.82	---	1.30	2.11	2.75
3/4" x 3-1/2"	Inst	LF	C-1	200	0.99	---	0.84	1.83	2.24	C-1	130	1.26	---	1.30	2.56	3.20
3/4" x 4-1/4"	Inst	LF	C-1	200	1.31	---	0.84	2.16	2.57	C-1	130	1.68	---	1.30	2.98	3.61
3/4" x 5-1/4"	Inst	LF	C-1	200	1.69	---	0.84	2.53	2.94	C-1	130	2.16	---	1.30	3.46	4.09
Half round																
1/4" x 1/2"	Inst	LF	C-1	266	0.10	---	0.63	0.73	1.04	C-1	173	0.13	---	0.97	1.10	1.58
5/16" x 5/8"	Inst	LF	C-1	266	0.15	---	0.63	0.78	1.09	C-1	173	0.19	---	0.97	1.17	1.64
3/8" x 3/4"	Inst	LF	C-1	266	0.19	---	0.63	0.82	1.13	C-1	173	0.24	---	0.97	1.21	1.69
1/2" x 1"	Inst	LF	C-1	266	0.29	---	0.63	0.92	1.23	C-1	173	0.37	---	0.97	1.34	1.82
3/4" x 1-1/2"	Inst	LF	C-1	266	0.51	---	0.63	1.15	1.46	C-1	173	0.66	---	0.97	1.63	2.11

Description	Oper	Unit	Costs Based On Large Volume							Costs Based On Small Volume						
			Crew Size	Avg Day Prod	Avg Mat'l Unit Cost	Avg Equip Unit Cost	Avg Labor Unit Cost	Avg Total Unit Cost	Avg Price Incl O&P	Crew Size	Avg Day Prod	Avg Mat'l Unit Cost	Avg Equip Unit Cost	Avg Labor Unit Cost	Avg Total Unit Cost	Avg Price Incl O&P

Molding (continued)

Mullion casing

Description	Oper	Unit	Crew Size	Avg Day Prod	Avg Mat'l Unit Cost	Avg Equip Unit Cost	Avg Labor Unit Cost	Avg Total Unit Cost	Avg Price Incl O&P	Crew Size	Avg Day Prod	Avg Mat'l Unit Cost	Avg Equip Unit Cost	Avg Labor Unit Cost	Avg Total Unit Cost	Avg Price Incl O&P
1/4" x 3-1/2"	Inst	LF	C-1	266	0.61	---	0.63	1.25	1.56	C-1	173	0.78	---	0.97	1.76	2.24
Parting bead																
3/4" x 1-5/8"	Inst	LF	C-1	266	0.21	---	0.63	0.85	1.16	C-1	173	0.27	---	0.97	1.25	1.72
Picture mold																
3/4" x 1-5/8"	Inst	LF	C-1	200	0.54	---	0.84	1.38	1.79	C-1	130	0.69	---	1.30	1.98	2.62
Quarter round																
1/4" x 1/4"	Inst	LF	C-1	266	0.10	---	0.63	0.73	1.04	C-1	173	0.13	---	0.97	1.10	1.58
3/8" x 3/8"	Inst	LF	C-1	266	0.14	---	0.63	0.77	1.08	C-1	173	0.18	---	0.97	1.15	1.63
1/2" x 1/2"	Inst	LF	C-1	266	0.16	---	0.63	0.80	1.11	C-1	173	0.21	---	0.97	1.18	1.66
5/8" x 5/8"	Inst	LF	C-1	266	0.23	---	0.63	0.86	1.17	C-1	173	0.29	---	0.97	1.26	1.74
3/4" x 3/4"	Inst	LF	C-1	266	0.29	---	0.63	0.92	1.23	C-1	173	0.37	---	0.97	1.34	1.82
1" x 1"	Inst	LF	C-1	266	0.60	---	0.63	1.23	1.54	C-1	173	0.77	---	0.97	1.74	2.22
Squares																
1/2" x 1/2"	Inst	LF	C-1	266	0.20	---	0.63	0.83	1.14	C-1	173	0.26	---	0.97	1.23	1.71
3/4" x 3/4"	Inst	LF	C-1	266	0.29	---	0.63	0.92	1.23	C-1	173	0.37	---	0.97	1.34	1.82
2" x 2"	Inst	LF	C-1	266	1.04	---	0.63	1.67	1.98	C-1	173	1.33	---	0.97	2.30	2.78
Stops, rounded edge																
3/8" x 1/2"	Inst	LF	C-1	266	0.15	---	0.63	0.78	1.09	C-1	173	0.19	---	0.97	1.17	1.64
3/8" x 3/4"	Inst	LF	C-1	266	0.20	---	0.63	0.83	1.14	C-1	173	0.26	---	0.97	1.23	1.71
3/8" x 1"	Inst	LF	C-1	266	0.25	---	0.63	0.88	1.19	C-1	173	0.32	---	0.97	1.29	1.77
3/8" x 1-1/4"	Inst	LF	C-1	266	0.25	---	0.63	0.88	1.19	C-1	173	0.32	---	0.97	1.29	1.77
3/8" x 1-5/8"	Inst	LF	C-1	266	0.35	---	0.63	0.98	1.29	C-1	173	0.45	---	0.97	1.42	1.90
1/2" x 3/4"	Inst	LF	C-1	266	0.24	---	0.63	0.87	1.18	C-1	173	0.30	---	0.97	1.28	1.76
1/2" x 1"	Inst	LF	C-1	266	0.29	---	0.63	0.92	1.23	C-1	173	0.37	---	0.97	1.34	1.82
1/2" x 1-1/4"	Inst	LF	C-1	266	0.36	---	0.63	1.00	1.31	C-1	173	0.46	---	0.97	1.44	1.92
1/2" x 1-5/8"	Inst	LF	C-1	266	0.43	---	0.63	1.06	1.37	C-1	173	0.54	---	0.97	1.52	2.00

Redwood

Band molding

Description	Oper	Unit	Crew Size	Avg Day Prod	Avg Mat'l Unit Cost	Avg Equip Unit Cost	Avg Labor Unit Cost	Avg Total Unit Cost	Avg Price Incl O&P	Crew Size	Avg Day Prod	Avg Mat'l Unit Cost	Avg Equip Unit Cost	Avg Labor Unit Cost	Avg Total Unit Cost	Avg Price Incl O&P
3/4" x 1-5/8"	Inst	LF	C-1	266	0.55	---	0.63	1.18	1.49	C-1	173	0.70	---	0.97	1.68	2.16
Brick mold																
1-5/8" x 1-5/8"	Inst	LF	C-1	200	1.13	---	0.84	1.97	2.38	C-1	130	1.44	---	1.30	2.74	3.37

			Costs Based On Large Volume							Costs Based On Small Volume						
Description	Oper	Unit	Crew Size	Avg Day Prod	Avg Mat'l Unit Cost	Avg Equip Unit Cost	Avg Labor Unit Cost	Avg Total Unit Cost	Avg Price Incl O&P	Crew Size	Avg Day Prod	Avg Mat'l Unit Cost	Avg Equip Unit Cost	Avg Labor Unit Cost	Avg Total Unit Cost	Avg Price Incl O&P

Molding (continued)

Lattice and battens

Description	Oper	Unit	Crew Size	Avg Day Prod	Avg Mat'l Unit Cost	Avg Equip Unit Cost	Avg Labor Unit Cost	Avg Total Unit Cost	Avg Price Incl O&P	Crew Size	Avg Day Prod	Avg Mat'l Unit Cost	Avg Equip Unit Cost	Avg Labor Unit Cost	Avg Total Unit Cost	Avg Price Incl O&P
5/16" x 1-1/4"	Inst	LF	C-1	266	0.23	---	0.63	0.86	1.17	C-1	173	0.29	---	0.97	1.26	1.74
5/16" x 1-5/8"	Inst	LF	C-1	266	0.28	---	0.63	0.91	1.22	C-1	173	0.35	---	0.97	1.33	1.80
5/16" x 2-1/2"	Inst	LF	C-1	266	0.36	---	0.63	1.00	1.31	C-1	173	0.46	---	0.97	1.44	1.92
5/16" x 3-1/2"	Inst	LF	C-1	266	0.50	---	0.63	1.13	1.44	C-1	173	0.64	---	0.97	1.61	2.09

Stucco mold

Description	Oper	Unit	Crew Size	Avg Day Prod	Avg Mat'l Unit Cost	Avg Equip Unit Cost	Avg Labor Unit Cost	Avg Total Unit Cost	Avg Price Incl O&P	Crew Size	Avg Day Prod	Avg Mat'l Unit Cost	Avg Equip Unit Cost	Avg Labor Unit Cost	Avg Total Unit Cost	Avg Price Incl O&P
7/8" x 1-3/8"	Inst	LF	C-1	200	0.48	---	0.84	1.32	1.73	C-1	130	0.61	---	1.30	1.90	2.54
7/8" x 1-5/8"	Inst	LF	C-1	200	0.54	---	0.84	1.38	1.79	C-1	130	0.69	---	1.30	1.98	2.62
7/8" x 2-1/2"	Inst	LF	C-1	200	0.66	---	0.84	1.51	1.92	C-1	130	0.85	---	1.30	2.14	2.78

Spindles
western hemlock, clear, kiln dried, turned for decorative applications

Planter design

1-11/16" x 1-11/16"

Description	Oper	Unit	Crew Size	Avg Day Prod	Avg Mat'l Unit Cost	Avg Equip Unit Cost	Avg Labor Unit Cost	Avg Total Unit Cost	Avg Price Incl O&P	Crew Size	Avg Day Prod	Avg Mat'l Unit Cost	Avg Equip Unit Cost	Avg Labor Unit Cost	Avg Total Unit Cost	Avg Price Incl O&P
3'-0" H	Inst	EA	C-1	24	4.63	---	7.02	11.65	15.09	C-1	16	5.92	---	10.80	16.72	22.02
4'-0" H	Inst	EA	C-1	24	5.94	---	7.02	12.96	16.40	C-1	16	7.60	---	10.80	18.40	23.70
2-3/8" x 2-3/8"																
3'-0" H	Inst	EA	C-1	24	9.63	---	7.02	16.65	20.09	C-1	16	12.32	---	10.80	23.12	28.42
4'-0" H	Inst	EA	C-1	24	13.19	---	7.02	20.21	23.65	C-1	16	16.88	---	10.80	27.68	32.98
5'-0" H	Inst	EA	C-1	24	16.00	---	7.02	23.02	26.46	C-1	16	20.48	---	10.80	31.28	36.58
6'-0" H	Inst	EA	C-1	20	20.38	---	8.43	28.80	32.93	C-1	13	26.08	---	12.97	39.05	45.40
8'-0" H	Inst	EA	C-1	20	39.63	---	8.43	48.05	52.18	C-1	13	50.72	---	12.97	63.69	70.04
3-1/4" x 3-1/4"																
3'-0" H	Inst	EA	C-1	24	18.25	---	7.02	25.27	28.71	C-1	16	23.36	---	10.80	34.16	39.46
4'-0" H	Inst	EA	C-1	24	24.38	---	7.02	31.40	34.84	C-1	16	31.20	---	10.80	42.00	47.30
5'-0" H	Inst	EA	C-1	24	30.88	---	7.02	37.90	41.34	C-1	16	39.52	---	10.80	50.32	55.62
6'-0" H	Inst	EA	C-1	20	36.94	---	8.43	45.36	49.49	C-1	13	47.28	---	12.97	60.25	66.60
8'-0" H	Inst	EA	C-1	20	53.31	---	8.43	61.74	65.87	C-1	13	68.24	---	12.97	81.21	87.56

Colonial design

1-11/16" x 1-11/16"

Description	Oper	Unit	Crew Size	Avg Day Prod	Avg Mat'l Unit Cost	Avg Equip Unit Cost	Avg Labor Unit Cost	Avg Total Unit Cost	Avg Price Incl O&P	Crew Size	Avg Day Prod	Avg Mat'l Unit Cost	Avg Equip Unit Cost	Avg Labor Unit Cost	Avg Total Unit Cost	Avg Price Incl O&P
1'-0" H	Inst	EA	C-1	28	1.50	---	6.02	7.52	10.47	C-1	18	1.92	---	9.26	11.18	15.72
1'-6" H	Inst	EA	C-1	28	2.69	---	6.02	8.71	11.66	C-1	18	3.44	---	9.26	12.70	17.24
2'-0" H	Inst	EA	C-1	28	3.44	---	6.02	9.46	12.41	C-1	18	4.40	---	9.26	13.66	18.20

			Costs Based On Large Volume							Costs Based On Small Volume					
Description	Oper	Unit	Crew Size	Avg Day Prod	Avg Mat'l Unit Cost	Avg Equip Unit Cost	Avg Labor Unit Cost	Avg Total Unit Cost	Avg Price Incl O&P	Crew Size	Avg Day Prod	Avg Mat'l Unit Cost	Avg Equip Unit Cost	Avg Labor Unit Cost	Avg Total Unit Cost	Avg Price Incl O&P

Molding (continued)

Description	Oper	Unit	Crew Size	Avg Day Prod	Avg Mat'l Unit Cost	Avg Equip Unit Cost	Avg Labor Unit Cost	Avg Total Unit Cost	Avg Price Incl O&P	Crew Size	Avg Day Prod	Avg Mat'l Unit Cost	Avg Equip Unit Cost	Avg Labor Unit Cost	Avg Total Unit Cost	Avg Price Incl O&P
2'-4" H	Inst	EA	C-1	24	3.56	---	7.02	10.59	14.03	C-1	16	4.56	---	10.80	15.36	20.66
2'-8" H	Inst	EA	C-1	24	4.00	---	7.02	11.02	14.46	C-1	16	5.12	---	10.80	15.92	21.22
3'-0" H	Inst	EA	C-1	24	4.75	---	7.02	11.77	15.21	C-1	16	6.08	---	10.80	16.88	22.18
2-3/8" x 2-3/8"																
1'-0" H	Inst	EA	C-1	28	3.06	---	6.02	9.08	12.03	C-1	18	3.92	---	9.26	13.18	17.72
1'-6" H	Inst	EA	C-1	28	4.25	---	6.02	10.27	13.22	C-1	18	5.44	---	9.26	14.70	19.24
2'-0" H	Inst	EA	C-1	28	6.44	---	6.02	12.46	15.41	C-1	18	8.24	---	9.26	17.50	22.04
2'-4" H	Inst	EA	C-1	24	6.88	---	7.02	13.90	17.34	C-1	16	8.80	---	10.80	19.60	24.90
2'-8" H	Inst	EA	C-1	24	8.75	---	7.02	15.77	19.21	C-1	16	11.20	---	10.80	22.00	27.30
3'-0" H	Inst	EA	C-1	24	9.63	---	7.02	16.65	20.09	C-1	16	12.32	---	10.80	23.12	28.42
3-1/4" x 3-1/4"																
1'-6" H	Inst	EA	C-1	28	7.19	---	6.02	13.21	16.16	C-1	18	9.20	---	9.26	18.46	23.00
2'-0" H	Inst	EA	C-1	28	9.75	---	6.02	15.77	18.72	C-1	18	12.48	---	9.26	21.74	26.28
2'-4" H	Inst	EA	C-1	24	11.44	---	7.02	18.46	21.90	C-1	16	14.64	---	10.80	25.44	30.74
2'-8" H	Inst	EA	C-1	24	12.94	---	7.02	19.96	23.40	C-1	16	16.56	---	10.80	27.36	32.66
3'-0" H	Inst	EA	C-1	24	18.25	---	7.02	25.27	28.71	C-1	16	23.36	---	10.80	34.16	39.46
8'-0" H	Inst	EA	C-1	20	53.31	---	8.43	61.74	65.87	C-1	13	68.24	---	12.97	81.21	87.56
Mediterranean design																
1-11/16" x 1-11/16"																
1'-0" H	Inst	EA	C-1	28	1.56	---	6.02	7.58	10.53	C-1	18	2.00	---	9.26	11.26	15.80
1'-6" H	Inst	EA	C-1	28	2.56	---	6.02	8.58	11.53	C-1	18	3.28	---	9.26	12.54	17.08
2'-0" H	Inst	EA	C-1	28	2.81	---	6.02	8.83	11.78	C-1	18	3.60	---	9.26	12.86	17.40
2'-4" H	Inst	EA	C-1	24	3.31	---	7.02	10.34	13.78	C-1	16	4.24	---	10.80	15.04	20.34
2'-8" H	Inst	EA	C-1	24	3.69	---	7.02	10.71	14.15	C-1	16	4.72	---	10.80	15.52	20.82
3'-0" H	Inst	EA	C-1	24	4.63	---	7.02	11.65	15.09	C-1	16	5.92	---	10.80	16.72	22.02
4'-0" H	Inst	EA	C-1	24	5.94	---	7.02	12.96	16.40	C-1	16	7.60	---	10.80	18.40	23.70
5'-0" H	Inst	EA	C-1	24	10.69	---	7.02	17.71	21.15	C-1	16	13.68	---	10.80	24.48	29.78
2-3/8" x 2-3/8"																
1'-0" H	Inst	EA	C-1	28	2.81	---	6.02	8.83	11.78	C-1	18	3.60	---	9.26	12.86	17.40
1'-6" H	Inst	EA	C-1	28	4.06	---	6.02	10.08	13.03	C-1	18	5.20	---	9.26	14.46	19.00
2'-0" H	Inst	EA	C-1	28	5.13	---	6.02	11.14	14.09	C-1	18	6.56	---	9.26	15.82	20.36
2'-4" H	Inst	EA	C-1	24	6.56	---	7.02	13.59	17.03	C-1	16	8.40	---	10.80	19.20	24.50
2'-8" H	Inst	EA	C-1	24	8.19	---	7.02	15.21	18.65	C-1	16	10.48	---	10.80	21.28	26.58
3'-0" H	Inst	EA	C-1	24	9.38	---	7.02	16.40	19.84	C-1	16	12.00	---	10.80	22.80	28.10

			Costs Based On Large Volume							Costs Based On Small Volume						
Description	Oper	Unit	Crew Size	Avg Day Prod	Avg Mat'l Unit Cost	Avg Equip Unit Cost	Avg Labor Unit Cost	Avg Total Unit Cost	Avg Price Incl O&P	Crew Size	Avg Day Prod	Avg Mat'l Unit Cost	Avg Equip Unit Cost	Avg Labor Unit Cost	Avg Total Unit Cost	Avg Price Incl O&P

Molding (continued)

Description	Oper	Unit	Crew Size	Avg Day Prod	Avg Mat'l Unit Cost	Avg Equip Unit Cost	Avg Labor Unit Cost	Avg Total Unit Cost	Avg Price Incl O&P	Crew Size	Avg Day Prod	Avg Mat'l Unit Cost	Avg Equip Unit Cost	Avg Labor Unit Cost	Avg Total Unit Cost	Avg Price Incl O&P
4'-0" H	Inst	EA	C-1	24	11.13	---	7.02	18.15	21.59	C-1	16	14.24	---	10.80	25.04	30.34
5'-0" H	Inst	EA	C-1	24	15.31	---	7.02	22.34	25.78	C-1	16	19.60	---	10.80	30.40	35.70
6'-0" H	Inst	EA	C-1	20	21.50	---	8.43	29.93	34.06	C-1	13	27.52	---	12.97	40.49	46.84
8'-0" H	Inst	EA	C-1	20	39.63	---	8.43	48.05	52.18	C-1	13	50.72	---	12.97	63.69	70.04
3-1/4" x 3-1/4"																
3'-0" H	Inst	EA	C-1	24	17.69	---	7.02	24.71	28.15	C-1	16	22.64	---	10.80	33.44	38.74
4'-0" H	Inst	EA	C-1	24	20.81	---	7.02	27.84	31.28	C-1	16	26.64	---	10.80	37.44	42.74
5'-0" H	Inst	EA	C-1	24	28.00	---	7.02	35.02	38.46	C-1	16	35.84	---	10.80	46.64	51.94
6'-0" H	Inst	EA	C-1	20	39.38	---	8.43	47.80	51.93	C-1	13	50.40	---	12.97	63.37	69.72
8'-0" H	Inst	EA	C-1	20	62.13	---	8.43	70.55	74.68	C-1	13	79.52	---	12.97	92.49	98.84
Spindle rails, 8'-0" H pieces																
For 1-11/16" spindles	Inst	EA	C-1	---	15.75	---	---	15.75	15.75	C-1	---	20.16	---	---	20.16	20.16
For 2-3/8" spindles	Inst	EA	C-1	---	22.75	---	---	22.75	22.75	C-1	---	29.12	---	---	29.12	29.12
For 3-1/4" spindles	Inst	EA	C-1	---	26.44	---	---	26.44	26.44	C-1	---	33.84	---	---	33.84	33.84

Painting & Cleaning

Interior. The crew output per day includes time calculations for normal handling and protection of furniture and other property not to be painted
Preparation. Excluding openings unless otherwise indicated

Description	Oper	Unit	Crew Size	Avg Day Prod	Avg Mat'l Unit Cost	Avg Equip Unit Cost	Avg Labor Unit Cost	Avg Total Unit Cost	Avg Price Incl O&P	Crew Size	Avg Day Prod	Avg Mat'l Unit Cost	Avg Equip Unit Cost	Avg Labor Unit Cost	Avg Total Unit Cost	Avg Price Incl O&P
Cleaning, wet																
Smooth finishes																
Plaster and drywall	Inst	SF	N-1	1680	0.01	---	0.09	0.10	0.14	N-1	1176	0.01	---	0.13	0.14	0.20
Paneling	Inst	SF	N-1	1800	0.01	---	0.09	0.10	0.14	N-1	1260	0.01	---	0.12	0.13	0.19
Millwork and trim	Inst	SF	N-1	1600	0.01	---	0.10	0.11	0.15	N-1	1120	0.01	---	0.14	0.15	0.21
Floors	Inst	SF	N-1	2240	0.01	---	0.07	0.08	0.11	N-1	1568	0.01	---	0.10	0.11	0.15
Sand finishes	Inst	SF	N-1	1160	0.01	---	0.13	0.14	0.21	N-1	812	0.01	---	0.19	0.20	0.29
Sheetrock or drywall; tape, fill, and finish only; tape kit includes 25 lb. bag taping compound and 250 ft. roll perforated tape; 735 SF per kit	Inst	SF	N-1	880	0.02	---	0.18	0.19	0.27	N-1	616	0.02	---	0.25	0.27	0.39
Sheetrock or drywall; thin coat plaster in lieu of taping (125 SF per 50 lb. bag)	Inst	SF	N-1	680	0.09	---	0.23	0.32	0.43	N-1	476	0.12	---	0.32	0.44	0.59

Painting

Interior and Exterior. There is a paint for almost every type of surface and surface condition. The large variety makes it impractical to consider each paint individually. For this reason, average output and average material cost/unit are based on the paints and prices listed below.

1. **Specifications.**

2. **Installation:** Paint can be applied by brush, roller or spray gun. Only application by brush or roller is considered in this section.

Characteristics - Interior

Type	Coverage SF/Gal.	Surface
Latex, flat	450	Plaster/drywall
Latex, enamel	450	Doors, windows, trim
Shellac	500	Woodwork
Varnish	500	Woodwork
Stain	500	Woodwork

Characteristics - Exterior

Type	Coverage SF/Gal.	Surface
Oil base*	300	Plain siding & stucco
Oil base*	450	Door, windows, trim
Oil base*	300	Shingle siding
Stain	200	Shingle siding
Latex, masonry	400	Stucco & masonry

*Certain latex paints may also be used on exterior work.

3. **Notes — Labor:** Average output, for both roller and brush, is based on what one painter can do in one day. The output for cleaning is also based on what one painter can do in one day.

4. **Estimating Technique:** Used to determine quantity prior to applying unit costs.

Interior.

a. Floors, walls and ceilings. Figure actual area. No deductions for openings.

b. Doors and windows. **Openings only to be painted.** Figure 36 SF or 4 SY for each side of each door and 27 SF for each side of each window. Based on doors 3'-0" x 7'-0" or smaller and windows 3'-0" x 4'-0" or smaller. Doors and windows. **Openings to be painted with walls.** Figure wall area plus 27 SF or 3 SY for each side of each door and 18 SF or 2 SY for each side of each window. Based on doors 3'-0" x 7'-0" or smaller and windows 3'-0" x 4'-0" or smaller.

For larger doors and windows, add 1'-0" to height and width and figure area.

c. Base, picture mold or chair rail. Less than 1'-0" wide, figure one SF/LF. On 1'-0" or larger, figure actual area.

d. Stairs (including treads, risers, cove and stringers). Add 2'-0" width (for treads, risers, etc.) times length plus 2'-0".

e. Balustrades. Add 1'-0" to height, figure two times area to paint two sides.

Exterior.

a. Walls. (No deductions for openings.)

Siding. Figure actual area plus 10%.

Shingles. Figure actual area plus 40%.

Brick, stucco, concrete and smooth wood surface. Figure actual area.

b. Doors and windows. See interior doors and windows.

c. Eaves (including soffit or exposed rafter ends and fascia).

Enclosed. If sidewalls are to be painted the same color, figure 1.5 times actual area. If sidewalls are to be painted a different color, figure 2.0 times actual area. If sidewalls are not to be painted, figure 2.5 times actual area.

Rafter ends exposed. If sidewalls are to be painted same color, figure 2.5 times actual area. If sidewalls are to be painted a different color, figure 3.0 times actual area. If sidewalls are not to be painted, figure 4.0 times actual area.

d. Porch rail. See interior — balustrades.

e. Gutter and downspouts. Figure two SF/LF or 2/9 SY/LF.

f. Latticework. Figure 2.0 times actual area for each side.

g. Fences.

Solid fence. Figure actual area of each side to be painted.

Basket weave. Figure 1.5 times actual area for each side to be painted.

Calculating Square Foot Coverage

Triangle

To find the number of square feet in any shape triangle or 3 sided surface, multiply the height by the width and divide the total by 2.

Square

Multiply the base measurement in feet times the height in feet.

Rectangle

Multiply the base measurement in feet times the height in feet.

Arch Roof

Multiply length (B) by width (A) and add one-half the total.

Circle

To find the number of square feet in a circle multiply the diameter (distance across) by itself and then multiply this total by .7854.

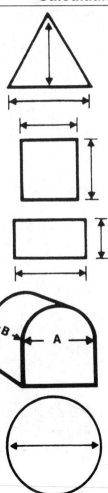

Cylinder

When the circumference (distance around the cylinder) is known, multiply height by circumference. When the diameter (distance across) is known, multiply diameter by 3.1416. This gives circumference. Then multiply by height.

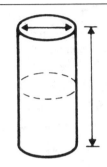

Gambrel Roof

Multiply length (B) by width (A) and add one-third of the total.

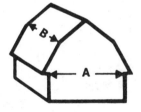

Cone

Determine area of base by multiplying 3.1416 times radius (A) in feet.
Determine the surface area of a cone by multiplying circumference of base (in feet) times one-half of the slant height (B) in feet.
Add the square foot area of the base to the square foot area of the cone for total square foot area.

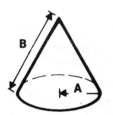

| | | | Costs Based On Large Volume | | | | | | | Costs Based On Small Volume | | | | | | |
Description	Oper	Unit	Crew Size	Avg Day Prod	Avg Mat'l Unit Cost	Avg Equip Unit Cost	Avg Labor Unit Cost	Avg Total Unit Cost	Avg Price Incl O&P	Crew Size	Avg Day Prod	Avg Mat'l Unit Cost	Avg Equip Unit Cost	Avg Labor Unit Cost	Avg Total Unit Cost	Avg Price Incl O&P

Painting (continued)

Light sanding

Description	Oper	Unit	Crew Size	Avg Day Prod	Avg Mat'l Unit Cost	Avg Equip Unit Cost	Avg Labor Unit Cost	Avg Total Unit Cost	Avg Price Incl O&P	Crew Size	Avg Day Prod	Avg Mat'l Unit Cost	Avg Equip Unit Cost	Avg Labor Unit Cost	Avg Total Unit Cost	Avg Price Incl O&P
Before first coat	Inst	SF	N-1	1600	0.01	---	0.10	0.11	0.15	N-1	1120	0.01	---	0.14	0.15	0.21
Before second coat	Inst	SF	N-1	1760	0.01	---	0.09	0.10	0.14	N-1	1232	0.01	---	0.13	0.14	0.19
Before third coat	Inst	SF	N-1	1920	0.01	---	0.08	0.09	0.13	N-1	1344	0.01	---	0.11	0.12	0.18
Liquid removal of paint or varnish																
Paneling (170 SF/gal)	Inst	SF	N-1	320	0.11	---	0.48	0.59	0.81	N-1	224	0.13	---	0.69	0.82	1.14
Millwork & trim (170 SF/gal)	Inst	SF	N-1	280	0.11	---	0.55	0.66	0.92	N-1	196	0.13	---	0.79	0.92	1.29
Floors (170 SF/gal)	Inst	SF	N-1	480	0.11	---	0.32	0.43	0.58	N-1	336	0.13	---	0.46	0.59	0.81
Burning off paint	Inst	SF	N-1	200	0.01	---	0.77	0.78	1.14	N-1	140	0.01	---	1.10	1.11	1.63

One coat application. Excluding openings unless otherwise indicated

Sizing, on sheetrock or plaster

Smooth finish

Description	Oper	Unit	Crew Size	Avg Day Prod	Avg Mat'l Unit Cost	Avg Equip Unit Cost	Avg Labor Unit Cost	Avg Total Unit Cost	Avg Price Incl O&P	Crew Size	Avg Day Prod	Avg Mat'l Unit Cost	Avg Equip Unit Cost	Avg Labor Unit Cost	Avg Total Unit Cost	Avg Price Incl O&P
Brush (650 SF/gal)	Inst	SF	N-1	2400	0.02	---	0.06	0.08	0.11	N-1	1680	0.02	---	0.09	0.11	0.16
Roller (625 SF/gal)	Inst	SF	N-1	3200	0.02	---	0.05	0.07	0.09	N-1	2240	0.02	---	0.07	0.09	0.13
Sand finish																
Brush (550 SF/gal)	Inst	SF	N-1	1720	0.02	---	0.09	0.11	0.15	N-1	1204	0.03	---	0.13	0.16	0.22
Roller (525 SF/gal)	Inst	SF	N-1	2320	0.02	---	0.07	0.09	0.12	N-1	1624	0.03	---	0.09	0.12	0.17

Sealer
Sheetrock or plaster

Smooth finish

Description	Oper	Unit	Crew Size	Avg Day Prod	Avg Mat'l Unit Cost	Avg Equip Unit Cost	Avg Labor Unit Cost	Avg Total Unit Cost	Avg Price Incl O&P	Crew Size	Avg Day Prod	Avg Mat'l Unit Cost	Avg Equip Unit Cost	Avg Labor Unit Cost	Avg Total Unit Cost	Avg Price Incl O&P
Brush (300 SF/gal)	Inst	SF	N-1	1440	0.04	---	0.11	0.15	0.20	N-1	1008	0.05	---	0.15	0.21	0.28
Roller (285 SF/gal)	Inst	SF	N-1	1840	0.05	---	0.08	0.13	0.17	N-1	1288	0.06	---	0.12	0.18	0.23
Spray (250 SF/gal)	Inst	SF	N-3	2600	0.05	0.01	0.06	0.12	0.15	N-3	1820	0.07	0.01	0.09	0.16	0.20
Sand finish																
Brush (250 SF/gal)	Inst	SF	N-1	1040	0.05	---	0.15	0.20	0.27	N-1	728	0.07	---	0.21	0.28	0.38
Roller (235 SF/gal)	Inst	SF	N-1	1360	0.06	---	0.11	0.17	0.22	N-1	952	0.07	---	0.16	0.23	0.31
Spray (210 SF/gal)	Inst	SF	N-3	2600	0.06	0.01	0.06	0.13	0.16	N-3	1820	0.08	0.01	0.09	0.18	0.22

Painting (continued)

Description	Oper	Unit	Crew Size	Avg Day Prod	Avg Mat'l Unit Cost	Avg Equip Unit Cost	Avg Labor Unit Cost	Avg Total Unit Cost	Avg Price Incl O&P	Crew Size	Avg Day Prod	Avg Mat'l Unit Cost	Avg Equip Unit Cost	Avg Labor Unit Cost	Avg Total Unit Cost	Avg Price Incl O&P
Acoustical tile or panels																
Brush (225 SF/gal)	Inst	SF	N-1	1240	0.06	---	0.12	0.18	0.24	N-1	868	0.07	---	0.18	0.25	0.33
Roller (200 SF/gal)	Inst	SF	N-1	1600	0.07	---	0.10	0.16	0.21	N-1	1120	0.08	---	0.14	0.22	0.28
Spray (160 SF/gal)	Inst	SF	N-3	2200	0.08	0.01	0.07	0.16	0.20	N-3	1540	0.10	0.01	0.10	0.22	0.27
Latex																
Drywall or plaster, latex flat																
Smooth finish																
Brush (300 SF/gal)	Inst	SF	N-1	1320	0.05	---	0.12	0.16	0.22	N-1	924	0.06	---	0.17	0.23	0.30
Roller (285 SF/gal)	Inst	SF	N-1	1680	0.05	---	0.09	0.14	0.18	N-1	1176	0.06	---	0.13	0.19	0.25
Spray (260 SF/gal)	Inst	SF	N-3	2400	0.05	0.01	0.07	0.13	0.16	N-3	1680	0.07	0.01	0.09	0.17	0.22
Sand finish																
Brush (250 SF/gal)	Inst	SF	N-1	960	0.06	---	0.16	0.22	0.29	N-1	672	0.07	---	0.23	0.30	0.41
Roller (235 SF/gal)	Inst	SF	N-1	1200	0.06	---	0.13	0.19	0.25	N-1	840	0.07	---	0.18	0.26	0.34
Spray (210 SF/gal)	Inst	SF	N-3	2400	0.07	0.01	0.07	0.14	0.17	N-3	1680	0.08	0.01	0.09	0.19	0.23
Texture or stipple applied to drywall, one coat																
Brush (125 SF/gal)	Inst	SF	N-1	1200	0.11	---	0.13	0.24	0.30	N-1	840	0.14	---	0.18	0.32	0.41
Roller (120 SF/gal)	Inst	SF	N-1	1600	0.12	---	0.10	0.21	0.26	N-1	1120	0.15	---	0.14	0.28	0.35
Paneling, latex enamel																
Brush (300 SF/gal)	Inst	SF	N-1	1320	0.05	---	0.12	0.16	0.22	N-1	924	0.06	---	0.17	0.23	0.30
Roller (285 SF/gal)	Inst	SF	N-1	1680	0.05	---	0.09	0.14	0.18	N-1	1176	0.06	---	0.13	0.19	0.25
Spray (260 SF/gal)	Inst	SF	N-3	2400	0.05	0.01	0.07	0.13	0.16	N-3	1680	0.07	0.01	0.09	0.17	0.22
Acoustical tile or panels, latex flat																
Brush (225 SF/gal)	Inst	SF	N-1	1120	0.06	---	0.14	0.20	0.26	N-1	784	0.08	---	0.20	0.27	0.37
Roller (210 SF/gal)	Inst	SF	N-1	1440	0.07	---	0.11	0.17	0.22	N-1	1008	0.08	---	0.15	0.24	0.31
Spray (185 SF/gal)	Inst	SF	N-3	2000	0.08	0.01	0.08	0.17	0.20	N-3	1400	0.09	0.01	0.11	0.22	0.28
Millwork and trim, latex enamel																
Doors and windows																
Roller and/or brush (360 SF/gal)	Inst	SF	N-1	640	0.04	---	0.24	0.28	0.39	N-1	448	0.05	---	0.34	0.39	0.55
Spray, doors only (325 SF/gal)	Inst	SF	N-3	1360	0.04	0.02	0.12	0.18	0.23	N-3	952	0.05	0.02	0.17	0.24	0.32

			Costs Based On Large Volume							Costs Based On Small Volume						
Description	Oper	Unit	Crew Size	Avg Day Prod	Avg Mat'l Unit Cost	Avg Equip Unit Cost	Avg Labor Unit Cost	Avg Total Unit Cost	Avg Price Incl O&P	Crew Size	Avg Day Prod	Avg Mat'l Unit Cost	Avg Equip Unit Cost	Avg Labor Unit Cost	Avg Total Unit Cost	Avg Price Incl O&P

Painting (continued)

Cabinets

Description	Oper	Unit	Crew Size	Avg Day Prod	Avg Mat'l Unit Cost	Avg Equip Unit Cost	Avg Labor Unit Cost	Avg Total Unit Cost	Avg Price Incl O&P	Crew Size	Avg Day Prod	Avg Mat'l Unit Cost	Avg Equip Unit Cost	Avg Labor Unit Cost	Avg Total Unit Cost	Avg Price Incl O&P
Roller and/or brush (360 SF/gal)	Inst	SF	N-1	600	0.04	---	0.26	0.30	0.42	N-1	420	0.05	---	0.37	0.42	0.59
Spray, doors only (325 SF/gal)	Inst	SF	N-3	1300	0.04	0.02	0.12	0.18	0.24	N-3	910	0.05	0.02	0.17	0.25	0.33
Louvers, spray (300 SF/gal)	Inst	SF	N-3	440	0.05	0.05	0.36	0.46	0.63	N-3	308	0.06	0.07	0.52	0.64	0.89

Picture molding, chair rail, base, ceiling mold etc., less than 6" H, (Note SF equals LF on trim less than 6" H)

Description	Oper	Unit	Crew Size	Avg Day Prod	Avg Mat'l Unit Cost	Avg Equip Unit Cost	Avg Labor Unit Cost	Avg Total Unit Cost	Avg Price Incl O&P	Crew Size	Avg Day Prod	Avg Mat'l Unit Cost	Avg Equip Unit Cost	Avg Labor Unit Cost	Avg Total Unit Cost	Avg Price Incl O&P
Brush (900 SF/gal)	Inst	SF	N-1	960	0.02	---	0.16	0.18	0.25	N-1	672	0.02	---	0.23	0.25	0.36

Floors, wood

Description	Oper	Unit	Crew Size	Avg Day Prod	Avg Mat'l Unit Cost	Avg Equip Unit Cost	Avg Labor Unit Cost	Avg Total Unit Cost	Avg Price Incl O&P	Crew Size	Avg Day Prod	Avg Mat'l Unit Cost	Avg Equip Unit Cost	Avg Labor Unit Cost	Avg Total Unit Cost	Avg Price Incl O&P
Brush (405 SF/gal)	Inst	SF	N-1	2000	0.03	---	0.08	0.11	0.15	N-1	1400	0.04	---	0.11	0.15	0.21
Roller (385 SF/gal)	Inst	SF	N-1	2400	0.04	---	0.06	0.10	0.13	N-1	1680	0.05	---	0.09	0.14	0.18
For custom colors, ADD	Inst	%	---	---	10.0	---	---	---	—	---	---	10.0	---	---	---	—

Floor seal

Description	Oper	Unit	Crew Size	Avg Day Prod	Avg Mat'l Unit Cost	Avg Equip Unit Cost	Avg Labor Unit Cost	Avg Total Unit Cost	Avg Price Incl O&P	Crew Size	Avg Day Prod	Avg Mat'l Unit Cost	Avg Equip Unit Cost	Avg Labor Unit Cost	Avg Total Unit Cost	Avg Price Incl O&P
Brush (450 SF/gal)	Inst	SF	N-1	2800	0.04	---	0.06	0.09	0.12	N-1	1960	0.04	---	0.08	0.12	0.16
Roller (430 SF/gal)	Inst	SF	N-1	3200	0.04	---	0.05	0.09	0.11	N-1	2240	0.05	---	0.07	0.12	0.15

Penetrating stainwax hardwood floors

Description	Oper	Unit	Crew Size	Avg Day Prod	Avg Mat'l Unit Cost	Avg Equip Unit Cost	Avg Labor Unit Cost	Avg Total Unit Cost	Avg Price Incl O&P	Crew Size	Avg Day Prod	Avg Mat'l Unit Cost	Avg Equip Unit Cost	Avg Labor Unit Cost	Avg Total Unit Cost	Avg Price Incl O&P
Brush (450 SF/gal)	Inst	SF	N-1	2000	0.04	---	0.08	0.11	0.15	N-1	1400	0.04	---	0.11	0.15	0.21
Roller (425 SF/gal)	Inst	SF	N-1	2400	0.04	---	0.06	0.10	0.13	N-1	1680	0.05	---	0.09	0.14	0.18

Natural finishes

Paneling brush work unless otherwise indicated

Stain, brush on - wipe off

Description	Oper	Unit	Crew Size	Avg Day Prod	Avg Mat'l Unit Cost	Avg Equip Unit Cost	Avg Labor Unit Cost	Avg Total Unit Cost	Avg Price Incl O&P	Crew Size	Avg Day Prod	Avg Mat'l Unit Cost	Avg Equip Unit Cost	Avg Labor Unit Cost	Avg Total Unit Cost	Avg Price Incl O&P
(360 SF/gal)	Inst	SF	N-1	560	0.04	---	0.28	0.32	0.45	N-1	392	0.06	---	0.39	0.45	0.63
Varnish (380 SF/gal)	Inst	SF	N-1	1360	0.05	---	0.11	0.16	0.21	N-1	952	0.06	---	0.16	0.22	0.30
Shellac (630 SF/gal)	Inst	SF	N-1	1600	0.03	---	0.10	0.13	0.17	N-1	1120	0.04	---	0.14	0.18	0.24
Lacquer																
Brush (450 SF/gal)	Inst	SF	N-1	1920	0.04	---	0.08	0.12	0.15	N-1	1344	0.04	---	0.11	0.16	0.21
Spray (300 SF/gal)	Inst	SF	N-3	2400	0.05	0.01	0.07	0.13	0.16	N-3	1680	0.07	0.01	0.09	0.17	0.22

Doors and windows, brush work unless otherwise indicated

Stain, brush on - wipe off

Description	Oper	Unit	Crew Size	Avg Day Prod	Avg Mat'l Unit Cost	Avg Equip Unit Cost	Avg Labor Unit Cost	Avg Total Unit Cost	Avg Price Incl O&P	Crew Size	Avg Day Prod	Avg Mat'l Unit Cost	Avg Equip Unit Cost	Avg Labor Unit Cost	Avg Total Unit Cost	Avg Price Incl O&P
(450 SF/gal)	Inst	SF	N-1	250	0.04	---	0.62	0.65	0.94	N-1	175	0.04	---	0.88	0.93	1.34
Varnish (550 SF/gal)	Inst	SF	N-1	340	0.03	---	0.45	0.49	0.70	N-1	238	0.04	---	0.65	0.69	0.99
Shellac (550 SF/gal)	Inst	SF	N-1	365	0.04	---	0.42	0.46	0.66	N-1	256	0.05	---	0.60	0.65	0.93

Painting (continued)

Description	Oper	Unit	Costs Based On Large Volume							Costs Based On Small Volume						
			Crew Size	Avg Day Prod	Avg Mat'l Unit Cost	Avg Equip Unit Cost	Avg Labor Unit Cost	Avg Total Unit Cost	Avg Price Incl O&P	Crew Size	Avg Day Prod	Avg Mat'l Unit Cost	Avg Equip Unit Cost	Avg Labor Unit Cost	Avg Total Unit Cost	Avg Price Incl O&P
Lacquer																
Brush (550 SF/gal)	Inst	SF	N-1	405	0.03	---	0.38	0.41	0.59	N-1	284	0.04	---	0.54	0.58	0.84
Spray doors (300 SF/gal)	Inst	SF	N-3	800	0.05	0.03	0.20	0.28	0.37	N-3	560	0.07	0.04	0.28	0.39	0.52
Cabinets, brush work unless otherwise indicated																
Stain, brush on-wipe off																
(450 SF/gal)	Inst	SF	N-1	235	0.04	---	0.66	0.69	1.00	N-1	165	0.04	---	0.94	0.98	1.42
Varnish (550 SF/gal)	Inst	SF	N-1	320	0.03	---	0.48	0.51	0.74	N-1	224	0.04	---	0.69	0.73	1.05
Shellac (550 SF/gal)	Inst	SF	N-1	345	0.04	---	0.45	0.48	0.69	N-1	242	0.05	---	0.64	0.68	0.98
Lacquer																
Brush (550 SF/gal)	Inst	SF	N-1	385	0.03	---	0.40	0.43	0.62	N-1	270	0.04	---	0.57	0.61	0.88
Spray (300 SF/gal)	Inst	SF	N-3	750	0.05	0.03	0.21	0.29	0.39	N-3	525	0.07	0.04	0.30	0.41	0.55
Louvers, lacquer, spray (300 SF/gal)	Inst	SF	N-3	480	0.05	0.04	0.33	0.43	0.58	N-3	336	0.07	0.06	0.47	0.60	0.82
Picture molding, chair rail, base, ceiling mold etc., less than 6" H, (Note SF equals LF on trim less than 6" H)																
Varnish, brush (900 SF/Gal)	Inst	SF	N-1	1040	0.02	---	0.15	0.17	0.24	N-1	728	0.03	---	0.21	0.24	0.34
Shellac, brush (900 SF/Gal)	Inst	SF	N-1	1040	0.02	---	0.15	0.17	0.24	N-1	728	0.03	---	0.21	0.24	0.34
Lacquer, spray (700 SF/Gal)	Inst	SF	N-1	1280	0.02	---	0.12	0.14	0.20	N-1	896	0.03	---	0.17	0.20	0.28
Floors, wood. Brush work unless otherwise indicated																
Shellac (450 SF/gal)	Inst	SF	N-1	2000	0.04	---	0.08	0.12	0.16	N-1	1400	0.06	---	0.11	0.17	0.22
Varnish (500 SF/gal)	Inst	SF	N-1	2000	0.04	---	0.08	0.11	0.15	N-1	1400	0.05	---	0.11	0.16	0.21
Buffing, by machine	Inst	SF	N-1	2800	0.01	---	0.06	0.06	0.09	N-1	1960	0.01	---	0.08	0.09	0.13
Waxing and polishing, by hand (1,000 SF/gal)	Inst	SF	N-1	1520	0.01	---	0.10	0.11	0.16	N-1	1064	0.01	---	0.14	0.16	0.23

Two coat application. Excluding openings unless otherwise indicated

For sizing or sealer, see One Coat Application

Latex. Note: All spray work, see One Coat Application

Sheetrock or plaster, latex flat Smooth finish																
Brush (170 SF/gal)	Inst	SF	N-1	880	0.08	---	0.18	0.26	0.34	N-1	616	0.10	---	0.25	0.35	0.47
Roller (160 SF/gal)	Inst	SF	N-1	1040	0.09	---	0.15	0.24	0.31	N-1	728	0.11	---	0.21	0.32	0.42

Description	Oper	Unit	Costs Based On Large Volume							Costs Based On Small Volume						
			Crew Size	Avg Day Prod	Avg Mat'l Unit Cost	Avg Equip Unit Cost	Avg Labor Unit Cost	Avg Total Unit Cost	Avg Price Incl O&P	Crew Size	Avg Day Prod	Avg Mat'l Unit Cost	Avg Equip Unit Cost	Avg Labor Unit Cost	Avg Total Unit Cost	Avg Price Incl O&P

Painting (continued)

Sand finish

Description	Oper	Unit	Crew Size	Avg Day Prod	Avg Mat'l Unit Cost	Avg Equip Unit Cost	Avg Labor Unit Cost	Avg Total Unit Cost	Avg Price Incl O&P	Crew Size	Avg Day Prod	Avg Mat'l Unit Cost	Avg Equip Unit Cost	Avg Labor Unit Cost	Avg Total Unit Cost	Avg Price Incl O&P
Brush (170 SF/gal)	Inst	SF	N-1	600	0.08	---	0.26	0.34	0.46	N-1	420	0.10	---	0.37	0.47	0.64
Roller (160 SF/gal)	Inst	SF	N-1	760	0.09	---	0.20	0.29	0.39	N-1	532	0.11	---	0.29	0.40	0.54

Paneling latex enamel

Description	Oper	Unit	Crew Size	Avg Day Prod	Avg Mat'l Unit Cost	Avg Equip Unit Cost	Avg Labor Unit Cost	Avg Total Unit Cost	Avg Price Incl O&P	Crew Size	Avg Day Prod	Avg Mat'l Unit Cost	Avg Equip Unit Cost	Avg Labor Unit Cost	Avg Total Unit Cost	Avg Price Incl O&P
Brush (170 SF/gal)	Inst	SF	N-1	880	0.08	---	0.18	0.26	0.34	N-1	616	0.10	---	0.25	0.35	0.47
Roller (160 SF/gal)	Inst	SF	N-1	1040	0.09	---	0.15	0.24	0.31	N-1	728	0.11	---	0.21	0.32	0.42

Acoustical tile or panels, latex flat

Description	Oper	Unit	Crew Size	Avg Day Prod	Avg Mat'l Unit Cost	Avg Equip Unit Cost	Avg Labor Unit Cost	Avg Total Unit Cost	Avg Price Incl O&P	Crew Size	Avg Day Prod	Avg Mat'l Unit Cost	Avg Equip Unit Cost	Avg Labor Unit Cost	Avg Total Unit Cost	Avg Price Incl O&P
Brush (130 SF/gal)	Inst	SF	N-1	680	0.11	---	0.23	0.33	0.44	N-1	476	0.13	---	0.32	0.46	0.61
Roller (120 SF/gal)	Inst	SF	N-1	880	0.12	---	0.18	0.29	0.37	N-1	616	0.15	---	0.25	0.40	0.51

Millwork and trim, enamel

Doors and windows
Roller and/or brush

Description	Oper	Unit	Crew Size	Avg Day Prod	Avg Mat'l Unit Cost	Avg Equip Unit Cost	Avg Labor Unit Cost	Avg Total Unit Cost	Avg Price Incl O&P	Crew Size	Avg Day Prod	Avg Mat'l Unit Cost	Avg Equip Unit Cost	Avg Labor Unit Cost	Avg Total Unit Cost	Avg Price Incl O&P
(200 SF/gal)	Inst	SF	N-1	335	0.07	---	0.46	0.53	0.75	N-1	235	0.09	---	0.66	0.74	1.05

Cabinets
Roller and/or brush

Description	Oper	Unit	Crew Size	Avg Day Prod	Avg Mat'l Unit Cost	Avg Equip Unit Cost	Avg Labor Unit Cost	Avg Total Unit Cost	Avg Price Incl O&P	Crew Size	Avg Day Prod	Avg Mat'l Unit Cost	Avg Equip Unit Cost	Avg Labor Unit Cost	Avg Total Unit Cost	Avg Price Incl O&P
(200 SF/gal)	Inst	SF	N-1	315	0.07	---	0.49	0.56	0.79	N-1	221	0.09	---	0.70	0.79	1.12

Louvers, see One Coat Application

Picture molding, chair rail, base, ceiling mold etc., less than
6" H, (Note SF equals LF on trim less than 6" H)

Description	Oper	Unit	Crew Size	Avg Day Prod	Avg Mat'l Unit Cost	Avg Equip Unit Cost	Avg Labor Unit Cost	Avg Total Unit Cost	Avg Price Incl O&P	Crew Size	Avg Day Prod	Avg Mat'l Unit Cost	Avg Equip Unit Cost	Avg Labor Unit Cost	Avg Total Unit Cost	Avg Price Incl O&P
Brush (510 SF/gal)	Inst	SF	N-1	500	0.03	---	0.31	0.34	0.48	N-1	350	0.03	---	0.44	0.47	0.68

Floors, wood

Description	Oper	Unit	Crew Size	Avg Day Prod	Avg Mat'l Unit Cost	Avg Equip Unit Cost	Avg Labor Unit Cost	Avg Total Unit Cost	Avg Price Incl O&P	Crew Size	Avg Day Prod	Avg Mat'l Unit Cost	Avg Equip Unit Cost	Avg Labor Unit Cost	Avg Total Unit Cost	Avg Price Incl O&P
Brush (230 SF/gal)	Inst	SF	N-1	1120	0.06	---	0.14	0.20	0.26	N-1	784	0.08	---	0.20	0.27	0.37
Roller (220 SF/gal)	Inst	SF	N-1	1280	0.06	---	0.12	0.18	0.24	N-1	896	0.08	---	0.17	0.25	0.33
For custom colors, ADD	Inst	%	---	---	10.00	---	---	---	---	---	---	10.00	---	---	---	---

For floor seal, penetrating stainwax, or natural finish, see One Coat Application

Exterior. The crew output per day includes time calculations for normal handling and protection of other property not to be painted

Preparation. Excluding openings unless otherwise indicated
Cleaning, wet

Description	Oper	Unit	Crew Size	Avg Day Prod	Avg Mat'l Unit Cost	Avg Equip Unit Cost	Avg Labor Unit Cost	Avg Total Unit Cost	Avg Price Incl O&P	Crew Size	Avg Day Prod	Avg Mat'l Unit Cost	Avg Equip Unit Cost	Avg Labor Unit Cost	Avg Total Unit Cost	Avg Price Incl O&P
Plain siding	Inst	SF	N-1	1480	0.01	---	0.10	0.12	0.17	N-1	1036	0.02	---	0.15	0.17	0.24

			Costs Based On Large Volume							Costs Based On Small Volume					
Description	Oper	Unit	Crew Size	Avg Day Prod	Avg Mat'l Unit Cost	Avg Equip Unit Cost	Avg Labor Unit Cost	Avg Total Unit Cost	Avg Price Incl O&P	Crew Size	Avg Day Prod	Avg Mat'l Unit Cost	Avg Equip Unit Cost	Avg Labor Unit Cost	Avg Total Unit Cost	Avg Price Incl O&P

Painting (continued)

Exterior doors and trim (Note: SF equals LF on trim

less than 6" high)	Inst	SF	N-1	1400	0.01	---	0.11	0.12	0.18	N-1	980	0.02	---	0.16	0.17	0.25
Windows, wash and clean glass	Inst	SF	N-1	1280	0.01	---	0.12	0.13	0.19	N-1	896	0.02	---	0.17	0.19	0.27
Porch floors and steps	Inst	SF	N-1	2240	0.01	---	0.07	0.08	0.12	N-1	1568	0.02	---	0.10	0.12	0.16

Acid wash

Gutters and downspouts	Inst	SF	N-1	1040	0.01	---	0.15	0.16	0.23	N-1	728	0.02	---	0.21	0.23	0.33

Sanding, light

Porch floors and steps	Inst	SF	N-1	1760	0.01	---	0.09	0.10	0.14	N-1	1232	0.02	---	0.13	0.14	0.20

Sanding and putting

Plain siding	Inst	SF	N-1	1440	0.01	---	0.11	0.12	0.17	N-1	1008	0.02	---	0.15	0.17	0.24

Exterior doors and trim (Note: SF equals LF on trim

less than 6" high)	Inst	SF	N-1	720	0.01	---	0.21	0.23	0.33	N-1	504	0.02	---	0.31	0.32	0.47

Puttying sash or reglazing
Windows (30 SF glass/lb.)

glazing compound	Inst	SF	N-1	240	0.47	---	0.64	1.11	1.41	N-1	168	0.58	---	0.92	1.50	1.93

One coat application. Excluding openings unless otherwise indicated

Latex, flat (unless otherwise indicated)
Plain siding

Brush (300 SF/gal)	Inst	SF	N-1	800	0.05	---	0.19	0.24	0.33	N-1	560	0.06	---	0.28	0.33	0.46
Roller (275 SF/gal)	Inst	SF	N-1	1040	0.05	---	0.15	0.20	0.27	N-1	728	0.06	---	0.21	0.28	0.37
Spray (325 SF/gal)	Inst	SF	N-3	1750	0.04	0.01	0.09	0.15	0.19	N-3	1225	0.05	0.02	0.13	0.20	0.26

Shingle siding

Brush (270 SF/gal)	Inst	SF	N-1	880	0.05	---	0.18	0.23	0.31	N-1	616	0.06	---	0.25	0.32	0.43
Roller (260 SF/gal)	Inst	SF	N-1	1200	0.05	---	0.13	0.18	0.24	N-1	840	0.07	---	0.18	0.25	0.34
Spray (300 SF/gal)	Inst	SF	N-3	1600	0.05	0.01	0.10	0.16	0.21	N-3	1120	0.06	0.02	0.14	0.22	0.29

Stucco

Brush (135 SF/gal)	Inst	SF	N-1	800	0.10	---	0.19	0.30	0.39	N-1	560	0.13	---	0.28	0.40	0.53
Roller (125 SF/gal)	Inst	SF	N-1	1200	0.11	---	0.13	0.24	0.30	N-1	840	0.14	---	0.18	0.32	0.41
Spray (150 SF/gal)	Inst	SF	N-3	1600	0.09	0.01	0.10	0.21	0.25	N-3	1120	0.12	0.02	0.14	0.28	0.34

Cement walls. See Cement base paint.

| | | | Costs Based On Large Volume | | | | | | | Costs Based On Small Volume | | | | | | |
Description	Oper	Unit	Crew Size	Avg Day Prod	Avg Mat'l Unit Cost	Avg Equip Unit Cost	Avg Labor Unit Cost	Avg Total Unit Cost	Avg Price Incl O&P	Crew Size	Avg Day Prod	Avg Mat'l Unit Cost	Avg Equip Unit Cost	Avg Labor Unit Cost	Avg Total Unit Cost	Avg Price Incl O&P

Painting (continued)

Masonry block, brick, tile; masonry latex

Description	Oper	Unit	Crew Size	Avg Day Prod	Mat'l	Equip	Labor	Total	Price O&P	Crew Size	Avg Day Prod	Mat'l	Equip	Labor	Total	Price O&P
Brush (180 SF/gal)	Inst	SF	N-1	880	0.08	---	0.18	0.25	0.34	N-1	616	0.10	---	0.25	0.35	0.47
Roller (125 SF/gal)	Inst	SF	N-1	1440	0.11	---	0.11	0.22	0.27	N-1	1008	0.14	---	0.15	0.29	0.36
Spray (160 SF/gal)	Inst	SF	N-3	1840	0.09	0.01	0.09	0.19	0.23	N-3	1288	0.11	0.02	0.12	0.25	0.31
Doors, exterior side only																
Brush (375 SF/gal)	Inst	SF	N-1	640	0.04	---	0.24	0.28	0.39	N-1	448	0.05	---	0.34	0.39	0.55
Roller (375 SF/gal)	Inst	SF	N-1	865	0.04	---	0.18	0.22	0.30	N-1	606	0.05	---	0.25	0.30	0.42
Windows, exterior side only, brush work (1500 SF/gal)	Inst	SF	N-1	640	0.01	---	0.24	0.25	0.36	N-1	448	0.01	---	0.34	0.36	0.52
Trim, less than 6" H, brush (Note: SF equals LF on trim less than 6" H) High gloss (300 SF/gal)	Inst	SF	N-1	900	0.05	---	0.17	0.22	0.30	N-1	630	0.06	---	0.24	0.30	0.42
Screens, full; high gloss																
Paint applied to wood only, brush work (700 SF/gal)	Inst	SF	N-1	540	0.02	---	0.29	0.31	0.44	N-1	378	0.03	---	0.41	0.43	0.62
Paint applied to wood (brush) and wire (spray) (475 SF/gal)	Inst	SF	N-3	450	0.03	0.05	0.35	0.43	0.60	N-3	315	0.04	0.07	0.51	0.61	0.85
Storm windows and doors, 2 lites, brush work (340 SF/gal)	Inst	SF	N-1	340	0.04	---	0.45	0.49	0.71	N-1	238	0.05	---	0.65	0.70	1.00
Blinds or shutters																
Brush (120 SF/gal)	Inst	SF	N-1	130	0.12	---	1.19	1.30	1.86	N-1	91	0.15	---	1.69	1.84	2.64
Spray (300 SF/gal)	Inst	SF	N-3	400	0.05	0.05	0.40	0.50	0.68	N-3	280	0.06	0.07	0.57	0.70	0.97
Gutters and downspouts, brush work (225 LF/gal), galvanized	Inst	SF	N-1	600	0.06	---	0.26	0.32	0.44	N-1	420	0.08	---	0.37	0.44	0.62
Porch floors and steps, wood																
Brush (340 SF/gal)	Inst	SF	N-1	1520	0.04	---	0.10	0.14	0.19	N-1	1064	0.05	---	0.14	0.20	0.26
Roller (325 SF/gal)	Inst	SF	N-1	1920	0.04	---	0.08	0.12	0.16	N-1	1344	0.05	---	0.11	0.17	0.22
Shingle roofs																
Brush (135 SF/gal)	Inst	SF	N-1	1040	0.10	---	0.15	0.25	0.32	N-1	728	0.13	---	0.21	0.34	0.44
Roller (125 SF/gal)	Inst	SF	N-1	1360	0.11	---	0.11	0.23	0.28	N-1	952	0.14	---	0.16	0.30	0.38
Spray (150 SF/gal)	Inst	SF	N-3	2080	0.09	0.01	0.08	0.18	0.22	N-3	1456	0.12	0.01	0.11	0.24	0.29

Painting (continued)

Description	Oper	Unit	Costs Based On Large Volume							Costs Based On Small Volume						
			Crew Size	Avg Day Prod	Avg Mat'l Unit Cost	Avg Equip Unit Cost	Avg Labor Unit Cost	Avg Total Unit Cost	Avg Price Incl O&P	Crew Size	Avg Day Prod	Avg Mat'l Unit Cost	Avg Equip Unit Cost	Avg Labor Unit Cost	Avg Total Unit Cost	Avg Price Incl O&P
For custom colors, ADD	Inst	%	---	---	10.0	---	---	---	—	---	---	10.0	---	---	---	—
Cement base paint (epoxy Concrete enamel)																
Cement walls, smooth finish																
Brush (120 SF/gal)	Inst	SF	N-1	1320	0.17	---	0.12	0.28	0.34	N-1	924	0.21	---	0.17	0.38	0.45
Roller (110 SF/gal)	Inst	SF	N-1	1920	0.18	---	0.08	0.26	0.30	N-1	1344	0.23	---	0.11	0.34	0.40
Concrete porch floors and steps																
Brush (400 SF/gal)	Inst	SF	N-1	2000	0.05	---	0.08	0.13	0.16	N-1	1400	0.06	---	0.11	0.17	0.22
Roller (375 SF/gal)	Inst	SF	N-1	2400	0.05	---	0.06	0.12	0.15	N-1	1680	0.07	---	0.09	0.16	0.20
Stain																
Shingle siding																
Brush (180 SF/gal)	Inst	SF	N-1	800	0.09	---	0.19	0.28	0.37	N-1	560	0.11	---	0.28	0.39	0.52
Roller (170 SF/gal)	Inst	SF	N-1	1120	0.09	---	0.14	0.23	0.30	N-1	784	0.12	---	0.20	0.31	0.41
Spray (200 SF/gal)	Inst	SF	N-3	1520	0.08	0.01	0.10	0.20	0.25	N-3	1064	0.10	0.02	0.15	0.27	0.34
Shingle roofs																
Brush (180 SF/gal)	Inst	SF	N-1	960	0.09	---	0.16	0.25	0.32	N-1	672	0.11	---	0.23	0.34	0.45
Roller (170 SF/gal)	Inst	SF	N-1	1280	0.09	---	0.12	0.21	0.27	N-1	896	0.12	---	0.17	0.29	0.37
Spray (200 SF/gal)	Inst	SF	N-3	1920	0.08	0.01	0.08	0.17	0.21	N-3	1344	0.10	0.02	0.12	0.23	0.29
Two coat application (excluding openings unless otherwise indicated)																
Latex, flat (unless otherwise indicated)																
Plain siding																
Brush (170 SF/gal)	Inst	SF	N-1	440	0.08	---	0.35	0.43	0.60	N-1	308	0.10	---	0.50	0.60	0.84
Roller (155 SF/gal)	Inst	SF	N-1	560	0.09	---	0.28	0.37	0.49	N-1	392	0.11	---	0.39	0.51	0.69
Spray (185 SF/gal)	Inst	SF	N-3	960	0.08	0.02	0.17	0.26	0.34	N-3	672	0.09	0.03	0.24	0.36	0.47
Shingle siding																
Brush (150 SF/gal)	Inst	SF	N-1	480	0.09	---	0.32	0.41	0.57	N-1	336	0.12	---	0.46	0.58	0.79
Roller (150 SF/gal)	Inst	SF	N-1	640	0.09	---	0.24	0.33	0.45	N-1	448	0.12	---	0.34	0.46	0.62
Spray (170 SF/gal)	Inst	SF	N-3	880	0.08	0.02	0.18	0.29	0.37	N-3	616	0.10	0.03	0.26	0.40	0.52

Description	Oper	Unit	Costs Based On Large Volume							Costs Based On Small Volume						
			Crew Size	Avg Day Prod	Avg Mat'l Unit Cost	Avg Equip Unit Cost	Avg Labor Unit Cost	Avg Total Unit Cost	Avg Price Incl O&P	Crew Size	Avg Day Prod	Avg Mat'l Unit Cost	Avg Equip Unit Cost	Avg Labor Unit Cost	Avg Total Unit Cost	Avg Price Incl O&P

Painting (continued)

Stucco

Description	Oper	Unit	Crew Size	Avg Day Prod	Avg Mat'l Unit Cost	Avg Equip Unit Cost	Avg Labor Unit Cost	Avg Total Unit Cost	Avg Price Incl O&P	Crew Size	Avg Day Prod	Avg Mat'l Unit Cost	Avg Equip Unit Cost	Avg Labor Unit Cost	Avg Total Unit Cost	Avg Price Incl O&P
Brush (90 SF/gal)	Inst	SF	N-1	480	0.16	---	0.32	0.48	0.63	N-1	336	0.19	---	0.46	0.65	0.87
Roller (85 SF/gal)	Inst	SF	N-1	680	0.16	---	0.23	0.39	0.50	N-1	476	0.21	---	0.32	0.53	0.68
Spray (100 SF/gal)	Inst	SF	N-3	880	0.14	0.02	0.18	0.34	0.43	N-3	616	0.18	0.03	0.26	0.47	0.59
Masonry block, brick, tile; masonry latex																
Brush (120 SF/gal)	Inst	SF	N-1	520	0.12	---	0.30	0.41	0.55	N-1	364	0.15	---	0.42	0.57	0.77
Roller (85 SF/gal)	Inst	SF	N-1	760	0.16	---	0.20	0.37	0.46	N-1	532	0.21	---	0.29	0.50	0.63
Spray (105 SF/gal)	Inst	SF	N-3	1000	0.13	0.02	0.16	0.31	0.39	N-3	700	0.17	0.03	0.23	0.42	0.53
Doors, exterior side only																
Brush (215 SF/gal)	Inst	SF	N-1	335	0.07	---	0.46	0.53	0.74	N-1	235	0.08	---	0.66	0.74	1.05
Roller (215 SF/gal)	Inst	SF	N-1	455	0.07	---	0.34	0.40	0.56	N-1	319	0.08	---	0.48	0.57	0.79
Windows, exterior side only, brush work (850 SF/gal)	Inst	SF	N-1	335	0.02	---	0.46	0.48	0.69	N-1	235	0.02	---	0.66	0.68	0.99
Trim, less than 6" H, brush (Note: SF equals LF on trim less than 6" H) High gloss (230 SF/gal)	Inst	SF	N-1	480	0.06	---	0.32	0.38	0.53	N-1	336	0.08	---	0.46	0.53	0.75
Screens, full; high gloss																
Paint applied to wood only, brush work (400 SF/gal)	Inst	SF	N-1	285	0.04	---	0.54	0.58	0.83	N-1	200	0.04	---	0.77	0.82	1.18
Paint applied to wood (brush) and wire (spray) (270 SF/gal)	Inst	SF	N-3	240	0.05	0.09	0.66	0.80	1.11	N-3	168	0.06	0.12	0.95	1.14	1.58
Storm windows and doors, 2 lites, brush work (195 SF/gal)	Inst	SF	N-1	180	0.07	---	0.86	0.93	1.33	N-1	126	0.09	---	1.22	1.31	1.89
Blinds or shutters																
Brush (65 SF/gal)	Inst	SF	N-1	70	0.22	---	2.20	2.42	3.45	N-1	49	0.27	---	3.15	3.41	4.89
Spray (170 SF/gal)	Inst	SF	N-3	210	0.08	0.10	0.76	0.94	1.30	N-3	147	0.10	0.14	1.08	1.33	1.84
Gutters and downspouts, brush work (130 LF/gal), galvanized	Inst	SF	N-1	315	0.11	---	0.49	0.60	0.83	N-1	221	0.13	---	0.70	0.83	1.16

					Costs Based On Large Volume							Costs Based On Small Volume					
Description	Oper	Unit	Crew Size	Avg Day Prod	Avg Mat'l Unit Cost	Avg Equip Unit Cost	Avg Labor Unit Cost	Avg Total Unit Cost	Avg Price Incl O&P	Crew Size	Avg Day Prod	Avg Mat'l Unit Cost	Avg Equip Unit Cost	Avg Labor Unit Cost	Avg Total Unit Cost	Avg Price Incl O&P

Painting (continued)

Porch floors and steps, wood

Description	Oper	Unit	Crew Size	Avg Day Prod	Mat'l	Equip	Labor	Total	O&P	Crew Size	Avg Day Prod	Mat'l	Equip	Labor	Total	O&P
Brush (195 SF/gal)	Inst	SF	N-1	800	0.07	---	0.19	0.26	0.36	N-1	560	0.09	---	0.28	0.36	0.49
Roller (185 SF/gal)	Inst	SF	N-1	1000	0.08	---	0.15	0.23	0.30	N-1	700	0.09	---	0.22	0.31	0.42
Shingle roofs																
Brush (75 SF/gal)	Inst	SF	N-1	640	0.19	---	0.24	0.43	0.54	N-1	448	0.23	---	0.34	0.58	0.74
Roller (70 SF/gal)	Inst	SF	N-1	800	0.20	---	0.19	0.39	0.48	N-1	560	0.25	---	0.28	0.53	0.65
Spray (85 SF/gal)	Inst	SF	N-3	1000	0.16	0.02	0.16	0.34	0.42	N-3	700	0.21	0.03	0.23	0.46	0.57
For custom colors, ADD	Inst	%	---	---	10.0	---	---	---	---	---	---	10.0	---	---	---	---

Cement base paint (epoxy concrete enamel)

Cement walls, smooth finish

Brush (80 SF/gal)	Inst	SF	N-1	720	0.25	---	0.21	0.46	0.56	N-1	504	0.31	---	0.31	0.62	0.76
Roller (75 SF/gal)	Inst	SF	N-1	1040	0.27	---	0.15	0.41	0.48	N-1	728	0.33	---	0.21	0.55	0.64
Concrete porch floors and steps																
Brush (225 SF/gal)	Inst	SF	N-1	1040	0.09	---	0.15	0.24	0.31	N-1	728	0.11	---	0.21	0.32	0.42
Roller (210 SF/gal)	Inst	SF	N-1	1000	0.10	---	0.15	0.25	0.32	N-1	700	0.12	---	0.22	0.34	0.44

Stain

Shingle siding

Brush (105 SF/gal)	Inst	SF	N-1	480	0.15	---	0.32	0.47	0.62	N-1	336	0.19	---	0.46	0.65	0.86
Roller (100 SF/gal)	Inst	SF	N-1	600	0.16	---	0.26	0.42	0.54	N-1	420	0.20	---	0.37	0.57	0.74
Spray (115 SF/gal)	Inst	SF	N-3	880	0.14	0.02	0.18	0.34	0.43	N-3	616	0.17	0.03	0.26	0.47	0.59
Shingle roofs																
Brush (105 SF/gal)	Inst	SF	N-1	560	0.15	---	0.28	0.43	0.56	N-1	392	0.19	---	0.39	0.58	0.77
Roller (100 SF/gal)	Inst	SF	N-1	760	0.16	---	0.20	0.36	0.46	N-1	532	0.20	---	0.29	0.49	0.63
Spray (115 SF/gal)	Inst	SF	N-3	1120	0.14	0.02	0.14	0.30	0.37	N-3	784	0.17	0.03	0.20	0.40	0.50

Paneling, Hardboard and Plywood

Removal

Sheets, plywood or hardboard	Demo	SF	L-2	1850	---	---	0.14	0.14	0.21	L-2	1295	---	---	0.20	0.20	0.30
Boards, wood	Demo	SF	L-2	1650	---	---	0.16	0.16	0.23	L-2	1155	---	---	0.23	0.23	0.33

			Costs Based On Large Volume							Costs Based On Small Volume					
Description	Oper	Unit	Crew Size	Avg Day Prod	Avg Mat'l Unit Cost	Avg Equip Unit Cost	Avg Labor Unit Cost	Avg Total Unit Cost	Avg Price Incl O&P	Crew Size	Avg Day Prod	Avg Mat'l Unit Cost	Avg Equip Unit Cost	Avg Labor Unit Cost	Avg Total Unit Cost	Avg Price Incl O&P

Paneling (continued)

Installation
Waste, nails, and adhesives not included

Economy hardboard

Presdwood, 4' x 8' sheets

Standard

Description	Oper	Unit	Crew Size	Avg Day Prod	Avg Mat'l Unit Cost	Avg Equip Unit Cost	Avg Labor Unit Cost	Avg Total Unit Cost	Avg Price Incl O&P	Crew Size	Avg Day Prod	Avg Mat'l Unit Cost	Avg Equip Unit Cost	Avg Labor Unit Cost	Avg Total Unit Cost	Avg Price Incl O&P
1/8" T	Inst	SF	C-11	1025	0.23	---	0.36	0.59	0.77	C-11	718	0.35	---	0.52	0.87	1.12
1/4" T	Inst	SF	C-11	1025	0.38	---	0.36	0.74	0.92	C-11	718	0.57	---	0.52	1.08	1.34
Tempered																
1/8" T	Inst	SF	C-11	1025	0.30	---	0.36	0.66	0.84	C-11	718	0.45	---	0.52	0.96	1.22
1/4" T	Inst	SF	C-11	1025	0.42	---	0.36	0.78	0.96	C-11	718	0.63	---	0.52	1.14	1.39
Duolux, 4' x 8' sheets																
Standard																
1/8" T	Inst	SF	C-11	1025	0.26	---	0.36	0.62	0.80	C-11	718	0.39	---	0.52	0.91	1.16
1/4" T	Inst	SF	C-11	1025	0.40	---	0.36	0.77	0.94	C-11	718	0.61	---	0.52	1.12	1.37
Tempered																
1/8" T	Inst	SF	C-11	1025	0.47	---	0.36	0.83	1.01	C-11	718	0.71	---	0.52	1.22	1.47
1/4" T	Inst	SF	C-11	1025	0.52	---	0.36	0.88	1.06	C-11	718	0.78	---	0.52	1.30	1.55
Particleboard, 40 lb. interior underlayment																
Nailed to floors																
3/8" T	Inst	SF	C-11	1535	0.22	---	0.24	0.46	0.58	C-11	1075	0.33	---	0.34	0.68	0.84
1/2" T	Inst	SF	C-11	1535	0.27	---	0.24	0.51	0.63	C-11	1075	0.41	---	0.34	0.75	0.92
5/8" T	Inst	SF	C-11	1535	0.35	---	0.24	0.59	0.71	C-11	1075	0.53	---	0.34	0.87	1.04
3/4" T	Inst	SF	C-11	1535	0.40	---	0.24	0.65	0.76	C-11	1075	0.61	---	0.34	0.95	1.12
Nailed to walls																
3/8" T	Inst	SF	C-11	1440	0.22	---	0.26	0.48	0.60	C-11	1008	0.33	---	0.37	0.70	0.88
1/2" T	Inst	SF	C-11	1440	0.27	---	0.26	0.53	0.66	C-11	1008	0.41	---	0.37	0.78	0.96
5/8" T	Inst	SF	C-11	1440	0.35	---	0.26	0.61	0.73	C-11	1008	0.53	---	0.37	0.90	1.07
3/4" T	Inst	SF	C-11	1440	0.40	---	0.26	0.66	0.79	C-11	1008	0.61	---	0.37	0.97	1.15
Masonite prefinished 4' x 8' panels																
1/4" T; oak and maple designs	Inst	SF	C-11	800	0.56	---	0.46	1.02	1.25	C-11	560	0.84	---	0.66	1.50	1.82
1/4" T; nutwood designs	Inst	SF	C-11	800	0.64	---	0.46	1.10	1.33	C-11	560	0.96	---	0.66	1.62	1.94
1/4" T; weathered white	Inst	SF	C-11	800	0.64	---	0.46	1.10	1.33	C-11	560	0.96	---	0.66	1.62	1.94
1/4" T; brick or stone designs	Inst	SF	C-11	800	0.91	---	0.46	1.37	1.60	C-11	560	1.36	---	0.66	2.02	2.34

			Costs Based On Large Volume							Costs Based On Small Volume					
Description	Oper	Unit	Crew Size	Avg Day Prod	Avg Mat'l Unit Cost	Avg Equip Unit Cost	Avg Labor Unit Cost	Avg Total Unit Cost	Avg Price Incl O&P	Crew Size	Avg Day Prod	Avg Mat'l Unit Cost	Avg Equip Unit Cost	Avg Labor Unit Cost	Avg Total Unit Cost	Avg Price Incl O&P

Paneling (continued)

Pegboard, 4' x 8' sheets
Presdwood, tempered

Description	Oper	Unit	Crew Size	Avg Day Prod	Avg Mat'l Unit Cost	Avg Equip Unit Cost	Avg Labor Unit Cost	Avg Total Unit Cost	Avg Price Incl O&P	Crew Size	Avg Day Prod	Avg Mat'l Unit Cost	Avg Equip Unit Cost	Avg Labor Unit Cost	Avg Total Unit Cost	Avg Price Incl O&P
1/8" T	Inst	SF	C-11	1025	0.48	---	0.36	0.84	1.02	C-11	718	0.73	---	0.52	1.24	1.49
1/4" T	Inst	SF	C-11	1025	0.64	---	0.36	1.00	1.18	C-11	718	0.96	---	0.52	1.48	1.73
Duolux, tempered																
1/8" T	Inst	SF	C-11	1025	0.50	---	0.36	0.86	1.03	C-11	718	0.74	---	0.52	1.26	1.51
1/4" T	Inst	SF	C-11	1025	0.71	---	0.36	1.07	1.24	C-11	718	1.06	---	0.52	1.58	1.83
Unfinished hardwood plywood, applied with nails (nail heads filled)																
Ash-Sen, flush face																
1/8" x 4' x 7' x 8'	Inst	SF	C-17	900	0.82	---	0.52	1.34	1.60	C-17	630	1.24	---	0.74	1.98	2.34
3/16" x 4' x 8'	Inst	SF	C-17	900	0.98	---	0.52	1.50	1.76	C-17	630	1.47	---	0.74	2.21	2.58
1/4" x 4' x 8'	Inst	SF	C-17	900	1.14	---	0.52	1.66	1.91	C-17	630	1.71	---	0.74	2.45	2.81
1/4" x 4' x 10'	Inst	SF	C-17	900	1.28	---	0.52	1.80	2.06	C-17	630	1.93	---	0.74	2.67	3.03
1/2" x 4' x 10'	Inst	SF	C-17	900	2.53	---	0.52	3.05	3.30	C-17	630	3.80	---	0.74	4.54	4.90
3/4" x 4' x 8' vertical core	Inst	SF	C-17	900	2.89	---	0.52	3.40	3.66	C-17	630	4.33	---	0.74	5.07	5.43
3/4" x 4' x 8' lumber core	Inst	SF	C-17	900	3.57	---	0.52	4.09	4.34	C-17	630	5.35	---	0.74	6.09	6.46
Ash, V-grooved																
3/16" x 4' x 8'	Inst	SF	C-17	900	0.90	---	0.52	1.42	1.68	C-17	630	1.36	---	0.74	2.10	2.46
1/4" x 4' x 8'	Inst	SF	C-17	900	1.03	---	0.52	1.55	1.81	C-17	630	1.55	---	0.74	2.29	2.66
1/4" x 4' x 10'	Inst	SF	C-17	900	1.18	---	0.52	1.70	1.95	C-17	630	1.77	---	0.74	2.51	2.87
Birch, natural, "A" grade face																
Flush face																
1/8" x 4' x 8'	Inst	SF	C-17	900	0.88	---	0.52	1.40	1.65	C-17	630	1.32	---	0.74	2.06	2.42
3/16" x 4' x 8'	Inst	SF	C-17	900	1.02	---	0.52	1.54	1.80	C-17	630	1.53	---	0.74	2.27	2.64
1/4" x 4' x 8'	Inst	SF	C-17	900	1.15	---	0.52	1.67	1.93	C-17	630	1.73	---	0.74	2.47	2.83
1/4" x 4' x 10'	Inst	SF	C-17	900	1.30	---	0.52	1.82	2.07	C-17	630	1.95	---	0.74	2.69	3.05
3/8" x 4' x 8'	Inst	SF	C-17	900	1.87	---	0.52	2.39	2.65	C-17	630	2.81	---	0.74	3.55	3.92
1/2" x 4' x 8'	Inst	SF	C-17	900	2.44	---	0.52	2.96	3.21	C-17	630	3.66	---	0.74	4.40	4.76
3/4" x 4' x 8' lumber core	Inst	SF	C-17	900	3.52	---	0.52	4.03	4.29	C-17	630	5.27	---	0.74	6.01	6.38
V-grooved																
1/4" x 4' x 8', mismatched	Inst	SF	C-17	900	1.26	---	0.52	1.78	2.03	C-17	630	1.89	---	0.74	2.63	2.99
1/4" x 4' x 8'	Inst	SF	C-17	900	1.18	---	0.52	1.70	1.95	C-17	630	1.77	---	0.74	2.51	2.87
1/4" x 4' x 10'	Inst	SF	C-17	900	1.32	---	0.52	1.84	2.10	C-17	630	1.99	---	0.74	2.73	3.09
Birch, select red,																
1/4" x 4' x 8'	Inst	SF	C-17	900	1.45	---	0.52	1.97	2.23	C-17	630	2.18	---	0.74	2.92	3.29

			Costs Based On Large Volume							Costs Based On Small Volume					
Description	Oper	Unit	Crew Size	Avg Day Prod	Avg Mat'l Unit Cost	Avg Equip Unit Cost	Avg Labor Unit Cost	Avg Total Unit Cost	Avg Price Incl O&P	Crew Size	Avg Day Prod	Avg Mat'l Unit Cost	Avg Equip Unit Cost	Avg Labor Unit Cost	Avg Total Unit Cost	Avg Price Incl O&P

Paneling (continued)

Birch, select white,

Description	Oper	Unit	Crew Size	Avg Day Prod	Avg Mat'l Unit Cost	Avg Equip Unit Cost	Avg Labor Unit Cost	Avg Total Unit Cost	Avg Price Incl O&P	Crew Size	Avg Day Prod	Avg Mat'l Unit Cost	Avg Equip Unit Cost	Avg Labor Unit Cost	Avg Total Unit Cost	Avg Price Incl O&P
1/4" x 4' x 8'	Inst	SF	C-17	900	1.45	---	0.52	1.97	2.23	C-17	630	2.18	---	0.74	2.92	3.29
Oak, flush face																
1/8" x 4' x 8'	Inst	SF	C-17	900	0.94	---	0.52	1.46	1.72	C-17	630	1.41	---	0.74	2.16	2.52
1/4" x 4' x 8'	Inst	SF	C-17	900	1.42	---	0.52	1.93	2.19	C-17	630	2.12	---	0.74	2.86	3.23
1/2" x 4' x 8'	Inst	SF	C-17	900	1.95	---	0.52	2.47	2.73	C-17	630	2.93	---	0.74	3.67	4.04
Philippine mahogany																
Rotary cut																
1/8" x 4' x 8'	Inst	SF	C-17	900	0.59	---	0.52	1.11	1.36	C-17	630	0.88	---	0.74	1.62	1.99
3/16" x 4' x 8'	Inst	SF	C-17	900	0.61	---	0.52	1.13	1.39	C-17	630	0.92	---	0.74	1.66	2.03
1/4" x 4' x 8'	Inst	SF	C-17	900	0.65	---	0.52	1.17	1.43	C-17	630	0.98	---	0.74	1.72	2.09
1/4" x 4' x 10'	Inst	SF	C-17	900	0.80	---	0.52	1.32	1.57	C-17	630	1.20	---	0.74	1.94	2.30
1/2" x 4' x 8'	Inst	SF	C-17	900	1.45	---	0.52	1.97	2.23	C-17	630	2.18	---	0.74	2.92	3.29
3/4" x 4' x 8' vertical core	Inst	SF	C-17	900	1.81	---	0.52	2.33	2.58	C-17	630	2.71	---	0.74	3.46	3.82
V-grooved																
3/16" x 4' x 8'	Inst	SF	C-17	900	0.64	---	0.52	1.16	1.41	C-17	630	0.96	---	0.74	1.70	2.07
1/4" x 4' x 8'	Inst	SF	C-17	900	0.77	---	0.52	1.29	1.55	C-17	630	1.16	---	0.74	1.90	2.26
1/4" x 4' x 10'	Inst	SF	C-17	900	0.85	---	0.52	1.37	1.62	C-17	630	1.28	---	0.74	2.02	2.38

Plaster and Stucco

Remove plaster or stucco from studs or sheathing

Lath (wood or metal) and plaster, walls and ceilings:

Description	Oper	Unit	Crew Size	Avg Day Prod	Avg Mat'l Unit Cost	Avg Equip Unit Cost	Avg Labor Unit Cost	Avg Total Unit Cost	Avg Price Incl O&P	Crew Size	Avg Day Prod	Avg Mat'l Unit Cost	Avg Equip Unit Cost	Avg Labor Unit Cost	Avg Total Unit Cost	Avg Price Incl O&P
2 coats	Demo	SY	L-2	130	---	---	2.00	2.00	2.96	L-2	98	---	---	2.67	2.67	3.95
3 coats	Demo	SY	L-2	120	---	---	2.17	2.17	3.21	L-2	90	---	---	2.89	2.89	4.28

Lath. Nails included

Gypsum lath, 16" x 48", applied with nails to ceilings or walls, 5% waste included, nails included (.067 lbs. per CY)

Description	Oper	Unit	Crew Size	Avg Day Prod	Avg Mat'l Unit Cost	Avg Equip Unit Cost	Avg Labor Unit Cost	Avg Total Unit Cost	Avg Price Incl O&P	Crew Size	Avg Day Prod	Avg Mat'l Unit Cost	Avg Equip Unit Cost	Avg Labor Unit Cost	Avg Total Unit Cost	Avg Price Incl O&P
3/8" T, perforated or plain	Inst	SY	P-1	80	2.17	---	2.05	4.22	5.20	P-1	60	2.71	---	2.73	5.44	6.75
1/2" T, perforated or plain	Inst	SY	P-1	80	2.43	---	2.05	4.48	5.46	P-1	60	3.04	---	2.73	5.77	7.08
3/8" T, insulating, aluminum foil back	Inst	SY	P-1	80	3.22	---	2.05	5.27	6.25	P-1	60	4.03	---	2.73	6.75	8.07
1/2" T, insulating, aluminum foil back	Inst	SY	P-1	80	3.48	---	2.05	5.53	6.51	P-1	60	4.35	---	2.73	7.08	8.39

Plaster & Stucco

Dimensions

1. Lath. Can be gypsum, wire or wood.

   a. Gypsum. Plain, perforated and insulating. Normally, each lath is 16" x 48" in 3/8" or 1/2" thicknesses; a five-piece bundle covers 3 SY.

   b. Wire. Only diamond and riblash are discussed here. Diamond lath is furnished in 27" x 96"-wide sheets covering 16 SY and 20 SY respectively.

   c. Wood. Can be fir, pine, redwood, spruce, etc. Bundles may consist of 50 or 100 pieces of 3/8" x 1-1/2" x 48" lath covering 3.4 SY and 6.8 SY respectively. There is usually a 3/8" gap between lath.

2. Plaster. Only two and three coat gypsum cement plaster are discussed here.

Installation.

1. Lath. Laths are nailed. The types and size of nails vary with the type and thickness of lath. Quantity will vary with o.c. spacing of studs or joists. Only lath applied to wood will be considered here.

   a. Nails for gypsum lath. Common type is 13 gauge blued 19/64" flathead, 1-1/8" long, spaced approximately 4" o.c., and 1-1/4" long spaced approximately 5" o.c. for 3/8" and 1/2" lath respectively.

   b. Nails for wire lath. For ceiling, common is 1-1/2" long 11 gauge barbed, galvanized with a 7/16" head diameter.

   c. Nails for wood lath. 3d fine common.

   Gypsum lath may be attached by the use of air-driven staples. Wire lath may be tied to support, usually with 18 gauge wire.

2. Plaster. Quantities of materials used to plaster vary with the type of lath and thickness of plaster.

   a. Two coat. Brown and finish coat.

   b. Three coat. Scratch, brown and finish coat.

   For types and quantities of material used, see **Notes on Material Pricing.**

Notes — Labor.

1. Lath. Output is based on what one lather can do in one day.

2. Plaster. Output is based on what two plasterers and one laborer can do in one day.

Stucco

1. Dimensions of material discussed.

   a. 18 gauge wire

   b. 15 lb. felt paper

   c. 1" x 18 gauge galvanized netting

   d. Mortar of 1:3 mix

2. Installation.

   a. Lathing consists of the following:

      1) 18 gauge wire stretched taut horizontally across studs at approximately 8" o.c.

      2) 15 lb. felt paper placed over wire.

      3) 1" x 18 gauge galvanized netting placed over felt.

   b. Mortar.

      1) The mortar mix used in this section is a 1:3 mix.

      2) One CY of mortar is comprised of one (1) CY sand, nine (9) CF portland cement and one hundred (100) lbs. hydrated lime.

      3) Mortar requirements for 100 SY of stucco are as follows:

3. Estimating technique. Determine area and deduct area of window and door openings. No waste has been included in the following figures unless otherwise noted. For waste, add 10% to total area.

| | Cubic Yards Per CSY | |
Stucco Thickness	On Masonry	On Netting
1/2"	1.50	1.75
5/8"	1.90	2.20
3/4"	2.20	2.60
1"	2.90	3.40

			Costs Based On Large Volume							Costs Based On Small Volume					
Description	Oper	Unit	Crew Size	Avg Day Prod	Avg Mat'l Unit Cost	Avg Equip Unit Cost	Avg Labor Unit Cost	Avg Total Unit Cost	Avg Price Incl O&P	Crew Size	Avg Day Prod	Avg Mat'l Unit Cost	Avg Equip Unit Cost	Avg Labor Unit Cost	Avg Total Unit Cost	Avg Price Incl O&P

Plaster and Stucco (continued)

For installation with staples, ADD OUTPUT, DEDUCT COST	Inst	SY	P-1	25	---	---	0.49	0.49	0.72	P-1	19	---	---	0.65	0.65	0.96

Metal lath, nailed to ceilings or walls, 5% waste and nails (0.067 lbs. per SY) included
Diamond lath (junior mesh), 27" x 96" sheets, nailed to wood members @ 16" o.c.

3.4 lb black painted	Inst	SY	P-1	120	3.24	---	1.36	4.60	5.26	P-1	90	4.05	---	1.82	5.87	6.74
3.4 lb galvanized	Inst	SY	P-1	120	3.77	---	1.36	5.13	5.79	P-1	90	4.71	---	1.82	6.53	7.40

Riblath, 3/8" high rib, 27" x 96" sheets, nailed to wood members @ 24" o.c.

3.4 lb painted	Inst	SY	P-1	160	3.77	---	1.02	4.79	5.28	P-1	120	4.71	---	1.36	6.07	6.73
3.4 lb galvanized	Inst	SY	P-1	160	4.82	---	1.02	5.84	6.33	P-1	120	6.02	---	1.36	7.38	8.04

Wood lath, nailed to ceilings or walls. 5% waste and nails
included, redwood, "A" grade and better

3/8" x 1-1/2" x 48" @3/8" spacing	Inst	SY	P-1	80	3.78	---	2.05	5.83	6.81	P-1	60	4.73	---	2.73	7.45	8.77

Labor adjustments, lath
For gypsum or wood lath above second floor, DEDUCT

OUTPUT, ADD COST	Inst	SY	P-1	12	---	---	0.36	0.36	0.53	P-1	9	---	---	0.48	0.48	0.71

For metal lath above second floor, DEDUCT OUTPUT,

ADD COST	Inst	SY	P-1	25	---	---	0.19	0.19	0.28	P-1	19	---	---	0.25	0.25	0.37

Plaster (only), applied to ceilings and walls; 10% waste included

Note: Material price includes gypsum plaster, sand, hydrated lime and gauging plaster

Two coats gypsum plaster

On gypsum lath	Inst	SY	P-4	115	1.66	0.31	3.99	5.96	7.87	P-4	86	2.08	0.41	5.32	7.80	10.36
On unit masonry, no lath	Inst	SY	P-4	112	1.79	0.32	4.09	6.20	8.16	P-4	84	2.23	0.42	5.46	8.11	10.73

Three coats gypsum plaster

On gypsum lath	Inst	SY	P-4	83	1.77	0.43	5.52	7.72	10.37	P-4	62	2.21	0.57	7.37	10.14	13.68
On unit masonry, no lath	Inst	SY	P-4	83	1.91	0.43	5.52	7.86	10.51	P-4	62	2.38	0.57	7.37	10.32	13.85
On wire lath	Inst	SY	P-4	80	2.66	0.44	5.73	8.83	11.59	P-4	60	3.33	0.59	7.64	11.56	15.23
On wood lath	Inst	SY	P-4	80	1.87	0.44	5.73	8.04	10.79	P-4	60	2.34	0.59	7.64	10.57	14.23

Labor adjustments, plaster
For plaster above second floor,

DEDUCT OUTPUT, ADD COST	Inst	SY	P-4	16	---	---	1.32	1.32	1.95	P-4	12	---	---	1.76	1.76	2.60

Description	Oper	Unit	Costs Based On Large Volume							Costs Based On Small Volume						
			Crew Size	Avg Day Prod	Avg Mat'l Unit Cost	Avg Equip Unit Cost	Avg Labor Unit Cost	Avg Total Unit Cost	Avg Price Incl O&P	Crew Size	Avg Day Prod	Avg Mat'l Unit Cost	Avg Equip Unit Cost	Avg Labor Unit Cost	Avg Total Unit Cost	Avg Price Incl O&P

Plaster and Stucco (continued)

Thin coat plaster over sheetrock

Description	Oper	Unit	Crew Size	Avg Day Prod	Avg Mat'l Unit Cost	Avg Equip Unit Cost	Avg Labor Unit Cost	Avg Total Unit Cost	Avg Price Incl O&P	Crew Size	Avg Day Prod	Avg Mat'l Unit Cost	Avg Equip Unit Cost	Avg Labor Unit Cost	Avg Total Unit Cost	Avg Price Incl O&P
(in lieu of taping)	Inst	SY	P-4	125	0.67	---	3.67	4.34	6.10	P-4	94	0.84	---	4.89	5.73	8.08

Stucco, exterior walls
Netting, galvanized, 1" x 20 ga. x 36",

Description	Oper	Unit	Crew Size	Avg Day Prod	Avg Mat'l Unit Cost	Avg Equip Unit Cost	Avg Labor Unit Cost	Avg Total Unit Cost	Avg Price Incl O&P	Crew Size	Avg Day Prod	Avg Mat'l Unit Cost	Avg Equip Unit Cost	Avg Labor Unit Cost	Avg Total Unit Cost	Avg Price Incl O&P
with 18 ga. wire and 15 lb felt	Inst	SY	P-1	60	2.91	---	2.73	5.64	6.95	P-1	45	3.64	---	3.64	7.28	9.03
Steel-Tex, 49" W x 11-1/2' L																
rolls, with felt backing	Inst	SY	P-1	80	3.23	---	2.05	5.28	6.26	P-1	60	4.02	---	2.73	6.75	8.06
1 coat work with float finish																
over masonry	Inst	SY	P-4	115	0.45	0.31	3.99	4.75	6.66	P-4	86	0.56	0.41	5.32	6.29	8.84
2 coat work with float finish																
over masonry	Inst	SY	P-4	60	1.36	0.59	7.64	9.59	13.26	P-4	45	1.70	0.78	10.19	12.67	17.56
over metal netting	Inst	SY	P-5	85	1.90	0.42	8.86	11.18	15.43	P-4	64	2.38	0.55	11.81	14.74	20.41
3 coat work with float finish																
over metal netting	Inst	SY	P-5	65	2.63	0.54	11.58	14.75	20.31	P-5	49	3.29	0.72	15.44	19.45	26.87

Plumbing, accessories and fixtures

Bath accessories, average quality

Description	Oper	Unit	Crew Size	Avg Day Prod	Avg Mat'l Unit Cost	Avg Equip Unit Cost	Avg Labor Unit Cost	Avg Total Unit Cost	Avg Price Incl O&P	Crew Size	Avg Day Prod	Avg Mat'l Unit Cost	Avg Equip Unit Cost	Avg Labor Unit Cost	Avg Total Unit Cost	Avg Price Incl O&P
Cup holder; chrome;																
surface mounted	Inst	EA	C-1	24	5.40	---	7.02	12.42	15.86	C-1	18	6.75	---	9.36	16.11	20.70
Cup holder; chrome; recessed	Inst	EA	C-1	16	10.80	---	10.53	21.33	26.50	C-1	12	13.50	---	14.05	27.55	34.43
Cup and toothbrush holder																
Chrome	Inst	EA	C-1	24	7.20	---	7.02	14.22	17.66	C-1	18	9.00	---	9.36	18.36	22.95
Antique brass	Inst	EA	C-1	24	19.56	---	7.02	26.58	30.02	C-1	18	24.45	---	9.36	33.81	38.40
White and gold	Inst	EA	C-1	24	10.80	---	7.02	17.82	21.26	C-1	18	13.50	---	9.36	22.86	27.45
Cup, toothbrush, and soapholder,																
chrome recessed	Inst	EA	C-1	16	14.40	---	10.53	24.93	30.10	C-1	12	18.00	---	14.05	32.05	38.93
Glass shelf, 24" long																
Chrome	Inst	EA	C-1	20	18.00	---	8.43	26.43	30.56	C-1	15	22.50	---	11.24	33.74	39.24
Antique brass	Inst	EA	C-1	20	54.00	---	8.43	62.43	66.56	C-1	15	67.50	---	11.24	78.74	84.24
Polished chrome	Inst	EA	C-1	20	42.00	---	8.43	50.43	54.56	C-1	15	52.50	---	11.24	63.74	69.24
Stainless steel	Inst	EA	C-1	20	24.00	---	8.43	32.43	36.56	C-1	15	30.00	---	11.24	41.24	46.74

Plumbing

The material cost of an item includes the fixture, water supply, and trim (includes fittings and faucets). The labor cost of an item includes installation of the fixture and connection of water supply and/or electricity, but no demolition or clean-up included. Average rough-in of pipe, waste, and vent is an "add" item shown below each major grouping of fixtures, unless noted otherwise.

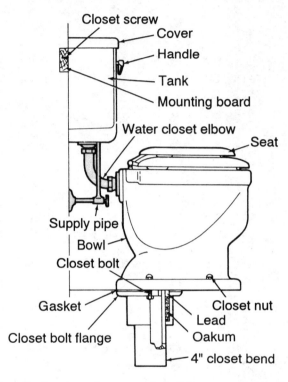

Detail of water closet and wall-hung tank in place

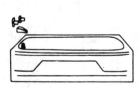

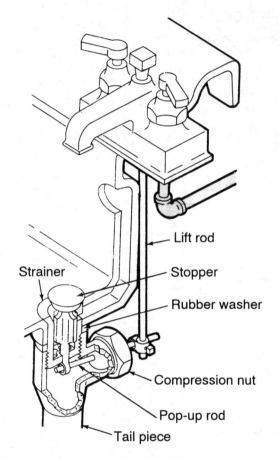

Detail of pop-up drain

From: *Basic Plumbing with Illustrations*
Craftsman Book Company

			Costs Based On Large Volume							Costs Based On Small Volume					
Description	Oper	Unit	Crew Size	Avg Day Prod	Avg Mat'l Unit Cost	Avg Equip Unit Cost	Avg Labor Unit Cost	Avg Total Unit Cost	Avg Price Incl O&P	Crew Size	Avg Day Prod	Avg Mat'l Unit Cost	Avg Equip Unit Cost	Avg Labor Unit Cost	Avg Total Unit Cost	Avg Price Incl O&P

Plumbing (continued)

Grab bars, stainless steel

Description	Oper	Unit	Crew Size	Avg Day Prod	Avg Mat'l Unit Cost	Avg Equip Unit Cost	Avg Labor Unit Cost	Avg Total Unit Cost	Avg Price Incl O&P	Crew Size	Avg Day Prod	Avg Mat'l Unit Cost	Avg Equip Unit Cost	Avg Labor Unit Cost	Avg Total Unit Cost	Avg Price Incl O&P
16" long	Inst	EA	C-1	20	31.20	---	8.43	39.63	43.76	C-1	15	39.00	---	11.24	50.24	55.74
24" long	Inst	EA	C-1	20	36.00	---	8.43	44.43	48.56	C-1	15	45.00	---	11.24	56.24	61.74
32" long	Inst	EA	C-1	20	43.20	---	8.43	51.63	55.76	C-1	15	54.00	---	11.24	65.24	70.74
Angle bar, 16" L x 32" H	Inst	EA	C-1	20	60.00	---	8.43	68.43	72.56	C-1	15	75.00	---	11.24	86.24	91.74
Robe hooks, single or double	Inst	EA	C-1	32	4.80	---	5.27	10.07	12.65	C-1	24	6.00	---	7.02	13.02	16.46
Shower curtain rod 1" dia. x 5.5' L with adjacent rod holder	Inst	EA	C-1	32	16.80	---	5.27	22.07	24.65	C-1	24	21.00	---	7.02	28.02	31.46
Shower equipment. See Plumbing fixtures, shower stalls below																
Soap holder/dish, chrome	Inst	EA	C-1	32	5.40	---	5.27	10.67	13.25	C-1	24	6.75	---	7.02	13.77	17.21
Soap holder with drain																
Chrome	Inst	EA	C-1	32	6.30	---	5.27	11.57	14.15	C-1	24	7.88	---	7.02	14.90	18.34
Antique brass	Inst	EA	C-1	32	19.20	---	5.27	24.47	27.05	C-1	24	24.00	---	7.02	31.02	34.46
Soap holder with tray																
Chrome	Inst	EA	C-1	32	7.20	---	5.27	12.47	15.05	C-1	24	9.00	---	7.02	16.02	19.46
Antique brass	Inst	EA	C-1	32	12.00	---	5.27	17.27	19.85	C-1	24	15.00	---	7.02	22.02	25.46
Soap holder; recessed; chrome	Inst	EA	C-1	24	14.40	---	7.02	21.42	24.86	C-1	18	18.00	---	9.36	27.36	31.95
Soap holder and utility bar; recessed																
Chrome	Inst	EA	C-1	24	16.80	---	7.02	23.82	27.26	C-1	18	21.00	---	9.36	30.36	34.95
Antique brass	Inst	EA	C-1	24	31.20	---	7.02	38.22	41.66	C-1	18	39.00	---	9.36	48.36	52.95
Toilet roll holder																
Chrome	Inst	EA	C-1	32	12.00	---	5.27	17.27	19.85	C-1	24	15.00	---	7.02	22.02	25.46
Antique brass	Inst	EA	C-1	32	26.40	---	5.27	31.67	34.25	C-1	24	33.00	---	7.02	40.02	43.46
Toilet roll holder; recessed																
Chrome	Inst	EA	C-1	24	14.40	---	7.02	21.42	24.86	C-1	18	18.00	---	9.36	27.36	31.95
Antique brass	Inst	EA	C-1	24	31.20	---	7.02	38.22	41.66	C-1	18	39.00	---	9.36	48.36	52.95
Toilet roll holder with hood; recessed																
Chrome	Inst	EA	C-1	24	28.80	---	7.02	35.82	39.26	C-1	18	36.00	---	9.36	45.36	49.95
Antique brass	Inst	EA	C-1	24	42.00	---	7.02	49.02	52.46	C-1	18	52.50	---	9.36	61.86	66.45
Towel bars, round																
Chrome																
18"	Inst	EA	C-1	32	19.20	---	5.27	24.47	27.05	C-1	24	24.00	---	7.02	31.02	34.46
24"	Inst	EA	C-1	32	20.40	---	5.27	25.67	28.25	C-1	24	25.50	---	7.02	32.52	35.96
30"	Inst	EA	C-1	32	21.60	---	5.27	26.87	29.45	C-1	24	27.00	---	7.02	34.02	37.46
36"	Inst	EA	C-1	32	22.80	---	5.27	28.07	30.65	C-1	24	28.50	---	7.02	35.52	38.96

			Costs Based On Large Volume							Costs Based On Small Volume						
Description	Oper	Unit	Crew Size	Avg Day Prod	Avg Mat'l Unit Cost	Avg Equip Unit Cost	Avg Labor Unit Cost	Avg Total Unit Cost	Avg Price Incl O&P	Crew Size	Avg Day Prod	Avg Mat'l Unit Cost	Avg Equip Unit Cost	Avg Labor Unit Cost	Avg Total Unit Cost	Avg Price Incl O&P

Plumbing (continued)

Description	Oper	Unit	Crew Size	Avg Day Prod	Avg Mat'l Unit Cost	Avg Equip Unit Cost	Avg Labor Unit Cost	Avg Total Unit Cost	Avg Price Incl O&P	Crew Size	Avg Day Prod	Avg Mat'l Unit Cost	Avg Equip Unit Cost	Avg Labor Unit Cost	Avg Total Unit Cost	Avg Price Incl O&P
Antique brass																
30"	Inst	EA	C-1	32	36.00	---	5.27	41.27	43.85	C-1	24	45.00	---	7.02	52.02	55.46
36"	Inst	EA	C-1	32	42.00	---	5.27	47.27	49.85	C-1	24	52.50	---	7.02	59.52	62.96
Towel pin, chrome	Inst	EA	C-1	36	4.80	---	4.68	9.48	11.78	C-1	27	6.00	---	6.24	12.24	15.30
Towel ring																
Chrome	Inst	EA	C-1	36	7.20	---	4.68	11.88	14.18	C-1	27	9.00	---	6.24	15.24	18.30
Antique brass	Inst	EA	C-1	36	12.00	---	4.68	16.68	18.98	C-1	27	15.00	---	6.24	21.24	24.30
Towel ladder, 4 arms																
Antique brass	Inst	EA	C-1	24	76.80	---	7.02	83.82	87.26	C-1	18	96.00	---	9.36	105.36	109.95
Polished chrome	Inst	EA	C-1	24	74.40	---	7.02	81.42	84.86	C-1	18	93.00	---	9.36	102.36	106.95
Towel supply shelf																
18" long	Inst	EA	C-1	32	36.00	---	5.27	41.27	43.85	C-1	24	45.00	---	7.02	52.02	55.46
24" long	Inst	EA	C-1	32	38.40	---	5.27	43.67	46.25	C-1	24	48.00	---	7.02	55.02	58.46
Towel three arm swing bar, chrome	Inst	EA	C-1	36	14.40	---	4.68	19.08	21.38	C-1	27	18.00	---	6.24	24.24	27.30
Vanity cabinets. See Cabinets.																
Wall to floor angle bar with flange with bolt, washer and screws	Inst	EA	C-1	18	57.60	---	9.36	66.96	71.55	C-1	14	72.00	---	12.49	84.49	90.60

Medicine cabinets. No electrical work included; for wall outlet cost, see Electrical

Description	Oper	Unit	Crew Size	Avg Day Prod	Avg Mat'l Unit Cost	Avg Equip Unit Cost	Avg Labor Unit Cost	Avg Total Unit Cost	Avg Price Incl O&P	Crew Size	Avg Day Prod	Avg Mat'l Unit Cost	Avg Equip Unit Cost	Avg Labor Unit Cost	Avg Total Unit Cost	Avg Price Incl O&P
Surface mounting, no wall opening																
Swing door cabinets																
Vinyl shutter door, 18" x 36" x 5-1/4"	Inst	EA	C-1	10	60.00	---	16.85	76.85	85.11	C-1	8	75.00	---	22.47	97.47	108.48
Sliding door cabinets, bi-passing mirror doors																
Unlighted																
24" x 16", budget model	Inst	EA	C-1	10	36.00	---	16.85	52.85	61.11	C-1	8	45.00	---	22.47	67.47	78.48
Decorator models, walnut, antique white, soft gold																
38" x 31" x 6", two adjustable shelves	Inst	EA	C-1	10	192.00	---	16.85	208.85	217.11	C-1	8	240.00	---	22.47	262.47	273.48
Lighted																
Budget models, four 60 watt incandescent bulbs																
24" x 19" x 8", one fixed shelf	Inst	EA	C-1	7	54.00	---	24.08	78.08	89.88	C-1	5	67.50	---	32.10	99.60	115.34
28" x 19" x 8", one fixed shelf	Inst	EA	C-1	7	60.00	---	24.08	84.08	95.88	C-1	5	75.00	---	32.10	107.10	122.84

			Costs Based On Large Volume							Costs Based On Small Volume						
Description	Oper	Unit	Crew Size	Avg Day Prod	Avg Mat'l Unit Cost	Avg Equip Unit Cost	Avg Labor Unit Cost	Avg Total Unit Cost	Avg Price Incl O&P	Crew Size	Avg Day Prod	Avg Mat'l Unit Cost	Avg Equip Unit Cost	Avg Labor Unit Cost	Avg Total Unit Cost	Avg Price Incl O&P

Plumbing (continued)

Deluxe models, four 60 watt incandescent bulbs
24" x 23" x 8-1/2", two

Description	Oper	Unit	Crew Size	Avg Day Prod	Avg Mat'l	Avg Equip	Avg Labor	Avg Total	Avg Price O&P	Crew Size	Avg Day Prod	Avg Mat'l	Avg Equip	Avg Labor	Avg Total	Avg Price O&P
adjustable shelves	Inst	EA	C-1	7	96.00	---	24.08	120.08	131.88	C-1	5	120.00	---	32.10	152.10	167.84

28" x 23" x 8-1/2", two

adjustable shelves	Inst	EA	C-1	7	120.00	---	24.08	144.08	155.88	C-1	5	150.00	---	32.10	182.10	197.84

36" x 23" x 8-1/2", two

adjustable shelves	Inst	EA	C-1	7	132.00	---	24.08	156.08	167.88	C-1	5	165.00	---	32.10	197.10	212.84

Decorator models, self-contained incandescent four bulb fixture antique gold
29" x 25" x 7-1/2", two

fixed shelves	Inst	EA	C-1	7	138.00	---	24.08	162.08	173.88	C-1	5	172.50	---	32.10	204.60	220.34

24" x 25" x 7-1/2", two

fixed shelves	Inst	EA	C-1	7	132.00	---	24.08	156.08	167.88	C-1	5	165.00	---	32.10	197.10	212.84

Three piece ensemble
Unlighted, decorator models
two cabinets and center mirror (24" x 36") all framed
antique gold 48" x 36" x 5", three

adjustable shelves	Inst	EA	C-1	5	210.00	---	33.71	243.71	260.23	C-1	4	262.50	---	44.95	307.45	329.47

Lighted, budget models
Mirror, vanity box with box with mirror doors, and incandescent light fixture

24" x 35", 4-bulb	Inst	EA	C-1	6	90.00	---	28.09	118.09	131.86	C-1	5	112.50	---	37.46	149.96	168.31
36" x 35", 4-bulb	Inst	EA	C-1	6	114.00	---	28.09	142.09	155.86	C-1	5	142.50	---	37.46	179.96	198.31

Recessed mounting, overall sizes
Swing door cabinets
Unlighted
Budget models
14" x 20", mirror door,

two fixed shelves	Inst	EA	C-1	8	24.00	---	21.07	45.07	55.39	C-1	6	30.00	---	28.09	58.09	71.86

16" x 22" x 3", mirror doors,

two fixed shelves	Inst	EA	C-1	8	28.80	---	21.07	49.87	60.19	C-1	6	36.00	---	28.09	64.09	77.86

17" x 24" x 3", vinyl shutter

adjustable shelves	Inst	EA	C-1	8	30.00	---	21.07	51.07	61.39	C-1	6	37.50	---	28.09	65.59	79.36

16" x 24" x 3", wood shutter,

adjustable shelves	Inst	EA	C-1	8	40.80	---	21.07	61.87	72.19	C-1	6	51.00	---	28.09	79.09	92.86

			Costs Based On Large Volume							Costs Based On Small Volume						
Description	Oper	Unit	Crew Size	Avg Day Prod	Avg Mat'l Unit Cost	Avg Equip Unit Cost	Avg Labor Unit Cost	Avg Total Unit Cost	Avg Price Incl O&P	Crew Size	Avg Day Prod	Avg Mat'l Unit Cost	Avg Equip Unit Cost	Avg Labor Unit Cost	Avg Total Unit Cost	Avg Price Incl O&P

Plumbing (continued)

Decorator models, two adjustable shelves, mirror door
18" x 24" x.3", antique gold

Description	Oper	Unit	Crew Size	Avg Day Prod	Avg Mat'l Unit Cost	Avg Equip Unit Cost	Avg Labor Unit Cost	Avg Total Unit Cost	Avg Price Incl O&P	Crew Size	Avg Day Prod	Avg Mat'l Unit Cost	Avg Equip Unit Cost	Avg Labor Unit Cost	Avg Total Unit Cost	Avg Price Incl O&P
or walnut	Inst	EA	C-1	8	67.20	---	21.07	88.27	98.59	C-1	6	84.00	---	28.09	112.09	125.86

21" x 31" x 3", antique gold

Description	Oper	Unit	Crew Size	Avg Day Prod	Avg Mat'l Unit Cost	Avg Equip Unit Cost	Avg Labor Unit Cost	Avg Total Unit Cost	Avg Price Incl O&P	Crew Size	Avg Day Prod	Avg Mat'l Unit Cost	Avg Equip Unit Cost	Avg Labor Unit Cost	Avg Total Unit Cost	Avg Price Incl O&P
oval mirror	Inst	EA	C-1	8	58.80	---	21.07	79.87	90.19	C-1	6	73.50	---	28.09	101.59	115.36

Oak framed mirror door with stained glass trim, enameled steel body with 2 adjustable

Description	Oper	Unit	Crew Size	Avg Day Prod	Avg Mat'l Unit Cost	Avg Equip Unit Cost	Avg Labor Unit Cost	Avg Total Unit Cost	Avg Price Incl O&P	Crew Size	Avg Day Prod	Avg Mat'l Unit Cost	Avg Equip Unit Cost	Avg Labor Unit Cost	Avg Total Unit Cost	Avg Price Incl O&P
shelves, 26" x 18" x 5"	Inst	EA	C-1	8	216.00	---	21.07	237.07	247.39	C-1	6	270.00	---	28.09	298.09	311.86
4 lights, oak mounted, ADD	Inst	LS	E-1	4	142.80	---	48.98	191.78	213.83	E-1	3	178.50	---	65.31	243.81	273.20

Lighted
Deluxe model, two adjustable shelves, mirror door, 4 slide out plastic utility

Description	Oper	Unit	Crew Size	Avg Day Prod	Avg Mat'l Unit Cost	Avg Equip Unit Cost	Avg Labor Unit Cost	Avg Total Unit Cost	Avg Price Incl O&P	Crew Size	Avg Day Prod	Avg Mat'l Unit Cost	Avg Equip Unit Cost	Avg Labor Unit Cost	Avg Total Unit Cost	Avg Price Incl O&P
drawers, 17" x 24" x 3"	Inst	EA	C-1	6	102.00	---	28.09	130.09	143.86	C-1	5	127.50	---	37.46	164.96	183.31

Decorator models, two attached lights adjustable shelves, antique gold or
walnut framed mirror door

Description	Oper	Unit	Crew Size	Avg Day Prod	Avg Mat'l Unit Cost	Avg Equip Unit Cost	Avg Labor Unit Cost	Avg Total Unit Cost	Avg Price Incl O&P	Crew Size	Avg Day Prod	Avg Mat'l Unit Cost	Avg Equip Unit Cost	Avg Labor Unit Cost	Avg Total Unit Cost	Avg Price Incl O&P
31" x 24" x 3"	Inst	EA	C-1	6	132.00	---	28.09	160.09	173.86	C-1	5	165.00	---	37.46	202.46	220.81
34" x 31" x 3", oval mirror door	Inst	EA	C-1	6	174.00	---	28.09	202.09	215.86	C-1	5	217.50	---	37.46	254.96	273.31

Related materials
Wall mounted decorator lights, 5" diameter
16" H, 10" extension, antique gold

Description	Oper	Unit	Crew Size	Avg Day Prod	Avg Mat'l Unit Cost	Avg Equip Unit Cost	Avg Labor Unit Cost	Avg Total Unit Cost	Avg Price Incl O&P	Crew Size	Avg Day Prod	Avg Mat'l Unit Cost	Avg Equip Unit Cost	Avg Labor Unit Cost	Avg Total Unit Cost	Avg Price Incl O&P
a) 4 bulb	Inst	PR	E-1	4	74.40	---	48.98	123.38	145.43	E-1	3	93.00	---	65.31	158.31	187.70
b) 3 bulb	Inst	PR	E-1	4	66.00	---	48.98	114.98	137.03	E-1	3	82.50	---	65.31	147.81	177.20

Plumbing fixtures

**Bathtubs; good quality fittings and faucets included in material cost. Labor
cost for installation only of fixture, fittings, faucets.
For rough-in, see Adjustments under bathtub section.**

Free-standing
Kohler products, enameled

Description	Oper	Unit	Crew Size	Avg Day Prod	Avg Mat'l Unit Cost	Avg Equip Unit Cost	Avg Labor Unit Cost	Avg Total Unit Cost	Avg Price Incl O&P	Crew Size	Avg Day Prod	Avg Mat'l Unit Cost	Avg Equip Unit Cost	Avg Labor Unit Cost	Avg Total Unit Cost	Avg Price Incl O&P
cast iron, 72" x 37-1/2"	Inst	EA	S-2	3.0	2237	---	110	2347	2397	S-2	2.4	2582	---	137	2719	2782

Crane products, acrylic plastic

Description	Oper	Unit	Crew Size	Avg Day Prod	Avg Mat'l Unit Cost	Avg Equip Unit Cost	Avg Labor Unit Cost	Avg Total Unit Cost	Avg Price Incl O&P	Crew Size	Avg Day Prod	Avg Mat'l Unit Cost	Avg Equip Unit Cost	Avg Labor Unit Cost	Avg Total Unit Cost	Avg Price Incl O&P
with cast brass legs, 66" x 30"	Inst	EA	S-2	3.0	1242	---	110	1351	1402	S-2	2.4	1433	---	137	1570	1633

			Costs Based On Large Volume							Costs Based On Small Volume						
Description	Oper	Unit	Crew Size	Avg Day Prod	Avg Mat'l Unit Cost	Avg Equip Unit Cost	Avg Labor Unit Cost	Avg Total Unit Cost	Avg Price Incl O&P	Crew Size	Avg Day Prod	Avg Mat'l Unit Cost	Avg Equip Unit Cost	Avg Labor Unit Cost	Avg Total Unit Cost	Avg Price Incl O&P

Plumbing (continued)

Recessed; American Standard Products
Cast-iron, enameled
60" L x 42" W x 15" H (Roma)

Description	Oper	Unit	Crew Size	Avg Day Prod	Avg Mat'l Unit Cost	Avg Equip Unit Cost	Avg Labor Unit Cost	Avg Total Unit Cost	Avg Price Incl O&P	Crew Size	Avg Day Prod	Avg Mat'l Unit Cost	Avg Equip Unit Cost	Avg Labor Unit Cost	Avg Total Unit Cost	Avg Price Incl O&P
White	Inst	EA	S-2	2.4	1247	---	137	1384	1447	S-2	1.9	1439	---	171	1610	1689
Colors	Inst	EA	S-2	2.4	1300	---	137	1437	1500	S-2	1.9	1500	---	171	1671	1750

Bathtub with whirlpool
Plumbing installation

Description	Oper	Unit	Crew Size	Avg Day Prod	Avg Mat'l Unit Cost	Avg Equip Unit Cost	Avg Labor Unit Cost	Avg Total Unit Cost	Avg Price Incl O&P	Crew Size	Avg Day Prod	Avg Mat'l Unit Cost	Avg Equip Unit Cost	Avg Labor Unit Cost	Avg Total Unit Cost	Avg Price Incl O&P
White	Inst	EA	S-2	1.2	2692	---	274	2966	3093	S-2	1.0	3107	---	343	3449	3607
Colors	Inst	EA	S-2	1.2	2746	---	274	3020	3146	S-2	1.0	3168	---	343	3511	3668
Electrical installation	Inst	EA	E-1	1.3	195	---	151	346	414	E-1	1.0	225	---	188	413	498

60" L x 32" W x 15" H (Spectra)

Description	Oper	Unit	Crew Size	Avg Day Prod	Avg Mat'l Unit Cost	Avg Equip Unit Cost	Avg Labor Unit Cost	Avg Total Unit Cost	Avg Price Incl O&P	Crew Size	Avg Day Prod	Avg Mat'l Unit Cost	Avg Equip Unit Cost	Avg Labor Unit Cost	Avg Total Unit Cost	Avg Price Incl O&P
White	Inst	EA	S-2	2.4	538	---	137	675	738	S-2	1.9	621	---	171	792	871
Colors	Inst	EA	S-2	2.4	633	---	137	770	833	S-2	1.9	731	---	171	902	981

Bathtub with whirlpool
Plumbing installation

Description	Oper	Unit	Crew Size	Avg Day Prod	Avg Mat'l Unit Cost	Avg Equip Unit Cost	Avg Labor Unit Cost	Avg Total Unit Cost	Avg Price Incl O&P	Crew Size	Avg Day Prod	Avg Mat'l Unit Cost	Avg Equip Unit Cost	Avg Labor Unit Cost	Avg Total Unit Cost	Avg Price Incl O&P
White	Inst	EA	S-2	1.2	2159	---	274	2433	2560	S-2	1.0	2492	---	343	2834	2992
Colors	Inst	EA	S-2	1.2	2213	---	274	2487	2613	S-2	1.0	2553	---	343	2896	3053
Electrical installation	Inst	EA	E-1	1.3	195	---	151	346	414	E-1	1.0		---	188	188	273

Grab bars (recessed) in bathtub

Description	Oper	Unit	Crew Size	Avg Day Prod	Avg Mat'l Unit Cost	Avg Equip Unit Cost	Avg Labor Unit Cost	Avg Total Unit Cost	Avg Price Incl O&P	Crew Size	Avg Day Prod	Avg Mat'l Unit Cost	Avg Equip Unit Cost	Avg Labor Unit Cost	Avg Total Unit Cost	Avg Price Incl O&P
White	Inst	EA	S-2	2.4	797	---	137	934	997	S-2	1.9	920	---	171	1091	1170
Colors	Inst	EA	S-2	2.4	924	---	137	1061	1124	S-2	1.9	1067	---	171	1238	1317

54" L x 30" W x 15" H (Bildor)

Description	Oper	Unit	Crew Size	Avg Day Prod	Avg Mat'l Unit Cost	Avg Equip Unit Cost	Avg Labor Unit Cost	Avg Total Unit Cost	Avg Price Incl O&P	Crew Size	Avg Day Prod	Avg Mat'l Unit Cost	Avg Equip Unit Cost	Avg Labor Unit Cost	Avg Total Unit Cost	Avg Price Incl O&P
White	Inst	EA	S-2	2.4	556	---	137	693	757	S-2	1.9	642	---	171	813	892
Colors	Inst	EA	S-2	2.4	684	---	137	821	884	S-2	1.9	789	---	171	960	1039

60" L x 30" W x 15" H (Bildor)

Description	Oper	Unit	Crew Size	Avg Day Prod	Avg Mat'l Unit Cost	Avg Equip Unit Cost	Avg Labor Unit Cost	Avg Total Unit Cost	Avg Price Incl O&P	Crew Size	Avg Day Prod	Avg Mat'l Unit Cost	Avg Equip Unit Cost	Avg Labor Unit Cost	Avg Total Unit Cost	Avg Price Incl O&P
White	Inst	EA	S-2	2.4	394	---	137	531	594	S-2	1.9	455	---	171	626	705
Colors	Inst	EA	S-2	2.4	489	---	137	626	689	S-2	1.9	564	---	171	735	814

66" L x 32" W x 15" H (Contour)

Description	Oper	Unit	Crew Size	Avg Day Prod	Avg Mat'l Unit Cost	Avg Equip Unit Cost	Avg Labor Unit Cost	Avg Total Unit Cost	Avg Price Incl O&P	Crew Size	Avg Day Prod	Avg Mat'l Unit Cost	Avg Equip Unit Cost	Avg Labor Unit Cost	Avg Total Unit Cost	Avg Price Incl O&P
White	Inst	EA	S-2	2.0	768	---	164	933	1008	S-2	1.6	887	---	206	1092	1187
Colors	Inst	EA	S-2	2.0	959	---	164	1124	1200	S-2	1.6	1107	---	206	1313	1407

72" L x 36" W x 18" H (Fontaine)

Description	Oper	Unit	Crew Size	Avg Day Prod	Avg Mat'l Unit Cost	Avg Equip Unit Cost	Avg Labor Unit Cost	Avg Total Unit Cost	Avg Price Incl O&P	Crew Size	Avg Day Prod	Avg Mat'l Unit Cost	Avg Equip Unit Cost	Avg Labor Unit Cost	Avg Total Unit Cost	Avg Price Incl O&P
White	Inst	EA	S-2	2.0	1379	---	164	1544	1619	S-2	1.6	1592	---	206	1797	1892
Colors	Inst	EA	S-2	2.0	1433	---	164	1597	1673	S-2	1.6	1653	---	206	1859	1953

| | | | Costs Based On Large Volume | | | | | | | Costs Based On Small Volume | | | | | |
Description	Oper	Unit	Crew Size	Avg Day Prod	Avg Mat'l Unit Cost	Avg Equip Unit Cost	Avg Labor Unit Cost	Avg Total Unit Cost	Avg Price Incl O&P	Crew Size	Avg Day Prod	Avg Mat'l Unit Cost	Avg Equip Unit Cost	Avg Labor Unit Cost	Avg Total Unit Cost	Avg Price Incl O&P
Plumbing (continued)																
Bathtub with whirlpool																
Plumbing installation																
White	Inst	EA	S-2	1.0	3077	---	329	3406	3557	S-2	0.8	3551	---	411	3962	4151
Colors	Inst	EA	S-2	1.0	3130	---	329	3459	3611	S-2	0.8	3612	---	411	4023	4212
Electrical installation	Inst	EA	E-1	1.3	195	---	151	346	414	E-1	1.0	225	---	188	413	498
Steel, enameled																
60" L x 30" W x 15" H (Salem)																
White, slip resistant	Inst	EA	S-2	2.4	285	---	137	422	485	S-2	1.9	329	---	171	500	579
Colors, slip resistant	Inst	EA	S-2	2.4	290	---	137	427	490	S-2	1.9	335	---	171	506	585
Fiberglass																
54" L x 42" W x 34" H (Roma Pool)																
White	Inst	EA	S-2	2.4	1460	---	137	1597	1660	S-2	1.9	1685	---	171	1856	1935
Colors	Inst	EA	S-2	2.4	1513	---	137	1650	1713	S-2	1.9	1746	---	171	1917	1996
Bathtub with whirlpool																
Plumbing installation																
White	Inst	EA	S-2	1.2	2595	---	274	2869	2995	S-2	1.0	2994	---	343	3337	3494
Colors	Inst	EA	S-2	1.2	2648	---	274	2922	3048	S-2	1.0	3056	---	343	3398	3556
Electrical installation	Inst	EA	E-1	1.3	195	---	151	346	414	E-1	1.0	225	---	188	413	498
66" L x 36" W x 15" H (Roma)																
White	Inst	EA	S-2	2.0	1418	---	164	1583	1658	S-2	1.6	1637	---	206	1842	1937
Colors	Inst	EA	S-2	2.0	1472	---	164	1636	1712	S-2	1.6	1698	---	206	1904	1998
Bathtub with whirlpool																
Plumbing installation																
White	Inst	EA	S-2	1.2	2232	---	274	2506	2632	S-2	1.0	2576	---	343	2918	3076
Colors	Inst	EA	S-2	1.2	2285	---	274	2560	2686	S-2	1.0	2637	---	343	2980	3137
Electrical installation	Inst	EA	E-1	1.3	195	---	151	346	414	E-1	1.0	225	---	188	413	498
72" L x 44" W x 15" H (Ellisse)																
White	Inst	EA	S-2	2.0	1568	---	164	1732	1808	S-2	1.6	1809	---	206	2015	2109
Colors	Inst	EA	S-2	2.0	1621	---	164	1786	1861	S-2	1.6	1871	---	206	2076	2171
Bathtub with whirlpool																
Plumbing installation																
White	Inst	EA	S-2	1.0	2512	---	329	2841	2992	S-2	0.8	2898	---	411	3309	3498
Colors	Inst	EA	S-2	1.0	2565	---	329	2894	3045	S-2	0.8	2960	---	411	3371	3560
Electrical installation	Inst	EA	E-1	1.3	195	---	151	346	414	E-1	1.0	225	---	188	413	498

			Costs Based On Large Volume						Costs Based On Small Volume						
Description	Oper	Unit	Crew Size	Avg Day Prod	Avg Mat'l Unit Cost	Avg Equip Unit Cost	Avg Labor Unit Cost	Avg Total Unit Cost	Avg Price Incl O&P	Crew Size	Avg Day Prod	Avg Mat'l Unit Cost	Avg Equip Unit Cost	Avg Labor Unit Cost	Avg Total Unit Cost	Avg Price Incl O&P

Plumbing (continued)

72" L x 60" W x 15" H (Elisse Grande)

Description	Oper	Unit	Crew Size	Avg Day Prod	Avg Mat'l Unit Cost	Avg Equip Unit Cost	Avg Labor Unit Cost	Avg Total Unit Cost	Avg Price Incl O&P	Crew Size	Avg Day Prod	Avg Mat'l Unit Cost	Avg Equip Unit Cost	Avg Labor Unit Cost	Avg Total Unit Cost	Avg Price Incl O&P
White	Inst	EA	S-2	2.0	1880	---	164	2044	2120	S-2	1.6	2169	---	206	2375	2469
Colors	Inst	EA	S-2	2.0	1933	---	164	2098	2173	S-2	1.6	2231	---	206	2436	2531
Bathtub with whirlpool																
Plumbing installation																
White	Inst	EA	S-2	1.0	3038	---	329	3367	3518	S-2	0.8	3506	---	411	3917	4106
Colors	Inst	EA	S-2	1.0	3091	---	329	3420	3572	S-2	0.8	3567	---	411	3978	4167
Electric installation	Inst	EA	E-1	1.3	195	---	151	346	414	E-1	1.0	225	---	188	413	498
Adjustments																
Remove & reset tub	Reset	EA	S-2	3.2	---	---	103	103	150	S-2	2.6	---	---	129	129	188
Install rough-in	Inst	EA	S-2	1.6	---	---	206	206	300	S-2	1.3	---	---	257	257	375
Shower fixture over tub with																
mixer valve, ADD	Inst	EA	S-1	---	130	---	---	130	130	S-1	---	150	---	---	150	150

Dishwashers, (high quality units) Labor includes rough-in

Built-in front loading, no front

Description	Oper	Unit	Crew Size	Avg Day Prod	Avg Mat'l Unit Cost	Avg Equip Unit Cost	Avg Labor Unit Cost	Avg Total Unit Cost	Avg Price Incl O&P	Crew Size	Avg Day Prod	Avg Mat'l Unit Cost	Avg Equip Unit Cost	Avg Labor Unit Cost	Avg Total Unit Cost	Avg Price Incl O&P
Six cycles	Inst	EA	S-3	2.4	780	---	165	945	1020	S-3	1.9	900	---	206	1106	1200
Five cycles	Inst	EA	S-3	2.4	718	---	165	882	958	S-3	1.9	828	---	206	1034	1128
Four cycles	Inst	EA	S-3	2.4	676	---	165	841	916	S-3	1.9	780	---	206	986	1080
Three cycles	Inst	EA	S-3	2.4	645	---	165	809	885	S-3	1.9	744	---	206	950	1044
Adjustments																
To only remove and reset																
dishwasher	Reset	EA	S-1	6.0	---	---	33	33	48	S-1	4.8	---	---	41	41	61
Front and side panel kits																
White or prime finish	Inst	EA	---	---	38	---	---	38	38	---	---	44	---	---	44	44
Regular color finish	Inst	EA	---	---	38	---	---	38	38	---	---	44	---	---	44	44
Brushed chrome finish	Inst	EA	---	---	47	---	---	47	47	---	---	54	---	---	54	54
Stainless steel finish	Inst	EA	---	---	62	---	---	62	62	---	---	72	---	---	72	72
Black-glass acrylic																
finish with trim kit	Inst	EA	---	---	103	---	---	103	103	---	---	119	---	---	119	119
Stainless steel trim kit	Inst	EA	---	---	47	---	---	47	47	---	---	54	---	---	54	54

			Costs Based On Large Volume							Costs Based On Small Volume					
Description	Oper	Unit	Crew Size	Avg Day Prod	Avg Mat'l Unit Cost	Avg Equip Unit Cost	Avg Labor Unit Cost	Avg Total Unit Cost	Avg Price Incl O&P	Crew Size	Avg Day Prod	Avg Mat'l Unit Cost	Avg Equip Unit Cost	Avg Labor Unit Cost	Avg Total Unit Cost	Avg Price Incl O&P

Plumbing (continued)

Convertible portables with hardwood tops, fronts, and side panels; front loading

Six cycles

White	Inst	EA	S-3	4.0	832	---	99	931	976	S-3	3.2	960	---	123	1083	1140
Colors	Inst	EA	S-3	4.0	865	---	99	963	1009	S-3	3.2	998	---	123	1121	1178

Four cycles

White	Inst	EA	S-3	4.0	770	---	99	868	914	S-3	3.2	888	---	123	1011	1068
Colors	Inst	EA	S-3	4.0	802	---	99	901	946	S-3	3.2	926	---	123	1049	1106

Three cycles

White	Inst	EA	S-3	4.0	728	---	99	827	872	S-3	3.2	840	---	123	963	1020
Colors	Inst	EA	S-3	4.0	761	---	99	859	905	S-3	3.2	878	---	123	1001	1058

Front loading portables

Three cycles with hardwood top

White	Inst	EA	S-3	4.0	686	---	99	785	831	S-3	3.2	792	---	123	915	972
Colors	Inst	EA	S-3	4.0	719	---	99	818	863	S-3	3.2	830	---	123	953	1010

Two cycles with porcelain top

White	Inst	EA	S-3	4.0	645	---	99	744	789	S-3	3.2	744	---	123	867	924
Colors	Inst	EA	S-3	4.0	677	---	99	776	821	S-3	3.2	782	---	123	905	962

Dishwasher - sink combination, 48" cabinet, with good quality fittings, faucets, and water supply kit

Six cycles

White	Inst	EA	S-3	1.5	1238	---	263	1501	1622	S-3	1.2	1428	---	329	1757	1908
Colors	Inst	EA	S-3	1.5	1270	---	263	1533	1654	S-3	1.2	1466	---	329	1795	1946

Three cycles

White	Inst	EA	S-3	1.5	1144	---	263	1407	1528	S-3	1.2	1320	---	329	1649	1800

To only remove and reset combination

dishwasher - sink	Reset	EA	S-1	3.6	---	---	55	55	81	S-1	2.9	---	---	69	69	101

Material adjustments

For standard quality, DEDUCT	Inst	%	---	---	20.0	---	---	---	—	---	---	20.0	---	---	---	—
For economy quality, DEDUCT	Inst	%	---	---	30.0	---	---	---	—	---	---	30.0	---	---	---	—

Garbage disposers (In-Sink-Erator Products), with wall switch

Labor includes rough-in. See also Trash compactors

Model "Badger 1," 1/3 hp, continuous feed, 1 yr. parts

protection	Inst	EA	S-3	2.2	100	---	179	280	362	S-3	1.8	116	---	224	340	443

			Costs Based On Large Volume							Costs Based On Small Volume					
Description	Oper	Unit	Crew Size	Avg Day Prod	Avg Mat'l Unit Cost	Avg Equip Unit Cost	Avg Labor Unit Cost	Avg Total Unit Cost	Avg Price Incl O&P	Crew Size	Avg Day Prod	Avg Mat'l Unit Cost	Avg Equip Unit Cost	Avg Labor Unit Cost	Avg Total Unit Cost	Avg Price Incl O&P

Plumbing (continued)

Description	Oper	Unit	Crew Size	Avg Day Prod	Avg Mat'l Unit Cost	Avg Equip Unit Cost	Avg Labor Unit Cost	Avg Total Unit Cost	Avg Price Incl O&P	Crew Size	Avg Day Prod	Avg Mat'l Unit Cost	Avg Equip Unit Cost	Avg Labor Unit Cost	Avg Total Unit Cost	Avg Price Incl O&P
Model "Badger V," 1/2 hp, continuous feed, 1 yr. parts protection	Inst	EA	S-3	2.2	108	---	179	287	370	S-3	1.8	125	---	224	349	452
Model 333," 1/2 hp, continuous feed, 1 yr. parts protection	Inst	EA	S-3	2.2	124	---	179	303	386	S-3	1.8	143	---	224	367	470
With stainless steel construction	Inst	EA	S-3	2.2	152	---	179	332	414	S-3	1.8	176	---	224	400	503
Model 77, 1/2 hp, automatic reversing feed, 5 year parts protection, stainless steel construction	Inst	EA	S-3	2.2	192	---	179	372	454	S-3	1.8	222	---	224	446	550
Classic, 1/2 hp, automatic reversing feed, 5 year parts protection, stainless steel construction	Inst	EA	S-3	2.2	280	---	179	459	542	S-3	1.8	323	---	224	547	650
Model 17, 1/2 hp, batch feed auto reversing feed, 5 year parts protection, stainless steel construction	Inst	EA	S-3	2.2	229	---	179	408	491	S-3	1.8	264	---	224	488	592
Classic/LC, 1/2 hp, batch feed auto reversing, 5-year parts protection, stainless steel construction	Inst	EA	S-3	2.2	320	---	179	499	582	S-3	1.8	369	---	224	593	697
Adjustments																
To only remove and reset garbage disposer	Reset	EA	S-1	8.0	---	---	25	25	36	S-1	6.4	---	---	31	31	45

Lavatories (bathroom sinks), white, with good quality fittings and faucets, concealed overflow (American Standard Products)
See also Vanity Units, this section

Description	Oper	Unit	Crew Size	Avg Day Prod	Avg Mat'l Unit Cost	Avg Equip Unit Cost	Avg Labor Unit Cost	Avg Total Unit Cost	Avg Price Incl O&P	Crew Size	Avg Day Prod	Avg Mat'l Unit Cost	Avg Equip Unit Cost	Avg Labor Unit Cost	Avg Total Unit Cost	Avg Price Incl O&P
Vitreous china																
Countertop units																
30" x 20" overall, with 17" x 11" x 6" deep oval bowl	Inst	EA	S-1	3.0	480	---	66	546	577	S-1	2.4	554	---	83	636	675
26" x 20" overall, with 17" x 11" x 6" deep oval bowl and ledge back	Inst	EA	S-1	3.0	363	---	66	429	460	S-1	2.4	419	---	83	501	540
22" x 19" overall, with 14" x 10" x 6" deep oval bowl and ledge back	Inst	EA	S-1	3.0	244	---	66	311	341	S-1	2.4	282	---	83	365	403
19" x 19" round, 6" deep bowl self rimming	Inst	EA	S-1	3.0	195	---	66	261	292	S-1	2.4	225	---	83	308	346
19" x 16" oval, 6" deep bowl	Inst	EA	S-1	3.0	166	---	66	233	263	S-1	2.4	192	---	83	275	313
21" x 17" oval, 6" deep bowl	Inst	EA	S-1	3.0	194	---	66	260	291	S-1	2.4	224	---	83	306	345
Wall hung units																
27" x 22" overall, pedestal model, 6" deep sculptured sculptured bowl	Inst	EA	S-1	2.6	563	---	77	639	675	S-1	2.1	650	---	96	745	789

			Costs Based On Large Volume							Costs Based On Small Volume						
Description	Oper	Unit	Crew Size	Avg Day Prod	Avg Mat'l Unit Cost	Avg Equip Unit Cost	Avg Labor Unit Cost	Avg Total Unit Cost	Avg Price Incl O&P	Crew Size	Avg Day Prod	Avg Mat'l Unit Cost	Avg Equip Unit Cost	Avg Labor Unit Cost	Avg Total Unit Cost	Avg Price Incl O&P

Plumbing (continued)

Description	Oper	Unit	Crew Size	Avg Day Prod	Avg Mat'l Unit Cost	Avg Equip Unit Cost	Avg Labor Unit Cost	Avg Total Unit Cost	Avg Price Incl O&P	Crew Size	Avg Day Prod	Avg Mat'l Unit Cost	Avg Equip Unit Cost	Avg Labor Unit Cost	Avg Total Unit Cost	Avg Price Incl O&P
22" x 21" overall, pedestal model, 6" deep square bowl	Inst	EA	S-1	2.6	355	---	77	431	467	S-1	2.1	410	---	96	505	549
20" x 18" overall, ledge on back and sides, 6" deep	Inst	EA	S-1	3.0	190	---	66	256	287	S-1	2.4	219	---	83	302	340
24" x 20" overall, ledge on back and sides, 6" deep	Inst	EA	S-1	3.0	199	---	66	265	296	S-1	2.4	230	---	83	312	351
Adjustments																
To only remove and reset lavatory	Reset	EA	S-1	4.0	---	---	50	50	73	S-1	3.2	---	---	62	62	91
To install rough-in	Inst	EA	S-1	1.2	150	---	166	315	392	S-1	1.0	173	---	207	380	475
For colors, ADD	Inst	%	---	---	30.0	---	---	---	—	---	---	30.0	---	---	---	—
For two 15" L towel bars, ADD	Inst	PR	---	---	26	---	---	26	26	---	---	30	---	---	30	30
Enameled cast-iron																
Countertop units																
20" x 18" overall, 6" deep square bowl	Inst	EA	S-1	3.0	185	---	66	251	281	S-1	2.4	213	---	83	296	334
18" x 18" overall, 6" deep circular bowl	Inst	EA	S-1	3.0	176	---	66	242	272	S-1	2.4	203	---	83	285	324
19" x 19" overall, 6" deep circular self-rimming bowl	Inst	EA	S-1	3.0	179	---	66	246	276	S-1	2.4	207	---	83	290	328
20" x 17" overall, 6" deep oval self rimming bowl	Inst	EA	S-1	3.0	189	---	66	255	285	S-1	2.4	218	---	83	300	339
Wall hung units; with ledge on back and sides																
19" x 17" overall, 6" deep square bowl	Inst	EA	S-1	3.0	199	---	66	265	296	S-1	2.4	230	---	83	312	351
20" x 18" overall, 6" deep square bowl	Inst	EA	S-1	3.0	205	---	66	272	302	S-1	2.4	237	---	83	320	358
Adjustments																
To only remove and reset lavatory	Reset	EA	S-1	4.0	---	---	50	50	73	S-1	3.2	---	---	62	62	91
To install rough-in	Inst	EA	S-1	1.2	150	---	166	315	392	S-1	1.0	173	---	207	380	475
For colors, ADD	Inst	%	---	---	25.0	---	---	---	—	---	---	25.0	---	---	---	—
For two 15" L towel bars, ADD	Inst	PR	---	---	26	---	---	26	26	---	---	30	---	---	30	30

			Costs Based On Large Volume							Costs Based On Small Volume						
Description	Oper	Unit	Crew Size	Avg Day Prod	Avg Mat'l Unit Cost	Avg Equip Unit Cost	Avg Labor Unit Cost	Avg Total Unit Cost	Avg Price Incl O&P	Crew Size	Avg Day Prod	Avg Mat'l Unit Cost	Avg Equip Unit Cost	Avg Labor Unit Cost	Avg Total Unit Cost	Avg Price Incl O&P

Plumbing (continued)

Enameled steel
Countertop Units
19" x 19" overall, 6" deep

Description	Oper	Unit	Crew Size	Avg Day Prod	Avg Mat'l Unit Cost	Avg Equip Unit Cost	Avg Labor Unit Cost	Avg Total Unit Cost	Avg Price Incl O&P	Crew Size	Avg Day Prod	Avg Mat'l Unit Cost	Avg Equip Unit Cost	Avg Labor Unit Cost	Avg Total Unit Cost	Avg Price Incl O&P
circular self-rimming bowl	Inst	EA	S-1	3.0	133	---	66	199	229	S-1	2.4	153	---	83	236	274
20" x 17" overall, 6" deep oval self rimming bowl	Inst	EA	S-1	3.0	134	---	66	200	231	S-1	2.4	155	---	83	237	276
Adjustments																
To only remove and reset lavatory	Reset	EA	S-1	4.0	---	---	50	50	73	S-1	3.2	---	---	62	62	91
To install rough-in	Inst	EA	S-1	1.2	150	---	166	315	392	S-1	1.0	173	---	207	380	475
For colors, ADD	Inst	%	---	---	15.0	---	---	---	—	---	---	15.0	---	---	---	—
Duramel plastic																
Countertop Units																
19" x 19" overall, 6" deep circular self-rimming bowl	Inst	EA	S-1	3.0	130	---	66	196	226.82	S-1	2.4	150	---	83	233	271
20" x 17" overall, 6" deep oval self rimming bowl	Inst	EA	S-1	3.0	134	---	66	200	230.72	S-1	2.4	155	---	83	237	276
Adjustments																
To only remove and reset lavatory	Reset	EA	S-1	4.0	---	---	50	50	72.61	S-1	3.2	---	---	62	62	91
To install rough-in	Inst	EA	S-1	1.2	150	---	166	315	391.54	S-1	1.0	173	---	207	380	475
For colors, ADD	Inst	%	---	---	12.0	---	---	---	—	---	---	12.0	---	---	---	—

Shower, tub, and shower/tub stall combinations, fiberglass reinforced plastic integral bath/shower and wall surrounds (Dura Glass), white, with good quality fittings, single control faucets, and shower head sprayer

Shower stall, no door, with plastic drain, one piece
Three wall model

Description	Oper	Unit	Crew Size	Avg Day Prod	Avg Mat'l Unit Cost	Avg Equip Unit Cost	Avg Labor Unit Cost	Avg Total Unit Cost	Avg Price Incl O&P	Crew Size	Avg Day Prod	Avg Mat'l Unit Cost	Avg Equip Unit Cost	Avg Labor Unit Cost	Avg Total Unit Cost	Avg Price Incl O&P
32" x 32" x 80" H	Inst	EA	S-2	1.8	325	---	183	508	592	S-2	1.4	375	---	228	603	709
36" x 37" x 74" H	Inst	EA	S-2	1.8	332	---	183	514	598	S-2	1.4	383	---	228	611	716
42" x 35" x 74" H	Inst	EA	S-2	1.6	338	---	206	544	638	S-2	1.3	390	---	257	647	765
48" x 35" x 74" H, w/ integral seat	Inst	EA	S-2	1.5	351	---	219	570	671	S-2	1.2	405	---	274	679	805
60" x 35" x 74" H, w/ integral seat	Inst	EA	S-2	1.4	371	---	235	605	714	S-2	1.1	428	---	294	721	856

			Costs Based On Large Volume								Costs Based On Small Volume					
Description	Oper	Unit	Crew Size	Avg Day Prod	Avg Mat'l Unit Cost	Avg Equip Unit Cost	Avg Labor Unit Cost	Avg Total Unit Cost	Avg Price Incl O&P	Crew Size	Avg Day Prod	Avg Mat'l Unit Cost	Avg Equip Unit Cost	Avg Labor Unit Cost	Avg Total Unit Cost	Avg Price Incl O&P

Plumbing (continued)

Description	Oper	Unit	Crew Size	Avg Day Prod	Avg Mat'l Unit Cost	Avg Equip Unit Cost	Avg Labor Unit Cost	Avg Total Unit Cost	Avg Price Incl O&P	Crew Size	Avg Day Prod	Avg Mat'l Unit Cost	Avg Equip Unit Cost	Avg Labor Unit Cost	Avg Total Unit Cost	Avg Price Incl O&P
Two wall model																
32" x 32" x 81" H	Inst	EA	S-2	1.8	403	---	183	586	670	S-2	1.4	465	---	228	693	799
36" x 36" x 81" H	Inst	EA	S-2	1.8	455	---	183	638	722	S-2	1.4	525	---	228	753	859
39" x 39" x 81" H	Inst	EA	S-2	1.6	397	---	206	602	697	S-2	1.3	458	---	257	715	833
Shower/Tub combination, 3 wall model, one piece, no door																
60" x 34" x 74" H	Inst	EA	S-2	1.4	449	---	235	683	792	S-2	1.1	518	---	294	811	946
60" x 34" x 76" H	Inst	EA	S-2	1.4	468	---	235	703	811	S-2	1.1	540	---	294	834	969
66" x 36" x 74" H	Inst	EA	S-2	1.4	611	---	235	846	954	S-2	1.1	705	---	294	999	1134
Tub with integral recessed step into tub																
Roman tub 60" x 48" x 17"	Inst	EA	S-2	1.4	449	---	235	683	792	S-2	1.1	518	---	294	811	946
Double tub model																
60" x 46" x 20" H	Inst	EA	S-2	1.4	598	---	235	833	941	S-2	1.1	690	---	294	984	1119
72" x 46" x 20" H	Inst	EA	S-2	1.4	735	---	235	969	1078	S-2	1.1	848	---	294	1141	1276
Sunken tub model																
60" x 40" x 15" H	Inst	EA	S-2	1.4	540	---	235	774	883	S-2	1.1	623	---	294	916	1051
80" x 51" x 24" H	Inst	EA	S-2	1.4	891	---	235	1125	1234	S-2	1.1	1028	---	294	1321	1456
Oval tub model																
60" x 36" x 22" H	Inst	EA	S-2	1.4	592	---	235	826	935	S-2	1.1	683	---	294	976	1111
72" x 36" x 22" H	Inst	EA	S-2	1.4	676	---	235	911	1019	S-2	1.1	780	---	294	1074	1209
Adjustments																
To remove only																
Shower stall	Demo	EA	S-2	6.5	---	---	51	51	74	S-2	5.2	---	---	63	63	92
Shower/tub	Demo	EA	S-2	6.0	---	---	55	55	80	S-2	4.8	---	---	69	69	100
Tub	Demo	EA	S-2	6.5	---	---	51	51	74	S-2	5.2	---	---	63	63	92
To remove and reset																
Shower stall	Reset	EA	S-2	3.2	---	---	103	103	150	S-2	2.6	---	---	129	129	188
Shower/tub	Reset	EA	S-2	3.0	---	---	110	110	160	S-2	2.4	---	---	137	137	200
Tub	Reset	EA	S-2	3.2	---	---	103	103	150	S-2	2.6	---	---	129	129	188
To remove and replace fiberglass shower stall receptor (floors)																
32" x 32"	R&R	EA	S-2	3.2	91	---	103	194	241	S-2	2.6	105	---	129	234	293
36" x 36"	R&R	EA	S-2	3.2	104	---	103	207	254	S-2	2.6	120	---	129	249	308
42" x 34"	R&R	EA	S-2	3.2	124	---	103	226	274	S-2	2.6	143	---	129	271	330
48" x 34"	R&R	EA	S-2	3.2	130	---	103	233	280	S-2	2.6	150	---	129	279	338
60" x 34"	R&R	EA	S-2	3.2	169	---	103	272	319	S-2	2.6	195	---	129	324	383
For colors, ADD	Inst	%	---	---	7.0	---	---	---	---	---	---	10.0	---	---	---	---

			Costs Based On Large Volume							Costs Based On Small Volume						
Description	Oper	Unit	Crew Size	Avg Day Prod	Avg Mat'l Unit Cost	Avg Equip Unit Cost	Avg Labor Unit Cost	Avg Total Unit Cost	Avg Price Incl O&P	Crew Size	Avg Day Prod	Avg Mat'l Unit Cost	Avg Equip Unit Cost	Avg Labor Unit Cost	Avg Total Unit Cost	Avg Price Incl O&P

Plumbing (continued)

To install rough-in, ADD

Description	Oper	Unit	Crew Size	Avg Day Prod	Avg Mat'l Unit Cost	Avg Equip Unit Cost	Avg Labor Unit Cost	Avg Total Unit Cost	Avg Price Incl O&P	Crew Size	Avg Day Prod	Avg Mat'l Unit Cost	Avg Equip Unit Cost	Avg Labor Unit Cost	Avg Total Unit Cost	Avg Price Incl O&P
Shower stall	Inst	EA	S-2	1.2	---	---	274	274	400	S-2	1.0	---	---	343	343	500
Shower/tub	Inst	EA	S-2	1.2	---	---	274	274	400	S-2	1.0	---	---	343	343	500
Tub	Inst	EA	S-2	1.6	---	---	206	206	300	S-2	1.3	---	---	257	257	375

Shower and tub doors

Hinged shower doors, anodized aluminum frame, tempered hammered glass, hardware included
64" door

Description	Oper	Unit	Crew Size	Avg Day Prod	Avg Mat'l Unit Cost	Avg Equip Unit Cost	Avg Labor Unit Cost	Avg Total Unit Cost	Avg Price Incl O&P	Crew Size	Avg Day Prod	Avg Mat'l Unit Cost	Avg Equip Unit Cost	Avg Labor Unit Cost	Avg Total Unit Cost	Avg Price Incl O&P
23" x 25" wide opening	Inst	EA	C-1	6.0	98	---	28	126	139	C-1	4.8	113	---	35	148	165
26" x 28" wide opening	Inst	EA	C-1	6.0	104	---	28	132	146	C-1	4.8	120	---	35	155	172

64" H door, with attached panel

Description	Oper	Unit	Crew Size	Avg Day Prod	Avg Mat'l Unit Cost	Avg Equip Unit Cost	Avg Labor Unit Cost	Avg Total Unit Cost	Avg Price Incl O&P	Crew Size	Avg Day Prod	Avg Mat'l Unit Cost	Avg Equip Unit Cost	Avg Labor Unit Cost	Avg Total Unit Cost	Avg Price Incl O&P
39" wide opening	Inst	EA	C-1	5.5	156	---	31	187	202	C-1	4.4	180	---	38	218	237

Sliding shower doors, 2 bi-passing panels, anodized aluminum frame, tempered hammered glass, hardware included
70" H doors

Description	Oper	Unit	Crew Size	Avg Day Prod	Avg Mat'l Unit Cost	Avg Equip Unit Cost	Avg Labor Unit Cost	Avg Total Unit Cost	Avg Price Incl O&P	Crew Size	Avg Day Prod	Avg Mat'l Unit Cost	Avg Equip Unit Cost	Avg Labor Unit Cost	Avg Total Unit Cost	Avg Price Incl O&P
46" to 48" wide opening	Inst	EA	C-1	5.5	208	---	31	239	254	C-1	4.4	240	---	38	278	297
46" to 48" wide opening, with etched design on panels	Inst	EA	C-1	5.5	234	---	31	265	280	C-1	4.4	270	---	38	308	327

Sliding tub doors, 2 bipassing panels, anodized aluminum frame, hardware included
tempered hammered glass, 58" H doors

Description	Oper	Unit	Crew Size	Avg Day Prod	Avg Mat'l Unit Cost	Avg Equip Unit Cost	Avg Labor Unit Cost	Avg Total Unit Cost	Avg Price Incl O&P	Crew Size	Avg Day Prod	Avg Mat'l Unit Cost	Avg Equip Unit Cost	Avg Labor Unit Cost	Avg Total Unit Cost	Avg Price Incl O&P
52" to 55" wide opening	Inst	EA	C-1	5.5	195	---	31	226	241	C-1	4.4	225	---	38	263	282
56" to 60" wide opening	Inst	EA	C-1	5.5	215	---	31	245	260	C-1	4.4	248	---	38	286	305
56" to 60" wide opening, with etched design on panels	Inst	EA	C-1	5.5	234	---	31	265	280	C-1	4.4	270	---	38	308	327

Vinyl plastic panels, 58" H doors

Description	Oper	Unit	Crew Size	Avg Day Prod	Avg Mat'l Unit Cost	Avg Equip Unit Cost	Avg Labor Unit Cost	Avg Total Unit Cost	Avg Price Incl O&P	Crew Size	Avg Day Prod	Avg Mat'l Unit Cost	Avg Equip Unit Cost	Avg Labor Unit Cost	Avg Total Unit Cost	Avg Price Incl O&P
60" wide opening	Inst	EA	C-1	5.5	104	---	31	135	150	C-1	4.4	120	---	38	158	177

Sinks (kitchen and utility), with good quality fittings, faucets, and sprayer (American Standard)

Enameled pressed steel, white

Description	Oper	Unit	Crew Size	Avg Day Prod	Avg Mat'l Unit Cost	Avg Equip Unit Cost	Avg Labor Unit Cost	Avg Total Unit Cost	Avg Price Incl O&P	Crew Size	Avg Day Prod	Avg Mat'l Unit Cost	Avg Equip Unit Cost	Avg Labor Unit Cost	Avg Total Unit Cost	Avg Price Incl O&P
Single bowl, 24" x 21"	Inst	EA	S-1	2.0	130	---	99	229	275	S-1	1.6	150	---	124	274	332
Double bowl bowl, 32" x 21"	Inst	EA	S-1	1.6	143	---	124	267	325	S-1	1.3	165	---	155	320	392

Enameled cast iron, white
Kitchen countertop

Description	Oper	Unit	Crew Size	Avg Day Prod	Avg Mat'l Unit Cost	Avg Equip Unit Cost	Avg Labor Unit Cost	Avg Total Unit Cost	Avg Price Incl O&P	Crew Size	Avg Day Prod	Avg Mat'l Unit Cost	Avg Equip Unit Cost	Avg Labor Unit Cost	Avg Total Unit Cost	Avg Price Incl O&P
12" x 15" single bowl bar sink	Inst	EA	S-1	2.0	185	---	99	284	330	S-1	1.6	213	---	124	337	395
24" x 21" single bowl	Inst	EA	S-1	2.0	189	---	99	288	334	S-1	1.6	218	---	124	342	399

			Costs Based On Large Volume							Costs Based On Small Volume					
Description	Oper	Unit	Crew Size	Avg Day Prod	Avg Mat'l Unit Cost	Avg Equip Unit Cost	Avg Labor Unit Cost	Avg Total Unit Cost	Avg Price Incl O&P	Crew Size	Avg Day Prod	Avg Mat'l Unit Cost	Avg Equip Unit Cost	Avg Labor Unit Cost	Avg Total Unit Cost	Avg Price Incl O&P

Plumbing (continued)

Description	Oper	Unit	Crew Size	Avg Day Prod	Avg Mat'l Unit Cost	Avg Equip Unit Cost	Avg Labor Unit Cost	Avg Total Unit Cost	Avg Price Incl O&P	Crew Size	Avg Day Prod	Avg Mat'l Unit Cost	Avg Equip Unit Cost	Avg Labor Unit Cost	Avg Total Unit Cost	Avg Price Incl O&P
30" x 21" single bowl	Inst	EA	S-1	2.0	209	---	99	309	355	S-1	1.6	242	---	124	366	423
32" x 21" double bowl	Inst	EA	S-1	1.6	226	---	124	351	408	S-1	1.3	261	---	155	416	488
42" x 21" double bowl	Inst	EA	S-1	1.6	356	---	124	481	538	S-1	1.3	411	---	155	566	638
Utility/Service																
14" x 20", with drilled back and rim guard	Inst	EA	S-1	2.0	286	---	99	385	431	S-1	1.6	330	---	124	454	512
22" x 18", with drilled back and rim guard	Inst	EA	S-1	2.0	342	---	99	441	487	S-1	1.6	395	---	124	519	576
48" x 20" double bowl	Inst	EA	S-1	1.6	406	---	124	530	587	S-1	1.3	468	---	155	623	695
Stainless steel (Polar Ware)																
Single bowl, self-rimming, 20 gauge																
Flat rim																
13" x 17" x 6-1/2" D	Inst	EA	S-1	2.0	147	---	99	246	292	S-1	1.6	170	---	124	294	351
13" x 17" x 9-3/4" D	Inst	EA	S-1	2.0	125	---	99	224	270	S-1	1.6	144	---	124	268	326
19" x 17" x 6-1/2" D	Inst	EA	S-1	2.0	142	---	99	241	287	S-1	1.6	164	---	124	288	345
22" x 19" x 6-1/2" D	Inst	EA	S-1	2.0	153	---	99	253	299	S-1	1.6	177	---	124	301	359
25" x 19" x 7" D	Inst	EA	S-1	2.0	164	---	99	263	309	S-1	1.6	189	---	124	313	371
31" x 19" x 7" D	Inst	EA	S-1	2.0	185	---	99	284	330	S-1	1.6	213	---	124	337	395
Ledge back																
13" x 17" x 6" D	Inst	EA	S-1	2.0	120	---	99	219	265	S-1	1.6	138	---	124	262	320
17" x 22" x 7" D	Inst	EA	S-1	2.0	142	---	99	241	287	S-1	1.6	164	---	124	288	345
22" x 22" x 7" D	Inst	EA	S-1	2.0	148	---	99	248	293	S-1	1.6	171	---	124	295	353
24" x 20" x 6-1/2" D	Inst	EA	S-1	2.0	130	---	99	229	275	S-1	1.6	150	---	124	274	332
25" x 22" x 7-1/4" D	Inst	EA	S-1	2.0	157	---	99	257	303	S-1	1.6	182	---	124	306	363
31" x 22" x 7" D	Inst	EA	S-1	2.0	191	---	99	291	336	S-1	1.6	221	---	124	345	402
Round/oval, flat rim																
13-1/2" round x 5-1/4" D	Inst	EA	S-1	2.0	104	---	99	203	249	S-1	1.6	120	---	124	244	302
13-1/2" x 10-1/2" x 5" D	Inst	EA	S-1	2.0	105	---	99	205	251	S-1	1.6	122	---	124	246	303
Double bowl, self-rimming, 20 gauge																
Flat rim																
25" x 17" x 9-3/4" D	Inst	EA	S-1	1.6	211	---	124	335	392	S-1	1.3	243	---	155	398	470
33" x 18" x 6-1/2" D	Inst	EA	S-1	1.6	140	---	124	265	322	S-1	1.3	162	---	155	317	389

Description	Oper	Unit	Costs Based On Large Volume							Costs Based On Small Volume						
			Crew Size	Avg Day Prod	Avg Mat'l Unit Cost	Avg Equip Unit Cost	Avg Labor Unit Cost	Avg Total Unit Cost	Avg Price Incl O&P	Crew Size	Avg Day Prod	Avg Mat'l Unit Cost	Avg Equip Unit Cost	Avg Labor Unit Cost	Avg Total Unit Cost	Avg Price Incl O&P

Plumbing (continued)

Description	Oper	Unit	Crew Size	Avg Day Prod	Avg Mat'l	Avg Equip	Avg Labor	Avg Total	Avg Price O&P	Crew Size	Avg Day Prod	Avg Mat'l	Avg Equip	Avg Labor	Avg Total	Avg Price O&P
Ledge back																
25" x 17" x 6-1/2" D	Inst	EA	S-1	1.6	144	---	124	269	326	S-1	1.3	167	---	155	322	393
33" x 22" x 6-1/2" D	Inst	EA	S-1	1.6	139	---	124	263	321	S-1	1.3	161	---	155	316	387
33" x 22" x 7-1/4" D, contoured	Inst	EA	S-1	1.6	179	---	124	304	361	S-1	1.3	207	---	155	362	434
43" x 22" x 7" D	Inst	EA	S-1	1.6	231	---	124	356	413	S-1	1.3	267	---	155	422	494
Corner unit, 6-1/2" D	Inst	EA	S-1	1.6	229	---	124	353	410	S-1	1.3	264	---	155	419	491
Triple bowl, self-rimming, 20 gauge																
Ledge back																
48" x 22" x 6-1/2" D	Inst	EA	S-1	1.4	356	---	142	498	564	S-1	1.1	411	---	178	589	670
48" x 22" x 7" D	Inst	EA	S-1	1.4	391	---	142	533	599	S-1	1.1	452	---	178	629	711
Adjustments																
To only remove and reset sink																
Single bowl	Reset	EA	S-1	4.0	---	---	50	50	73	S-1	3.2	---	---	62	62	91
Double bowl	Reset	EA	S-1	3.8	---	---	52	52	76	S-1	3.0	---	---	65	65	96
Triple bowl	Reset	EA	S-1	3.5	---	---	57	57	83	S-1	2.8	---	---	71	71	104
To install rough-in	Inst	EA	S-1	1.2	150	---	166	315	392	S-1	1.0	173	---	207	380	475
For 18 gauge steel, ADD	Inst	%	---	---	40.0	---	---	---	—	---	---	40.0	---	---	---	—
Spas, with good quality fittings and faucets																
Whirlpool baths (jacuzzi), colors chrome trim																
Adonis model, 62" x 51" x 24"	Inst	EA	S-2	0.8	2548	---	411	2959	3148	S-2	0.6	2940	---	514	3454	3690
Athena model, 72" x 56" x 23"	Inst	EA	S-2	0.8	2662	---	411	3074	3263	S-2	0.6	3072	---	514	3586	3822
Nova model, 60" x 42" x 18" with magic touch switch	Inst	EA	S-2	0.8	1565	---	411	1976	2166	S-2	0.6	1806	---	514	2320	2556
Cara model, 60" x 34" x 20"	Inst	EA	S-2	0.8	1700	---	411	2112	2301	S-2	0.6	1962	---	514	2476	2712
Lusso model, 60" x 40-1/2" x 30"	Inst	EA	S-2	0.8	2309	---	411	2720	2909	S-2	0.6	2664	---	514	3178	3414
Omni V bath and shower model, 60" x 34" x 78" (21" deep)	Inst	EA	S-2	0.6	2350	---	514	2864	3101	S-2	0.5	2712	---	643	3355	3650
Prima model, 72" x 36" x 18"	Inst	EA	S-2	0.8	1804	---	411	2216	2405	S-2	0.6	2082	---	514	2596	2832
Whirlpool spas (Jacuzzi); acrylic shells reinforced with fiberglass; plumbed and assembled at factory																
Celeste model, 72" x 66" x 30" 2 speed electric	Inst	EA	S-2	0.4	2568	---	822	3390	3768	S-2	0.3	2963	---	1028	3991	4463
Sienna model, 84" x 76" x 34"	Inst	EA	S-2	0.4	3416	---	822	4239	4617	S-2	0.3	3942	---	1028	4970	5443
Meridian, 84" x 84" x 37"	Inst	EA	S-2	0.4	3650	---	822	4473	4851	S-2	0.3	4212	---	1028	5240	5713

			Costs Based On Large Volume							Costs Based On Small Volume						

Description	Oper	Unit	Crew Size	Avg Day Prod	Avg Mat'l Unit Cost	Avg Equip Unit Cost	Avg Labor Unit Cost	Avg Total Unit Cost	Avg Price Incl O&P	Crew Size	Avg Day Prod	Avg Mat'l Unit Cost	Avg Equip Unit Cost	Avg Labor Unit Cost	Avg Total Unit Cost	Avg Price Incl O&P
Plumbing (continued)																
Espree model (portable), 64" x 70" x 28"	Inst	EA	S-2	0.4	2802	---	822	3624	**4002**	S-2	0.3	3233	---	1028	4261	**4733**
Cambio model (portable), 72" x 66" x 30"	Inst	EA	S-2	0.4	3920	---	822	4742	**5120**	S-2	0.3	4523	---	1028	5551	**6023**
Quanta model w/ integrated power system, 84" x 84" x 37"	Inst	EA	S-2	0.4	5148	---	822	5970	**6349**	S-2	0.3	5940	---	1028	6968	**7441**
Adjustments/related operations																
To remove and reset																
Whirlpool bath	Reset	EA	S-2	2.4	---	---	137	137	**200**	S-2	1.9	---	---	171	171	**250**
Whirlpool spa	Reset	EA	S-2	2.0	---	---	164	164	**240**	S-2	1.6	---	---	206	206	**300**
To rough-in																
Whirlpool bath	Inst	EA	S-2	1.6	228	---	206	433	**528**	S-2	1.3	263	---	257	520	**638**
Whirlpool spa	Inst	EA	S-2	1.6	228	---	206	433	**528**	S-2	1.3	263	---	257	520	**638**
Toilets (water closets), vitreous china, white, includes seat, shut- off valve, connectors, flanges, water supply valve (American Standard)																
One piece, floor mounted																
Low-line, silhoutte (Roma)	Inst	EA	S-1	1.8	793	---	111	904	**954**	S-1	1.4	915	---	138	1053	**1117**
Water saver model																
Elisse model	Inst	EA	S-1	1.8	658	---	111	768	**819**	S-1	1.4	759	---	138	897	**961**
Lexington model	Inst	EA	S-1	1.8	478	---	111	589	**640**	S-1	1.4	552	---	138	690	**754**
Two piece, close coupled, floor mounted, water saver models																
Economy model (Plebe)	Inst	EA	S-1	1.8	91	---	111	202	**252**	S-1	1.4	105	---	138	243	**307**
Average quality, 12" round																
Cadet rounded	Inst	EA	S-1	1.8	143	---	111	254	**304**	S-1	1.4	165	---	138	303	**367**
Cadet elongated	Inst	EA	S-1	1.8	170	---	111	281	**332**	S-1	1.4	197	---	138	335	**398**
Cadet elongated, 18" rim height	Inst	EA	S-1	1.8	260	---	111	371	**421**	S-1	1.4	300	---	138	438	**502**
Bidets, wall mounted																
Roma model	Inst	EA	S-1	2.2	659	---	90	750	**791**	S-1	1.8	761	---	113	874	**926**
Madval model	Inst	EA	S-1	2.2	605	---	90	695	**737**	S-1	1.8	698	---	113	811	**863**
Adjustments																
To remove and reset toilet or bidet	Reset	EA	S-1	4.0	---	---	50	50	**73**	S-1	3.2	---	---	62	62	**91**

			Costs Based On Large Volume							Costs Based On Small Volume						
Description	Oper	Unit	Crew Size	Avg Day Prod	Avg Mat'l Unit Cost	Avg Equip Unit Cost	Avg Labor Unit Cost	Avg Total Unit Cost	Avg Price Incl O&P	Crew Size	Avg Day Prod	Avg Mat'l Unit Cost	Avg Equip Unit Cost	Avg Labor Unit Cost	Avg Total Unit Cost	Avg Price Incl O&P

Plumbing (continued)

To install rough-in																
Toilet	Inst	EA	S-1	0.9	195	---	226	421	525	S-1	0.7	225	---	283	508	638
Bidet	Inst	EA	S-1	1.0	163	---	199	361	453	S-1	0.8	188	---	249	436	551
For colors, ADD	Inst	%	---	---	30.0	---	---	---	---	---	---	30.0	---	---	---	---

Vanity units. See Cabinets

Water heaters (A.O. Smith/National Products)

Electric, glass lined, 2 element heater, baked enamel, installation and connection only

Standard model

5 year Energy Saver, foam insulated

30 gal round	Inst	EA	S-1	1.8	198	---	111	308	359	S-1	1.4	228	---	138	366	430
40 gal round	Inst	EA	S-1	1.8	216	---	111	326	377	S-1	1.4	249	---	138	387	451
50 gal round	Inst	EA	S-1	1.7	241	---	117	358	411	S-1	1.4	278	---	146	424	491
66 gal round	Inst	EA	S-1	1.7	322	---	117	439	493	S-1	1.4	372	---	146	518	586
80 gal round	Inst	EA	S-1	1.7	373	---	117	490	544	S-1	1.4	431	---	146	577	644
120 gal round	Inst	EA	S-1	1.6	588	---	124	712	769	S-1	1.3	678	---	155	833	905
40 gal round, table top model	Inst	EA	S-1	1.8	226	---	111	337	388	S-1	1.4	261	---	138	399	463

Deluxe model

10-year Series II

30 gal round	Inst	EA	S-1	1.8	270	---	111	381	432	S-1	1.4	312	---	138	450	514
40 gal round	Inst	EA	S-1	1.8	296	---	111	407	458	S-1	1.4	342	---	138	480	544
50 gal round	Inst	EA	S-1	1.7	328	---	117	445	498	S-1	1.4	378	---	146	524	592
66 gal round	Inst	EA	S-1	1.7	429	---	117	546	600	S-1	1.4	495	---	146	641	709
80 gal round	Inst	EA	S-1	1.7	490	---	117	607	661	S-1	1.4	566	---	146	712	779

Gas, 3 stage automatic flame control, baked enamel, installation and connection only

Standard model

5 year Series I

30 gal round	Inst	EA	S-1	1.8	178	---	111	289	339	S-1	1.4	206	---	138	344	407
40 gal round	Inst	EA	S-1	1.8	195	---	111	306	356	S-1	1.4	225	---	138	363	427
50 gal round	Inst	EA	S-1	1.7	242	---	117	359	413	S-1	1.4	279	---	146	425	493
5 year Series II Fuel Saver, required in Calif., ADD	Inst	%	---	---	20.0	---	---	---	---	---	---	20.0	---	---	---	---

			Costs Based On Large Volume							Costs Based On Small Volume						
Description	Oper	Unit	Crew Size	Avg Day Prod	Avg Mat'l Unit Cost	Avg Equip Unit Cost	Avg Labor Unit Cost	Avg Total Unit Cost	Avg Price Incl O&P	Crew Size	Avg Day Prod	Avg Mat'l Unit Cost	Avg Equip Unit Cost	Avg Labor Unit Cost	Avg Total Unit Cost	Avg Price Incl O&P

Plumbing (continued)

Deluxe model
10-year Series II

30 gal round	Inst	EA	S-1	1.8	274	---	111	385	**436**	S-1	1.4	317	---	138	455	**518**
40 gal round	Inst	EA	S-1	1.8	304	---	111	415	**466**	S-1	1.4	351	---	138	489	**553**
50 gal round	Inst	EA	S-1	1.7	380	---	117	497	**550**	S-1	1.4	438	---	146	584	**652**

Solar water heating systems

Complete closed loop solar system with solar electric water heater with exchanger, differential control, heating element, circulator, tank and panel sensor, and collector

82 gallon capacity collector

One deluxe collector	---	LS	---	---	2548	---	---	---	—	---	---	2940	---	---	---	—
Two deluxe collectors																
Economy collectors	---	LS	---	---	2821	---	---	---	—	---	---	3255	---	---	---	—
Standard collectors	---	LS	---	---	3042	---	---	---	—	---	---	3510	---	---	---	—
Deluxe collectors	---	LS	---	---	3223	---	---	---	—	---	---	3719	---	---	---	—
Three collectors																
Economy collectors	---	LS	---	---	3380	---	---	---	—	---	---	3900	---	---	---	—
Standard collectors	---	LS	---	---	3743	---	---	---	—	---	---	4319	---	---	---	—
Four collectors																
Economy collectors	---	LS	---	---	3926	---	---	---	—	---	---	4530	---	---	---	—
Standard collectors	---	LS	---	---	4394	---	---	---	—	---	---	5070	---	---	---	—

120 gallon capacity system

Three collectors																
Economy collectors	---	LS	---	---	3978	---	---	---	—	---	---	4590	---	---	---	—
Standard collectors	---	LS	---	---	4290	---	---	---	—	---	---	4950	---	---	---	—
Four standard collectors	---	LS	---	---	4615	---	---	---	—	---	---	5325	---	---	---	—
Five collectors																
Economy collectors	---	LS	---	---	4810	---	---	---	—	---	---	5550	---	---	---	—
Standard collectors	---	LS	---	---	5200	---	---	---	—	---	---	6000	---	---	---	—
Six collectors																
Economy collectors	---	LS	---	---	5330	---	---	---	—	---	---	6150	---	---	---	—
Standard collectors	---	LS	---	---	5850	---	---	---	—	---	---	6750	---	---	---	—

			Costs Based On Large Volume							Costs Based On Small Volume					
Description	Oper	Unit	Crew Size	Avg Day Prod	Avg Mat'l Unit Cost	Avg Equip Unit Cost	Avg Labor Unit Cost	Avg Total Unit Cost	Avg Price Incl O&P	Crew Size	Avg Day Prod	Avg Mat'l Unit Cost	Avg Equip Unit Cost	Avg Labor Unit Cost	Avg Total Unit Cost	Avg Price Incl O&P

Plumbing (continued)

Material adjustments
Collector mounting kits, one required per collector panel

Description	Oper	Unit	Crew Size	Avg Day Prod	Avg Mat'l Unit Cost	Avg Equip Unit Cost	Avg Labor Unit Cost	Avg Total Unit Cost	Avg Price Incl O&P	Crew Size	Avg Day Prod	Avg Mat'l Unit Cost	Avg Equip Unit Cost	Avg Labor Unit Cost	Avg Total Unit Cost	Avg Price Incl O&P
Adjustable position hinge	---	EA	---	---	74.10	---	---	---	---	---	---	85.50	---	---	---	---
Integral flange	---	EA	---	---	6.50	---	---	---	---	---	---	7.50	---	---	---	---
Additional panels																
Economy collectors	---	EA	---	---	456.30	---	---	---	---	---	---	526.50	---	---	---	---
Standard collectors	---	EA	---	---	559.00	---	---	---	---	---	---	645.00	---	---	---	---
Deluxe collectors	---	EA	---	---	698.10	---	---	---	---	---	---	805.50	---	---	---	---
Individual components																
Solar storage tanks, glass lined with fiberglass insulation																
66 gallon	---	EA	---	---	325.00	---	---	---	---	---	---	375.00	---	---	---	---
82 gallon	---	EA	---	---	361.40	---	---	---	---	---	---	417.00	---	---	---	---
120 gallon	---	EA	---	---	546.00	---	---	---	---	---	---	630.00	---	---	---	---
Solar electric water heaters, glass lined with fiberglass insulation, heating element with thermostat																
66 gallon, 4.5 kw	---	EA	---	---	315.90	---	---	---	---	---	---	364.50	---	---	---	---
82 gallon, 4.5 kw	---	EA	---	---	353.60	---	---	---	---	---	---	408.00	---	---	---	---
120 gallon, 4.5 kw	---	EA	---	---	538.20	---	---	---	---	---	---	621.00	---	---	---	---
Additional element	---	EA	---	---	33.80	---	---	---	---	---	---	39.00	---	---	---	---
Solar electric water heaters with heat exchangers, glass lined with fiberglass insulation, two copper exchangers, powered circulator, differential control, adjustable thermostat																
82 gallon, 4.5 kw, 1/20 hp	---	EA	---	---	889.20	---	---	---	---	---	---	1026.00	---	---	---	---
120 gallon, 4.5 kw, 1/20 hp	---	EA	---	---	1041.30	---	---	---	---	---	---	1201.50	---	---	---	---

Water softeners (Bruner/Calgon)

Automatic water softeners, complete with yoke with 3/4" I.P.S. supply
Single tank units
8000 grain exchange capacity,

Description	Oper	Unit	Crew Size	Avg Day Prod	Avg Mat'l Unit Cost	Avg Equip Unit Cost	Avg Labor Unit Cost	Avg Total Unit Cost	Avg Price Incl O&P	Crew Size	Avg Day Prod	Avg Mat'l Unit Cost	Avg Equip Unit Cost	Avg Labor Unit Cost	Avg Total Unit Cost	Avg Price Incl O&P
160 lbs salt storage, 6 gpm	---	EA	---	---	468.00	---	---	---	---	---	---	540.00	---	---	---	---
15000 grain exchange capacity,																
200 lbs salt storage, 8.8 gpm	---	EA	---	---	516.10	---	---	---	---	---	---	595.50	---	---	---	---
25000 grain exchange capacity,																
175 lbs salt storage, 11.3 gpm	---	EA	---	---	620.10	---	---	---	---	---	---	715.50	---	---	---	---

Description	Oper	Unit	Costs Based On Large Volume							Costs Based On Small Volume						
			Crew Size	Avg Day Prod	Avg Mat'l Unit Cost	Avg Equip Unit Cost	Avg Labor Unit Cost	Avg Total Unit Cost	Avg Price Incl O&P	Crew Size	Avg Day Prod	Avg Mat'l Unit Cost	Avg Equip Unit Cost	Avg Labor Unit Cost	Avg Total Unit Cost	Avg Price Incl O&P

Plumbing (continued)

Two tank units, side-by-side 5000 grain exchange capacity,

Description	Oper	Unit	Crew Size	Avg Day Prod	Avg Mat'l Unit Cost	Avg Equip Unit Cost	Avg Labor Unit Cost	Avg Total Unit Cost	Avg Price Incl O&P	Crew Size	Avg Day Prod	Avg Mat'l Unit Cost	Avg Equip Unit Cost	Avg Labor Unit Cost	Avg Total Unit Cost	Avg Price Incl O&P
200 lbs salt strorage 8.8 gpm	---	EA	---	---	520.00	---	---	---	—	---	---	600.00	---	---	---	—
30000 grain exchange capacity,																
200 lbs salt strorage, 9 gpm	---	EA	---	---	611.00	---	---	---	—	---	---	705.00	---	---	---	—
45000 grain exchange capacity,																
200 lbs salt storage, 9.5 gpm	---	EA	---	---	708.50	---	---	---	—	---	---	817.50	---	---	---	—
Softy 1; for 5 persons with up to 50 grains																
per gallon of hardness	---	EA	---	---	377.00	---	---	---	—	---	---	435.00	---	---	---	—
Softy 2; above plus taste & odor																
removal capabilities	---	EA	---	---	396.50	---	---	---	—	---	---	457.50	---	---	---	—
Softy 3; for 6 persons with up to 73 grains																
per gallon of hardness	---	EA	---	---	429.00	---	---	---	—	---	---	495.00	---	---	---	—
Automatic water filters, complete with yoke and media with 3/4" I.P.S. supply																
Two tank units, side by side																
Eliminate rotten egg odor																
and rust	---	EA	---	---	592.80	---	---	---	—	---	---	684.00	---	---	---	—
Eliminate chlorine taste	---	EA	---	---	497.90	---	---	---	—	---	---	574.50	---	---	---	—
Clear up water discoloration	---	EA	---	---	445.90	---	---	---	—	---	---	514.50	---	---	---	—
Clear up corroding pipes	---	EA	---	---	464.10	---	---	---	—	---	---	535.50	---	---	---	—
Manual water filters, complete fiberglass tank with 3/4" pipe supplies with mineral packs																
Eliminate taste and odor	---	EA	---	---	442.00	---	---	---	—	---	---	510.00	---	---	---	—
Eliminate acid water	---	EA	---	---	375.70	---	---	---	—	---	---	433.50	---	---	---	—
Eliminate iron in solution	---	EA	---	---	455.00	---	---	---	—	---	---	525.00	---	---	---	—
Eliminate sediment	---	EA	---	---	375.70	---	---	---	—	---	---	433.50	---	---	---	—
Chemical feed pumps																
115 volt, 9 gal/day	---	EA	---	---	262.60	---	---	---	—	---	---	303.00	---	---	---	—
230 volt, 9 gal/day	---	EA	---	---	300.30	---	---	---	—	---	---	346.50	---	---	---	—

Plywood. See Framing.

Posts. See Framing.

Quarry Tile. See Masonry.

Rafters. See Framing.

					Costs Based On Large Volume						Costs Based On Small Volume					
Description	Oper	Unit	Crew Size	Avg Day Prod	Avg Mat'l Unit Cost	Avg Equip Unit Cost	Avg Labor Unit Cost	Avg Total Unit Cost	Avg Price Incl O&P	Crew Size	Avg Day Prod	Avg Mat'l Unit Cost	Avg Equip Unit Cost	Avg Labor Unit Cost	Avg Total Unit Cost	Avg Price Incl O&P

Range hoods

Metal finishes

Labor includes wiring and connection by electrician and installation by a carpenter in stud-exposed structure only.
Economy model, UL approved; mitered, welded construction; completely assembled and wired;
includes fan, motor, washable aluminum filter, and light

Description	Oper	Unit	Crew Size	Avg Day Prod	Avg Mat'l Unit Cost	Avg Equip Unit Cost	Avg Labor Unit Cost	Avg Total Unit Cost	Avg Price Incl O&P	Crew Size	Avg Day Prod	Avg Mat'l Unit Cost	Avg Equip Unit Cost	Avg Labor Unit Cost	Avg Total Unit Cost	Avg Price Incl O&P
24" wide																
3-1/4" x 10" duct, 160 CFM	Inst	EA	E-4	4.0	87	---	91	178	219	E-4	2.6	109	---	140	249	312
Non-ducted, 160 CFM	Inst	EA	E-4	4.0	91	---	91	182	223	E-4	2.6	114	---	140	254	317
30" wide																
7" round duct, 240 CFM	Inst	EA	E-4	4.0	83	---	91	174	215	E-4	2.6	103	---	140	243	307
3-1/4" x 10" duct, 160 CFM	Inst	EA	E-4	4.0	90	---	91	181	222	E-4	2.6	112	---	140	252	315
Non-ducted, 160 CFM	Inst	EA	E-4	4.0	95	---	91	186	227	E-4	2.6	119	---	140	259	322
36" wide																
7" round duct, 240 CFM	Inst	EA	E-4	4.0	85	---	91	177	218	E-4	2.6	107	---	140	247	310
3-1/4" x 10" duct, 160 CFM	Inst	EA	E-4	4.0	91	---	91	182	223	E-4	2.6	114	---	140	254	317
Non-ducted, 160 CFM	Inst	EA	E-4	4.0	99	---	91	191	232	E-4	2.6	124	---	140	264	328

Standard model, UL approved; deluxe mitered styling; solid state fan control for infinite speed
settings; wrap around, removable filter; built-in damper and dual light assembly

Description	Oper	Unit	Crew Size	Avg Day Prod	Avg Mat'l Unit Cost	Avg Equip Unit Cost	Avg Labor Unit Cost	Avg Total Unit Cost	Avg Price Incl O&P	Crew Size	Avg Day Prod	Avg Mat'l Unit Cost	Avg Equip Unit Cost	Avg Labor Unit Cost	Avg Total Unit Cost	Avg Price Incl O&P
30" wide x 9" deep																
3-1/4" x 10" duct	Inst	EA	E-4	4.0	162	---	91	254	295	E-4	2.6	203	---	140	343	406
Non-ducted	Inst	EA	E-4	4.0	176	---	91	268	309	E-4	2.6	221	---	140	361	424
36" wide x 9" deep																
3-1/4" x 10" duct	Inst	EA	E-4	4.0	167	---	91	258	299	E-4	2.6	208	---	140	348	412
Non-ducted	Inst	EA	E-4	4.0	183	---	91	275	316	E-4	2.6	229	---	140	369	433
42" wide x 9" deep																
3-1/4" x 10" duct	Inst	EA	E-4	3.0	171	---	121	292	347	E-4	2.0	214	---	187	400	485
Non-ducted	Inst	EA	E-4	3.0	188	---	121	309	364	E-4	2.0	235	---	187	421	506

Decorator/Designer model; solid state fan control for infinite speed settings; aluminum mesh
grease filters for easy cleaning; enclosed light assembly and switches

Description	Oper	Unit	Crew Size	Avg Day Prod	Avg Mat'l Unit Cost	Avg Equip Unit Cost	Avg Labor Unit Cost	Avg Total Unit Cost	Avg Price Incl O&P	Crew Size	Avg Day Prod	Avg Mat'l Unit Cost	Avg Equip Unit Cost	Avg Labor Unit Cost	Avg Total Unit Cost	Avg Price Incl O&P
Single faced hood-fan exhausts horizontally or vertically, 24" D canopy x 21" front to back																
30" wide	Inst	EA	E-4	3.0	360	---	121	481	536	E-4	2.0	450	---	187	637	721
36" wide	Inst	EA	E-4	3.0	372	---	121	494	549	E-4	2.0	466	---	187	652	737

209

			Costs Based On Large Volume							Costs Based On Small Volume						
Description	Oper	Unit	Crew Size	Avg Day Prod	Avg Mat'l Unit Cost	Avg Equip Unit Cost	Avg Labor Unit Cost	Avg Total Unit Cost	Avg Price Incl O&P	Crew Size	Avg Day Prod	Avg Mat'l Unit Cost	Avg Equip Unit Cost	Avg Labor Unit Cost	Avg Total Unit Cost	Avg Price Incl O&P

Range hoods (continued)

Single faced, contemporary style, duct horizontally or vertically
9" D canopy
30" wide

Description	Oper	Unit	Crew Size	Avg Day Prod	Avg Mat'l Unit Cost	Avg Equip Unit Cost	Avg Labor Unit Cost	Avg Total Unit Cost	Avg Price Incl O&P	Crew Size	Avg Day Prod	Avg Mat'l Unit Cost	Avg Equip Unit Cost	Avg Labor Unit Cost	Avg Total Unit Cost	Avg Price Incl O&P
with 330 cfm power unit	Inst	EA	E-4	3.0	333	---	121	455	509	E-4	2.0	417	---	187	603	688
with 410 cfm power unit	Inst	EA	E-4	3.0	371	---	121	492	547	E-4	2.0	464	---	187	651	735
36" wide																
with 330 cfm power unit	Inst	EA	E-4	3.0	342	---	121	463	518	E-4	2.0	427	---	187	614	698
with 410 cfm power unit	Inst	EA	E-4	3.0	381	---	121	502	557	E-4	2.0	476	---	187	663	747
Material adjustments																
Stainless steel, ADD	Inst	%	---	---	33.0	---	---	---	---	---	---	33.0	---	---	---	---

Reinforcing steel. See Concrete, Masonry, etc.

Resilient flooring

Description	Oper	Unit	Crew Size	Avg Day Prod	Avg Mat'l Unit Cost	Avg Equip Unit Cost	Avg Labor Unit Cost	Avg Total Unit Cost	Avg Price Incl O&P	Crew Size	Avg Day Prod	Avg Mat'l Unit Cost	Avg Equip Unit Cost	Avg Labor Unit Cost	Avg Total Unit Cost	Avg Price Incl O&P
Adhesive set sheet products	Demo	SY	L-2	160	---	---	1.63	1.63	2.41	L-2	112	---	---	2.32	2.32	3.44
Adhesive set tile products	Demo	SF	L-2	1500	---	---	0.17	0.17	0.26	L-2	1050	---	---	0.25	0.25	0.37

Install over smooth concrete subfloor

Tile
Asphalt, 9" x 9", 1/8"

Description	Oper	Unit	Crew Size	Avg Day Prod	Avg Mat'l Unit Cost	Avg Equip Unit Cost	Avg Labor Unit Cost	Avg Total Unit Cost	Avg Price Incl O&P	Crew Size	Avg Day Prod	Avg Mat'l Unit Cost	Avg Equip Unit Cost	Avg Labor Unit Cost	Avg Total Unit Cost	Avg Price Incl O&P
Group B colors (dark marbleized patterns)	Inst	SF	F-2	640	1.19	---	0.25	1.44	1.56	F-2	448	1.49	---	0.36	1.85	2.01
Group C colors (light marbleized patterns)	Inst	SF	F-2	640	1.36	---	0.25	1.61	1.73	F-2	448	1.70	---	0.36	2.06	2.22
Vinyl asbestos, 12" x 12"																
Peel and stick																
1/16" T	Inst	SF	F-2	960	1.48	---	0.17	1.64	1.72	F-2	672	1.85	---	0.24	2.09	2.20
0.080" T	Inst	SF	F-2	960	1.63	---	0.17	1.80	1.88	F-2	672	2.04	---	0.24	2.28	2.39
Plain patterns																
1/16" T	Inst	SF	F-2	960	1.04	---	0.17	1.21	1.29	F-2	672	1.31	---	0.24	1.55	1.66
1/8" T	Inst	SF	F-2	960	1.10	---	0.17	1.27	1.35	F-2	672	1.38	---	0.24	1.62	1.73
Solids																
1/16" T	Inst	SF	F-2	960	1.14	---	0.17	1.31	1.39	F-2	672	1.43	---	0.24	1.67	1.78
1/8" T	Inst	SF	F-2	960	1.21	---	0.17	1.38	1.46	F-2	672	1.52	---	0.24	1.76	1.87
Travertine/quarry																
1/8" T	Inst	SF	F-2	960	1.34	---	0.17	1.51	1.59	F-2	672	1.68	---	0.24	1.92	2.03

Resilient flooring

1. **Sheet Products.** Linoleum or vinyl sheet are normally installed either over a wood subfloor or a smooth concrete subfloor. When laid over wood, a layer of felt must first be laid in paste. This keeps irregularities in the wood from showing through. The amount of paste required to bond both felt and sheet is approximately 16 gallons (5% waste included) per 100 SY. When laid over smooth concrete subfloor, the sheet products can be bonded directly to the floor. However, the concrete subfloor should not be in direct contact with the ground because of excessive moisture. For bonding to concrete, 8 gallons of paste is required (5% waste included) per 100 SY. After laying the flooring over concrete or wood, wax is applied using 0.5 gallon per 100 SY. Paste, wax, felt (as needed), and 10% sheet waste are included in material costs.

2. **Tile Products.** All resilient tile can be bonded the same way as sheet products, either to smooth concrete or to felt over wood subfloor. When bonded to smooth concrete, a concrete primer (0.5 gal/100 SF) is first applied to seal the concrete. The tiles are then bonded to the sealed floor with resilient tile cement (0.6 gal/100 SF). On wood subfloors, felt is first laid in paste (0.9 gal/100 SF) and then the tiles are bonded to the felt with resilient tile emulsion (0.7 gal/100 SF). Bonding materials, felt (as needed), and 10% tile waste are included in material costs.

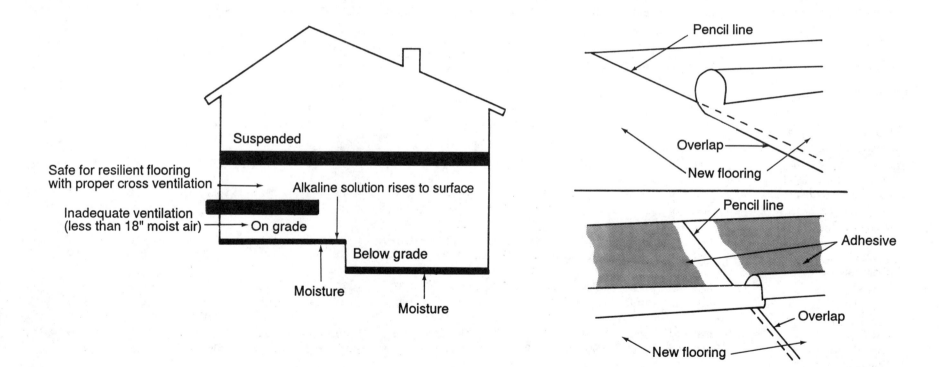

211

			Costs Based On Large Volume							Costs Based On Small Volume					
Description	Oper	Unit	Crew Size	Avg Day Prod	Avg Mat'l Unit Cost	Avg Equip Unit Cost	Avg Labor Unit Cost	Avg Total Unit Cost	Avg Price Incl O&P	Crew Size	Avg Day Prod	Avg Mat'l Unit Cost	Avg Equip Unit Cost	Avg Labor Unit Cost	Avg Total Unit Cost	Avg Price Incl O&P

Resilient flooring (continued)

Sheet
Linoleum, 6' wide
  Plain patterns

Description	Oper	Unit	Crew Size	Avg Day Prod	Avg Mat'l Unit Cost	Avg Equip Unit Cost	Avg Labor Unit Cost	Avg Total Unit Cost	Avg Price Incl O&P	Crew Size	Avg Day Prod	Avg Mat'l Unit Cost	Avg Equip Unit Cost	Avg Labor Unit Cost	Avg Total Unit Cost	Avg Price Incl O&P
3/32" T	Inst	SY	F-2	70	8.02	---	2.32	10.33	11.40	F-2	49	10.02	---	3.31	13.33	14.85
1/8" T	Inst	SY	F-2	70	8.72	---	2.32	11.04	12.11	F-2	49	10.91	---	3.31	14.21	15.74
Designer patterns																
3/32" T	Inst	SY	F-2	70	9.85	---	2.32	12.17	13.23	F-2	49	12.32	---	3.31	15.62	17.15
1/8" T	Inst	SY	F-2	70	10.76	---	2.32	13.08	14.15	F-2	49	13.46	---	3.31	16.76	18.29
Vinyl, 6' wide																
Plain patterns																
0.070" T	Inst	SY	F-2	70	11.12	---	2.32	13.44	14.51	F-2	49	13.91	---	3.31	17.21	18.74
0.093" T	Inst	SY	F-2	70	12.80	---	2.32	15.12	16.19	F-2	49	16.01	---	3.31	19.31	20.84

Install over wood subfloor

Tile
Asphalt, 9" x 9", 1/8"
  Group B colors (dark

Description	Oper	Unit	Crew Size	Avg Day Prod	Avg Mat'l Unit Cost	Avg Equip Unit Cost	Avg Labor Unit Cost	Avg Total Unit Cost	Avg Price Incl O&P	Crew Size	Avg Day Prod	Avg Mat'l Unit Cost	Avg Equip Unit Cost	Avg Labor Unit Cost	Avg Total Unit Cost	Avg Price Incl O&P
marbleized patterns)	Inst	SF	F-2	540	1.30	---	0.30	1.60	1.73	F-2	378	1.62	---	0.43	2.05	2.25
Group C colors (light																
marbleized patterns)	Inst	SF	F-2	540	1.46	---	0.30	1.76	1.90	F-2	378	1.83	---	0.43	2.26	2.46
Vinyl asbestos, 12" x 12"																
Peel and stick																
1/16" T	Inst	SF	F-2	750	1.57	---	0.22	1.79	1.89	F-2	525	1.97	---	0.31	2.27	2.42
0.080" T	Inst	SF	F-2	750	1.73	---	0.22	1.94	2.04	F-2	525	2.16	---	0.31	2.47	2.61
Plain patterns																
1/16" T	Inst	SF	F-2	750	1.15	---	0.22	1.37	1.47	F-2	525	1.44	---	0.31	1.75	1.89
1/8" T	Inst	SF	F-2	750	1.21	---	0.22	1.43	1.53	F-2	525	1.52	---	0.31	1.82	1.97
Solids																
1/16" T	Inst	SF	F-2	750	1.25	---	0.22	1.46	1.56	F-2	525	1.56	---	0.31	1.87	2.01
1/8" T	Inst	SF	F-2	750	1.32	---	0.22	1.54	1.64	F-2	525	1.65	---	0.31	1.96	2.10
Travertine/quarry																
1/8" T	Inst	SF	F-2	750	1.45	---	0.22	1.67	1.77	F-2	525	1.82	---	0.31	2.12	2.27

Description	Oper	Unit	Costs Based On Large Volume							Costs Based On Small Volume						
			Crew Size	Avg Day Prod	Avg Mat'l Unit Cost	Avg Equip Unit Cost	Avg Labor Unit Cost	Avg Total Unit Cost	Avg Price Incl O&P	Crew Size	Avg Day Prod	Avg Mat'l Unit Cost	Avg Equip Unit Cost	Avg Labor Unit Cost	Avg Total Unit Cost	Avg Price Incl O&P

Resilient flooring (continued)

Sheet
Linoleum, 6' wide
Plain patterns

Description	Oper	Unit	Crew	Day	Mat'l	Equip	Labor	Total	Price	Crew	Day	Mat'l	Equip	Labor	Total	Price
3/32" T	Inst	SY	F-2	60	8.89	---	2.70	11.59	12.84	F-2	42	11.12	---	3.86	14.98	16.75
1/8" T	Inst	SY	F-2	60	9.60	---	2.70	12.30	13.55	F-2	42	12.00	---	3.86	15.86	17.64
Designer patterns 3/32" T	Inst	SY	F-2	60	10.73	---	2.70	13.43	14.67	F-2	42	13.41	---	3.86	17.27	19.05
1/8" T	Inst	SY	F-2	60	11.64	---	2.70	14.34	15.59	F-2	42	14.55	---	3.86	18.41	20.19
Vinyl, 6' wide Plain patterns 0.070" T	Inst	SY	F-2	60	12.00	---	2.70	14.70	15.95	F-2	42	15.00	---	3.86	18.86	20.64
0.093" T	Inst	SY	F-2	60	13.68	---	2.70	16.38	17.63	F-2	42	17.10	---	3.86	20.96	22.74

Related materials and operations
Top set base / Vinyl / 2-1/2" H

Colors	Inst	LF	F-2	500	0.46	---	0.32	0.78	0.93	F-2	350	0.57	---	0.46	1.03	1.25
Wood grain	Inst	LF	F-2	500	0.61	---	0.32	0.94	1.09	F-2	350	0.77	---	0.46	1.23	1.44
4" H Colors	Inst	LF	F-2	500	0.53	---	0.32	0.85	1.00	F-2	350	0.66	---	0.46	1.12	1.34
Wood grain	Inst	LF	F-2	500	0.72	---	0.32	1.04	1.19	F-2	350	0.90	---	0.46	1.36	1.58
Linoleum cove, 7/16" x 1-1/4" Softwood	Inst	LF	F-2	350	0.53	---	0.46	0.99	1.20	F-2	245	0.66	---	0.66	1.32	1.63

Retaining walls. See Concrete or Masonry.

Roofing

Aluminum, nailed to wood
Corrugated (2-1/2"), 26" W, with

3-3/4" side lap and 6" end lap	Demo	SQ	L-2	10	---	---	26.01	26.01	38.49	L-2	9.5	---	---	27.38	27.38	40.52

Corrugated (2-1/2"), 26" W, with 3-3/4" side lap and 6" end lap
0.0175" thick

Natural	Inst	SQ	U-3	15	104.51	---	42.70	147.21	167.28	U-3	13	114.97	---	49.27	164.23	187.39
Painted	Inst	SQ	U-3	15	117.69	---	42.70	160.39	180.46	U-3	13	129.46	---	49.27	178.73	201.89

Roofing

225 lb. fiberglass, three tab strip shingles. Three bundle/square; 20 year.

1. **Dimensions.** Each shingle is 12" x 36". With a 5" exposure to the weather, 80 shingles are required to cover one square (100 SF).

2. **Installation.**

   a. Over wood. After scatter-nailing one ply of 15 lb. felt, the shingles are installed. Four nails per shingle is customary.

   b. Over existing roofing. 15 lb. felt is not required. Shingles are installed the same as over wood, except longer nails are used, i.e., 1-1/4" in lieu of 1".

3. **Estimating technique.** Determine roof area and add a percentage for starters, ridge, and valleys. The percent to be added varies, but generally:

   a. For plain gable and hip — add 10%.

   b. For gable or hip with dormers or intersecting roof(s) — add 15%.

   c. For gable or hip with dormers and intersecting roof(s) — add 20%.

When ridge or hip shingles are special ordered, the above percentages should be reduced by approximately 50%. Unit costs are then applied to the area calculated (incl. allowance above).

90 lb. mineral surfaced roll roofing.

1. **Dimensions.** A roll is 36" wide x 36'-0" long, or 108 SF. 8 SF is included for head and end laps. It covers one square (100 SF).

2. **Installation.** Usually, lap cement and 7/8" galvanized nails are furnished with each roll.

   a. Over wood. Roll roofing is usually applied directly to sheathing with 7/8" galvanized nails. End and head laps are normally 6" and 2" or 3" respectively.

   b. Over existing roofing. Applied the same as over wood, except nails are not less than 1-1/4" long.

   c. Roll roofing may be installed by the exposed nail or the concealed nail method. In the exposed nail method, nails are visible at laps, and head laps are usually 2". In the concealed nail method, no nails are visible, the head lap is a minimum of 3", and there is a 9"-wide strip of roll installed along rakes and eaves.

3. **Estimating technique.** Determine the area and add a percentage for hip and/or ridge. The percentage to be added does vary, but generally:

   a. For plain gable or hip — add 5%.

   b. For cut-up roof — add 10%.

If metal ridge and/or hip are used, the above percentages should be reduced by approximately 50%. Unit costs are then applied to the area calculated (incl. allowance above).

Wood Shingles

1. **Dimensions.** Shingles are available in 16", 18", and 24" lengths and in uniform widths of 4", 5", or 6". But they are commonly furnished in random widths averaging 4" wide.

2. **Installation.** The normal exposure to weather for 16", 18" and 24" shingles on roofs with 1/4 or steeper pitch is 5", 5-1/2", and 7-1/2" respectively. Where the slope is less than 1/4, the exposure is usually reduced. Generally, 3d commons are used when shingles are applied to strip sheathing. 5d commons are used when shingles are applied over existing shingles. Two nails per shingle is the general rule, but on some roofs, one nail per shingle is used.

3. **Estimating technique.** Determine the roof area and add a percentage for waste, starters, ridge, and hip shingles.

   a. For material cost.

Shingle Length	Exposure	% of Sq. Required To Cover 100 SF	Lbs. of Nails Per Square (2 Nails/Shingle)	
			3d	5d
24"	7-1/2"	100	2.3	2.7
24"	7"	107	2.4	2.9
24"	6-1/2"	115	2.6	3.2
24"	6"	125	2.8	3.4
24"	5-3/4"	130	2.9	3.6
18"	5-1/2"	100	3.1	3.7
18"	5"	110	3.4	4.1
18"	4-1/2"	122	3.8	4.6
16"	5"	100	3.4	4.1
16"	4-1/2"	111	3.8	4.6
16"	4"	125	4.2	5.1
16"	3-3/4"	133	4.5	5.5

1) Wood shingles. The amount of exposure determines the percentage of one square of shingles required to cover 100 SF of roof. Multiply the material cost per square of shingles by the appropriate percent (given on previous page) to determine the cost to cover 100 SF of roof with wood shingles. The following table does not include cutting waste: NOTE: Nails and the exposure factors in this table have already been calculated into the Average Unit Material Costs:

2) Nails. The weight of nails required per square varies with the size of the nail and with the shingle exposure. Multiply cost per lb by the appropriate lbs per sq. In the above table, lbs of nails per sq. are based on two nails per shingle. For one nail per shingle, deduct 50%.

3) The percentage to be added for starters, hip, and ridge shingles does vary, but generally:

    a) Plain gable and hip — add 10%.

    b) Gable or hip with dormers or intersecting roof(s) — add 15%.

    c) Gable or hip with dormers and intersecting roof(s) — add 20%.

When ridge and/or hip shingles are special ordered, the above percentages should be reduced by 50%. Unit costs are then applied to the area calculated (incl. allowance above).

Built-Up Roofing

1. **Dimensions.**

    a. 15 lb. felt. Rolls are 36" wide x 144'-0" long or 432 SF — four (4) square per roll. 32 SF (or 8 SF/sq) is for laps.

    b. 30 lb felt. rolls are 36" wide x 72'-0" long or 216 SF — two (2) square per roll. 16 SF (or 8 SF/sq) is for laps.

    c. Asphalt. Available in 100 lb cartons. Average asphalt usage as follows:

        1) Top coat without gravel — 35 lbs per square.

        2) Top coat with gravel — 60 lbs per square.

        3) Each coat (except top coat) — 25 lbs per square.

    d. Tar. Available in 550 lb kegs. Average tar usage as follows:

        1) Top coat without gravel — 40 lbs per square.

        2) Top coat without gravel — 75 lbs per square.

        3) Each coat (except top coat) — 30 lbs per square.

    e. Gravel or slag. Normally, 400 lbs of gravel or 275 lbs of slag are used to cover one square.

2. **Installation.**

    a. Over wood. Normally, one or two plies of felt are applied by scatter nailing before hot mopping is commenced. Subsequent plies of felt and gravel or slag are imbedded in hot asphalt or tar.

    b. Over concrete. Every ply of felt and gravel or slag is imbedded in hot asphalt or tar.

3. **Estimating technique.**

    a. For buildings with parapet walls, use outside dimensions of building in determining area. The area, thus determined, is usually sufficient to include the flashing.

    b. For buildings without parapet walls, determine area.

    c. No opening less than 10'-0" x 10'-0" (100 SF) should be deducted. 50% of the opening may be deducted for openings larger than 100 SF but smaller than 300 SF, and 100% of the openings exceeding 300 SF.

36"

12"

7" lap

5" to weather

Tab

Tab notch

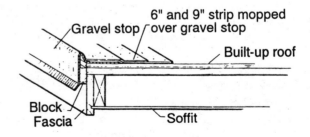

6" and 9" strip mopped over gravel stop

Gravel stop

Built-up roof

Block
Fascia

Soffit

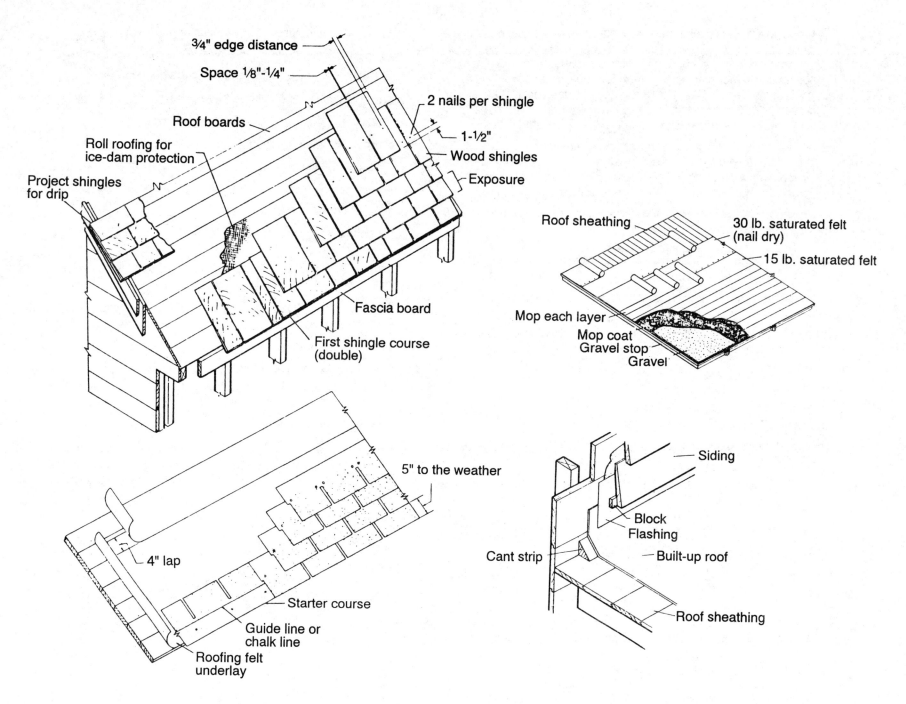

¾" edge distance

Space ⅛"-¼"

Roof boards

Roll roofing for
ice-dam protection

2 nails per shingle

1-½"

Wood shingles

Exposure

Project shingles
for drip

Fascia board

First shingle course
(double)

Roof sheathing

30 lb. saturated felt
(nail dry)

15 lb. saturated felt

Mop each layer

Mop coat
Gravel stop
Gravel

5" to the weather

4" lap

Starter course

Guide line or
chalk line

Roofing felt
underlay

Siding

Block
Flashing

Built-up roof

Cant strip

Roof sheathing

216

Description	Oper	Unit	Costs Based On Large Volume							Costs Based On Small Volume						
			Crew Size	Avg Day Prod	Avg Mat'l Unit Cost	Avg Equip Unit Cost	Avg Labor Unit Cost	Avg Total Unit Cost	Avg Price Incl O&P	Crew Size	Avg Day Prod	Avg Mat'l Unit Cost	Avg Equip Unit Cost	Avg Labor Unit Cost	Avg Total Unit Cost	Avg Price Incl O&P

Roofing (continued)

0.019" thick

Description	Oper	Unit	Crew Size	Avg Day Prod	Avg Mat'l Unit Cost	Avg Equip Unit Cost	Avg Labor Unit Cost	Avg Total Unit Cost	Avg Price Incl O&P	Crew Size	Avg Day Prod	Avg Mat'l Unit Cost	Avg Equip Unit Cost	Avg Labor Unit Cost	Avg Total Unit Cost	Avg Price Incl O&P
Natural	Inst	SQ	U-3	15	108.91	---	42.70	151.61	171.68	U-3	13	119.80	---	49.27	169.07	192.22
Painted	Inst	SQ	U-3	15	122.08	---	42.70	164.78	184.85	U-3	13	134.29	---	49.27	183.56	206.72

Composition. Asphalt or Fiberglass

Shingle, 240#/SQ., strip, 3 tab,

Description	Oper	Unit	Crew Size	Avg Day Prod	Avg Mat'l Unit Cost	Avg Equip Unit Cost	Avg Labor Unit Cost	Avg Total Unit Cost	Avg Price Incl O&P	Crew Size	Avg Day Prod	Avg Mat'l Unit Cost	Avg Equip Unit Cost	Avg Labor Unit Cost	Avg Total Unit Cost	Avg Price Incl O&P
5" exp.	Demo	SQ	L-2	16	---	---	16.26	16.26	24.06	L-2	14.5	---	---	17.94	17.94	26.55

Shingles nailed; 3 tab strips, 12" x 36" with 5" exp.; 240#/SQ, seal down type

Over existing roofing

Description	Oper	Unit	Crew Size	Avg Day Prod	Avg Mat'l Unit Cost	Avg Equip Unit Cost	Avg Labor Unit Cost	Avg Total Unit Cost	Avg Price Incl O&P	Crew Size	Avg Day Prod	Avg Mat'l Unit Cost	Avg Equip Unit Cost	Avg Labor Unit Cost	Avg Total Unit Cost	Avg Price Incl O&P
Gable, plain	Inst	SQ	R-6	15.0	33.05	---	24.68	57.73	71.05	R-6	13.5	36.36	---	27.42	63.78	78.58
Gable with dormers	Inst	SQ	R-6	13.5	33.65	---	27.42	61.07	75.88	R-6	12.2	37.02	---	30.34	67.36	83.74
Gable with intersecting roofs	Inst	SQ	R-6	13.5	33.65	---	27.42	61.07	75.88	R-6	12.2	37.02	---	30.34	67.36	83.74
Gable with dormers & intersections	Inst	SQ	R-6	12.0	33.95	---	30.85	64.80	81.46	R-6	10.8	37.35	---	34.28	71.62	90.13
Hip, plain	Inst	SQ	R-6	14.3	33.35	---	25.83	59.18	73.13	R-6	13.0	36.69	---	28.48	65.16	80.54
Hip with dormers	Inst	SQ	R-6	12.8	33.95	---	28.92	62.87	78.49	R-6	11.5	37.35	---	32.19	69.53	86.92
Hip with intersecting roofs	Inst	SQ	R-6	12.8	33.95	---	28.92	62.87	78.49	R-6	11.5	37.35	---	32.19	69.53	86.92
Hip with dormers & intersections	Inst	SQ	R-6	11.3	34.25	---	32.76	67.01	84.70	R-6	10.2	37.68	---	36.29	73.97	93.56

Over wood decks

Description	Oper	Unit	Crew Size	Avg Day Prod	Avg Mat'l Unit Cost	Avg Equip Unit Cost	Avg Labor Unit Cost	Avg Total Unit Cost	Avg Price Incl O&P	Crew Size	Avg Day Prod	Avg Mat'l Unit Cost	Avg Equip Unit Cost	Avg Labor Unit Cost	Avg Total Unit Cost	Avg Price Incl O&P
Gable, plain	Inst	SQ	R-6	14.0	36.76	---	26.44	63.20	77.48	R-6	12.6	40.44	---	29.38	69.82	85.68
Gable with dormers	Inst	SQ	R-6	12.5	37.36	---	29.61	66.98	82.97	R-6	11.3	41.10	---	32.76	73.86	91.55
Gable with intersecting roofs	Inst	SQ	R-6	12.5	37.36	---	29.61	66.98	82.97	R-6	11.3	41.10	---	32.76	73.86	91.55
Gable with dormers & intersections	Inst	SQ	R-6	11.0	37.66	---	33.65	71.31	89.49	R-6	10.0	41.43	---	37.02	78.45	98.44
Hip, plain	Inst	SQ	R-6	13.3	37.06	---	27.77	64.83	79.83	R-6	12.0	40.77	---	30.85	71.62	88.27
Hip with dormers	Inst	SQ	R-6	11.8	37.66	---	31.37	69.03	85.97	R-6	10.6	41.43	---	34.92	76.35	95.21
Hip with intersecting roofs	Inst	SQ	R-6	11.8	37.66	---	31.37	69.03	85.97	R-6	10.6	41.43	---	34.92	76.35	95.21
Hip with dormers & intersections	Inst	SQ	R-6	10.3	37.96	---	35.94	73.90	93.31	R-6	9.3	41.76	---	39.80	81.56	103.06

Related materials and operations

Ridge or Hip roll, 90# mineral

Description	Oper	Unit	Crew Size	Avg Day Prod	Avg Mat'l Unit Cost	Avg Equip Unit Cost	Avg Labor Unit Cost	Avg Total Unit Cost	Avg Price Incl O&P	Crew Size	Avg Day Prod	Avg Mat'l Unit Cost	Avg Equip Unit Cost	Avg Labor Unit Cost	Avg Total Unit Cost	Avg Price Incl O&P
surfaced	Inst	LF	R-2	800	0.06	---	0.38	0.44	0.65	R-2	735	0.07	---	0.42	0.48	0.71
Valley roll, 90# mineral surfaced	Inst	LF	R-2	800	0.14	---	0.38	0.52	0.73	R-2	735	0.16	---	0.42	0.57	0.80

			Costs Based On Large Volume							Costs Based On Small Volume					
Description	Oper	Unit	Crew Size	Avg Day Prod	Avg Mat'l Unit Cost	Avg Equip Unit Cost	Avg Labor Unit Cost	Avg Total Unit Cost	Avg Price Incl O&P	Crew Size	Avg Day Prod	Avg Mat'l Unit Cost	Avg Equip Unit Cost	Avg Labor Unit Cost	Avg Total Unit Cost	Avg Price Incl O&P

Roofing (continued)

Ridge or Hip shingles, 9" x 12"
with 5" exp.; 80 pieces/bundle @ 33.3 LF/bundle

Description	Oper	Unit	Crew Size	Avg Day Prod	Avg Mat'l Unit Cost	Avg Equip Unit Cost	Avg Labor Unit Cost	Avg Total Unit Cost	Avg Price Incl O&P	Crew Size	Avg Day Prod	Avg Mat'l Unit Cost	Avg Equip Unit Cost	Avg Labor Unit Cost	Avg Total Unit Cost	Avg Price Incl O&P
Over existing roofing	Inst	LF	R-2	700	0.47	---	0.44	0.91	1.14	R-2	645	0.52	---	0.47	0.99	1.25
Over wood decks	Inst	LF	R-2	660	0.47	---	0.46	0.93	1.18	R-2	605	0.52	---	0.50	1.02	1.30

Built-up or membrane roofing over existing roofing materials

Description	Oper	Unit	Crew Size	Avg Day Prod	Avg Mat'l Unit Cost	Avg Equip Unit Cost	Avg Labor Unit Cost	Avg Total Unit Cost	Avg Price Incl O&P	Crew Size	Avg Day Prod	Avg Mat'l Unit Cost	Avg Equip Unit Cost	Avg Labor Unit Cost	Avg Total Unit Cost	Avg Price Incl O&P
One mop coat over smooth surface																
with asphalt	Inst	SQ	R-17	135	6.75	---	6.76	13.51	17.16	R-17	120	7.43	---	7.60	15.03	19.13
with tar	Inst	SQ	R-17	135	11.40	---	6.76	18.16	21.81	R-17	120	12.54	---	7.60	20.14	24.25
Remove gravel, mop one coat, redistribute old gravel																
with asphalt	Inst	SQ	R-17	53	10.50	---	17.21	27.71	37.01	R-17	45	11.55	---	20.27	31.82	42.77
with tar	Inst	SQ	R-17	53	17.60	---	17.21	34.81	44.11	R-17	45	19.36	---	20.27	39.63	50.58
Mop in one 30# cap sheet, plus smooth top mop coat																
with asphalt	Inst	SQ	R-17	57	18.43	---	16.01	34.43	43.07	R-17	50	20.27	---	18.25	38.51	48.37
with tar	Inst	SQ	R-17	57	25.73	---	16.01	41.73	50.37	R-17	50	28.30	---	18.25	46.54	56.40
Remove gravel, mop in one 30# cap sheet, mop in and redistribute old gravel																
with asphalt	Inst	SQ	R-17	33	22.18	---	27.65	49.82	64.75	R-17	30	24.39	---	30.41	54.80	71.23
with tar	Inst	SQ	R-17	33	31.93	---	27.65	59.57	74.50	R-17	30	35.12	---	30.41	65.53	81.95
Mop in two-ply 15# felt, plus smooth top mop coat																
with asphalt	Inst	SQ	R-17	36	22.18	---	25.34	47.52	61.20	R-17	33	24.39	---	27.65	52.04	66.97
with tar	Inst	SQ	R-17	36	32.13	---	25.34	57.47	71.15	R-17	33	35.34	---	27.65	62.98	77.91
Remove gravel, mop in two-ply 15# felt, mop in and redistribute old gravel																
with asphalt	Inst	SQ	R-17	25	25.93		36.49	62.42	82.13	R-17	23	28.52	---	39.67	68.18	89.60
with tar	Inst	SQ	R-17	25	38.33	---	36.49	74.82	94.53	R-17	23	42.16	---	39.67	81.82	103.24

Built-up or membrane roofing over smooth wood or concrete deck

Description	Oper	Unit	Crew Size	Avg Day Prod	Avg Mat'l Unit Cost	Avg Equip Unit Cost	Avg Labor Unit Cost	Avg Total Unit Cost	Avg Price Incl O&P	Crew Size	Avg Day Prod	Avg Mat'l Unit Cost	Avg Equip Unit Cost	Avg Labor Unit Cost	Avg Total Unit Cost	Avg Price Incl O&P
3 ply																
with gravel	Demo	SQ	L-2	11	---	---	23.64	23.64	34.99	L-2	10	---	---	26.01	26.01	38.49
without gravel	Demo	SQ	L-2	14	---	---	18.58	18.58	27.49	L-2	13	---	---	20.01	20.01	29.61
5 ply																
with gravel	Demo	SQ	L-2	10	---	---	26.01	26.01	38.49	L-2	9	---	---	28.90	28.90	42.77
without gravel	Demo	SQ	L-2	13	---	---	20.01	20.01	29.61	L-2	12	---	---	21.67	21.67	32.08

Costs Based On Large Volume   Costs Based On Small Volume

Description	Oper	Unit	Crew Size	Avg Day Prod	Avg Mat'l Unit Cost	Avg Equip Unit Cost	Avg Labor Unit Cost	Avg Total Unit Cost	Avg Price Incl O&P	Crew Size	Avg Day Prod	Avg Mat'l Unit Cost	Avg Equip Unit Cost	Avg Labor Unit Cost	Avg Total Unit Cost	Avg Price Incl O&P

Roofing (continued)

Installed over smooth wood decks

Nail one-ply 15# felt, plus mop in one 90# mineral surfaced ply

Description	Oper	Unit	Crew Size	Avg Day Prod	Avg Mat'l Unit Cost	Avg Equip Unit Cost	Avg Labor Unit Cost	Avg Total Unit Cost	Avg Price Incl O&P	Crew Size	Avg Day Prod	Avg Mat'l Unit Cost	Avg Equip Unit Cost	Avg Labor Unit Cost	Avg Total Unit Cost	Avg Price Incl O&P
with asphalt	Inst	SQ	R-17	80	27.46	---	11.40	38.87	45.03	R-17	70	30.21	---	13.03	43.24	50.28
without tar	Inst	SQ	R-17	80	30.56	---	11.40	41.97	48.13	R-17	70	33.62	---	13.03	46.65	53.69

Nail one and mop in two-ply 15# felt, plus smooth top mop coat

Description	Oper	Unit	Crew Size	Avg Day Prod	Avg Mat'l Unit Cost	Avg Equip Unit Cost	Avg Labor Unit Cost	Avg Total Unit Cost	Avg Price Incl O&P	Crew Size	Avg Day Prod	Avg Mat'l Unit Cost	Avg Equip Unit Cost	Avg Labor Unit Cost	Avg Total Unit Cost	Avg Price Incl O&P
with asphalt	Inst	SQ	R-17	33	27.14	---	27.65	54.78	69.71	R-17	30	29.85	---	30.41	60.26	76.68
without tar	Inst	SQ	R-17	33	37.09	---	27.65	64.73	79.66	R-17	30	40.80	---	30.41	71.21	87.63

Nail one and mop in two-ply 15# felt, plus mop in and distribute gravel

Description	Oper	Unit	Crew Size	Avg Day Prod	Avg Mat'l Unit Cost	Avg Equip Unit Cost	Avg Labor Unit Cost	Avg Total Unit Cost	Avg Price Incl O&P	Crew Size	Avg Day Prod	Avg Mat'l Unit Cost	Avg Equip Unit Cost	Avg Labor Unit Cost	Avg Total Unit Cost	Avg Price Incl O&P
with asphalt	Inst	SQ	R-17	25	36.29	---	36.49	72.78	92.49	R-17	23	39.92	---	39.67	79.58	101.00
without tar	Inst	SQ	R-17	25	48.69	---	36.49	85.18	104.89	R-17	23	53.56	---	39.67	93.22	114.64

Nail one and mop in three-ply 15#, plus mop in and distribute gravel

Description	Oper	Unit	Crew Size	Avg Day Prod	Avg Mat'l Unit Cost	Avg Equip Unit Cost	Avg Labor Unit Cost	Avg Total Unit Cost	Avg Price Incl O&P	Crew Size	Avg Day Prod	Avg Mat'l Unit Cost	Avg Equip Unit Cost	Avg Labor Unit Cost	Avg Total Unit Cost	Avg Price Incl O&P
with asphalt	Inst	SQ	R-17	20	44.00	---	45.62	89.62	114.25	R-17	18	48.40	---	50.69	99.09	126.46
without tar	Inst	SQ	R-17	20	59.05	---	45.62	104.67	129.30	R-17	18	64.96	---	50.69	115.64	143.01

Nail one and mop in four-ply 15#, plus mop in and distribute gravel

Description	Oper	Unit	Crew Size	Avg Day Prod	Avg Mat'l Unit Cost	Avg Equip Unit Cost	Avg Labor Unit Cost	Avg Total Unit Cost	Avg Price Incl O&P	Crew Size	Avg Day Prod	Avg Mat'l Unit Cost	Avg Equip Unit Cost	Avg Labor Unit Cost	Avg Total Unit Cost	Avg Price Incl O&P
with asphalt	Inst	SQ	R-17	16	51.71	---	57.02	108.73	139.53	R-17	14	56.88	---	65.17	122.05	157.24
without tar	Inst	SQ	R-17	16	69.41	---	57.02	126.43	157.23	R-17	14	76.35	---	65.17	141.52	176.71

Installed over smooth concrete decks

Mop in one-ply 15# felt, plus mop in one 90# mineral surfaced ply

Description	Oper	Unit	Crew Size	Avg Day Prod	Avg Mat'l Unit Cost	Avg Equip Unit Cost	Avg Labor Unit Cost	Avg Total Unit Cost	Avg Price Incl O&P	Crew Size	Avg Day Prod	Avg Mat'l Unit Cost	Avg Equip Unit Cost	Avg Labor Unit Cost	Avg Total Unit Cost	Avg Price Incl O&P
with asphalt	Inst	SQ	R-17	62	31.21	---	14.72	45.93	53.87	R-17	55	34.33	---	16.59	50.92	59.88
without tar	Inst	SQ	R-17	62	36.76	---	14.72	51.48	59.42	R-17	55	40.44	---	16.59	57.03	65.98

Mop in three-ply 15# felt, plus smooth top mop coat

Description	Oper	Unit	Crew Size	Avg Day Prod	Avg Mat'l Unit Cost	Avg Equip Unit Cost	Avg Labor Unit Cost	Avg Total Unit Cost	Avg Price Incl O&P	Crew Size	Avg Day Prod	Avg Mat'l Unit Cost	Avg Equip Unit Cost	Avg Labor Unit Cost	Avg Total Unit Cost	Avg Price Incl O&P
with asphalt	Inst	SQ	R-17	27	29.89	---	33.79	63.68	81.92	R-17	25	32.88	---	36.49	69.37	89.08
without tar	Inst	SQ	R-17	27	42.49	---	33.79	76.28	94.52	R-17	25	46.74	---	36.49	83.23	102.94

Mop in three-ply 15# felt, plus mop in and distribute gravel

Description	Oper	Unit	Crew Size	Avg Day Prod	Avg Mat'l Unit Cost	Avg Equip Unit Cost	Avg Labor Unit Cost	Avg Total Unit Cost	Avg Price Incl O&P	Crew Size	Avg Day Prod	Avg Mat'l Unit Cost	Avg Equip Unit Cost	Avg Labor Unit Cost	Avg Total Unit Cost	Avg Price Incl O&P
with asphalt	Inst	SQ	R-17	21	39.04	---	43.44	82.48	105.94	R-17	19	42.94	---	48.02	90.96	116.89
without tar	Inst	SQ	R-17	21	54.09	---	43.44	97.53	120.99	R-17	19	59.50	---	48.02	107.51	133.44

Mop in four-ply 15# felt, plus mop in and distribute gravel

Description	Oper	Unit	Crew Size	Avg Day Prod	Avg Mat'l Unit Cost	Avg Equip Unit Cost	Avg Labor Unit Cost	Avg Total Unit Cost	Avg Price Incl O&P	Crew Size	Avg Day Prod	Avg Mat'l Unit Cost	Avg Equip Unit Cost	Avg Labor Unit Cost	Avg Total Unit Cost	Avg Price Incl O&P
with asphalt	Inst	SQ	R-17	17	46.75	---	53.67	100.42	129.40	R-17	15	51.43	---	60.82	112.25	145.09
without tar	Inst	SQ	R-17	17	64.45	---	53.67	118.12	147.10	R-17	15	70.90	---	60.82	131.72	164.56

Mop in five-ply 15# felt, plus mop in and distribute gravel

Description	Oper	Unit	Crew Size	Avg Day Prod	Avg Mat'l Unit Cost	Avg Equip Unit Cost	Avg Labor Unit Cost	Avg Total Unit Cost	Avg Price Incl O&P	Crew Size	Avg Day Prod	Avg Mat'l Unit Cost	Avg Equip Unit Cost	Avg Labor Unit Cost	Avg Total Unit Cost	Avg Price Incl O&P
with asphalt	Inst	SQ	R-17	15	54.46	---	60.82	115.29	148.13	R-17	13	59.91	---	70.18	130.09	167.99
without tar	Inst	SQ	R-17	15	74.81	---	60.82	135.64	168.48	R-17	13	82.29	---	70.18	152.47	190.37

Description	Oper	Unit	Costs Based On Large Volume							Costs Based On Small Volume						
			Crew Size	Avg Day Prod	Avg Mat'l Unit Cost	Avg Equip Unit Cost	Avg Labor Unit Cost	Avg Total Unit Cost	Avg Price Incl O&P	Crew Size	Avg Day Prod	Avg Mat'l Unit Cost	Avg Equip Unit Cost	Avg Labor Unit Cost	Avg Total Unit Cost	Avg Price Incl O&P

Roofing (continued)

Clay tile
2 piece interlocking	Demo	SQ	L-2	10	---	---	26.01	26.01	38.49	L-2	9	---	---	28.90	28.90	42.77
1 piece	Demo	SQ	L-2	11	---	---	23.64	23.64	34.99	L-2	10	---	---	26.01	26.01	38.49

Clay tile; over wood; includes felt
Interlocking tile shingles

Early American	Inst	SQ	R-7	5.4	126.43	---	80.59	207.02	245.70	R-7	5.0	139.07	---	87.04	226.11	267.89
Lanai	Inst	SQ	R-7	5.4	155.30	---	80.59	235.89	274.58	R-7	5.0	170.83	---	87.04	257.87	299.65

Tile Shingles

Mission	Inst	SQ	R-7	3.0	74.45	---	145.07	219.52	289.15	R-7	2.8	81.90	---	155.43	237.32	311.93
Spanish	Inst	SQ	R-7	3.0	97.55	---	145.07	242.62	312.25	R-7	2.8	107.31	---	155.43	262.73	337.34

Mineral surfaced roll
Single coverage 90#/SQ roll, with

6" end lap and 2" head lap	Demo	SQ	L-2	32	---	---	8.13	8.13	12.03	L-2	30	---	---	8.67	8.67	12.83

Single coverage roll on
Plain gable over
Existing roofing with

Nails concealed	Inst	SQ	R-4	23	21.95	---	14.68	36.63	43.67	R-4	21	24.14	---	16.08	40.22	47.94
Nails exposed	Inst	SQ	R-4	25	19.95	---	13.51	33.46	39.94	R-4	23	21.95	---	14.68	36.63	43.67

Wood deck with

Nails concealed	Inst	SQ	R-4	22	22.57	---	15.35	37.92	45.29	R-4	20	24.83	---	16.88	41.71	49.82
Nails exposed	Inst	SQ	R-4	24	20.52	---	14.07	34.59	41.34	R-4	22	22.57	---	15.35	37.92	45.29

Plain hip over
Existing roofing with

Nails concealed	Inst	SQ	R-4	22	22.57	---	15.35	37.92	45.29	R-4	20	24.83	---	16.88	41.71	49.82
Nails exposed	Inst	SQ	R-4	24	20.52	---	14.07	34.59	41.34	R-4	22	22.57	---	15.35	37.92	45.29

Wood deck with

Nails concealed	Inst	SQ	R-4	20	22.99	---	16.88	39.87	47.98	R-4	18	25.29	---	18.76	44.05	53.05
Nails exposed	Inst	SQ	R-4	23	20.90	---	14.68	35.58	42.63	R-4	21	22.99	---	16.08	39.07	46.79

Double coverage selvage roll, with

19" lap and 17" exposure	Demo	SQ	L-2	22	---	---	11.82	11.82	17.50	L-2	20	---	---	13.00	13.00	19.25

			Costs Based On Large Volume							Costs Based On Small Volume						
Description	Oper	Unit	Crew Size	Avg Day Prod	Avg Mat'l Unit Cost	Avg Equip Unit Cost	Avg Labor Unit Cost	Avg Total Unit Cost	Avg Price Incl O&P	Crew Size	Avg Day Prod	Avg Mat'l Unit Cost	Avg Equip Unit Cost	Avg Labor Unit Cost	Avg Total Unit Cost	Avg Price Incl O&P

Roofing (continued)

Double coverage selvage roll, with 19" lap and 17" exposure (2 rolls/SQ)
Plain gable over
Wood deck with

Description	Oper	Unit	Crew Size	Avg Day Prod	Avg Mat'l Unit Cost	Avg Equip Unit Cost	Avg Labor Unit Cost	Avg Total Unit Cost	Avg Price Incl O&P	Crew Size	Avg Day Prod	Avg Mat'l Unit Cost	Avg Equip Unit Cost	Avg Labor Unit Cost	Avg Total Unit Cost	Avg Price Incl O&P
Nails exposed	Inst	SQ	R-4	14	45.89	---	24.12	70.00	81.58	R-4	13	50.47	---	25.97	76.45	88.92

Related materials and operations
Starter strips (36' L rolls) along eaves and up rakes and gable ends on new roofing over wood decks

Description	Oper	Unit	Crew Size	Avg Day Prod	Avg Mat'l Unit Cost	Avg Equip Unit Cost	Avg Labor Unit Cost	Avg Total Unit Cost	Avg Price Incl O&P	Crew Size	Avg Day Prod	Avg Mat'l Unit Cost	Avg Equip Unit Cost	Avg Labor Unit Cost	Avg Total Unit Cost	Avg Price Incl O&P
9" W	Inst	LF	---	---	0.22	---	---	0.22	0.22	---	---	0.24	---	---	0.24	0.24
12" W	Inst	LF	---	---	0.28	---	---	0.28	0.28	---	---	0.31	---	---	0.31	0.31
18" W	Inst	LF	---	---	0.33	---	---	0.33	0.33	---	---	0.37	---	---	0.37	0.37
24" W	Inst	LF	---	---	0.44	---	---	0.44	0.44	---	---	0.49	---	---	0.49	0.49

Wood

Shakes
24" L with 10" exposure

Description	Oper	Unit	Crew Size	Avg Day Prod	Avg Mat'l Unit Cost	Avg Equip Unit Cost	Avg Labor Unit Cost	Avg Total Unit Cost	Avg Price Incl O&P	Crew Size	Avg Day Prod	Avg Mat'l Unit Cost	Avg Equip Unit Cost	Avg Labor Unit Cost	Avg Total Unit Cost	Avg Price Incl O&P
1/2" to 3/4" T	Demo	SQ	L-2	25.0	---	---	10.40	10.40	15.40	L-2	23.0	---	---	11.31	11.31	16.74
3/4" to 5/4" T	Demo	SQ	L-2	22.5	---	---	11.56	11.56	17.11	L-2	21.0	---	---	12.38	12.38	18.33

Shakes, over wood deck; 2 nails/shake, 24" L with 10" exp., 6" W (avg.), red cedar, sawn one side
Gable

Description	Oper	Unit	Crew Size	Avg Day Prod	Avg Mat'l Unit Cost	Avg Equip Unit Cost	Avg Labor Unit Cost	Avg Total Unit Cost	Avg Price Incl O&P	Crew Size	Avg Day Prod	Avg Mat'l Unit Cost	Avg Equip Unit Cost	Avg Labor Unit Cost	Avg Total Unit Cost	Avg Price Incl O&P
1/2" to 3/4" T	Inst	SQ	R-14	12.6	87.85	---	34.47	122.32	138.86	R-14	11.6	96.64	---	37.47	134.10	152.08
3/4" to 5/4" T	Inst	SQ	R-15	11.3	132.05	---	39.87	171.92	191.06	R-15	10.4	145.26	---	43.34	188.59	209.40
Gable with dormers																
1/2" to 3/4" T	Inst	SQ	R-14	12.0	89.49	---	36.19	125.68	143.05	R-14	11.0	98.44	---	39.34	137.78	156.66
3/4" to 5/4" T	Inst	SQ	R-15	10.7	134.51	---	42.11	176.62	196.83	R-15	9.8	147.96	---	45.77	193.73	215.70
Gable with valleys																
1/2" to 3/4" T	Inst	SQ	R-14	12.0	89.49	---	36.19	125.68	143.05	R-14	11.0	98.44	---	39.34	137.78	156.66
3/4" to 5/4" T	Inst	SQ	R-15	10.7	134.51	---	42.11	176.62	196.83	R-15	9.8	147.96	---	45.77	193.73	215.70
Hip																
1/2" to 3/4" T	Inst	SQ	R-14	12.0	89.49	---	36.19	125.68	143.05	R-14	11.0	98.44	---	39.34	137.78	156.66
3/4" to 5/4" T	Inst	SQ	R-15	10.7	134.51	---	42.11	176.62	196.83	R-15	9.8	147.96	---	45.77	193.73	215.70
Hip with valleys																
1/2" to 3/4" T	Inst	SQ	R-14	11.3	91.13	---	38.43	129.56	148.01	R-14	10.4	100.24	---	41.78	142.02	162.07
3/4" to 5/4" T	Inst	SQ	R-15	10.2	136.97	---	44.17	181.14	202.34	R-15	9.4	150.67	---	48.01	198.68	221.73

			Costs Based On Large Volume							Costs Based On Small Volume						
Description	Oper	Unit	Crew Size	Avg Day Prod	Avg Mat'l Unit Cost	Avg Equip Unit Cost	Avg Labor Unit Cost	Avg Total Unit Cost	Avg Price Incl O&P	Crew Size	Avg Day Prod	Avg Mat'l Unit Cost	Avg Equip Unit Cost	Avg Labor Unit Cost	Avg Total Unit Cost	Avg Price Incl O&P

Roofing (continued)

Related materials and operations
Ridge/Hip units, 10" W with

Description	Oper	Unit	Crew Size	Avg Day Prod	Avg Mat'l Unit Cost	Avg Equip Unit Cost	Avg Labor Unit Cost	Avg Total Unit Cost	Avg Price Incl O&P	Crew Size	Avg Day Prod	Avg Mat'l Unit Cost	Avg Equip Unit Cost	Avg Labor Unit Cost	Avg Total Unit Cost	Avg Price Incl O&P
10" exp., 20 pcs/bundle	Inst	LF	R-9	900	0.99	---	0.37	1.37	1.55	R-9	825	1.09	---	0.41	1.50	1.70
Roll valley, galvanized, unpainted, 28 gauge, 50' L rolls																
14" W	Inst	LF	U-1	120	0.74	---	1.59	2.32	3.07	U-1	110	0.81	---	1.73	2.54	3.36
20" W	Inst	LF	U-1	120	1.16	---	1.59	2.74	3.49	U-1	110	1.28	---	1.73	3.00	3.82
Rosin sized sheathing paper 36" W, 500 SF/roll; nailed																
Over open sheathing	Inst	SQ	R-9	95	2.84	---	3.54	6.38	8.08	R-9	85	3.12	---	3.96	7.08	8.99
Over solid sheathing	Inst	SQ	R-9	140	2.84	---	2.41	5.24	6.40	R-9	125	3.12	---	2.69	5.82	7.11
Shingles																
16" L with 5" exposure	Demo	SQ	L-2	12.0	---	---	21.67	21.67	32.08	L-2	11.0	---	---	23.56	23.56	34.87
18" L with 5-1/2" exposure	Demo	SQ	L-2	12.6	---	---	20.64	20.64	30.55	L-2	11.6	---	---	22.44	22.44	33.21
24" L with 7-1/2" exposure	Demo	SQ	L-2	18.0	---	---	14.45	14.45	21.38	L-2	16.6	---	---	15.71	15.71	23.24
Shingles, red cedar, No. 1 perfect, 4" W (avg.), 2 nails per shingle																
Over existing roofing materials on																
Gable, plain																
16" L with 5" exposure	Inst	SQ	R-11	5.6	137.85	---	68.84	206.69	239.74	R-11	5.2	151.64	---	74.83	226.47	262.39
18" L with 5-1/2" exposure	Inst	SQ	R-11	6.2	157.47	---	62.18	219.65	249.50	R-11	5.7	173.22	---	67.59	240.81	273.25
24" L with 7-1/2" exposure	Inst	SQ	R-11	8.6	140.93	---	44.83	185.76	207.28	R-11	7.9	155.03	---	48.73	203.75	227.14
Gable with dormers																
16" L with 5" exposure	Inst	SQ	R-11	5.4	139.05	---	71.39	210.44	244.71	R-11	5.0	152.96	---	77.60	230.56	267.81
18" L with 5-1/2" exposure	Inst	SQ	R-11	6.0	158.86	---	64.26	223.11	253.95	R-11	5.5	174.74	---	69.84	244.58	278.11
24" L with 7-1/2" exposure	Inst	SQ	R-11	6.3	142.16	---	61.20	203.36	232.73	R-11	5.8	156.38	---	66.52	222.89	254.82
Gable with intersecting roofs																
16" L with 5" exposure	Inst	SQ	R-11	5.4	139.05	---	71.39	210.44	244.71	R-11	5.0	152.96	---	77.60	230.56	267.81
18" L with 5-1/2" exposure	Inst	SQ	R-11	6.0	158.86	---	64.26	223.11	253.95	R-11	5.5	174.74	---	69.84	244.58	278.11
24" L with 7-1/2" exposure	Inst	SQ	R-11	6.3	142.16	---	61.20	203.36	232.73	R-11	5.8	156.38	---	66.52	222.89	254.82
Gable with dormers & intersecting roofs																
16" L with 5" exposure	Inst	SQ	R-11	5.2	141.45	---	74.14	215.59	251.18	R-11	4.8	155.60	---	80.59	236.18	274.86
18" L with 5-1/2" exposure	Inst	SQ	R-11	5.8	161.62	---	66.47	228.09	260.00	R-11	5.3	177.78	---	72.25	250.03	284.72
24" L with 7-1/2" exposure	Inst	SQ	R-11	8.1	144.62	---	47.60	192.21	215.06	R-11	7.5	159.08	---	51.74	210.81	235.65

			Costs Based On Large Volume							Costs Based On Small Volume					
Description	Oper	Unit	Crew Size	Avg Day Prod	Avg Mat'l Unit Cost	Avg Equip Unit Cost	Avg Labor Unit Cost	Avg Total Unit Cost	Avg Price Incl O&P	Crew Size	Avg Day Prod	Avg Mat'l Unit Cost	Avg Equip Unit Cost	Avg Labor Unit Cost	Avg Total Unit Cost	Avg Price Incl O&P

Roofing (continued)

Hip, plain

Description	Oper	Unit	Crew Size	Avg Day Prod	Avg Mat'l Unit Cost	Avg Equip Unit Cost	Avg Labor Unit Cost	Avg Total Unit Cost	Avg Price Incl O&P	Crew Size	Avg Day Prod	Avg Mat'l Unit Cost	Avg Equip Unit Cost	Avg Labor Unit Cost	Avg Total Unit Cost	Avg Price Incl O&P
16" L with 5" exposure	Inst	SQ	R-11	5.4	139.05	---	71.39	210.44	244.71	R-11	5.0	152.96	---	77.60	230.56	267.81
18" L with 5-1/2" exposure	Inst	SQ	R-11	6.0	158.86	---	64.26	223.11	253.95	R-11	5.5	174.74	---	69.84	244.58	278.11
24" L with 7-1/2" exposure	Inst	SQ	R-11	8.3	142.16	---	46.45	188.61	210.91	R-11	7.6	156.38	---	50.49	206.86	231.10
Hip with dormers																
16" L with 5" exposure	Inst	SQ	R-11	5.2	140.25	---	74.14	214.39	249.98	R-11	4.8	154.28	---	80.59	234.86	273.54
18" L with 5-1/2" exposure	Inst	SQ	R-11	5.7	160.24	---	67.64	227.88	260.34	R-11	5.2	176.26	---	73.52	249.78	285.07
24" L with 7-1/2" exposure	Inst	SQ	R-11	7.9	143.39	---	48.80	192.19	215.61	R-11	7.3	157.73	---	53.05	210.77	236.23
Hip with intersecting roofs																
16" L with 5" exposure	Inst	SQ	R-11	5.2	140.25	---	74.14	214.39	249.98	R-11	4.8	154.28	---	80.59	234.86	273.54
18" L with 5-1/2" exposure	Inst	SQ	R-11	5.7	160.24	---	67.64	227.88	260.34	R-11	5.2	176.26	---	73.52	249.78	285.07
24" L with 7-1/2" exposure	Inst	SQ	R-11	7.9	143.39	---	48.80	192.19	215.61	R-11	7.3	157.73	---	53.05	210.77	236.23
Hip with dormers & intersecting roofs																
16" L with 5" exposure	Inst	SQ	R-11	5.0	142.65	---	77.11	219.76	256.77	R-11	4.6	156.92	---	83.81	240.73	280.96
18" L with 5-1/2" exposure	Inst	SQ	R-11	5.5	163.00	---	70.10	233.10	266.75	R-11	5.1	179.31	---	76.19	255.50	292.07
24" L with 7-1/2" exposure	Inst	SQ	R-11	7.7	145.84	---	50.07	195.91	219.95	R-11	7.1	160.43	---	54.42	214.85	240.98

Over wood decks on

Gable, plain

Description	Oper	Unit	Crew Size	Avg Day Prod	Avg Mat'l Unit Cost	Avg Equip Unit Cost	Avg Labor Unit Cost	Avg Total Unit Cost	Avg Price Incl O&P	Crew Size	Avg Day Prod	Avg Mat'l Unit Cost	Avg Equip Unit Cost	Avg Labor Unit Cost	Avg Total Unit Cost	Avg Price Incl O&P
16" L with 5" exposure	Inst	SQ	R-11	6.0	134.79	---	64.26	199.05	229.89	R-11	5.5	148.27	---	69.84	218.11	251.64
18" L with 5-1/2" exposure	Inst	SQ	R-11	6.6	154.69	---	58.41	213.11	241.14	R-11	6.1	170.16	---	63.49	233.65	264.13
24" L with 7-1/2" exposure	Inst	SQ	R-11	9.0	138.44	---	42.84	181.28	201.84	R-11	8.3	152.29	---	46.56	198.85	221.20
Gable with dormers																
16" L with 5" exposure	Inst	SQ	R-11	5.8	135.99	---	66.47	202.46	234.37	R-11	5.3	149.59	---	72.25	221.84	256.52
18" L with 5-1/2" exposure	Inst	SQ	R-11	6.4	156.08	---	60.24	216.31	245.23	R-11	5.9	171.68	---	65.48	237.16	268.59
24" L with 7-1/2" exposure	Inst	SQ	R-11	8.7	139.67	---	44.31	183.98	205.25	R-11	8.0	153.64	---	48.17	201.80	224.92
Gable with intersecting roofs																
16" L with 5" exposure	Inst	SQ	R-11	5.8	135.99	---	66.47	202.46	234.37	R-11	5.3	149.59	---	72.25	221.84	256.52
18" L with 5-1/2" exposure	Inst	SQ	R-11	6.4	156.08	---	60.24	216.31	245.23	R-11	5.9	171.68	---	65.48	237.16	268.59
24" L with 7-1/2" exposure	Inst	SQ	R-11	8.7	139.67	---	44.31	183.98	205.25	R-11	8.0	153.64	---	48.17	201.80	224.92
Gable with dormers & intersecting roofs																
16" L with 5" exposure	Inst	SQ	R-11	5.6	138.39	---	68.84	207.23	240.28	R-11	5.2	152.23	---	74.83	227.06	262.98
18" L with 5-1/2" exposure	Inst	SQ	R-11	6.2	158.84	---	62.18	221.02	250.87	R-11	5.7	174.73	---	67.59	242.32	274.76
24" L with 7-1/2" exposure	Inst	SQ	R-11	8.5	142.13	---	45.36	187.48	209.25	R-11	7.8	156.34	---	49.30	205.64	229.30

Description	Oper	Unit	Costs Based On Large Volume							Costs Based On Small Volume						
			Crew Size	Avg Day Prod	Avg Mat'l Unit Cost	Avg Equip Unit Cost	Avg Labor Unit Cost	Avg Total Unit Cost	Avg Price Incl O&P	Crew Size	Avg Day Prod	Avg Mat'l Unit Cost	Avg Equip Unit Cost	Avg Labor Unit Cost	Avg Total Unit Cost	Avg Price Incl O&P

Roofing (continued)

Hip, plain

Description	Oper	Unit	Crew Size	Avg Day Prod	Avg Mat'l Unit Cost	Avg Equip Unit Cost	Avg Labor Unit Cost	Avg Total Unit Cost	Avg Price Incl O&P	Crew Size	Avg Day Prod	Avg Mat'l Unit Cost	Avg Equip Unit Cost	Avg Labor Unit Cost	Avg Total Unit Cost	Avg Price Incl O&P
16" L with 5" exposure	Inst	SQ	R-11	5.8	135.99	---	66.47	202.46	234.37	R-11	5.3	149.59	---	72.25	221.84	256.52
18" L with 5-1/2" exposure	Inst	SQ	R-11	6.4	156.08	---	60.24	216.31	245.23	R-11	5.9	171.68	---	65.48	237.16	268.59
24" L with 7-1/2" exposure	Inst	SQ	R-11	8.7	139.67	---	44.31	183.98	205.25	R-11	8.0	153.64	---	48.17	201.80	224.92
Hip with dormers																
16" L with 5" exposure	Inst	SQ	R-11	5.6	137.19	---	68.84	206.03	239.08	R-11	5.2	150.91	---	74.83	225.74	261.66
18" L with 5-1/2" exposure	Inst	SQ	R-11	6.1	157.46	---	63.20	220.66	251.00	R-11	5.6	173.20	---	68.70	241.90	274.88
24" L with 7-1/2" exposure	Inst	SQ	R-11	8.3	140.90	---	46.45	187.35	209.64	R-11	7.6	154.99	---	50.49	205.48	229.71
Hip with intersecting roofs																
16" L with 5" exposure	Inst	SQ	R-11	5.6	137.19	---	68.84	206.03	239.08	R-11	5.2	150.91	---	74.83	225.74	261.66
18" L with 5-1/2" exposure	Inst	SQ	R-11	6.1	157.46	---	63.20	220.66	251.00	R-11	5.6	173.20	---	68.70	241.90	274.88
24" L with 7-1/2" exposure	Inst	SQ	R-11	8.3	140.90	---	46.45	187.35	209.64	R-11	7.6	154.99	---	50.49	205.48	229.71
Hip with dormers & intersecting roofs																
16" L with 5" exposure	Inst	SQ	R-11	5.4	139.59	---	71.39	210.98	245.25	R-11	5.0	153.55	---	77.60	231.15	268.40
18" L with 5-1/2" exposure	Inst	SQ	R-11	5.9	160.22	---	65.34	225.57	256.93	R-11	5.4	176.25	---	71.03	247.27	281.37
24" L with 7-1/2" exposure	Inst	SQ	R-11	8.1	143.35	---	47.60	190.95	213.80	R-11	7.5	157.69	---	51.74	209.43	234.26

Related materials and operations

Ridge/Hip units, 40 pcs/bundle

Over existing roofing

Description	Oper	Unit	Crew Size	Avg Day Prod	Avg Mat'l Unit Cost	Avg Equip Unit Cost	Avg Labor Unit Cost	Avg Total Unit Cost	Avg Price Incl O&P	Crew Size	Avg Day Prod	Avg Mat'l Unit Cost	Avg Equip Unit Cost	Avg Labor Unit Cost	Avg Total Unit Cost	Avg Price Incl O&P
5" exposure	Inst	LF	R-9	560	1.01	---	0.60	1.61	1.90	R-9	515	1.11	---	0.65	1.77	2.08
5-1/2" exposure	Inst	LF	R-9	590	0.93	---	0.57	1.50	1.77	R-9	540	1.02	---	0.62	1.64	1.94
7-1/2" exposure	Inst	LF	R-9	700	0.86	---	0.48	1.34	1.57	R-9	645	0.94	---	0.52	1.46	1.71
Over wood decks																
5" exposure	Inst	LF	R-9	600	1.01	---	0.56	1.57	1.84	R-9	550	1.11	---	0.61	1.73	2.02
5-1/2" exposure	Inst	LF	R-9	630	0.93	---	0.53	1.46	1.72	R-9	575	1.02	---	0.59	1.60	1.88
7-1/2" exposure	Inst	LF	R-9	750	0.86	---	0.45	1.30	1.52	R-9	690	2.68	---	0.49	3.17	3.41
Roll valley, galvanized, 28 gauge																
50' L rolls, unpainted																
14" W	Inst	LF	U-1	120	0.74	---	1.59	2.32	3.07	U-1	110	0.81	---	1.73	2.54	3.36
20" W	Inst	LF	U-1	120	1.16	---	1.59	2.74	3.49	U-1	110	1.28	---	1.73	3.00	3.82
Rosin sized sheathing paper, 36" W, 500 SF/roll; nailed																
Over open sheathing	Inst	SQ	R-9	95	2.84	---	3.54	6.38	8.08	R-9	85	3.12	---	3.96	7.08	8.99
Over solid sheathing	Inst	SQ	R-9	140	2.84	---	2.41	5.24	6.40	R-9	125	3.12	---	2.69	5.82	7.11

					Costs Based On Large Volume					Costs Based On Small Volume					
Description	Oper	Unit	Crew Size	Avg Day Prod	Avg Mat'l Unit Cost	Avg Equip Unit Cost	Avg Labor Unit Cost	Avg Total Unit Cost	Avg Price Incl O&P	Crew Size	Avg Day Prod	Avg Mat'l Unit Cost	Avg Equip Unit Cost	Avg Labor Unit Cost	Avg Total Unit Cost	Avg Price Incl O&P

Siding (continued)

Related materials and operations
Rosin sized sheathing paper,
36" W 500 SF/roll; nailed

Description	Oper	Unit	Crew Size	Avg Day Prod	Avg Mat'l Unit Cost	Avg Equip Unit Cost	Avg Labor Unit Cost	Avg Total Unit Cost	Avg Price Incl O&P	Crew Size	Avg Day Prod	Avg Mat'l Unit Cost	Avg Equip Unit Cost	Avg Labor Unit Cost	Avg Total Unit Cost	Avg Price Incl O&P
Over solid sheathing	Inst	SF	C-18	12500	0.03	---	0.03	0.06	0.07	C-18	11500	0.04	---	0.03	0.06	0.08

Site Work

Demolition, wreck and remove to dumpster

Concrete
Footings, with air tools; reinforced

Description	Oper	Unit	Crew Size	Avg Day Prod	Avg Mat'l Unit Cost	Avg Equip Unit Cost	Avg Labor Unit Cost	Avg Total Unit Cost	Avg Price Incl O&P	Crew Size	Avg Day Prod	Avg Mat'l Unit Cost	Avg Equip Unit Cost	Avg Labor Unit Cost	Avg Total Unit Cost	Avg Price Incl O&P
8" T x 12" W (.67 CF/LF)	Demo	LF	A-2	120	---	0.73	2.49	3.22	4.39	A-2	84	---	1.04	3.56	4.60	6.27
8" T x 16" W (.89 CF/LF)	Demo	LF	A-2	100	---	0.88	2.99	3.86	5.27	A-2	70	---	1.25	4.27	5.52	7.53
8" T x 20" W (1.11 CF/LF)	Demo	LF	A-2	90	---	0.97	3.32	4.29	5.86	A-2	63	---	1.39	4.75	6.13	8.36
12" T x 12" W (1.00 CF/LF)	Demo	LF	A-2	70	---	1.25	4.27	5.52	7.53	A-2	49	---	1.79	6.10	7.89	10.75
12" T x 16" W (1.33 CF/LF)	Demo	LF	A-2	60	---	1.46	4.98	6.44	8.78	A-2	42	---	2.08	7.12	9.20	12.55
12" T x 20" W (1.67 CF/LF)	Demo	LF	A-2	55	---	1.59	5.44	7.03	9.58	A-2	39	---	2.27	7.76	10.04	13.69
12"T x 24" W (2.00 CF/LF)	Demo	LF	A-2	50	---	1.75	5.98	7.73	10.54	A-2	35	---	2.50	8.54	11.04	15.06

Foundations and retaining walls, with air tools, per LF wall
With reinforcing
4'-0" H

Description	Oper	Unit	Crew Size	Avg Day Prod	Avg Mat'l Unit Cost	Avg Equip Unit Cost	Avg Labor Unit Cost	Avg Total Unit Cost	Avg Price Incl O&P	Crew Size	Avg Day Prod	Avg Mat'l Unit Cost	Avg Equip Unit Cost	Avg Labor Unit Cost	Avg Total Unit Cost	Avg Price Incl O&P
8" T (2.67 CF/LF)	Demo	SF	A-2	60	---	1.46	4.98	6.44	8.78	A-2	42	---	2.08	7.12	9.20	12.55
12" T (4.00 CF/LF)	Demo	SF	A-2	45	---	1.95	6.64	8.59	11.71	A-2	32	---	2.78	9.49	12.27	16.73

8'-0" H

Description	Oper	Unit	Crew Size	Avg Day Prod	Avg Mat'l Unit Cost	Avg Equip Unit Cost	Avg Labor Unit Cost	Avg Total Unit Cost	Avg Price Incl O&P	Crew Size	Avg Day Prod	Avg Mat'l Unit Cost	Avg Equip Unit Cost	Avg Labor Unit Cost	Avg Total Unit Cost	Avg Price Incl O&P
8" T (5.33 CF/LF)	Demo	SF	A-2	55	---	1.59	5.44	7.03	9.58	A-2	39	---	2.27	7.76	10.04	13.69
12" T (8.00 CF/LF)	Demo	SF	A-2	40	---	2.19	7.47	9.66	13.17	A-2	28	---	3.13	10.68	13.80	18.82

12'-0" H

Description	Oper	Unit	Crew Size	Avg Day Prod	Avg Mat'l Unit Cost	Avg Equip Unit Cost	Avg Labor Unit Cost	Avg Total Unit Cost	Avg Price Incl O&P	Crew Size	Avg Day Prod	Avg Mat'l Unit Cost	Avg Equip Unit Cost	Avg Labor Unit Cost	Avg Total Unit Cost	Avg Price Incl O&P
8" T (8.00 CF/LF)	Demo	SF	A-2	50	---	1.75	5.98	7.73	10.54	A-2	35	---	2.50	8.54	11.04	15.06
12" T (12.00 CF/LF)	Demo	SF	A-2	35	---	2.50	8.54	11.04	15.06	A-2	25	---	3.57	12.20	15.77	21.51

Without reinforcing
4'-0" H

Description	Oper	Unit	Crew Size	Avg Day Prod	Avg Mat'l Unit Cost	Avg Equip Unit Cost	Avg Labor Unit Cost	Avg Total Unit Cost	Avg Price Incl O&P	Crew Size	Avg Day Prod	Avg Mat'l Unit Cost	Avg Equip Unit Cost	Avg Labor Unit Cost	Avg Total Unit Cost	Avg Price Incl O&P
8" T (2.67 CF/LF)	Demo	SF	A-2	80	---	1.09	3.74	4.83	6.59	A-2	56	---	1.56	5.34	6.90	9.41
12" T (4.00 CF/LF)	Demo	SF	A-2	60	---	1.46	4.98	6.44	8.78	A-2	42	---	2.08	7.12	9.20	12.55

8'-0" H

Description	Oper	Unit	Crew Size	Avg Day Prod	Avg Mat'l Unit Cost	Avg Equip Unit Cost	Avg Labor Unit Cost	Avg Total Unit Cost	Avg Price Incl O&P	Crew Size	Avg Day Prod	Avg Mat'l Unit Cost	Avg Equip Unit Cost	Avg Labor Unit Cost	Avg Total Unit Cost	Avg Price Incl O&P
8" T (5.33 CF/LF)	Demo	SF	A-2	70	---	1.25	4.27	5.52	7.53	A-2	49	---	1.79	6.10	7.89	10.75
12" T (8.00 CF/LF)	Demo	SF	A-2	50	---	1.75	5.98	7.73	10.54	A-2	35	---	2.50	8.54	11.04	15.06

Site Work - Demolition

Average Output is based on what the designated crew can do in one day. In this section, the man might be a Laborer, a Carpenter, a Floor Layer, etc. Who does the wrecking depends on the quantity of wrecking to be done.

A Contractor might use Laborers exclusively on a large volume, whereas, on a small volume, the contractor might use a Carpenter(s) or a Carpenter and a Laborer.

The quantity of work that a man does in a day is greatly influenced by his selection of tools and/or equipment. A man can remove more brick with a compressor and pneumatic tool than he can with a sledge hammer or pry bar. In this section, the phrase "by hand" includes the use of hand tools, i.e.,

sledge hammers, wrecking bars, claw hammers, etc. When the Average Output is based on the use of equipment (not hand tools), a description of the equipment is provided.

The Average Output is not based on the use of "heavy" equipment such as bulldozers, cranes with wrecking balls, etc. The Average Output does include the labor involved in hauling wrecked material or debris to a dumpster located at the site. Average rental costs for dumpsters are: $175.00 for 40 CY (20' x 8' x 8' H); $150.00 for 30 CY (20' x 8' x 6' H). Rental period equals delivery and pickup when full.

			Costs Based On Large Volume							Costs Based On Small Volume					
Description	Oper	Unit	Crew Size	Avg Day Prod	Avg Mat'l Unit Cost	Avg Equip Unit Cost	Avg Labor Unit Cost	Avg Total Unit Cost	Avg Price Incl O&P	Crew Size	Avg Day Prod	Avg Mat'l Unit Cost	Avg Equip Unit Cost	Avg Labor Unit Cost	Avg Total Unit Cost	Avg Price Incl O&P

Site Work, demolition (continued)

12'-0" H

Description	Oper	Unit	Crew Size	Avg Day Prod	Avg Mat'l Unit Cost	Avg Equip Unit Cost	Avg Labor Unit Cost	Avg Total Unit Cost	Avg Price Incl O&P	Crew Size	Avg Day Prod	Avg Mat'l Unit Cost	Avg Equip Unit Cost	Avg Labor Unit Cost	Avg Total Unit Cost	Avg Price Incl O&P
8" T (8.00 CF/LF)	Demo	SF	A-2	60	---	1.46	4.98	6.44	8.78	A-2	42	---	2.08	7.12	9.20	12.55
12" T (12.00 CF/LF)	Demo	SF	A-2	40	---	2.19	7.47	9.66	13.17	A-2	28	---	3.13	10.68	13.80	18.82

Slabs with air tools

With reinforcing

Description	Oper	Unit	Crew Size	Avg Day Prod	Avg Mat'l Unit Cost	Avg Equip Unit Cost	Avg Labor Unit Cost	Avg Total Unit Cost	Avg Price Incl O&P	Crew Size	Avg Day Prod	Avg Mat'l Unit Cost	Avg Equip Unit Cost	Avg Labor Unit Cost	Avg Total Unit Cost	Avg Price Incl O&P
4" T	Demo	SF	A-2	350	---	0.25	0.85	1.10	1.51	A-2	245	---	0.36	1.22	1.58	2.15
5" T	Demo	SF	A-2	425	---	0.21	0.70	0.91	1.24	A-2	298	---	0.29	1.00	1.30	1.77
6" T	Demo	SF	A-2	370	---	0.24	0.81	1.04	1.42	A-2	259	---	0.34	1.15	1.49	2.03

Without reinforcing

Description	Oper	Unit	Crew Size	Avg Day Prod	Avg Mat'l Unit Cost	Avg Equip Unit Cost	Avg Labor Unit Cost	Avg Total Unit Cost	Avg Price Incl O&P	Crew Size	Avg Day Prod	Avg Mat'l Unit Cost	Avg Equip Unit Cost	Avg Labor Unit Cost	Avg Total Unit Cost	Avg Price Incl O&P
4" T	Demo	SF	A-2	500	---	0.18	0.60	0.77	1.05	A-2	350	---	0.25	0.85	1.10	1.51
5" T	Demo	SF	A-2	425	---	0.21	0.70	0.91	1.24	A-2	298	---	0.29	1.00	1.30	1.77
6" T	Demo	SF	A-2	370	---	0.24	0.81	1.04	1.42	A-2	259	---	0.34	1.15	1.49	2.03

Masonry

Brick

Chimneys

Description	Oper	Unit	Crew Size	Avg Day Prod	Avg Mat'l Unit Cost	Avg Equip Unit Cost	Avg Labor Unit Cost	Avg Total Unit Cost	Avg Price Incl O&P	Crew Size	Avg Day Prod	Avg Mat'l Unit Cost	Avg Equip Unit Cost	Avg Labor Unit Cost	Avg Total Unit Cost	Avg Price Incl O&P
4" T wall	Demo	VLF	L-2	25	---	---	10.40	10.40	15.40	L-2	18	---	---	14.86	14.86	22.00
8" T wall	Demo	VLF	L-2	10	---	---	26.01	26.01	38.49	L-2	7	---	---	37.15	37.15	54.99
Columns, 12" x 12" o.d.	Demo	VLF	L-2	50	---	---	5.20	5.20	7.70	L-2	35	---	---	7.43	7.43	11.00
Veneer, 4" T, with air tools	Demo	SF	A-2	270	---	0.32	1.11	1.43	1.95	A-2	189	---	0.46	1.58	2.04	2.79

Walls, with air tools

Description	Oper	Unit	Crew Size	Avg Day Prod	Avg Mat'l Unit Cost	Avg Equip Unit Cost	Avg Labor Unit Cost	Avg Total Unit Cost	Avg Price Incl O&P	Crew Size	Avg Day Prod	Avg Mat'l Unit Cost	Avg Equip Unit Cost	Avg Labor Unit Cost	Avg Total Unit Cost	Avg Price Incl O&P
8" T wall	Demo	SF	A-2	135	---	0.65	2.21	2.86	3.90	A-2	95	---	0.93	3.16	4.09	5.58
12" T wall	Demo	SF	A-2	95	---	0.92	3.15	4.07	5.55	A-2	67	---	1.32	4.50	5.81	7.92

Masonry (concrete) block; lightweight (haydite) or heavyweight blocks

Foundations and retaining walls; no excavation included

Without reinforcing or with only lateral reinforcing

With air tools

Description	Oper	Unit	Crew Size	Avg Day Prod	Avg Mat'l Unit Cost	Avg Equip Unit Cost	Avg Labor Unit Cost	Avg Total Unit Cost	Avg Price Incl O&P	Crew Size	Avg Day Prod	Avg Mat'l Unit Cost	Avg Equip Unit Cost	Avg Labor Unit Cost	Avg Total Unit Cost	Avg Price Incl O&P
8" W x 8" H x 16" L	Demo	SF	A-2	280	---	0.31	1.07	1.38	1.88	A-2	196	---	0.45	1.53	1.97	2.69
12" W x 8" H x 16" L	Demo	SF	A-2	240	---	0.36	1.25	1.61	2.20	A-2	168	---	0.52	1.78	2.30	3.14

Without air tools

Description	Oper	Unit	Crew Size	Avg Day Prod	Avg Mat'l Unit Cost	Avg Equip Unit Cost	Avg Labor Unit Cost	Avg Total Unit Cost	Avg Price Incl O&P	Crew Size	Avg Day Prod	Avg Mat'l Unit Cost	Avg Equip Unit Cost	Avg Labor Unit Cost	Avg Total Unit Cost	Avg Price Incl O&P
8" W x 8" H x 16" L	Demo	SF	L-2	225	---	---	1.16	1.16	1.71	L-2	158	---	---	1.65	1.65	2.44
12" W x 8" H x 16" L	Demo	SF	L-2	190	---	---	1.37	1.37	2.03	L-2	133	---	---	1.96	1.96	2.89

| | | | Costs Based On Large Volume | | | | | | Costs Based On Small Volume | | | | | | |
Description	Oper	Unit	Crew Size	Avg Day Prod	Avg Mat'l Unit Cost	Avg Equip Unit Cost	Avg Labor Unit Cost	Avg Total Unit Cost	Avg Price Incl O&P	Crew Size	Avg Day Prod	Avg Mat'l Unit Cost	Avg Equip Unit Cost	Avg Labor Unit Cost	Avg Total Unit Cost	Avg Price Incl O&P
Site Work, demolition (continued)																
With vertical reinforcing in every other core (2 core blocks) with core filled																
With air tools																
8" W x 8" H x 16" L	Demo	SF	A-2	170	---	0.51	1.76	2.27	3.10	A-2	119	---	0.74	2.51	3.25	4.43
12" W x 8" H x 16" L	Demo	SF	A-2	145	---	0.60	2.06	2.67	3.63	A-2	102	---	0.86	2.95	3.81	5.19
Exterior walls (above grade) and partitions; no shoring included																
Without reinforcing or with only lateral reinforcing																
With air tools																
8" W x 8" H x 16" L	Demo	SF	A-2	350	---	0.25	0.85	1.10	1.51	A-2	245	---	0.36	1.22	1.58	2.15
12" W x 8" H x 16" L	Demo	SF	A-2	300	---	0.29	1.00	1.29	1.76	A-2	210	---	0.42	1.42	1.84	2.51
Without air tools																
8" W x 8" H x 16" L	Demo	SF	L-2	280	---	---	0.93	0.93	1.37	L-2	196	---	---	1.33	1.33	1.96
12" W x 8" H x 16" L	Demo	SF	L-2	240	---	---	1.08	1.08	1.60	L-2	168	---	---	1.55	1.55	2.29
Fences, without reinforcing or with only lateral reinforcing																
With air tools																
6" W x 4" H x 16" L	Demo	SF	A-2	390	---	0.22	0.77	0.99	1.35	A-2	273	---	0.32	1.10	1.42	1.93
6" W x 6" H x 16" L	Demo	SF	A-2	370	---	0.24	0.81	1.04	1.42	A-2	259	---	0.34	1.15	1.49	2.03
8" W x 8" H x 16" L	Demo	SF	A-2	350	---	0.25	0.85	1.10	1.51	A-2	245	---	0.36	1.22	1.58	2.15
12" W x 8" H x 16" L	Demo	SF	A-2	300	---	0.29	1.00	1.29	1.76	A-2	210	---	0.42	1.42	1.84	2.51
Without air tools																
6" W x 4" H x 16" L	Demo	SF	L-2	310	---	---	0.84	0.84	1.24	L-2	217	---	---	1.20	1.20	1.77
6" W x 6" H x 16" L	Demo	SF	L-2	295	---	---	0.88	0.88	1.30	L-2	207	---	---	1.26	1.26	1.86
8" W x 8" H x 16" L	Demo	SF	L-2	280	---	---	0.93	0.93	1.37	L-2	196	---	---	1.33	1.33	1.96
12" W x 8" H x 16" L	Demo	SF	L-2	240	---	---	1.08	1.08	1.60	L-2	168	---	---	1.55	1.55	2.29
Quarry tile, 6" or 9" squares																
Floors																
Conventional mortar set	Demo	SF	L-2	445	---	---	0.58	0.58	0.86	L-2	312	---	---	0.83	0.83	1.24
Dry-set mortar	Demo	SF	L-2	515	---	---	0.51	0.51	0.75	L-2	361	---	---	0.72	0.72	1.07
Rough Carpentry (Framing)																
Dimension lumber																
Beams, set on steel columns																
Built-up from 2" lumber																
4" T x 10" W - 10' L (2 pcs)	Demo	LF	L-2	855	---	---	0.30	0.30	0.45	L-2	599	---	---	0.43	0.43	0.64
4" T x 12" W - 12' L (2 pcs)	Demo	LF	L-2	1025	---	---	0.25	0.25	0.38	L-2	718	---	---	0.36	0.36	0.54

			Costs Based On Large Volume							Costs Based On Small Volume						
Description	Oper	Unit	Crew Size	Avg Day Prod	Avg Mat'l Unit Cost	Avg Equip Unit Cost	Avg Labor Unit Cost	Avg Total Unit Cost	Avg Price Incl O&P	Crew Size	Avg Day Prod	Avg Mat'l Unit Cost	Avg Equip Unit Cost	Avg Labor Unit Cost	Avg Total Unit Cost	Avg Price Incl O&P

Site Work, demolition (continued)

Description	Oper	Unit	Crew Size	Avg Day Prod	Avg Mat'l Unit Cost	Avg Equip Unit Cost	Avg Labor Unit Cost	Avg Total Unit Cost	Avg Price Incl O&P	Crew Size	Avg Day Prod	Avg Mat'l Unit Cost	Avg Equip Unit Cost	Avg Labor Unit Cost	Avg Total Unit Cost	Avg Price Incl O&P
6" T x 10" W - 10' L (3 pcs)	Demo	LF	L-2	855	---	---	0.30	0.30	0.45	L-2	599	---	---	0.43	0.43	0.64
6" T x 12" W - 12' L (3 pcs)	Demo	LF	L-2	1025	---	---	0.25	0.25	0.38	L-2	718	---	---	0.36	0.36	0.54
Single member (solid lumber)																
3" T x 12" W - 12' L	Demo	LF	L-2	1025	---	---	0.25	0.25	0.38	L-2	718	---	---	0.36	0.36	0.54
4" T x 12" W - 12' L	Demo	LF	L-2	1025	---	---	0.25	0.25	0.38	L-2	718	---	---	0.36	0.36	0.54
Bracing, diagonal, notched-in, studs o.c.																
1" x 6" - 10'	Demo	LF	L-2	605	---	---	0.43	0.43	0.64	L-2	424	---	---	0.61	0.61	0.91
Bridging, "X" type, 1" x 3", (8", 10", 12" T) 16" o.c.	Demo	LF	L-2	350	---	---	0.74	0.74	1.10	L-2	245	---	---	1.06	1.06	1.57
Columns or posts																
4" x 4" -8' L	Demo	LF	L-2	770	---	---	0.34	0.34	0.50	L-2	539	---	---	0.48	0.48	0.71
6" x 6" -8' L	Demo	LF	L-2	770	---	---	0.34	0.34	0.50	L-2	539	---	---	0.48	0.48	0.71
6" x 8" -8' L	Demo	LF	L-2	650	---	---	0.40	0.40	0.59	L-2	455	---	---	0.57	0.57	0.85
8" x 8" -8' L	Demo	LF	L-2	615	---	---	0.42	0.42	0.63	L-2	431	---	---	0.60	0.60	0.89
Fascia																
1" x 4" - 12" L	Demo	LF	L-2	1315	---	---	0.20	0.20	0.29	L-2	921	---	---	0.28	0.28	0.42
Firestops or stiffeners																
2" x 4" - 16"	Demo	LF	L-2	475	---	---	0.55	0.55	0.81	L-2	333	---	---	0.78	0.78	1.16
2" x 6" - 16"	Demo	LF	L-2	475	---	---	0.55	0.55	0.81	L-2	333	---	---	0.78	0.78	1.16
Furring strips, 1" x 4" - 8' L																
Walls; strips 12" o.c.																
Studs 16" o.c.	Demo	SF	L-2	1215	---	---	0.21	0.21	0.32	L-2	851	---	---	0.31	0.31	0.45
Studs 24" o.c.	Demo	SF	L-2	1365	---	---	0.19	0.19	0.28	L-2	956	---	---	0.27	0.27	0.40
Masonry (concrete blocks)	Demo	SF	L-2	1095	---	---	0.24	0.24	0.35	L-2	767	---	---	0.34	0.34	0.50
Concrete	Demo	SF	L-2	645	---	---	0.40	0.40	0.60	L-2	452	---	---	0.58	0.58	0.85
Ceiling; joists 16" o.c.																
Strips 12" o.c.	Demo	SF	L-2	840	---	---	0.31	0.31	0.46	L-2	588	---	---	0.44	0.44	0.65
Strips 16" o.c.	Demo	SF	L-2	1075	---	---	0.24	0.24	0.36	L-2	753	---	---	0.35	0.35	0.51
Headers or lintels, (over openings)																
Built-up or single member																
4" T x 6" W - 4' L	Demo	LF	L-2	560	---	---	0.46	0.46	0.69	L-2	392	---	---	0.66	0.66	0.98
4" T x 8" W - 8' L	Demo	LF	L-2	720	---	---	0.36	0.36	0.53	L-2	504	---	---	0.52	0.52	0.76

			Costs Based On Large Volume							Costs Based On Small Volume					
Description	Oper	Unit	Crew Size	Avg Day Prod	Avg Mat'l Unit Cost	Avg Equip Unit Cost	Avg Labor Unit Cost	Avg Total Unit Cost	Avg Price Incl O&P	Crew Size	Avg Day Prod	Avg Mat'l Unit Cost	Avg Equip Unit Cost	Avg Labor Unit Cost	Avg Total Unit Cost	Avg Price Incl O&P

Site Work, demolition (continued)

4" T x 10" W - 10' L	Demo	LF	L-2	855	---	---	0.30	0.30	0.45	L-2	599	---	---	0.43	0.43	0.64
4" T x 12" W - 12' L	Demo	LF	L-2	1025	---	---	0.25	0.25	0.38	L-2	718	---	---	0.36	0.36	0.54
4" T x 14" W - 14' L	Demo	LF	L-2	1025	---	---	0.25	0.25	0.38	L-2	718	---	---	0.36	0.36	0.54
Joists																
Ceiling																
2" x 4" - 8' L	Demo	LF	L-2	1230	---	---	0.21	0.21	0.31	L-2	861	---	---	0.30	0.30	0.45
2" x 4" - 10' L	Demo	LF	L-2	1395	---	---	0.19	0.19	0.28	L-2	977	---	---	0.27	0.27	0.39
2" x 8" - 12' L	Demo	LF	L-2	1420	---	---	0.18	0.18	0.27	L-2	994	---	---	0.26	0.26	0.39
2" x 10" - 14' L	Demo	LF	L-2	1535	---	---	0.17	0.17	0.25	L-2	1075	---	---	0.24	0.24	0.36
2" x 12" - 16' L	Demo	LF	L-2	1585	---	---	0.16	0.16	0.24	L-2	1110	---	---	0.23	0.23	0.35
Floor; seated on sill plate																
2" x 8" - 12' L	Demo	LF	L-2	1600	---	---	0.16	0.16	0.24	L-2	1120	---	---	0.23	0.23	0.34
2" x 10" - 14' L	Demo	LF	L-2	1725	---	---	0.15	0.15	0.22	L-2	1208	---	---	0.22	0.22	0.32
2" x 12" - 16' L	Demo	LF	L-2	1755	---	---	0.15	0.15	0.22	L-2	1229	---	---	0.21	0.21	0.31
Ledgers																
Nailed, 2" x 6" - 12' L	Demo	LF	L-2	710	---	---	0.37	0.37	0.54	L-2	497	---	---	0.52	0.52	0.77
Bolted, 3" x 8" - 12' L	Demo	LF	L-2	970	---	---	0.27	0.27	0.40	L-2	679	---	---	0.38	0.38	0.57
Plates; 2" x 4" or 2" x 6"																
Double top nailed	Demo	LF	L-2	820	---	---	0.32	0.32	0.47	L-2	574	---	---	0.45	0.45	0.67
Sill, nailed	Demo	LF	L-2	1360	---	---	0.19	0.19	0.28	L-2	952	---	---	0.27	0.27	0.40
Sill or bottom, bolted	Demo	LF	L-2	685	---	---	0.38	0.38	0.56	L-2	480	---	---	0.54	0.54	0.80
Rafters																
Common																
2" x 4" - 14' L	Demo	LF	L-2	1795	---	---	0.14	0.14	0.21	L-2	1257	---	---	0.21	0.21	0.31
2" x 6" - 14' L	Demo	LF	L-2	1660	---	---	0.16	0.16	0.23	L-2	1162	---	---	0.22	0.22	0.33
2" x 8" - 14' L	Demo	LF	L-2	1435	---	---	0.18	0.18	0.27	L-2	1005	---	---	0.26	0.26	0.38
2" x 10" - 14' L	Demo	LF	L-2	1195	---	---	0.22	0.22	0.32	L-2	837	---	---	0.31	0.31	0.46
Hip and/or valley																
2" x 4" - 16' L	Demo	LF	L-2	2050	---	---	0.13	0.13	0.19	L-2	1435	---	---	0.18	0.18	0.27
2" x 6" - 16' L	Demo	LF	L-2	1895	---	---	0.14	0.14	0.20	L-2	1327	---	---	0.20	0.20	0.29
2" x 8" - 16' L	Demo	LF	L-2	1640	---	---	0.16	0.16	0.23	L-2	1148	---	---	0.23	0.23	0.34
2" x 10" - 16' L	Demo	LF	L-2	1360	---	---	0.19	0.19	0.28	L-2	952	---	---	0.27	0.27	0.40

			Costs Based On Large Volume							Costs Based On Small Volume					

Description	Oper	Unit	Crew Size	Avg Day Prod	Avg Mat'l Unit Cost	Avg Equip Unit Cost	Avg Labor Unit Cost	Avg Total Unit Cost	Avg Price Incl O&P	Crew Size	Avg Day Prod	Avg Mat'l Unit Cost	Avg Equip Unit Cost	Avg Labor Unit Cost	Avg Total Unit Cost	Avg Price Incl O&P
Site Work, demolition (continued)																
Jack																
2" x 4" - 6' L	Demo	LF	L-2	805	---	---	0.32	0.32	0.48	L-2	564	---	---	0.46	0.46	0.68
2" x 6" - 6' L	Demo	LF	L-2	740	---	---	0.35	0.35	0.52	L-2	518	---	---	0.50	0.50	0.74
2" x 8" - 6' L	Demo	LF	L-2	655	---	---	0.40	0.40	0.59	L-2	459	---	---	0.57	0.57	0.84
2" x 10" - 6' L	Demo	LF	L-2	540	---	---	0.48	0.48	0.71	L-2	378	---	---	0.69	0.69	1.02
Roof decking, solid T&G																
2" x 6" - 12' L	Demo	LF	L-2	645	---	---	0.40	0.40	0.60	L-2	452	---	---	0.58	0.58	0.85
2" x 8" - 12' L	Demo	LF	L-2	900	---	---	0.29	0.29	0.43	L-2	630	---	---	0.41	0.41	0.61
Studs																
2" x 4" - 8' L	Demo	LF	L-2	1360	---	---	0.19	0.19	0.28	L-2	952	---	---	0.27	0.27	0.40
2" x 6" - 8' L	Demo	LF	L-2	1360	---	---	0.19	0.19	0.28	L-2	952	---	---	0.27	0.27	0.40
Stud partitions, studs 16" o.c. with bottom and double top plates and firestops; per LF partition																
2" x 4" - 8' L	Demo	LF	L-2	100	---	---	2.60	2.60	3.85	L-2	70	---	---	3.72	3.72	5.50
2" x 6" - 8' L	Demo	LF	L-2	100	---	---	2.60	2.60	3.85	L-2	70	---	---	3.72	3.72	5.50
Boards																
Sheathing, regular or diagonal, 1" x 8"																
Roof	Demo	SF	L-2	1340	---	---	0.19	0.19	0.29	L-2	938	---	---	0.28	0.28	0.41
Sidewall	Demo	SF	L-2	1615	---	---	0.16	0.16	0.24	L-2	1131	---	---	0.23	0.23	0.34
Subflooring, regular or diagonal																
1" x 8" - 16' L	Demo	SF	L-2	1525	---	---	0.17	0.17	0.25	L-2	1068	---	---	0.24	0.24	0.36
1" x 10" - 16' L	Demo	SF	L-2	1930	---	---	0.13	0.13	0.20	L-2	1351	---	---	0.19	0.19	0.28
Plywood																
Sheathing																
Roof																
1/2" T, CDX	Demo	SF	L-2	1970	---	---	0.13	0.13	0.20	L-2	1379	---	---	0.19	0.19	0.28
5/8" T, CDX	Demo	SF	L-2	1935	---	---	0.13	0.13	0.20	L-2	1355	---	---	0.19	0.19	0.28
Wall																
3/8" or 1/2" T, CDX	Demo	SF	L-2	2520	---	---	0.10	0.10	0.15	L-2	1764	---	---	0.15	0.15	0.22
5/8" T, CDX	Demo	SF	L-2	2460	---	---	0.11	0.11	0.16	L-2	1722	---	---	0.15	0.15	0.22
Subflooring																
5/8", 3/4" T, CDX	Demo	SF	L-2	2305	---	---	0.11	0.11	0.17	L-2	1614	---	---	0.16	0.16	0.24
1-1/8" T, 2-4-1, T&G long edges	Demo	SF	L-2	1615	---	---	0.16	0.16	0.24	L-2	1131	---	---	0.23	0.23	0.34

			Costs Based On Large Volume						Costs Based On Small Volume					
Description	Oper	Unit	Crew Size	Avg Day Prod	Avg Mat'l Unit Cost	Avg Equip Unit Cost	Avg Labor Unit Cost	Avg Total Unit Cost	Avg Price Incl O&P	Crew Size	Avg Day Prod	Avg Mat'l Unit Cost	Avg Equip Unit Cost	Avg Labor Unit Cost	Avg Total Unit Cost	Avg Price Incl O&P

Site Work, demolition (continued)

Trusses, "W" pattern with gin pole, 24' to 30' spans

Description	Oper	Unit	Crew Size	Avg Day Prod	Avg Mat'l Unit Cost	Avg Equip Unit Cost	Avg Labor Unit Cost	Avg Total Unit Cost	Avg Price Incl O&P	Crew Size	Avg Day Prod	Avg Mat'l Unit Cost	Avg Equip Unit Cost	Avg Labor Unit Cost	Avg Total Unit Cost	Avg Price Incl O&P
3 - in - 12 slope	Demo	SF	L-2	26	---	---	10.00	10.00	14.80	L-2	18	---	---	14.29	14.29	21.15
5 - in - 12 slope	Demo	SF	L-2	26	---	---	10.00	10.00	14.80	L-2	18	---	---	14.29	14.29	21.15
Finish carpentry																
Bath accessories, screw on type	Demo	EA	L-2	125	---	---	2.08	2.08	3.08	L-2	88	---	---	2.97	2.97	4.40
Cabinets																
Kitchen, to 3' x 4', wood;																
base, wall, or peninsula	Demo	EA	L-2	25	---	---	10.40	10.40	15.40	L-2	18	---	---	14.86	14.86	22.00
Medicine, metal	Demo	EA	L-2	30	---	---	8.67	8.67	12.83	L-2	21	---	---	12.38	12.38	18.33
Vanity, cabinet and sink top																
Disconnect plumbing and remove to dumpster	Demo	EA	L-2	16	---	---	16.26	16.26	24.06	L-2	11	---	---	23.22	23.22	34.37
Remove old unit, replace with new unit reconnect plumbing	Demo	EA	S-2	7	---	---	47.00	47.00	68.62	S-2	5	---	---	67.14	67.14	98.02
Hardwood flooring (over wood subfloors)																
Block, set in mastic	Demo	SF	L-2	1200	---	---	0.22	0.22	0.32	L-2	840	---	---	0.31	0.31	0.46
Strip, nailed	Demo	SF	L-2	900	---	---	0.29	0.29	0.43	L-2	630	---	---	0.41	0.41	0.61
Marlite panels, 4' x 8', adhesive set	Demo	SF	L-2	1850	---	---	0.14	0.14	0.21	L-2	1295	---	---	0.20	0.20	0.30
Molding and trim																
At base (floor)	Demo	LF	L-2	1060	---	---	0.25	0.25	0.36	L-2	742	---	---	0.35	0.35	0.52
At ceiling	Demo	LF	L-2	1200	---	---	0.22	0.22	0.32	L-2	840	---	---	0.31	0.31	0.46
On walls or cabinets	Demo	LF	L-2	1600	---	---	0.16	0.16	0.24	L-2	1120	---	---	0.23	0.23	0.34
Paneling																
Plywood, prefinished	Demo	SF	L-2	1850	---	---	0.14	0.14	0.21	L-2	1295	---	---	0.20	0.20	0.30
Wood	Demo	SF	L-2	1650	---	---	0.16	0.16	0.23	L-2	1155	---	---	0.23	0.23	0.33
Weather protection																
Insulation																
Batt/Roll, with wall or ceiling finish already removed																
Joists, 16" or 24" o.c.	Demo	SF	L-2	2935	---	---	0.09	0.09	0.13	L-2	2055	---	---	0.13	0.13	0.19
Rafters, 16" or 24" o.c.	Demo	SF	L-2	2560	---	---	0.10	0.10	0.15	L-2	1792	---	---	0.15	0.15	0.21
Studs, 16" or 24" o.c.	Demo	SF	L-2	3285	---	---	0.08	0.08	0.12	L-2	2300	---	---	0.11	0.11	0.17

			Costs Based On Large Volume							Costs Based On Small Volume					
Description	Oper	Unit	Crew Size	Avg Day Prod	Avg Mat'l Unit Cost	Avg Equip Unit Cost	Avg Labor Unit Cost	Avg Total Unit Cost	Avg Price Incl O&P	Crew Size	Avg Day Prod	Avg Mat'l Unit Cost	Avg Equip Unit Cost	Avg Labor Unit Cost	Avg Total Unit Cost	Avg Price Incl O&P

Site Work, demolition (continued)

Loose, with ceiling finish already removed
Joists, 16" or 24" o.c.

Description	Oper	Unit	Crew Size	Avg Day Prod	Avg Mat'l Unit Cost	Avg Equip Unit Cost	Avg Labor Unit Cost	Avg Total Unit Cost	Avg Price Incl O&P	Crew Size	Avg Day Prod	Avg Mat'l Unit Cost	Avg Equip Unit Cost	Avg Labor Unit Cost	Avg Total Unit Cost	Avg Price Incl O&P
4" T	Demo	SF	L-2	3900	---	---	0.07	0.07	0.10	L-2	2730	---	---	0.10	0.10	0.14
6" T	Demo	SF	L-2	2340	---	---	0.11	0.11	0.16	L-2	1638	---	---	0.16	0.16	0.23
Rigid																
Roofs																
1/2" T	Demo	SQ	L-2	17	---	---	15.30	15.30	22.64	L-2	12	---	---	21.86	21.86	32.35
1" T	Demo	SQ	L-2	15	---	---	17.34	17.34	25.66	L-2	11	---	---	24.77	24.77	36.66
Walls, 1/2" T	Demo	SQ	L-2	2140	---	---	0.12	0.12	0.18	L-2	1498	---	---	0.17	0.17	0.26
Sheet metal																
Gutter and downspouts																
Aluminum	Demo	LF	L-2	850	---	---	0.31	0.31	0.45	L-2	595	---	---	0.44	0.44	0.65
Galvanized	Demo	LF	L-2	640	---	---	0.41	0.41	0.60	L-2	448	---	---	0.58	0.58	0.86

Roofing and siding

Aluminum
Roofing, nailed to wood
Corrugated (2-1/2"), 26" W with 3-3/4" side lap and

Description	Oper	Unit	Crew Size	Avg Day Prod	Avg Mat'l Unit Cost	Avg Equip Unit Cost	Avg Labor Unit Cost	Avg Total Unit Cost	Avg Price Incl O&P	Crew Size	Avg Day Prod	Avg Mat'l Unit Cost	Avg Equip Unit Cost	Avg Labor Unit Cost	Avg Total Unit Cost	Avg Price Incl O&P
6" end lap	Demo	SQ	L-2	10	---	---	26.01	26.01	38.49	L-2	7	---	---	37.15	37.15	54.99
Siding, nailed to wood																
Clapboard (i.e., lap drop)																
8" exposure	Demo	SF	L-2	1025	---	---	0.25	0.25	0.38	L-2	718	---	---	0.36	0.36	0.54
10" exposure	Demo	SF	L-2	1280	---	---	0.20	0.20	0.30	L-2	896	---	---	0.29	0.29	0.43
Corrugated (2-1/2"), 26" W with 2-1/2" side lap and 4" end lap	Demo	SF	L-2	1200	---	---	0.22	0.22	0.32	L-2	840	---	---	0.31	0.31	0.46
Panels, 4' x 8'	Demo	SF	L-2	2400	---	---	0.11	0.11	0.16	L-2	1680	---	---	0.15	0.15	0.23
Shingle, 24" L with 12" exposure	Demo	SF	L-2	1450	---	---	0.18	0.18	0.27	L-2	1015	---	---	0.26	0.26	0.38
Asbestos shingle																
Siding, nailed; 24" x 12" with 11" exposure	Demo	SF	L-2	2160	---	---	0.12	0.12	0.18	L-2	1512	---	---	0.17	0.17	0.25
Asphalt shingle roofing																
240 lb./SQ., strip, 3 tab 5" exposure	Demo	SQ	L-2	16	---	---	16.26	16.26	24.06	L-2	11	---	---	23.22	23.22	34.37
Built-up/hot roofing (to wood deck)																

Description	Oper	Unit	Costs Based On Large Volume							Costs Based On Small Volume						
			Crew Size	Avg Day Prod	Avg Mat'l Unit Cost	Avg Equip Unit Cost	Avg Labor Unit Cost	Avg Total Unit Cost	Avg Price Incl O&P	Crew Size	Avg Day Prod	Avg Mat'l Unit Cost	Avg Equip Unit Cost	Avg Labor Unit Cost	Avg Total Unit Cost	Avg Price Incl O&P

Site Work, demolition (continued)

Description	Oper	Unit	Crew Size	Avg Day Prod	Avg Mat'l Unit Cost	Avg Equip Unit Cost	Avg Labor Unit Cost	Avg Total Unit Cost	Avg Price Incl O&P	Crew Size	Avg Day Prod	Avg Mat'l Unit Cost	Avg Equip Unit Cost	Avg Labor Unit Cost	Avg Total Unit Cost	Avg Price Incl O&P
3 ply																
With gravel	Demo	SQ	L-2	11	---	---	23.64	23.64	34.99	L-2	8	---	---	33.78	33.78	49.99
Without gravel	Demo	SQ	L-2	14	---	---	18.58	18.58	27.49	L-2	10	---	---	26.54	26.54	39.28
5 ply																
With gravel	Demo	SQ	L-2	10	---	---	26.01	26.01	38.49	L-2	7	---	---	37.15	37.15	54.99
Without gravel	Demo	SQ	L-2	13	---	---	20.01	20.01	29.61	L-2	9	---	---	28.58	28.58	42.30
Clay tile roofing																
2 piece interlocking	Demo	SF	L-2	10	---	---	26.01	26.01	38.49	L-2	7	---	---	37.15	37.15	54.99
1 piece	Demo	SF	L-2	11	---	---	23.64	23.64	34.99	L-2	8	---	---	33.78	33.78	49.99
Hardboard siding																
Lap, 1/2" T x 12" W x 16' L, with 11" exposure	Demo	SF	L-2	1345	---	---	0.19	0.19	0.29	L-2	942	---	---	0.28	0.28	0.41
Panels, 7/16" T x 4' W x 8' H	Demo	SF	L-2	2520	---	---	0.10	0.10	0.15	L-2	1764	---	---	0.15	0.15	0.22
Mineral surfaced roll roofing																
Single coverage 90 lb./SQ. roll with 6" end lap and 2" headlap	Demo	SQ	L-2	32	---	---	8.13	8.13	12.03	L-2	22	---	---	11.61	11.61	17.18
Double coverage selvage roll, with 6" end lap and 17" exposure	Demo	SQ	L-2	22	---	---	11.82	11.82	17.50	L-2	15	---	---	16.89	16.89	25.00
Wood																
Roofing																
Shakes																
24" L with 10" exposure																
1/2" to 3/4" T	Demo	SQ	L-2	25	---	---	10.40	10.40	15.40	L-2	18	---	---	14.86	14.86	22.00
3/4" to 5/4" T	Demo	SQ	L-2	23	---	---	11.56	11.56	17.11	L-2	16	---	---	16.51	16.51	24.44
Shingles																
16" L with 5" exposure	Demo	SQ	L-2	12	---	---	21.67	21.67	32.08	L-2	8	---	---	30.96	30.96	45.82
18" L with 5-1/2" exposure	Demo	SQ	L-2	13	---	---	20.64	20.64	30.55	L-2	9	---	---	29.49	29.49	43.64
24" L with 7-1/2" exposure	Demo	SQ	L-2	18	---	---	14.45	14.45	21.38	L-2	13	---	---	20.64	20.64	30.55
Siding																
Bevel																
1/2" x 8" with 6-3/4" exp.	Demo	SF	L-2	910	---	---	0.29	0.29	0.42	L-2	637	---	---	0.41	0.41	0.60
5/8" x 10" with 8-3/4" exp.	Demo	SF	L-2	1180	---	---	0.22	0.22	0.33	L-2	826	---	---	0.31	0.31	0.47
3/4" x 12" with 10-3/4" exp.	Demo	SF	L-2	1450	---	---	0.18	0.18	0.27	L-2	1015	---	---	0.26	0.26	0.38

| | | | Costs Based On Large Volume | | | | | | Costs Based On Small Volume | | | | | |
Description	Oper	Unit	Crew Size	Avg Day Prod	Avg Mat'l Unit Cost	Avg Equip Unit Cost	Avg Labor Unit Cost	Avg Total Unit Cost	Avg Price Incl O&P	Crew Size	Avg Day Prod	Avg Mat'l Unit Cost	Avg Equip Unit Cost	Avg Labor Unit Cost	Avg Total Unit Cost	Avg Price Incl O&P
Site Work, demolition (continued)																
Drop (horizontal), 1/4" T&G																
1" x 8" with 7" exposure	Demo	SF	L-2	945	---	---	0.28	0.28	0.41	L-2	662	---	---	0.39	0.39	0.58
1" x 10" with 9" exposure	Demo	SF	L-2	1215	---	---	0.21	0.21	0.32	L-2	851	---	---	0.31	0.31	0.45
Board (1" x 12") and batten (1" x 2") @ 12" o.c.																
Horizontal	Demo	SF	L-2	1280	---	---	0.20	0.20	0.30	L-2	896	---	---	0.29	0.29	0.43
Vertical																
Standard	Demo	SF	L-2	1025	---	---	0.25	0.25	0.38	L-2	718	---	---	0.36	0.36	0.54
Reverse	Demo	SF	L-2	1065	---	---	0.24	0.24	0.36	L-2	746	---	---	0.35	0.35	0.52
Board on board (1" x 12" with 1-1/2" overlap), vertical	Demo	SF	L-2	950	---	---	0.27	0.27	0.41	L-2	665	---	---	0.39	0.39	0.58
Plywood (1/2" T) with battens (1" x 2")																
16" o.c. battens	Demo	SF	L-2	2255	---	---	0.12	0.12	0.17	L-2	1579	---	---	0.16	0.16	0.24
24" o.c. battens	Demo	SF	L-2	2365	---	---	0.11	0.11	0.16	L-2	1656	---	---	0.16	0.16	0.23
Shakes																
24" L with 11-1/2" exp.																
1/2" to 3/4" T	Demo	SF	L-2	1600	---	---	0.16	0.16	0.24	L-2	1120	---	---	0.23	0.23	0.34
3/4" to 5/4" T	Demo	SF	L-2	1440	---	---	0.18	0.18	0.27	L-2	1008	---	---	0.26	0.26	0.38
Shingles																
16" L with 7-1/2" exposure	Demo	SF	L-2	1000	---	---	0.26	0.26	0.38	L-2	700	---	---	0.37	0.37	0.55
18" L with 8-1/2" exposure	Demo	SF	L-2	1130	---	---	0.23	0.23	0.34	L-2	791	---	---	0.33	0.33	0.49
24" L with 11-1/2" exposure	Demo	SF	L-2	1530	---	---	0.17	0.17	0.25	L-2	1071	---	---	0.24	0.24	0.36
Doors, Windows, Glazing																
Doors with related trim and frame																
Closet, with track																
Folding, 4 doors	Demo	Set	L-2	12	---	---	21.67	21.67	32.08	L-2	8	---	---	30.96	30.96	45.82
Sliding, 2 or 3 doors	Demo	Set	L-2	12	---	---	21.67	21.67	32.08	L-2	8	---	---	30.96	30.96	45.82
Entry, 3' x 7'	Demo	EA	L-2	14	---	---	18.58	18.58	27.49	L-2	10	---	---	26.54	26.54	39.28
Fire, 3' x 7'	Demo	EA	L-2	14	---	---	18.58	18.58	27.49	L-2	10	---	---	26.54	26.54	39.28
Garage																
Wood, aluminum, or hardboard																
Single	Demo	EA	L-2	8	---	---	32.51	32.51	48.12	L-2	6	---	---	46.44	46.44	68.74
Double	Demo	EA	L-2	6	---	---	43.35	43.35	64.15	L-2	4	---	---	61.92	61.92	91.65

			Costs Based On Large Volume							Costs Based On Small Volume					
Description	Oper	Unit	Crew Size	Avg Day Prod	Avg Mat'l Unit Cost	Avg Equip Unit Cost	Avg Labor Unit Cost	Avg Total Unit Cost	Avg Price Incl O&P	Crew Size	Avg Day Prod	Avg Mat'l Unit Cost	Avg Equip Unit Cost	Avg Labor Unit Cost	Avg Total Unit Cost	Avg Price Incl O&P

Site Work, demolition (continued)

Description	Oper	Unit	Crew Size	Avg Day Prod	Avg Mat'l Unit Cost	Avg Equip Unit Cost	Avg Labor Unit Cost	Avg Total Unit Cost	Avg Price Incl O&P	Crew Size	Avg Day Prod	Avg Mat'l Unit Cost	Avg Equip Unit Cost	Avg Labor Unit Cost	Avg Total Unit Cost	Avg Price Incl O&P
Steel																
Single	Demo	EA	L-2	7	---	---	37.15	37.15	54.99	L-2	5	---	---	53.08	53.08	78.56
Double	Demo	EA	L-2	5	---	---	52.02	52.02	76.98	L-2	4	---	---	74.31	74.31	109.98
Glass sliding, with track																
2 lites wide	Demo	Set	L-2	8	---	---	32.51	32.51	48.12	L-2	6	---	---	46.44	46.44	68.74
3 lites wide	Demo	Set	L-2	6	---	---	43.35	43.35	64.15	L-2	4	---	---	61.92	61.92	91.65
4 lites wide	Demo	Set	L-2	4	---	---	65.02	65.02	96.23	L-2	3	---	---	92.89	92.89	137.47
Interior, 3' x 7'	Demo	EA	L-2	16	---	---	16.26	16.26	24.06	L-2	11	---	---	23.22	23.22	34.37
Screen, 3' x 7'	Demo	EA	L-2	20	---	---	13.00	13.00	19.25	L-2	14	---	---	18.58	18.58	27.49
Storm combination, 3' x 7'	Demo	EA	L-2	20	---	---	13.00	13.00	19.25	L-2	14	---	---	18.58	18.58	27.49
Windows, with related trim and frame																
To 12 SF																
Aluminum	Demo	EA	L-2	21	---	---	12.38	12.38	18.33	L-2	15	---	---	17.69	17.69	26.19
Wood	Demo	EA	L-2	16	---	---	16.26	16.26	24.06	L-2	11	---	---	23.22	23.22	34.37
13 SF to 50 SF																
Aluminum	Demo	EA	L-2	13	---	---	20.01	20.01	29.61	L-2	9	---	---	28.58	28.58	42.30
Wood	Demo	EA	L-2	10	---	---	26.01	26.01	38.49	L-2	7	---	---	37.15	37.15	54.99
Glazing, clean sash and remove old putty or rubber																
3/32" T float, putty or rubber																
8" x 12" (0.667 SF)	Demo	SF	G-1	25	---	---	6.66	6.66	9.79	G-1	18	---	---	9.51	9.51	13.98
12" x 16" (1.333 SF)	Demo	SF	G-1	45	---	---	3.70	3.70	5.44	G-1	32	---	---	5.28	5.28	7.77
14" x 20" (1.944 SF)	Demo	SF	G-1	55	---	---	3.03	3.03	4.45	G-1	39	---	---	4.32	4.32	6.36
16" x 24" (2.667 SF)	Demo	SF	G-1	70	---	---	2.38	2.38	3.50	G-1	49	---	---	3.40	3.40	4.99
24" x 26" (4.333 SF)	Demo	SF	G-1	95	---	---	1.75	1.75	2.58	G-1	67	---	---	2.50	2.50	3.68
36" x 24" (6.000 SF)	Demo	SF	G-1	125	---	---	1.33	1.33	1.96	G-1	88	---	---	1.90	1.90	2.80
1/8" T float, putty, steel sash																
12" x 16" (1.333 SF)	Demo	SF	G-1	45	---	---	3.70	3.70	5.44	G-1	32	---	---	5.28	5.28	7.77
16" x 20" (2.222 SF)	Demo	SF	G-1	65	---	---	2.56	2.56	3.76	G-1	46	---	---	3.66	3.66	5.38
16" x 24" (2.667 SF)	Demo	SF	G-1	70	---	---	2.38	2.38	3.50	G-1	49	---	---	3.40	3.40	4.99
24" x 26" (4.333 SF)	Demo	SF	G-1	95	---	---	1.75	1.75	2.58	G-1	67	---	---	2.50	2.50	3.68
28" x 32" (6.222 SF)	Demo	SF	G-1	130	---	---	1.28	1.28	1.88	G-1	91	---	---	1.83	1.83	2.69
36" x 36" (9.000 SF)	Demo	SF	G-1	160	---	---	1.04	1.04	1.53	G-1	112	---	---	1.49	1.49	2.18
36" x 48" (12.000 SF)	Demo	SF	G-1	190	---	---	0.88	0.88	1.29	G-1	133	---	---	1.25	1.25	1.84

			Costs Based On Large Volume							Costs Based On Small Volume					
Description	Oper	Unit	Crew Size	Avg Day Prod	Avg Mat'l Unit Cost	Avg Equip Unit Cost	Avg Labor Unit Cost	Avg Total Unit Cost	Avg Price Incl O&P	Crew Size	Avg Day Prod	Avg Mat'l Unit Cost	Avg Equip Unit Cost	Avg Labor Unit Cost	Avg Total Unit Cost	Avg Price Incl O&P

Site Work, demolition (continued)

1/4" T float
  Wood sash with putty

72" x 48" (24.0 SF)	Demo	SF	G-1	185	---	---	0.90	0.90	1.32	G-1	130	---	---	1.29	1.29	1.89

  Aluminum sash with aluminum channel and rigid neoprene rubber

48" x 96" (32.0 SF)	Demo	SF	G-1	175	---	---	0.95	0.95	1.40	G-1	123	---	---	1.36	1.36	2.00
96" x 96" (64.0 SF)	Demo	SF	G-1	180	---	---	0.92	0.92	1.36	G-1	126	---	---	1.32	1.32	1.94

1" T insulating glass; with 2 pieces 1/4" float and 1/2" air space

To 6.0 SF	Demo	SF	G-1	50	---	---	3.33	3.33	4.89	G-1	35	---	---	4.76	4.76	6.99
6.1 to 12.0 SF	Demo	SF	G-1	110	---	---	1.51	1.51	2.22	G-1	77	---	---	2.16	2.16	3.18
12.1 to 18.0 SF	Demo	SF	G-1	150	---	---	1.11	1.11	1.63	G-1	105	---	---	1.59	1.59	2.33
18.1 to 24.0 SF	Demo	SF	G-1	145	---	---	1.15	1.15	1.69	G-1	102	---	---	1.64	1.64	2.41

Aluminum sliding door glass with aluminum channel and rigid neoprene rubber
  34" x 76" (17.944 SF)
  5/8" T insulating glass with 2 pieces 5/32" T tempered with 1-1/4"

air space	Demo	SF	G-1	170	---	---	0.98	0.98	1.44	G-1	119	---	---	1.40	1.40	2.06
5/32" T tempered	Demo	SF	G-1	195	---	---	0.85	0.85	1.25	G-1	137	---	---	1.22	1.22	1.79

  46" x 76" (24.278 SF)
  5/8" T insulating glass with 2 pieces 5/32" T tempered with 1-1/4"

air space	Demo	SF	G-1	215	---	---	0.77	0.77	1.14	G-1	151	---	---	1.11	1.11	1.63
5/32" T tempered	Demo	SF	G-1	245	---	---	0.68	0.68	1.00	G-1	172	---	---	0.97	0.97	1.43

Finishes. Exterior and interior, with hand tools; no insulation removal included
Plaster and stucco; remove to studs or sheathing
  Lath (wood or metal) and plaster, walls and ceiling

2 coats	Demo	SY	L-2	130	---	---	2.00	2.00	2.96	L-2	91	---	---	2.86	2.86	4.23
3 coats	Demo	SY	L-2	120	---	---	2.17	2.17	3.21	L-2	84	---	---	3.10	3.10	4.58

  Stucco and metal netting

2 coats	Demo	SY	L-2	95	---	---	2.74	2.74	4.05	L-2	67	---	---	3.91	3.91	5.79
3 coats	Demo	SY	L-2	80	---	---	3.25	3.25	4.81	L-2	56	---	---	4.64	4.64	6.87

Wallboard, gypsum (drywall), walls and ceilings	Demo	SF	L-2	1540	---	---	0.17	0.17	0.25	L-2	1078	---	---	0.24	0.24	0.36

Ceramic, metal, plastic tile
  Floors, 1" x 1"

Adhesive or dry-set base	Demo	SF	L-2	550	---	---	0.47	0.47	0.70	L-2	385	---	---	0.68	0.68	1.00
Conventional mortar base	Demo	SF	L-2	475	---	---	0.55	0.55	0.81	L-2	333	---	---	0.78	0.78	1.16

			Costs Based On Large Volume							Costs Based On Small Volume						
Description	Oper	Unit	Crew Size	Avg Day Prod	Avg Mat'l Unit Cost	Avg Equip Unit Cost	Avg Labor Unit Cost	Avg Total Unit Cost	Avg Price Incl O&P	Crew Size	Avg Day Prod	Avg Mat'l Unit Cost	Avg Equip Unit Cost	Avg Labor Unit Cost	Avg Total Unit Cost	Avg Price Incl O&P

Site Work, demolition (continued)

Walls, 1" x 1" or 4-1/4" x 4 1/4"

Description	Oper	Unit	Crew Size	Avg Day Prod	Mat'l	Equip	Labor	Total	Price O&P	Crew Size	Avg Day Prod	Mat'l	Equip	Labor	Total	Price O&P
Adhesive or dry-set base	Demo	SF	L-2	480	---	---	0.54	0.54	0.80	L-2	336	---	---	0.77	0.77	1.15
Conventional mortar base	Demo	SF	L-2	400	---	---	0.65	0.65	0.96	L-2	280	---	---	0.93	0.93	1.37
Acoustical or insulating ceiling tile																
Adhesive set, tile only	Demo	SF	L-2	1300	---	---	0.20	0.20	0.30	L-2	910	---	---	0.29	0.29	0.42
Stapled, tile only	Demo	SF	L-2	1170	---	---	0.22	0.22	0.33	L-2	819	---	---	0.32	0.32	0.47
Stapled, tile and furring strips	Demo	SF	L-2	1540	---	---	0.17	0.17	0.25	L-2	1078	---	---	0.24	0.24	0.36
Suspended ceiling system; panels and grid system	Demo	SF	L-2	1780	---	---	0.15	0.15	0.22	L-2	1246	---	---	0.21	0.21	0.31
Resilient flooring, adhesive set																
Sheet products	Demo	SY	L-2	160	---	---	1.63	1.63	2.41	L-2	112	---	---	2.32	2.32	3.44
Tile products	Demo	SF	L-2	1500	---	---	0.17	0.17	0.26	L-2	1050	---	---	0.25	0.25	0.37
Wallpaper. NOTE: Average output is expressed in rolls (36.0 SF/ single roll)																
Single layer of paper from plaster with steaming equipment	Demo	Roll	L-1	20	---	---	6.63	6.63	9.82	L-1	14	---	---	9.48	9.48	14.03
Several layers of paper from plaster with steaming equipment	Demo	Roll	L-1	12	---	---	10.57	10.57	15.65	L-1	9	---	---	15.10	15.10	22.35
Vinyls (with non-woven, woven, or synthetic fiber backings) from plaster with steaming equipment	Demo	Roll	L-1	30	---	---	4.33	4.33	6.42	L-1	21	---	---	6.19	6.19	9.16
Single layer of paper from drywall with steaming equipment	Demo	Roll	L-1	20	---	---	6.63	6.63	9.82	L-1	14	---	---	9.48	9.48	14.03
Several layers of paper from drywall with steaming equipment	Demo	Roll	L-1	12	---	---	10.57	10.57	15.65	L-1	9	---	---	15.10	15.10	22.35
Vinyls (with synthetic fiber backing) from drywall with steaming equipment	Demo	Roll	L-1	30	---	---	4.33	4.33	6.42	L-1	21	---	---	6.19	6.19	9.16
Vinyls (with other backings), from drywall with steaming equipment	Demo	Roll	L-1	20	---	---	6.63	6.63	9.82	L-1	14	---	---	9.48	9.48	14.03

Excavation

Description	Oper	Unit	Crew Size	Avg Day Prod	Mat'l	Equip	Labor	Total	Price O&P	Crew Size	Avg Day Prod	Mat'l	Equip	Labor	Total	Price O&P
Digging out or trenching			L-2	475	---	---	0.55	0.55	0.81	L-2	333	---	---	0.78	0.78	1.16
Pits, medium earth piled																
With front end loader, track mounted, 1-1/2 CY capacity;																
55 CY per hour	Demo	CY	V-1	440	---	0.48	0.58	1.05	1.33	V-1	308	---	0.68	0.82	1.51	1.89

Description	Oper	Unit	Costs Based On Large Volume							Costs Based On Small Volume						
			Crew Size	Avg Day Prod	Avg Mat'l Unit Cost	Avg Equip Unit Cost	Avg Labor Unit Cost	Avg Total Unit Cost	Avg Price Incl O&P	Crew Size	Avg Day Prod	Avg Mat'l Unit Cost	Avg Equip Unit Cost	Avg Labor Unit Cost	Avg Total Unit Cost	Avg Price Incl O&P

Site Work, excavation (continued)

Description	Oper	Unit	Crew	Prod	Mat'l	Equip	Labor	Total	O&P	Crew	Prod	Mat'l	Equip	Labor	Total	O&P
By hand																
To 4'-0" D	Demo	CY	L-2	15	---	---	17.34	17.34	25.66	L-2	11	---	---	24.77	24.77	36.66
4'-0" to 6'-0" D	Demo	CY	L-2	10	---	---	26.01	26.01	38.49	L-2	7	---	---	37.15	37.15	54.99
6'-0" to 8'-0" D	Demo	CY	L-2	6	---	---	43.35	43.35	64.15	L-2	4	---	---	61.92	61.92	91.65
Continuous footing or trench, medium earth piled																
With tractor backhoe, 3/8 CY																
capacity (48 HP); 15 CY																
per hour	Demo	CY	V-2	120	---	1.65	2.11	3.76	4.75	V-2	84	---	2.35	3.01	5.37	6.78
By hand, to 4'-0" D	Demo	CY	L-2	15	---	---	17.34	17.34	25.66	L-2	11	---	---	24.77	24.77	36.66
Backfilling, by hand, medium soil																
Without compaction	Demo	CY	L-2	28	---	---	9.29	9.29	13.75	L-2	20	---	---	13.27	13.27	19.64
With hand compaction																
6" layers	Demo	CY	L-2	17	---	---	15.30	15.30	22.64	L-2	12	---	---	21.86	21.86	32.35
12" layers	Demo	CY	L-2	22	---	---	11.82	11.82	17.50	L-2	15	---	---	16.89	16.89	25.00
With vibrating plate compaction																
6" layers	Demo	CY	A-2	20	---	1.04	14.95	15.99	23.02	A-2	14	---	1.49	21.35	22.84	32.88
12" layers	Demo	CY	A-2	24	---	0.87	12.46	13.33	19.18	A-2	17	---	1.24	17.79	19.04	27.40

Stairs
Stair Parts

Description	Oper	Unit	Crew	Prod	Mat'l	Equip	Labor	Total	O&P	Crew	Prod	Mat'l	Equip	Labor	Total	O&P
Balusters, stock pine																
1-1/16" x 1-1/16"	Inst	LF	C-1	200	0.63	---	0.84	1.47	1.88	C-1	140	0.74	---	1.20	1.94	2.53
1-5/8" x 1-5/8"	Inst	LF	C-1	160	1.21	---	1.05	2.26	2.78	C-1	112	1.43	---	1.50	2.93	3.67
Balusters, turned,																
30" high																
Pine	Inst	EA	C-1	24	5.50	---	7.02	12.52	15.96	C-1	17	6.50	---	10.03	16.53	21.45
Birch	Inst	EA	C-1	24	6.60	---	7.02	13.62	17.06	C-1	17	7.80	---	10.03	17.83	22.75
42" high																
Pine	Inst	EA	C-1	20	6.60	---	8.43	15.03	19.16	C-1	14	7.80	---	12.04	19.84	25.74
Birch	Inst	EA	C-1	20	8.80	---	8.43	17.23	21.36	C-1	14	10.40	---	12.04	22.44	28.34
Newels, 3-1/4" wide																
Starting	Inst	EA	C-1	6	55.00	---	28.09	83.09	96.86	C-1	4	65.00	---	40.13	105.13	124.79
Landing	Inst	EA	C-1	4	82.50	---	42.14	124.64	145.28	C-1	3	97.50	---	60.20	157.70	187.19
Railings, Built-up oak	Inst	LF	C-1	50	8.80	---	3.37	12.17	13.82	C-1	58	10.40	---	2.93	13.33	14.77
Railings, Sub rail oak	Inst	LF	C-1	100	3.63	---	1.69	5.32	6.14	C-1	70	4.29	---	2.41	6.70	7.88

			Costs Based On Large Volume							Costs Based On Small Volume					
Description	Oper	Unit	Crew Size	Avg Day Prod	Avg Mat'l Unit Cost	Avg Equip Unit Cost	Avg Labor Unit Cost	Avg Total Unit Cost	Avg Price Incl O&P	Crew Size	Avg Day Prod	Avg Mat'l Unit Cost	Avg Equip Unit Cost	Avg Labor Unit Cost	Avg Total Unit Cost	Avg Price Incl O&P

Stairs (continued)

Risers, 3/4" x 7-1/2" high

Description	Oper	Unit	Crew Size	Avg Day Prod	Avg Mat'l Unit Cost	Avg Equip Unit Cost	Avg Labor Unit Cost	Avg Total Unit Cost	Avg Price Incl O&P	Crew Size	Avg Day Prod	Avg Mat'l Unit Cost	Avg Equip Unit Cost	Avg Labor Unit Cost	Avg Total Unit Cost	Avg Price Incl O&P
Beech	Inst	LF	C-1	60	4.57	---	2.81	7.37	8.75	C-1	42	5.40	---	4.01	9.41	11.37
Fir	Inst	LF	C-1	60	1.27	---	2.81	4.07	5.45	C-1	42	1.50	---	4.01	5.51	7.47
Oak	Inst	LF	C-1	60	4.18	---	2.81	6.99	8.37	C-1	42	4.94	---	4.01	8.95	10.92
Pine	Inst	LF	C-1	60	1.27	---	2.81	4.07	5.45	C-1	42	1.50	---	4.01	5.51	7.47
Skirt board, pine																
1" x 10"	Inst	LF	C-1	50	1.60	---	3.37	4.97	6.62	C-1	35	1.89	---	4.82	6.70	9.06
1" x 12"	Inst	LF	C-1	45	1.93	---	3.75	5.67	7.51	C-1	32	2.28	---	5.35	7.63	10.25
Treads, oak																
1-16" x 9-1/2" wide																
3' long	Inst	EA	C-1	16	18.59	---	10.53	29.12	34.29	C-1	11	21.97	---	15.05	37.02	44.39
4' long	Inst	EA	C-1	15	25.96	---	11.24	37.20	42.70	C-1	11	30.68	---	16.05	46.73	54.60
1-1/16" x 11-1/2" wide																
3' long	Inst	EA	C-1	16	22.28	---	10.53	32.81	37.97	C-1	11	26.33	---	15.05	41.37	48.75
6' long	Inst	EA	C-1	12	47.30	---	14.05	61.35	68.23	C-1	8	55.90	---	20.07	75.97	85.80
For beech treads, ADD	Inst	%	---	---	40.0	---	---	---	---	---	---	40.0	---	---	---	---
For mitered return nosings, ADD	Inst	LF	C-1	60	2.48	---	2.81	5.28	6.66	C-1	42	2.93				

Stairs, shop fabricated. Per flight

Basement stairs, soft wood, open risers

Description	Oper	Unit	Crew Size	Avg Day Prod	Avg Mat'l Unit Cost	Avg Equip Unit Cost	Avg Labor Unit Cost	Avg Total Unit Cost	Avg Price Incl O&P	Crew Size	Avg Day Prod	Avg Mat'l Unit Cost	Avg Equip Unit Cost	Avg Labor Unit Cost	Avg Total Unit Cost	Avg Price Incl O&P
3' wide, 8' high	Inst	Flt	C-18	4.0	93.50	---	84.27	177.77	219.07	C-18	2.8	110.50	---	120.39	230.89	289.88
Box stairs, no handrails, 3' wide																
Oak treads																
2' high	Inst	Flt	C-18	5.0	110.00	---	67.42	177.42	210.45	C-18	3.5	130.00	---	96.31	226.31	273.51
4' high	Inst	Flt	C-18	4.0	247.50	---	84.27	331.77	373.07	C-18	2.8	292.50	---	120.39	412.89	471.88
6' high	Inst	Flt	C-18	3.5	401.50	---	96.31	497.81	545.01	C-18	2.5	474.50	---	137.59	612.09	679.51
8' high	Inst	Flt	C-18	3.0	511.50	---	112.37	623.87	678.92	C-18	2.1	604.50	---	160.52	765.02	843.68
Pine treads for carpet																
2' high	Inst	Flt	C-18	5.0	82.50	---	67.42	149.92	182.95	C-18	3.5	97.50	---	96.31	193.81	241.01
4' high	Inst	Flt	C-18	4.0	154.00	---	84.27	238.27	279.57	C-18	2.8	182.00	---	120.39	302.39	361.38
6' high	Inst	Flt	C-18	3.5	242.00	---	96.31	338.31	385.51	C-18	2.5	286.00	---	137.59	423.59	491.01
8' high	Inst	Flt	C-18	3.0	297.00	---	112.37	409.37	464.42	C-18	2.1	351.00	---	160.52	511.52	590.18
Stair rail with balusters,																
5 risers	Inst	EA	C-18	15.0	143.00	---	22.47	165.47	176.48	C-18	10.5	169.00	---	32.10	201.10	216.84
For 4' wide stairs, ADD	Inst	%	C-18	---	25.0	5.0	---	---	---	C-18	---	25.0	5.0	---	---	---

Description	Oper	Unit	Costs Based On Large Volume							Costs Based On Small Volume						
			Crew Size	Avg Day Prod	Avg Mat'l Unit Cost	Avg Equip Unit Cost	Avg Labor Unit Cost	Avg Total Unit Cost	Avg Price Incl O&P	Crew Size	Avg Day Prod	Avg Mat'l Unit Cost	Avg Equip Unit Cost	Avg Labor Unit Cost	Avg Total Unit Cost	Avg Price Incl O&P

Siding (continued)

Related materials and operations
Rosin sized sheathing paper,
36" W 500 SF/roll; nailed

Description	Oper	Unit	Crew Size	Avg Day Prod	Avg Mat'l Unit Cost	Avg Equip Unit Cost	Avg Labor Unit Cost	Avg Total Unit Cost	Avg Price Incl O&P	Crew Size	Avg Day Prod	Avg Mat'l Unit Cost	Avg Equip Unit Cost	Avg Labor Unit Cost	Avg Total Unit Cost	Avg Price Incl O&P
Over solid sheathing	Inst	SF	C-18	12500	0.03	---	0.03	0.06	0.07	C-18	11500	0.04	---	0.03	0.06	0.08

Site Work

Demolition, wreck and remove to dumpster

Concrete
Footings, with air tools; reinforced

Description	Oper	Unit	Crew Size	Avg Day Prod	Avg Mat'l Unit Cost	Avg Equip Unit Cost	Avg Labor Unit Cost	Avg Total Unit Cost	Avg Price Incl O&P	Crew Size	Avg Day Prod	Avg Mat'l Unit Cost	Avg Equip Unit Cost	Avg Labor Unit Cost	Avg Total Unit Cost	Avg Price Incl O&P
8" T x 12" W (.67 CF/LF)	Demo	LF	A-2	120	---	0.73	2.49	3.22	4.39	A-2	84	---	1.04	3.56	4.60	6.27
8" T x 16" W (.89 CF/LF)	Demo	LF	A-2	100	---	0.88	2.99	3.86	5.27	A-2	70	---	1.25	4.27	5.52	7.53
8" T x 20" W (1.11 CF/LF)	Demo	LF	A-2	90	---	0.97	3.32	4.29	5.86	A-2	63	---	1.39	4.75	6.13	8.36
12" T x 12" W (1.00 CF/LF)	Demo	LF	A-2	70	---	1.25	4.27	5.52	7.53	A-2	49	---	1.79	6.10	7.89	10.75
12" T x 16" W (1.33 CF/LF)	Demo	LF	A-2	60	---	1.46	4.98	6.44	8.78	A-2	42	---	2.08	7.12	9.20	12.55
12" T x 20" W (1.67 CF/LF)	Demo	LF	A-2	55	---	1.59	5.44	7.03	9.58	A-2	39	---	2.27	7.76	10.04	13.69
12"T x 24" W (2.00 CF/LF)	Demo	LF	A-2	50	---	1.75	5.98	7.73	10.54	A-2	35	---	2.50	8.54	11.04	15.06

Foundations and retaining walls, with air tools, per LF wall

With reinforcing
4'-0" H

Description	Oper	Unit	Crew Size	Avg Day Prod	Avg Mat'l Unit Cost	Avg Equip Unit Cost	Avg Labor Unit Cost	Avg Total Unit Cost	Avg Price Incl O&P	Crew Size	Avg Day Prod	Avg Mat'l Unit Cost	Avg Equip Unit Cost	Avg Labor Unit Cost	Avg Total Unit Cost	Avg Price Incl O&P
8" T (2.67 CF/LF)	Demo	SF	A-2	60	---	1.46	4.98	6.44	8.78	A-2	42	---	2.08	7.12	9.20	12.55
12" T (4.00 CF/LF)	Demo	SF	A-2	45	---	1.95	6.64	8.59	11.71	A-2	32	---	2.78	9.49	12.27	16.73

8'-0" H

Description	Oper	Unit	Crew Size	Avg Day Prod	Avg Mat'l Unit Cost	Avg Equip Unit Cost	Avg Labor Unit Cost	Avg Total Unit Cost	Avg Price Incl O&P	Crew Size	Avg Day Prod	Avg Mat'l Unit Cost	Avg Equip Unit Cost	Avg Labor Unit Cost	Avg Total Unit Cost	Avg Price Incl O&P
8" T (5.33 CF/LF)	Demo	SF	A-2	55	---	1.59	5.44	7.03	9.58	A-2	39	---	2.27	7.76	10.04	13.69
12" T (8.00 CF/LF)	Demo	SF	A-2	40	---	2.19	7.47	9.66	13.17	A-2	28	---	3.13	10.68	13.80	18.82

12'-0" H

Description	Oper	Unit	Crew Size	Avg Day Prod	Avg Mat'l Unit Cost	Avg Equip Unit Cost	Avg Labor Unit Cost	Avg Total Unit Cost	Avg Price Incl O&P	Crew Size	Avg Day Prod	Avg Mat'l Unit Cost	Avg Equip Unit Cost	Avg Labor Unit Cost	Avg Total Unit Cost	Avg Price Incl O&P
8" T (8.00 CF/LF)	Demo	SF	A-2	50	---	1.75	5.98	7.73	10.54	A-2	35	---	2.50	8.54	11.04	15.06
12" T (12.00 CF/LF)	Demo	SF	A-2	35	---	2.50	8.54	11.04	15.06	A-2	25	---	3.57	12.20	15.77	21.51

Without reinforcing
4'-0" H

Description	Oper	Unit	Crew Size	Avg Day Prod	Avg Mat'l Unit Cost	Avg Equip Unit Cost	Avg Labor Unit Cost	Avg Total Unit Cost	Avg Price Incl O&P	Crew Size	Avg Day Prod	Avg Mat'l Unit Cost	Avg Equip Unit Cost	Avg Labor Unit Cost	Avg Total Unit Cost	Avg Price Incl O&P
8" T (2.67 CF/LF)	Demo	SF	A-2	80	---	1.09	3.74	4.83	6.59	A-2	56	---	1.56	5.34	6.90	9.41
12" T (4.00 CF/LF)	Demo	SF	A-2	60	---	1.46	4.98	6.44	8.78	A-2	42	---	2.08	7.12	9.20	12.55

8'-0" H

Description	Oper	Unit	Crew Size	Avg Day Prod	Avg Mat'l Unit Cost	Avg Equip Unit Cost	Avg Labor Unit Cost	Avg Total Unit Cost	Avg Price Incl O&P	Crew Size	Avg Day Prod	Avg Mat'l Unit Cost	Avg Equip Unit Cost	Avg Labor Unit Cost	Avg Total Unit Cost	Avg Price Incl O&P
8" T (5.33 CF/LF)	Demo	SF	A-2	70	---	1.25	4.27	5.52	7.53	A-2	49	---	1.79	6.10	7.89	10.75
12" T (8.00 CF/LF)	Demo	SF	A-2	50	---	1.75	5.98	7.73	10.54	A-2	35	---	2.50	8.54	11.04	15.06

Site Work - Demolition

Average Output is based on what the designated crew can do in one day. In this section, the man might be a Laborer, a Carpenter, a Floor Layer, etc. Who does the wrecking depends on the quantity of wrecking to be done.

A Contractor might use Laborers exclusively on a large volume, whereas, on a small volume, the contractor might use a Carpenter(s) or a Carpenter and a Laborer.

The quantity of work that a man does in a day is greatly influenced by his selection of tools and/or equipment. A man can remove more brick with a compressor and pneumatic tool than he can with a sledge hammer or pry bar. In this section, the phrase "by hand" includes the use of hand tools, i.e.,

sledge hammers, wrecking bars, claw hammers, etc. When the Average Output is based on the use of equipment (not hand tools), a description of the equipment is provided.

The Average Output is not based on the use of "heavy" equipment such as bulldozers, cranes with wrecking balls, etc. The Average Output does include the labor involved in hauling wrecked material or debris to a dumpster located at the site. Average rental costs for dumpsters are: $175.00 for 40 CY (20' x 8' x 8' H); $150.00 for 30 CY (20' x 8' x 6' H). Rental period equals delivery and pickup when full.

			Costs Based On Large Volume							Costs Based On Small Volume					
Description	Oper	Unit	Crew Size	Avg Day Prod	Avg Mat'l Unit Cost	Avg Equip Unit Cost	Avg Labor Unit Cost	Avg Total Unit Cost	Avg Price Incl O&P	Crew Size	Avg Day Prod	Avg Mat'l Unit Cost	Avg Equip Unit Cost	Avg Labor Unit Cost	Avg Total Unit Cost	Avg Price Incl O&P

Site Work, demolition (continued)

12'-0" H

Description	Oper	Unit	Crew Size	Avg Day Prod	Avg Mat'l Unit Cost	Avg Equip Unit Cost	Avg Labor Unit Cost	Avg Total Unit Cost	Avg Price Incl O&P	Crew Size	Avg Day Prod	Avg Mat'l Unit Cost	Avg Equip Unit Cost	Avg Labor Unit Cost	Avg Total Unit Cost	Avg Price Incl O&P
8" T (8.00 CF/LF)	Demo	SF	A-2	60	---	1.46	4.98	6.44	8.78	A-2	42	---	2.08	7.12	9.20	12.55
12" T (12.00 CF/LF)	Demo	SF	A-2	40	---	2.19	7.47	9.66	13.17	A-2	28	---	3.13	10.68	13.80	18.82
Slabs with air tools																
With reinforcing																
4" T	Demo	SF	A-2	350	---	0.25	0.85	1.10	1.51	A-2	245	---	0.36	1.22	1.58	2.15
5" T	Demo	SF	A-2	425	---	0.21	0.70	0.91	1.24	A-2	298	---	0.29	1.00	1.30	1.77
6" T	Demo	SF	A-2	370	---	0.24	0.81	1.04	1.42	A-2	259	---	0.34	1.15	1.49	2.03
Without reinforcing																
4" T	Demo	SF	A-2	500	---	0.18	0.60	0.77	1.05	A-2	350	---	0.25	0.85	1.10	1.51
5" T	Demo	SF	A-2	425	---	0.21	0.70	0.91	1.24	A-2	298	---	0.29	1.00	1.30	1.77
6" T	Demo	SF	A-2	370	---	0.24	0.81	1.04	1.42	A-2	259	---	0.34	1.15	1.49	2.03
Masonry																
Brick																
Chimneys																
4" T wall	Demo	VLF	L-2	25	---	---	10.40	10.40	15.40	L-2	18	---	---	14.86	14.86	22.00
8" T wall	Demo	VLF	L-2	10	---	---	26.01	26.01	38.49	L-2	7	---	---	37.15	37.15	54.99
Columns, 12" x 12" o.d.	Demo	VLF	L-2	50	---	---	5.20	5.20	7.70	L-2	35	---	---	7.43	7.43	11.00
Veneer, 4" T, with air tools	Demo	SF	A-2	270	---	0.32	1.11	1.43	1.95	A-2	189	---	0.46	1.58	2.04	2.79
Walls, with air tools																
8" T wall	Demo	SF	A-2	135	---	0.65	2.21	2.86	3.90	A-2	95	---	0.93	3.16	4.09	5.58
12" T wall	Demo	SF	A-2	95	---	0.92	3.15	4.07	5.55	A-2	67	---	1.32	4.50	5.81	7.92
Masonry (concrete) block; lightweight (haydite) or heavyweight blocks																
Foundations and retaining walls; no excavation included																
Without reinforcing or with only lateral reinforcing																
With air tools																
8" W x 8" H x 16" L	Demo	SF	A-2	280	---	0.31	1.07	1.38	1.88	A-2	196	---	0.45	1.53	1.97	2.69
12" W x 8" H x 16" L	Demo	SF	A-2	240	---	0.36	1.25	1.61	2.20	A-2	168	---	0.52	1.78	2.30	3.14
Without air tools																
8" W x 8" H x 16" L	Demo	SF	L-2	225	---	---	1.16	1.16	1.71	L-2	158	---	---	1.65	1.65	2.44
12" W x 8" H x 16" L	Demo	SF	L-2	190	---	---	1.37	1.37	2.03	L-2	133	---	---	1.96	1.96	2.89

			Costs Based On Large Volume							Costs Based On Small Volume					
Description	Oper	Unit	Crew Size	Avg Day Prod	Avg Mat'l Unit Cost	Avg Equip Unit Cost	Avg Labor Unit Cost	Avg Total Unit Cost	Avg Price Incl O&P	Crew Size	Avg Day Prod	Avg Mat'l Unit Cost	Avg Equip Unit Cost	Avg Labor Unit Cost	Avg Total Unit Cost	Avg Price Incl O&P

Site Work, demolition (continued)

With vertical reinforcing in every other core (2 core blocks) with core filled
  With air tools

Description	Oper	Unit	Crew Size	Avg Day Prod	Avg Mat'l Unit Cost	Avg Equip Unit Cost	Avg Labor Unit Cost	Avg Total Unit Cost	Avg Price Incl O&P	Crew Size	Avg Day Prod	Avg Mat'l Unit Cost	Avg Equip Unit Cost	Avg Labor Unit Cost	Avg Total Unit Cost	Avg Price Incl O&P
8" W x 8" H x 16" L	Demo	SF	A-2	170	---	0.51	1.76	2.27	3.10	A-2	119	---	0.74	2.51	3.25	4.43
12" W x 8" H x 16" L	Demo	SF	A-2	145	---	0.60	2.06	2.67	3.63	A-2	102	---	0.86	2.95	3.81	5.19

Exterior walls (above grade) and partitions; no shoring included
  Without reinforcing or with only lateral reinforcing
    With air tools

Description	Oper	Unit	Crew Size	Avg Day Prod	Avg Mat'l Unit Cost	Avg Equip Unit Cost	Avg Labor Unit Cost	Avg Total Unit Cost	Avg Price Incl O&P	Crew Size	Avg Day Prod	Avg Mat'l Unit Cost	Avg Equip Unit Cost	Avg Labor Unit Cost	Avg Total Unit Cost	Avg Price Incl O&P
8" W x 8" H x 16" L	Demo	SF	A-2	350	---	0.25	0.85	1.10	1.51	A-2	245	---	0.36	1.22	1.58	2.15
12" W x 8" H x 16" L	Demo	SF	A-2	300	---	0.29	1.00	1.29	1.76	A-2	210	---	0.42	1.42	1.84	2.51

Without air tools

Description	Oper	Unit	Crew Size	Avg Day Prod	Avg Mat'l Unit Cost	Avg Equip Unit Cost	Avg Labor Unit Cost	Avg Total Unit Cost	Avg Price Incl O&P	Crew Size	Avg Day Prod	Avg Mat'l Unit Cost	Avg Equip Unit Cost	Avg Labor Unit Cost	Avg Total Unit Cost	Avg Price Incl O&P
8" W x 8" H x 16" L	Demo	SF	L-2	280	---	---	0.93	0.93	1.37	L-2	196	---	---	1.33	1.33	1.96
12" W x 8" H x 16" L	Demo	SF	L-2	240	---	---	1.08	1.08	1.60	L-2	168	---	---	1.55	1.55	2.29

Fences, without reinforcing or with only lateral reinforcing
  With air tools

Description	Oper	Unit	Crew Size	Avg Day Prod	Avg Mat'l Unit Cost	Avg Equip Unit Cost	Avg Labor Unit Cost	Avg Total Unit Cost	Avg Price Incl O&P	Crew Size	Avg Day Prod	Avg Mat'l Unit Cost	Avg Equip Unit Cost	Avg Labor Unit Cost	Avg Total Unit Cost	Avg Price Incl O&P
6" W x 4" H x 16" L	Demo	SF	A-2	390	---	0.22	0.77	0.99	1.35	A-2	273	---	0.32	1.10	1.42	1.93
6" W x 6" H x 16" L	Demo	SF	A-2	370	---	0.24	0.81	1.04	1.42	A-2	259	---	0.34	1.15	1.49	2.03
8" W x 8" H x 16" L	Demo	SF	A-2	350	---	0.25	0.85	1.10	1.51	A-2	245	---	0.36	1.22	1.58	2.15
12" W x 8" H x 16" L	Demo	SF	A-2	300	---	0.29	1.00	1.29	1.76	A-2	210	---	0.42	1.42	1.84	2.51

Without air tools

Description	Oper	Unit	Crew Size	Avg Day Prod	Avg Mat'l Unit Cost	Avg Equip Unit Cost	Avg Labor Unit Cost	Avg Total Unit Cost	Avg Price Incl O&P	Crew Size	Avg Day Prod	Avg Mat'l Unit Cost	Avg Equip Unit Cost	Avg Labor Unit Cost	Avg Total Unit Cost	Avg Price Incl O&P
6" W x 4" H x 16" L	Demo	SF	L-2	310	---	---	0.84	0.84	1.24	L-2	217	---	---	1.20	1.20	1.77
6" W x 6" H x 16" L	Demo	SF	L-2	295	---	---	0.88	0.88	1.30	L-2	207	---	---	1.26	1.26	1.86
8" W x 8" H x 16" L	Demo	SF	L-2	280	---	---	0.93	0.93	1.37	L-2	196	---	---	1.33	1.33	1.96
12" W x 8" H x 16" L	Demo	SF	L-2	240	---	---	1.08	1.08	1.60	L-2	168	---	---	1.55	1.55	2.29

Quarry tile, 6" or 9" squares
  Floors

Description	Oper	Unit	Crew Size	Avg Day Prod	Avg Mat'l Unit Cost	Avg Equip Unit Cost	Avg Labor Unit Cost	Avg Total Unit Cost	Avg Price Incl O&P	Crew Size	Avg Day Prod	Avg Mat'l Unit Cost	Avg Equip Unit Cost	Avg Labor Unit Cost	Avg Total Unit Cost	Avg Price Incl O&P
Conventional mortar set	Demo	SF	L-2	445	---	---	0.58	0.58	0.86	L-2	312	---	---	0.83	0.83	1.24
Dry-set mortar	Demo	SF	L-2	515	---	---	0.51	0.51	0.75	L-2	361	---	---	0.72	0.72	1.07

Rough Carpentry (Framing)

Dimension lumber
  Beams, set on steel columns
    Built-up from 2" lumber

Description	Oper	Unit	Crew Size	Avg Day Prod	Avg Mat'l Unit Cost	Avg Equip Unit Cost	Avg Labor Unit Cost	Avg Total Unit Cost	Avg Price Incl O&P	Crew Size	Avg Day Prod	Avg Mat'l Unit Cost	Avg Equip Unit Cost	Avg Labor Unit Cost	Avg Total Unit Cost	Avg Price Incl O&P
4" T x 10" W - 10' L (2 pcs)	Demo	LF	L-2	855	---	---	0.30	0.30	0.45	L-2	599	---	---	0.43	0.43	0.64
4" T x 12" W - 12' L (2 pcs)	Demo	LF	L-2	1025	---	---	0.25	0.25	0.38	L-2	718	---	---	0.36	0.36	0.54

			Costs Based On Large Volume							Costs Based On Small Volume					
Description	Oper	Unit	Crew Size	Avg Day Prod	Avg Mat'l Unit Cost	Avg Equip Unit Cost	Avg Labor Unit Cost	Avg Total Unit Cost	Avg Price Incl O&P	Crew Size	Avg Day Prod	Avg Mat'l Unit Cost	Avg Equip Unit Cost	Avg Labor Unit Cost	Avg Total Unit Cost	Avg Price Incl O&P

Site Work, demolition (continued)

Description	Oper	Unit	Crew Size	Avg Day Prod	Avg Mat'l Unit Cost	Avg Equip Unit Cost	Avg Labor Unit Cost	Avg Total Unit Cost	Avg Price Incl O&P	Crew Size	Avg Day Prod	Avg Mat'l Unit Cost	Avg Equip Unit Cost	Avg Labor Unit Cost	Avg Total Unit Cost	Avg Price Incl O&P
6" T x 10" W - 10' L (3 pcs)	Demo	LF	L-2	855	---	---	0.30	0.30	0.45	L-2	599	---	---	0.43	0.43	0.64
6" T x 12" W - 12' L (3 pcs)	Demo	LF	L-2	1025	---	---	0.25	0.25	0.38	L-2	718	---	---	0.36	0.36	0.54
Single member (solid lumber)																
3" T x 12" W - 12' L	Demo	LF	L-2	1025	---	---	0.25	0.25	0.38	L-2	718	---	---	0.36	0.36	0.54
4" T x 12" W - 12' L	Demo	LF	L-2	1025	---	---	0.25	0.25	0.38	L-2	718	---	---	0.36	0.36	0.54
Bracing, diagonal, notched-in, studs o.c.																
1" x 6" - 10'	Demo	LF	L-2	605	---	---	0.43	0.43	0.64	L-2	424	---	---	0.61	0.61	0.91
Bridging, "X" type, 1" x 3", (8", 10", 12" T) 16" o.c.	Demo	LF	L-2	350	---	---	0.74	0.74	1.10	L-2	245	---	---	1.06	1.06	1.57
Columns or posts																
4" x 4" -8' L	Demo	LF	L-2	770	---	---	0.34	0.34	0.50	L-2	539	---	---	0.48	0.48	0.71
6" x 6" -8' L	Demo	LF	L-2	770	---	---	0.34	0.34	0.50	L-2	539	---	---	0.48	0.48	0.71
6" x 8" -8' L	Demo	LF	L-2	650	---	---	0.40	0.40	0.59	L-2	455	---	---	0.57	0.57	0.85
8" x 8" -8' L	Demo	LF	L-2	615	---	---	0.42	0.42	0.63	L-2	431	---	---	0.60	0.60	0.89
Fascia																
1" x 4" - 12" L	Demo	LF	L-2	1315	---	---	0.20	0.20	0.29	L-2	921	---	---	0.28	0.28	0.42
Firestops or stiffeners																
2" x 4" - 16"	Demo	LF	L-2	475	---	---	0.55	0.55	0.81	L-2	333	---	---	0.78	0.78	1.16
2" x 6" - 16"	Demo	LF	L-2	475	---	---	0.55	0.55	0.81	L-2	333	---	---	0.78	0.78	1.16
Furring strips, 1" x 4" - 8' L																
Walls; strips 12" o.c.																
Studs 16" o.c.	Demo	SF	L-2	1215	---	---	0.21	0.21	0.32	L-2	851	---	---	0.31	0.31	0.45
Studs 24" o.c.	Demo	SF	L-2	1365	---	---	0.19	0.19	0.28	L-2	956	---	---	0.27	0.27	0.40
Masonry (concrete blocks)	Demo	SF	L-2	1095	---	---	0.24	0.24	0.35	L-2	767	---	---	0.34	0.34	0.50
Concrete	Demo	SF	L-2	645	---	---	0.40	0.40	0.60	L-2	452	---	---	0.58	0.58	0.85
Ceiling; joists 16" o.c.																
Strips 12" o.c.	Demo	SF	L-2	840	---	---	0.31	0.31	0.46	L-2	588	---	---	0.44	0.44	0.65
Strips 16" o.c.	Demo	SF	L-2	1075	---	---	0.24	0.24	0.36	L-2	753	---	---	0.35	0.35	0.51
Headers or lintels, (over openings)																
Built-up or single member																
4" T x 6" W - 4' L	Demo	LF	L-2	560	---	---	0.46	0.46	0.69	L-2	392	---	---	0.66	0.66	0.98
4" T x 8" W - 8' L	Demo	LF	L-2	720	---	---	0.36	0.36	0.53	L-2	504	---	---	0.52	0.52	0.76

			Costs Based On Large Volume							Costs Based On Small Volume					
Description	Oper	Unit	Crew Size	Avg Day Prod	Avg Mat'l Unit Cost	Avg Equip Unit Cost	Avg Labor Unit Cost	Avg Total Unit Cost	Avg Price Incl O&P	Crew Size	Avg Day Prod	Avg Mat'l Unit Cost	Avg Equip Unit Cost	Avg Labor Unit Cost	Avg Total Unit Cost	Avg Price Incl O&P

Site Work, demolition (continued)

Description	Oper	Unit	Crew Size	Avg Day Prod	Avg Mat'l Unit Cost	Avg Equip Unit Cost	Avg Labor Unit Cost	Avg Total Unit Cost	Avg Price Incl O&P	Crew Size	Avg Day Prod	Avg Mat'l Unit Cost	Avg Equip Unit Cost	Avg Labor Unit Cost	Avg Total Unit Cost	Avg Price Incl O&P
4" T x 10" W - 10' L	Demo	LF	L-2	855	---	---	0.30	0.30	0.45	L-2	599	---	---	0.43	0.43	0.64
4" T x 12" W - 12' L	Demo	LF	L-2	1025	---	---	0.25	0.25	0.38	L-2	718	---	---	0.36	0.36	0.54
4" T x 14" W - 14' L	Demo	LF	L-2	1025	---	---	0.25	0.25	0.38	L-2	718	---	---	0.36	0.36	0.54
Joists																
Ceiling																
2" x 4" - 8' L	Demo	LF	L-2	1230	---	---	0.21	0.21	0.31	L-2	861	---	---	0.30	0.30	0.45
2" x 4" - 10' L	Demo	LF	L-2	1395	---	---	0.19	0.19	0.28	L-2	977	---	---	0.27	0.27	0.39
2" x 8" - 12' L	Demo	LF	L-2	1420	---	---	0.18	0.18	0.27	L-2	994	---	---	0.26	0.26	0.39
2" x 10" - 14' L	Demo	LF	L-2	1535	---	---	0.17	0.17	0.25	L-2	1075	---	---	0.24	0.24	0.36
2" x 12" - 16' L	Demo	LF	L-2	1585	---	---	0.16	0.16	0.24	L-2	1110	---	---	0.23	0.23	0.35
Floor; seated on sill plate																
2" x 8" - 12' L	Demo	LF	L-2	1600	---	---	0.16	0.16	0.24	L-2	1120	---	---	0.23	0.23	0.34
2" x 10" - 14' L	Demo	LF	L-2	1725	---	---	0.15	0.15	0.22	L-2	1208	---	---	0.22	0.22	0.32
2" x 12" - 16' L	Demo	LF	L-2	1755	---	---	0.15	0.15	0.22	L-2	1229	---	---	0.21	0.21	0.31
Ledgers																
Nailed, 2" x 6" - 12' L	Demo	LF	L-2	710	---	---	0.37	0.37	0.54	L-2	497	---	---	0.52	0.52	0.77
Bolted, 3" x 8" - 12' L	Demo	LF	L-2	970	---	---	0.27	0.27	0.40	L-2	679	---	---	0.38	0.38	0.57
Plates; 2" x 4" or 2" x 6"																
Double top nailed	Demo	LF	L-2	820	---	---	0.32	0.32	0.47	L-2	574	---	---	0.45	0.45	0.67
Sill, nailed	Demo	LF	L-2	1360	---	---	0.19	0.19	0.28	L-2	952	---	---	0.27	0.27	0.40
Sill or bottom, bolted	Demo	LF	L-2	685	---	---	0.38	0.38	0.56	L-2	480	---	---	0.54	0.54	0.80
Rafters																
Common																
2" x 4" - 14' L	Demo	LF	L-2	1795	---	---	0.14	0.14	0.21	L-2	1257	---	---	0.21	0.21	0.31
2" x 6" - 14' L	Demo	LF	L-2	1660	---	---	0.16	0.16	0.23	L-2	1162	---	---	0.22	0.22	0.33
2" x 8" - 14' L	Demo	LF	L-2	1435	---	---	0.18	0.18	0.27	L-2	1005	---	---	0.26	0.26	0.38
2" x 10" - 14' L	Demo	LF	L-2	1195	---	---	0.22	0.22	0.32	L-2	837	---	---	0.31	0.31	0.46
Hip and/or valley																
2" x 4" - 16' L	Demo	LF	L-2	2050	---	---	0.13	0.13	0.19	L-2	1435	---	---	0.18	0.18	0.27
2" x 6" - 16' L	Demo	LF	L-2	1895	---	---	0.14	0.14	0.20	L-2	1327	---	---	0.20	0.20	0.29
2" x 8" - 16' L	Demo	LF	L-2	1640	---	---	0.16	0.16	0.23	L-2	1148	---	---	0.23	0.23	0.34
2" x 10" - 16' L	Demo	LF	L-2	1360	---	---	0.19	0.19	0.28	L-2	952	---	---	0.27	0.27	0.40

Description	Oper	Unit	Costs Based On Large Volume							Costs Based On Small Volume						
			Crew Size	Avg Day Prod	Avg Mat'l Unit Cost	Avg Equip Unit Cost	Avg Labor Unit Cost	Avg Total Unit Cost	Avg Price Incl O&P	Crew Size	Avg Day Prod	Avg Mat'l Unit Cost	Avg Equip Unit Cost	Avg Labor Unit Cost	Avg Total Unit Cost	Avg Price Incl O&P

Site Work, demolition (continued)

Jack

Description	Oper	Unit	Crew Size	Avg Day Prod	Avg Mat'l Unit Cost	Avg Equip Unit Cost	Avg Labor Unit Cost	Avg Total Unit Cost	Avg Price Incl O&P	Crew Size	Avg Day Prod	Avg Mat'l Unit Cost	Avg Equip Unit Cost	Avg Labor Unit Cost	Avg Total Unit Cost	Avg Price Incl O&P
2" x 4" - 6' L	Demo	LF	L-2	805	---	---	0.32	0.32	0.48	L-2	564	---	---	0.46	0.46	0.68
2" x 6" - 6' L	Demo	LF	L-2	740	---	---	0.35	0.35	0.52	L-2	518	---	---	0.50	0.50	0.74
2" x 8" - 6' L	Demo	LF	L-2	655	---	---	0.40	0.40	0.59	L-2	459	---	---	0.57	0.57	0.84
2" x 10" - 6' L	Demo	LF	L-2	540	---	---	0.48	0.48	0.71	L-2	378	---	---	0.69	0.69	1.02

Roof decking, solid T&G

Description	Oper	Unit	Crew Size	Avg Day Prod	Avg Mat'l Unit Cost	Avg Equip Unit Cost	Avg Labor Unit Cost	Avg Total Unit Cost	Avg Price Incl O&P	Crew Size	Avg Day Prod	Avg Mat'l Unit Cost	Avg Equip Unit Cost	Avg Labor Unit Cost	Avg Total Unit Cost	Avg Price Incl O&P
2" x 6" - 12' L	Demo	LF	L-2	645	---	---	0.40	0.40	0.60	L-2	452	---	---	0.58	0.58	0.85
2" x 8" - 12' L	Demo	LF	L-2	900	---	---	0.29	0.29	0.43	L-2	630	---	---	0.41	0.41	0.61

Studs

Description	Oper	Unit	Crew Size	Avg Day Prod	Avg Mat'l Unit Cost	Avg Equip Unit Cost	Avg Labor Unit Cost	Avg Total Unit Cost	Avg Price Incl O&P	Crew Size	Avg Day Prod	Avg Mat'l Unit Cost	Avg Equip Unit Cost	Avg Labor Unit Cost	Avg Total Unit Cost	Avg Price Incl O&P
2" x 4" - 8' L	Demo	LF	L-2	1360	---	---	0.19	0.19	0.28	L-2	952	---	---	0.27	0.27	0.40
2" x 6" - 8' L	Demo	LF	L-2	1360	---	---	0.19	0.19	0.28	L-2	952	---	---	0.27	0.27	0.40

Stud partitions, studs 16" o.c. with bottom and double top plates and firestops; per LF partition

Description	Oper	Unit	Crew Size	Avg Day Prod	Avg Mat'l Unit Cost	Avg Equip Unit Cost	Avg Labor Unit Cost	Avg Total Unit Cost	Avg Price Incl O&P	Crew Size	Avg Day Prod	Avg Mat'l Unit Cost	Avg Equip Unit Cost	Avg Labor Unit Cost	Avg Total Unit Cost	Avg Price Incl O&P
2" x 4" - 8' L	Demo	LF	L-2	100	---	---	2.60	2.60	3.85	L-2	70	---	---	3.72	3.72	5.50
2" x 6" - 8' L	Demo	LF	L-2	100	---	---	2.60	2.60	3.85	L-2	70	---	---	3.72	3.72	5.50

Boards

Sheathing, regular or diagonal, 1" x 8"

Description	Oper	Unit	Crew Size	Avg Day Prod	Avg Mat'l Unit Cost	Avg Equip Unit Cost	Avg Labor Unit Cost	Avg Total Unit Cost	Avg Price Incl O&P	Crew Size	Avg Day Prod	Avg Mat'l Unit Cost	Avg Equip Unit Cost	Avg Labor Unit Cost	Avg Total Unit Cost	Avg Price Incl O&P
Roof	Demo	SF	L-2	1340	---	---	0.19	0.19	0.29	L-2	938	---	---	0.28	0.28	0.41
Sidewall	Demo	SF	L-2	1615	---	---	0.16	0.16	0.24	L-2	1131	---	---	0.23	0.23	0.34

Subflooring, regular or diagonal

Description	Oper	Unit	Crew Size	Avg Day Prod	Avg Mat'l Unit Cost	Avg Equip Unit Cost	Avg Labor Unit Cost	Avg Total Unit Cost	Avg Price Incl O&P	Crew Size	Avg Day Prod	Avg Mat'l Unit Cost	Avg Equip Unit Cost	Avg Labor Unit Cost	Avg Total Unit Cost	Avg Price Incl O&P
1" x 8" - 16' L	Demo	SF	L-2	1525	---	---	0.17	0.17	0.25	L-2	1068	---	---	0.24	0.24	0.36
1" x 10" - 16' L	Demo	SF	L-2	1930	---	---	0.13	0.13	0.20	L-2	1351	---	---	0.19	0.19	0.28

Plywood

Sheathing

Roof

Description	Oper	Unit	Crew Size	Avg Day Prod	Avg Mat'l Unit Cost	Avg Equip Unit Cost	Avg Labor Unit Cost	Avg Total Unit Cost	Avg Price Incl O&P	Crew Size	Avg Day Prod	Avg Mat'l Unit Cost	Avg Equip Unit Cost	Avg Labor Unit Cost	Avg Total Unit Cost	Avg Price Incl O&P
1/2" T, CDX	Demo	SF	L-2	1970	---	---	0.13	0.13	0.20	L-2	1379	---	---	0.19	0.19	0.28
5/8" T, CDX	Demo	SF	L-2	1935	---	---	0.13	0.13	0.20	L-2	1355	---	---	0.19	0.19	0.28

Wall

Description	Oper	Unit	Crew Size	Avg Day Prod	Avg Mat'l Unit Cost	Avg Equip Unit Cost	Avg Labor Unit Cost	Avg Total Unit Cost	Avg Price Incl O&P	Crew Size	Avg Day Prod	Avg Mat'l Unit Cost	Avg Equip Unit Cost	Avg Labor Unit Cost	Avg Total Unit Cost	Avg Price Incl O&P
3/8" or 1/2" T, CDX	Demo	SF	L-2	2520	---	---	0.10	0.10	0.15	L-2	1764	---	---	0.15	0.15	0.22
5/8" T, CDX	Demo	SF	L-2	2460	---	---	0.11	0.11	0.16	L-2	1722	---	---	0.15	0.15	0.22

Subflooring

Description	Oper	Unit	Crew Size	Avg Day Prod	Avg Mat'l Unit Cost	Avg Equip Unit Cost	Avg Labor Unit Cost	Avg Total Unit Cost	Avg Price Incl O&P	Crew Size	Avg Day Prod	Avg Mat'l Unit Cost	Avg Equip Unit Cost	Avg Labor Unit Cost	Avg Total Unit Cost	Avg Price Incl O&P
5/8", 3/4" T, CDX	Demo	SF	L-2	2305	---	---	0.11	0.11	0.17	L-2	1614	---	---	0.16	0.16	0.24
1-1/8" T, 2-4-1, T&G long edges	Demo	SF	L-2	1615	---	---	0.16	0.16	0.24	L-2	1131	---	---	0.23	0.23	0.34

			Costs Based On Large Volume							Costs Based On Small Volume						
Description	Oper	Unit	Crew Size	Avg Day Prod	Avg Mat'l Unit Cost	Avg Equip Unit Cost	Avg Labor Unit Cost	Avg Total Unit Cost	Avg Price Incl O&P	Crew Size	Avg Day Prod	Avg Mat'l Unit Cost	Avg Equip Unit Cost	Avg Labor Unit Cost	Avg Total Unit Cost	Avg Price Incl O&P

Site Work, demolition (continued)

Trusses, "W" pattern with gin pole, 24' to 30' spans

Description	Oper	Unit	Crew Size	Avg Day Prod	Avg Mat'l Unit Cost	Avg Equip Unit Cost	Avg Labor Unit Cost	Avg Total Unit Cost	Avg Price Incl O&P	Crew Size	Avg Day Prod	Avg Mat'l Unit Cost	Avg Equip Unit Cost	Avg Labor Unit Cost	Avg Total Unit Cost	Avg Price Incl O&P
3 - in - 12 slope	Demo	SF	L-2	26	---	---	10.00	10.00	14.80	L-2	18	---	---	14.29	14.29	21.15
5 - in - 12 slope	Demo	SF	L-2	26	---	---	10.00	10.00	14.80	L-2	18	---	---	14.29	14.29	21.15

Finish carpentry

Description	Oper	Unit	Crew Size	Avg Day Prod	Avg Mat'l Unit Cost	Avg Equip Unit Cost	Avg Labor Unit Cost	Avg Total Unit Cost	Avg Price Incl O&P	Crew Size	Avg Day Prod	Avg Mat'l Unit Cost	Avg Equip Unit Cost	Avg Labor Unit Cost	Avg Total Unit Cost	Avg Price Incl O&P
Bath accessories, screw on type	Demo	EA	L-2	125	---	---	2.08	2.08	3.08	L-2	88	---	---	2.97	2.97	4.40
Cabinets																
Kitchen, to 3' x 4', wood;																
base, wall, or peninsula	Demo	EA	L-2	25	---	---	10.40	10.40	15.40	L-2	18	---	---	14.86	14.86	22.00
Medicine, metal	Demo	EA	L-2	30	---	---	8.67	8.67	12.83	L-2	21	---	---	12.38	12.38	18.33
Vanity, cabinet and sink top																
Disconnect plumbing and remove to dumpster	Demo	EA	L-2	16	---	---	16.26	16.26	24.06	L-2	11	---	---	23.22	23.22	34.37
Remove old unit, replace with new unit reconnect plumbing	Demo	EA	S-2	7	---	---	47.00	47.00	68.62	S-2	5	---	---	67.14	67.14	98.02
Hardwood flooring (over wood subfloors)																
Block, set in mastic	Demo	SF	L-2	1200	---	---	0.22	0.22	0.32	L-2	840	---	---	0.31	0.31	0.46
Strip, nailed	Demo	SF	L-2	900	---	---	0.29	0.29	0.43	L-2	630	---	---	0.41	0.41	0.61
Marlite panels, 4' x 8', adhesive set	Demo	SF	L-2	1850	---	---	0.14	0.14	0.21	L-2	1295	---	---	0.20	0.20	0.30
Molding and trim																
At base (floor)	Demo	LF	L-2	1060	---	---	0.25	0.25	0.36	L-2	742	---	---	0.35	0.35	0.52
At ceiling	Demo	LF	L-2	1200	---	---	0.22	0.22	0.32	L-2	840	---	---	0.31	0.31	0.46
On walls or cabinets	Demo	LF	L-2	1600	---	---	0.16	0.16	0.24	L-2	1120	---	---	0.23	0.23	0.34
Paneling																
Plywood, prefinished	Demo	SF	L-2	1850	---	---	0.14	0.14	0.21	L-2	1295	---	---	0.20	0.20	0.30
Wood	Demo	SF	L-2	1650	---	---	0.16	0.16	0.23	L-2	1155	---	---	0.23	0.23	0.33

Weather protection

Insulation

Batt/Roll, with wall or ceiling finish already removed

Description	Oper	Unit	Crew Size	Avg Day Prod	Avg Mat'l Unit Cost	Avg Equip Unit Cost	Avg Labor Unit Cost	Avg Total Unit Cost	Avg Price Incl O&P	Crew Size	Avg Day Prod	Avg Mat'l Unit Cost	Avg Equip Unit Cost	Avg Labor Unit Cost	Avg Total Unit Cost	Avg Price Incl O&P
Joists, 16" or 24" o.c.	Demo	SF	L-2	2935	---	---	0.09	0.09	0.13	L-2	2055	---	---	0.13	0.13	0.19
Rafters, 16" or 24" o.c.	Demo	SF	L-2	2560	---	---	0.10	0.10	0.15	L-2	1792	---	---	0.15	0.15	0.21
Studs, 16" or 24" o.c.	Demo	SF	L-2	3285	---	---	0.08	0.08	0.12	L-2	2300	---	---	0.11	0.11	0.17

			Costs Based On Large Volume							Costs Based On Small Volume						
Description	Oper	Unit	Crew Size	Avg Day Prod	Avg Mat'l Unit Cost	Avg Equip Unit Cost	Avg Labor Unit Cost	Avg Total Unit Cost	Avg Price Incl O&P	Crew Size	Avg Day Prod	Avg Mat'l Unit Cost	Avg Equip Unit Cost	Avg Labor Unit Cost	Avg Total Unit Cost	Avg Price Incl O&P

Site Work, demolition (continued)

Loose, with ceiling finish already removed
Joists, 16" or 24" o.c.

Description	Oper	Unit	Crew Size	Avg Day Prod	Avg Mat'l Unit Cost	Avg Equip Unit Cost	Avg Labor Unit Cost	Avg Total Unit Cost	Avg Price Incl O&P	Crew Size	Avg Day Prod	Avg Mat'l Unit Cost	Avg Equip Unit Cost	Avg Labor Unit Cost	Avg Total Unit Cost	Avg Price Incl O&P
4" T	Demo	SF	L-2	3900	---	---	0.07	0.07	0.10	L-2	2730	---	---	0.10	0.10	0.14
6" T	Demo	SF	L-2	2340	---	---	0.11	0.11	0.16	L-2	1638	---	---	0.16	0.16	0.23
Rigid																
Roofs																
1/2" T	Demo	SQ	L-2	17	---	---	15.30	15.30	22.64	L-2	12	---	---	21.86	21.86	32.35
1" T	Demo	SQ	L-2	15	---	---	17.34	17.34	25.66	L-2	11	---	---	24.77	24.77	36.66
Walls, 1/2" T	Demo	SQ	L-2	2140	---	---	0.12	0.12	0.18	L-2	1498	---	---	0.17	0.17	0.26
Sheet metal																
Gutter and downspouts																
Aluminum	Demo	LF	L-2	850	---	---	0.31	0.31	0.45	L-2	595	---	---	0.44	0.44	0.65
Galvanized	Demo	LF	L-2	640	---	---	0.41	0.41	0.60	L-2	448	---	---	0.58	0.58	0.86

Roofing and siding

Aluminum
Roofing, nailed to wood
Corrugated (2-1/2"), 26" W with 3-3/4" side lap and

Description	Oper	Unit	Crew Size	Avg Day Prod	Avg Mat'l Unit Cost	Avg Equip Unit Cost	Avg Labor Unit Cost	Avg Total Unit Cost	Avg Price Incl O&P	Crew Size	Avg Day Prod	Avg Mat'l Unit Cost	Avg Equip Unit Cost	Avg Labor Unit Cost	Avg Total Unit Cost	Avg Price Incl O&P
6" end lap	Demo	SQ	L-2	10	---	---	26.01	26.01	38.49	L-2	7	---	---	37.15	37.15	54.99

Siding, nailed to wood
Clapboard (i.e., lap drop)

Description	Oper	Unit	Crew Size	Avg Day Prod	Avg Mat'l Unit Cost	Avg Equip Unit Cost	Avg Labor Unit Cost	Avg Total Unit Cost	Avg Price Incl O&P	Crew Size	Avg Day Prod	Avg Mat'l Unit Cost	Avg Equip Unit Cost	Avg Labor Unit Cost	Avg Total Unit Cost	Avg Price Incl O&P
8" exposure	Demo	SF	L-2	1025	---	---	0.25	0.25	0.38	L-2	718	---	---	0.36	0.36	0.54
10" exposure	Demo	SF	L-2	1280	---	---	0.20	0.20	0.30	L-2	896	---	---	0.29	0.29	0.43

Corrugated (2-1/2"), 26" W with 2-1/2" side lap and 4"

Description	Oper	Unit	Crew Size	Avg Day Prod	Avg Mat'l Unit Cost	Avg Equip Unit Cost	Avg Labor Unit Cost	Avg Total Unit Cost	Avg Price Incl O&P	Crew Size	Avg Day Prod	Avg Mat'l Unit Cost	Avg Equip Unit Cost	Avg Labor Unit Cost	Avg Total Unit Cost	Avg Price Incl O&P
end lap	Demo	SF	L-2	1200	---	---	0.22	0.22	0.32	L-2	840	---	---	0.31	0.31	0.46
Panels, 4' x 8'	Demo	SF	L-2	2400	---	---	0.11	0.11	0.16	L-2	1680	---	---	0.15	0.15	0.23

Shingle, 24" L with 12"

Description	Oper	Unit	Crew Size	Avg Day Prod	Avg Mat'l Unit Cost	Avg Equip Unit Cost	Avg Labor Unit Cost	Avg Total Unit Cost	Avg Price Incl O&P	Crew Size	Avg Day Prod	Avg Mat'l Unit Cost	Avg Equip Unit Cost	Avg Labor Unit Cost	Avg Total Unit Cost	Avg Price Incl O&P
exposure	Demo	SF	L-2	1450	---	---	0.18	0.18	0.27	L-2	1015	---	---	0.26	0.26	0.38

Asbestos shingle
Siding, nailed; 24" x 12"

Description	Oper	Unit	Crew Size	Avg Day Prod	Avg Mat'l Unit Cost	Avg Equip Unit Cost	Avg Labor Unit Cost	Avg Total Unit Cost	Avg Price Incl O&P	Crew Size	Avg Day Prod	Avg Mat'l Unit Cost	Avg Equip Unit Cost	Avg Labor Unit Cost	Avg Total Unit Cost	Avg Price Incl O&P
with 11" exposure	Demo	SF	L-2	2160	---	---	0.12	0.12	0.18	L-2	1512	---	---	0.17	0.17	0.25

Asphalt shingle roofing
240 lb./SQ., strip, 3 tab

Description	Oper	Unit	Crew Size	Avg Day Prod	Avg Mat'l Unit Cost	Avg Equip Unit Cost	Avg Labor Unit Cost	Avg Total Unit Cost	Avg Price Incl O&P	Crew Size	Avg Day Prod	Avg Mat'l Unit Cost	Avg Equip Unit Cost	Avg Labor Unit Cost	Avg Total Unit Cost	Avg Price Incl O&P
5" exposure	Demo	SQ	L-2	16	---	---	16.26	16.26	24.06	L-2	11	---	---	23.22	23.22	34.37

Built-up/hot roofing (to wood deck)

			Costs Based On Large Volume							Costs Based On Small Volume					
Description	Oper	Unit	Crew Size	Avg Day Prod	Avg Mat'l Unit Cost	Avg Equip Unit Cost	Avg Labor Unit Cost	Avg Total Unit Cost	Avg Price Incl O&P	Crew Size	Avg Day Prod	Avg Mat'l Unit Cost	Avg Equip Unit Cost	Avg Labor Unit Cost	Avg Total Unit Cost	Avg Price Incl O&P

Site Work, demolition (continued)

Description	Oper	Unit	Crew Size	Avg Day Prod	Avg Mat'l Unit Cost	Avg Equip Unit Cost	Avg Labor Unit Cost	Avg Total Unit Cost	Avg Price Incl O&P	Crew Size	Avg Day Prod	Avg Mat'l Unit Cost	Avg Equip Unit Cost	Avg Labor Unit Cost	Avg Total Unit Cost	Avg Price Incl O&P
3 ply																
With gravel	Demo	SQ	L-2	11	---	---	23.64	23.64	34.99	L-2	8	---	---	33.78	33.78	49.99
Without gravel	Demo	SQ	L-2	14	---	---	18.58	18.58	27.49	L-2	10	---	---	26.54	26.54	39.28
5 ply																
With gravel	Demo	SQ	L-2	10	---	---	26.01	26.01	38.49	L-2	7	---	---	37.15	37.15	54.99
Without gravel	Demo	SQ	L-2	13	---	---	20.01	20.01	29.61	L-2	9	---	---	28.58	28.58	42.30
Clay tile roofing																
2 piece interlocking	Demo	SF	L-2	10	---	---	26.01	26.01	38.49	L-2	7	---	---	37.15	37.15	54.99
1 piece	Demo	SF	L-2	11	---	---	23.64	23.64	34.99	L-2	8	---	---	33.78	33.78	49.99
Hardboard siding																
Lap, 1/2" T x 12" W x 16' L, with 11" exposure	Demo	SF	L-2	1345	---	---	0.19	0.19	0.29	L-2	942	---	---	0.28	0.28	0.41
Panels, 7/16" T x 4' W x 8' H	Demo	SF	L-2	2520	---	---	0.10	0.10	0.15	L-2	1764	---	---	0.15	0.15	0.22
Mineral surfaced roll roofing																
Single coverage 90 lb./SQ. roll with 6" end lap and 2" headlap	Demo	SQ	L-2	32	---	---	8.13	8.13	12.03	L-2	22	---	---	11.61	11.61	17.18
Double coverage selvage roll, with 6" end lap and 17" exposure	Demo	SQ	L-2	22	---	---	11.82	11.82	17.50	L-2	15	---	---	16.89	16.89	25.00
Wood																
Roofing																
Shakes																
24" L with 10" exposure																
1/2" to 3/4" T	Demo	SQ	L-2	25	---	---	10.40	10.40	15.40	L-2	18	---	---	14.86	14.86	22.00
3/4" to 5/4" T	Demo	SQ	L-2	23	---	---	11.56	11.56	17.11	L-2	16	---	---	16.51	16.51	24.44
Shingles																
16" L with 5" exposure	Demo	SQ	L-2	12	---	---	21.67	21.67	32.08	L-2	8	---	---	30.96	30.96	45.82
18" L with 5-1/2" exposure	Demo	SQ	L-2	13	---	---	20.64	20.64	30.55	L-2	9	---	---	29.49	29.49	43.64
24" L with 7-1/2" exposure	Demo	SQ	L-2	18	---	---	14.45	14.45	21.38	L-2	13	---	---	20.64	20.64	30.55
Siding																
Bevel																
1/2" x 8" with 6-3/4" exp.	Demo	SF	L-2	910	---	---	0.29	0.29	0.42	L-2	637	---	---	0.41	0.41	0.60
5/8" x 10" with 8-3/4" exp.	Demo	SF	L-2	1180	---	---	0.22	0.22	0.33	L-2	826	---	---	0.31	0.31	0.47
3/4" x 12" with 10-3/4" exp.	Demo	SF	L-2	1450	---	---	0.18	0.18	0.27	L-2	1015	---	---	0.26	0.26	0.38

			Costs Based On Large Volume							Costs Based On Small Volume					
Description	Oper	Unit	Crew Size	Avg Day Prod	Avg Mat'l Unit Cost	Avg Equip Unit Cost	Avg Labor Unit Cost	Avg Total Unit Cost	Avg Price Incl O&P	Crew Size	Avg Day Prod	Avg Mat'l Unit Cost	Avg Equip Unit Cost	Avg Labor Unit Cost	Avg Total Unit Cost	Avg Price Incl O&P

Site Work, demolition (continued)

Drop (horizontal), 1/4" T&G

Description	Oper	Unit	Crew Size	Avg Day Prod	Avg Mat'l Unit Cost	Avg Equip Unit Cost	Avg Labor Unit Cost	Avg Total Unit Cost	Avg Price Incl O&P	Crew Size	Avg Day Prod	Avg Mat'l Unit Cost	Avg Equip Unit Cost	Avg Labor Unit Cost	Avg Total Unit Cost	Avg Price Incl O&P
1" x 8" with 7" exposure	Demo	SF	L-2	945	---	---	0.28	0.28	0.41	L-2	662	---	---	0.39	0.39	0.58
1" x 10" with 9" exposure	Demo	SF	L-2	1215	---	---	0.21	0.21	0.32	L-2	851	---	---	0.31	0.31	0.45
Board (1" x 12") and batten (1" x 2") @ 12" o.c.																
Horizontal	Demo	SF	L-2	1280	---	---	0.20	0.20	0.30	L-2	896	---	---	0.29	0.29	0.43
Vertical																
Standard	Demo	SF	L-2	1025	---	---	0.25	0.25	0.38	L-2	718	---	---	0.36	0.36	0.54
Reverse	Demo	SF	L-2	1065	---	---	0.24	0.24	0.36	L-2	746	---	---	0.35	0.35	0.52
Board on board (1" x 12" with 1-1/2" overlap), vertical	Demo	SF	L-2	950	---	---	0.27	0.27	0.41	L-2	665	---	---	0.39	0.39	0.58
Plywood (1/2" T) with battens (1" x 2")																
16" o.c. battens	Demo	SF	L-2	2255	---	---	0.12	0.12	0.17	L-2	1579	---	---	0.16	0.16	0.24
24" o.c. battens	Demo	SF	L-2	2365	---	---	0.11	0.11	0.16	L-2	1656	---	---	0.16	0.16	0.23
Shakes																
24" L with 11-1/2" exp.																
1/2" to 3/4" T	Demo	SF	L-2	1600	---	---	0.16	0.16	0.24	L-2	1120	---	---	0.23	0.23	0.34
3/4" to 5/4" T	Demo	SF	L-2	1440	---	---	0.18	0.18	0.27	L-2	1008	---	---	0.26	0.26	0.38
Shingles																
16" L with 7-1/2" exposure	Demo	SF	L-2	1000	---	---	0.26	0.26	0.38	L-2	700	---	---	0.37	0.37	0.55
18" L with 8-1/2" exposure	Demo	SF	L-2	1130	---	---	0.23	0.23	0.34	L-2	791	---	---	0.33	0.33	0.49
24" L with 11-1/2" exposure	Demo	SF	L-2	1530	---	---	0.17	0.17	0.25	L-2	1071	---	---	0.24	0.24	0.36

Doors, Windows, Glazing

Doors with related trim and frame

Closet, with track

Description	Oper	Unit	Crew Size	Avg Day Prod	Avg Mat'l Unit Cost	Avg Equip Unit Cost	Avg Labor Unit Cost	Avg Total Unit Cost	Avg Price Incl O&P	Crew Size	Avg Day Prod	Avg Mat'l Unit Cost	Avg Equip Unit Cost	Avg Labor Unit Cost	Avg Total Unit Cost	Avg Price Incl O&P
Folding, 4 doors	Demo	Set	L-2	12	---	---	21.67	21.67	32.08	L-2	8	---	---	30.96	30.96	45.82
Sliding, 2 or 3 doors	Demo	Set	L-2	12	---	---	21.67	21.67	32.08	L-2	8	---	---	30.96	30.96	45.82
Entry, 3' x 7'	Demo	EA	L-2	14	---	---	18.58	18.58	27.49	L-2	10	---	---	26.54	26.54	39.28
Fire, 3' x 7'	Demo	EA	L-2	14	---	---	18.58	18.58	27.49	L-2	10	---	---	26.54	26.54	39.28
Garage																
Wood, aluminum, or hardboard																
Single	Demo	EA	L-2	8	---	---	32.51	32.51	48.12	L-2	6	---	---	46.44	46.44	68.74
Double	Demo	EA	L-2	6	---	---	43.35	43.35	64.15	L-2	4	---	---	61.92	61.92	91.65

				Costs Based On Large Volume						Costs Based On Small Volume					
Description	Oper	Unit	Crew Size	Avg Day Prod	Avg Mat'l Unit Cost	Avg Equip Unit Cost	Avg Labor Unit Cost	Avg Total Unit Cost	Avg Price Incl O&P	Crew Size	Avg Day Prod	Avg Mat'l Unit Cost	Avg Equip Unit Cost	Avg Labor Unit Cost	Avg Total Unit Cost	Avg Price Incl O&P

Site Work, demolition (continued)

Steel

| Single | Demo | EA | L-2 | 7 | --- | --- | 37.15 | 37.15 | 54.99 | L-2 | 5 | --- | --- | 53.08 | 53.08 | 78.56 |
| Double | Demo | EA | L-2 | 5 | --- | --- | 52.02 | 52.02 | 76.98 | L-2 | 4 | --- | --- | 74.31 | 74.31 | 109.98 |

Glass sliding, with track

2 lites wide	Demo	Set	L-2	8	---	---	32.51	32.51	48.12	L-2	6	---	---	46.44	46.44	68.74
3 lites wide	Demo	Set	L-2	6	---	---	43.35	43.35	64.15	L-2	4	---	---	61.92	61.92	91.65
4 lites wide	Demo	Set	L-2	4	---	---	65.02	65.02	96.23	L-2	3	---	---	92.89	92.89	137.47
Interior, 3' x 7'	Demo	EA	L-2	16	---	---	16.26	16.26	24.06	L-2	11	---	---	23.22	23.22	34.37
Screen, 3' x 7'	Demo	EA	L-2	20	---	---	13.00	13.00	19.25	L-2	14	---	---	18.58	18.58	27.49
Storm combination, 3' x 7'	Demo	EA	L-2	20	---	---	13.00	13.00	19.25	L-2	14	---	---	18.58	18.58	27.49

Windows, with related trim and frame

To 12 SF

| Aluminum | Demo | EA | L-2 | 21 | --- | --- | 12.38 | 12.38 | 18.33 | L-2 | 15 | --- | --- | 17.69 | 17.69 | 26.19 |
| Wood | Demo | EA | L-2 | 16 | --- | --- | 16.26 | 16.26 | 24.06 | L-2 | 11 | --- | --- | 23.22 | 23.22 | 34.37 |

13 SF to 50 SF

| Aluminum | Demo | EA | L-2 | 13 | --- | --- | 20.01 | 20.01 | 29.61 | L-2 | 9 | --- | --- | 28.58 | 28.58 | 42.30 |
| Wood | Demo | EA | L-2 | 10 | --- | --- | 26.01 | 26.01 | 38.49 | L-2 | 7 | --- | --- | 37.15 | 37.15 | 54.99 |

Glazing, clean sash and remove old putty or rubber

3/32" T float, putty or rubber

8" x 12" (0.667 SF)	Demo	SF	G-1	25	---	---	6.66	6.66	9.79	G-1	18	---	---	9.51	9.51	13.98
12" x 16" (1.333 SF)	Demo	SF	G-1	45	---	---	3.70	3.70	5.44	G-1	32	---	---	5.28	5.28	7.77
14" x 20" (1.944 SF)	Demo	SF	G-1	55	---	---	3.03	3.03	4.45	G-1	39	---	---	4.32	4.32	6.36
16" x 24" (2.667 SF)	Demo	SF	G-1	70	---	---	2.38	2.38	3.50	G-1	49	---	---	3.40	3.40	4.99
24" x 26" (4.333 SF)	Demo	SF	G-1	95	---	---	1.75	1.75	2.58	G-1	67	---	---	2.50	2.50	3.68
36" x 24" (6.000 SF)	Demo	SF	G-1	125	---	---	1.33	1.33	1.96	G-1	88	---	---	1.90	1.90	2.80

1/8" T float, putty, steel sash

12" x 16" (1.333 SF)	Demo	SF	G-1	45	---	---	3.70	3.70	5.44	G-1	32	---	---	5.28	5.28	7.77
16" x 20" (2.222 SF)	Demo	SF	G-1	65	---	---	2.56	2.56	3.76	G-1	46	---	---	3.66	3.66	5.38
16" x 24" (2.667 SF)	Demo	SF	G-1	70	---	---	2.38	2.38	3.50	G-1	49	---	---	3.40	3.40	4.99
24" x 26" (4.333 SF)	Demo	SF	G-1	95	---	---	1.75	1.75	2.58	G-1	67	---	---	2.50	2.50	3.68
28" x 32" (6.222 SF)	Demo	SF	G-1	130	---	---	1.28	1.28	1.88	G-1	91	---	---	1.83	1.83	2.69
36" x 36" (9.000 SF)	Demo	SF	G-1	160	---	---	1.04	1.04	1.53	G-1	112	---	---	1.49	1.49	2.18
36" x 48" (12.000 SF)	Demo	SF	G-1	190	---	---	0.88	0.88	1.29	G-1	133	---	---	1.25	1.25	1.84

			Costs Based On Large Volume							Costs Based On Small Volume					
Description	Oper	Unit	Crew Size	Avg Day Prod	Avg Mat'l Unit Cost	Avg Equip Unit Cost	Avg Labor Unit Cost	Avg Total Unit Cost	Avg Price Incl O&P	Crew Size	Avg Day Prod	Avg Mat'l Unit Cost	Avg Equip Unit Cost	Avg Labor Unit Cost	Avg Total Unit Cost	Avg Price Incl O&P

Site Work, demolition (continued)

1/4" T float
  Wood sash with putty

72" x 48" (24.0 SF)	Demo	SF	G-1	185	---	---	0.90	0.90	1.32	G-1	130	---	---	1.29	1.29	1.89

  Aluminum sash with aluminum channel and rigid neoprene rubber

48" x 96" (32.0 SF)	Demo	SF	G-1	175	---	---	0.95	0.95	1.40	G-1	123	---	---	1.36	1.36	2.00
96" x 96" (64.0 SF)	Demo	SF	G-1	180	---	---	0.92	0.92	1.36	G-1	126	---	---	1.32	1.32	1.94

  1" T insulating glass; with 2 pieces 1/4" float and 1/2" air space

To 6.0 SF	Demo	SF	G-1	50	---	---	3.33	3.33	4.89	G-1	35	---	---	4.76	4.76	6.99
6.1 to 12.0 SF	Demo	SF	G-1	110	---	---	1.51	1.51	2.22	G-1	77	---	---	2.16	2.16	3.18
12.1 to 18.0 SF	Demo	SF	G-1	150	---	---	1.11	1.11	1.63	G-1	105	---	---	1.59	1.59	2.33
18.1 to 24.0 SF	Demo	SF	G-1	145	---	---	1.15	1.15	1.69	G-1	102	---	---	1.64	1.64	2.41

  Aluminum sliding door glass with aluminum channel and rigid neoprene rubber
  34" x 76" (17.944 SF)
    5/8" T insulating glass with 2 pieces 5/32" T tempered with 1-1/4"

air space	Demo	SF	G-1	170	---	---	0.98	0.98	1.44	G-1	119	---	---	1.40	1.40	2.06
5/32" T tempered	Demo	SF	G-1	195	---	---	0.85	0.85	1.25	G-1	137	---	---	1.22	1.22	1.79

  46" x 76" (24.278 SF)
    5/8" T insulating glass with 2 pieces 5/32" T tempered with 1-1/4"

air space	Demo	SF	G-1	215	---	---	0.77	0.77	1.14	G-1	151	---	---	1.11	1.11	1.63
5/32" T tempered	Demo	SF	G-1	245	---	---	0.68	0.68	1.00	G-1	172	---	---	0.97	0.97	1.43

Finishes. Exterior and interior, with hand tools; no insulation removal included

  Plaster and stucco; remove to studs or sheathing
  Lath (wood or metal) and plaster, walls and ceiling

2 coats	Demo	SY	L-2	130	---	---	2.00	2.00	2.96	L-2	91	---	---	2.86	2.86	4.23
3 coats	Demo	SY	L-2	120	---	---	2.17	2.17	3.21	L-2	84	---	---	3.10	3.10	4.58

  Stucco and metal netting

2 coats	Demo	SY	L-2	95	---	---	2.74	2.74	4.05	L-2	67	---	---	3.91	3.91	5.79
3 coats	Demo	SY	L-2	80	---	---	3.25	3.25	4.81	L-2	56	---	---	4.64	4.64	6.87

Wallboard, gypsum (drywall), walls and ceilings	Demo	SF	L-2	1540	---	---	0.17	0.17	0.25	L-2	1078	---	---	0.24	0.24	0.36

  Ceramic, metal, plastic tile
  Floors, 1" x 1"

Adhesive or dry-set base	Demo	SF	L-2	550	---	---	0.47	0.47	0.70	L-2	385	---	---	0.68	0.68	1.00
Conventional mortar base	Demo	SF	L-2	475	---	---	0.55	0.55	0.81	L-2	333	---	---	0.78	0.78	1.16

			Costs Based On Large Volume						Costs Based On Small Volume					
Description	Oper	Unit	Crew Size	Avg Day Prod	Avg Mat'l Unit Cost	Avg Equip Unit Cost	Avg Labor Unit Cost	Avg Total Unit Cost	Avg Price Incl O&P	Crew Size	Avg Day Prod	Avg Mat'l Unit Cost	Avg Equip Unit Cost	Avg Labor Unit Cost	Avg Total Unit Cost	Avg Price Incl O&P

Site Work, demolition (continued)

Walls, 1" x 1" or 4-1/4" x 4 1/4"

Description	Oper	Unit	Crew Size	Avg Day Prod	Avg Mat'l Unit Cost	Avg Equip Unit Cost	Avg Labor Unit Cost	Avg Total Unit Cost	Avg Price Incl O&P	Crew Size	Avg Day Prod	Avg Mat'l Unit Cost	Avg Equip Unit Cost	Avg Labor Unit Cost	Avg Total Unit Cost	Avg Price Incl O&P
Adhesive or dry-set base	Demo	SF	L-2	480	---	---	0.54	0.54	0.80	L-2	336	---	---	0.77	0.77	1.15
Conventional mortar base	Demo	SF	L-2	400	---	---	0.65	0.65	0.96	L-2	280	---	---	0.93	0.93	1.37
Acoustical or insulating ceiling tile																
Adhesive set, tile only	Demo	SF	L-2	1300	---	---	0.20	0.20	0.30	L-2	910	---	---	0.29	0.29	0.42
Stapled, tile only	Demo	SF	L-2	1170	---	---	0.22	0.22	0.33	L-2	819	---	---	0.32	0.32	0.47
Stapled, tile and furring strips	Demo	SF	L-2	1540	---	---	0.17	0.17	0.25	L-2	1078	---	---	0.24	0.24	0.36
Suspended ceiling system; panels and grid system	Demo	SF	L-2	1780	---	---	0.15	0.15	0.22	L-2	1246	---	---	0.21	0.21	0.31
Resilient flooring, adhesive set																
Sheet products	Demo	SY	L-2	160	---	---	1.63	1.63	2.41	L-2	112	---	---	2.32	2.32	3.44
Tile products	Demo	SF	L-2	1500	---	---	0.17	0.17	0.26	L-2	1050	---	---	0.25	0.25	0.37
Wallpaper. NOTE: Average output is expressed in rolls (36.0 SF/ single roll)																
Single layer of paper from plaster with steaming equipment	Demo	Roll	L-1	20	---	---	6.63	6.63	9.82	L-1	14	---	---	9.48	9.48	14.03
Several layers of paper from plaster with steaming equipment	Demo	Roll	L-1	12	---	---	10.57	10.57	15.65	L-1	9	---	---	15.10	15.10	22.35
Vinyls (with non-woven, woven, or synthetic fiber backings) from plaster with steaming equipment	Demo	Roll	L-1	30	---	---	4.33	4.33	6.42	L-1	21	---	---	6.19	6.19	9.16
Single layer of paper from drywall with steaming equipment	Demo	Roll	L-1	20	---	---	6.63	6.63	9.82	L-1	14	---	---	9.48	9.48	14.03
Several layers of paper from drywall with steaming equipment	Demo	Roll	L-1	12	---	---	10.57	10.57	15.65	L-1	9	---	---	15.10	15.10	22.35
Vinyls (with synthetic fiber backing) from drywall with steaming equipment	Demo	Roll	L-1	30	---	---	4.33	4.33	6.42	L-1	21	---	---	6.19	6.19	9.16
Vinyls (with other backings), from drywall with steaming equipment	Demo	Roll	L-1	20	---	---	6.63	6.63	9.82	L-1	14	---	---	9.48	9.48	14.03

Excavation

Description	Oper	Unit	Crew Size	Avg Day Prod	Avg Mat'l Unit Cost	Avg Equip Unit Cost	Avg Labor Unit Cost	Avg Total Unit Cost	Avg Price Incl O&P	Crew Size	Avg Day Prod	Avg Mat'l Unit Cost	Avg Equip Unit Cost	Avg Labor Unit Cost	Avg Total Unit Cost	Avg Price Incl O&P
Digging out or trenching			L-2	475	---	---	0.55	0.55	0.81	L-2	333	---	---	0.78	0.78	1.16
Pits, medium earth piled																
With front end loader, track mounted, 1-1/2 CY capacity; 55 CY per hour	Demo	CY	V-1	440	---	0.48	0.58	1.05	1.33	V-1	308	---	0.68	0.82	1.51	1.89

Description	Oper	Unit	Costs Based On Large Volume							Costs Based On Small Volume						
			Crew Size	Avg Day Prod	Avg Mat'l Unit Cost	Avg Equip Unit Cost	Avg Labor Unit Cost	Avg Total Unit Cost	Avg Price Incl O&P	Crew Size	Avg Day Prod	Avg Mat'l Unit Cost	Avg Equip Unit Cost	Avg Labor Unit Cost	Avg Total Unit Cost	Avg Price Incl O&P

Site Work, excavation (continued)

Description	Oper	Unit	Crew Size	Avg Day Prod	Avg Mat'l Unit Cost	Avg Equip Unit Cost	Avg Labor Unit Cost	Avg Total Unit Cost	Avg Price Incl O&P	Crew Size	Avg Day Prod	Avg Mat'l Unit Cost	Avg Equip Unit Cost	Avg Labor Unit Cost	Avg Total Unit Cost	Avg Price Incl O&P
By hand																
To 4'-0" D	Demo	CY	L-2	15	---	---	17.34	17.34	25.66	L-2	11	---	---	24.77	24.77	36.66
4'-0" to 6'-0" D	Demo	CY	L-2	10	---	---	26.01	26.01	38.49	L-2	7	---	---	37.15	37.15	54.99
6'-0" to 8'-0" D	Demo	CY	L-2	6	---	---	43.35	43.35	64.15	L-2	4	---	---	61.92	61.92	91.65
Continuous footing or trench, medium earth piled																
With tractor backhoe, 3/8 CY																
capacity (48 HP); 15 CY																
per hour	Demo	CY	V-2	120	---	1.65	2.11	3.76	4.75	V-2	84	---	2.35	3.01	5.37	6.78
By hand, to 4'-0" D	Demo	CY	L-2	15	---	---	17.34	17.34	25.66	L-2	11	---	---	24.77	24.77	36.66
Backfilling, by hand, medium soil																
Without compaction	Demo	CY	L-2	28	---	---	9.29	9.29	13.75	L-2	20	---	---	13.27	13.27	19.64
With hand compaction																
6" layers	Demo	CY	L-2	17	---	---	15.30	15.30	22.64	L-2	12	---	---	21.86	21.86	32.35
12" layers	Demo	CY	L-2	22	---	---	11.82	11.82	17.50	L-2	15	---	---	16.89	16.89	25.00
With vibrating plate compaction																
6" layers	Demo	CY	A-2	20	---	1.04	14.95	15.99	23.02	A-2	14	---	1.49	21.35	22.84	32.88
12" layers	Demo	CY	A-2	24	---	0.87	12.46	13.33	19.18	A-2	17	---	1.24	17.79	19.04	27.40

Stairs
Stair Parts

Description	Oper	Unit	Crew Size	Avg Day Prod	Avg Mat'l Unit Cost	Avg Equip Unit Cost	Avg Labor Unit Cost	Avg Total Unit Cost	Avg Price Incl O&P	Crew Size	Avg Day Prod	Avg Mat'l Unit Cost	Avg Equip Unit Cost	Avg Labor Unit Cost	Avg Total Unit Cost	Avg Price Incl O&P
Balusters, stock pine																
1-1/16" x 1-1/16"	Inst	LF	C-1	200	0.63	---	0.84	1.47	1.88	C-1	140	0.74	---	1.20	1.94	2.53
1-5/8" x 1-5/8"	Inst	LF	C-1	160	1.21	---	1.05	2.26	2.78	C-1	112	1.43	---	1.50	2.93	3.67
Balusters, turned,																
30" high																
Pine	Inst	EA	C-1	24	5.50	---	7.02	12.52	15.96	C-1	17	6.50	---	10.03	16.53	21.45
Birch	Inst	EA	C-1	24	6.60	---	7.02	13.62	17.06	C-1	17	7.80	---	10.03	17.83	22.75
42" high																
Pine	Inst	EA	C-1	20	6.60	---	8.43	15.03	19.16	C-1	14	7.80	---	12.04	19.84	25.74
Birch	Inst	EA	C-1	20	8.80	---	8.43	17.23	21.36	C-1	14	10.40	---	12.04	22.44	28.34
Newels, 3-1/4" wide																
Starting	Inst	EA	C-1	6	55.00	---	28.09	83.09	96.86	C-1	4	65.00	---	40.13	105.13	124.79
Landing	Inst	EA	C-1	4	82.50	---	42.14	124.64	145.28	C-1	3	97.50	---	60.20	157.70	187.19
Railings, Built-up oak	Inst	LF	C-1	50	8.80	---	3.37	12.17	13.82	C-1	58	10.40	---	2.93	13.33	14.77
Railings, Sub rail oak	Inst	LF	C-1	100	3.63	---	1.69	5.32	6.14	C-1	70	4.29	---	2.41	6.70	7.88

Description	Oper	Unit	Costs Based On Large Volume							Costs Based On Small Volume						
			Crew Size	Avg Day Prod	Avg Mat'l Unit Cost	Avg Equip Unit Cost	Avg Labor Unit Cost	Avg Total Unit Cost	Avg Price Incl O&P	Crew Size	Avg Day Prod	Avg Mat'l Unit Cost	Avg Equip Unit Cost	Avg Labor Unit Cost	Avg Total Unit Cost	Avg Price Incl O&P

Stairs (continued)

Risers, 3/4" x 7-1/2" high

Description	Oper	Unit	Crew Size	Avg Day Prod	Avg Mat'l	Avg Equip	Avg Labor	Avg Total	Avg O&P	Crew Size	Avg Day Prod	Avg Mat'l	Avg Equip	Avg Labor	Avg Total	Avg O&P
Beech	Inst	LF	C-1	60	4.57	---	2.81	7.37	8.75	C-1	42	5.40	---	4.01	9.41	11.37
Fir	Inst	LF	C-1	60	1.27	---	2.81	4.07	5.45	C-1	42	1.50	---	4.01	5.51	7.47
Oak	Inst	LF	C-1	60	4.18	---	2.81	6.99	8.37	C-1	42	4.94	---	4.01	8.95	10.92
Pine	Inst	LF	C-1	60	1.27	---	2.81	4.07	5.45	C-1	42	1.50	---	4.01	5.51	7.47
Skirt board, pine																
1" x 10"	Inst	LF	C-1	50	1.60	---	3.37	4.97	6.62	C-1	35	1.89	---	4.82	6.70	9.06
1" x 12"	Inst	LF	C-1	45	1.93	---	3.75	5.67	7.51	C-1	32	2.28	---	5.35	7.63	10.25
Treads, oak																
1-16" x 9-1/2" wide																
3' long	Inst	EA	C-1	16	18.59	---	10.53	29.12	34.29	C-1	11	21.97	---	15.05	37.02	44.39
4' long	Inst	EA	C-1	15	25.96	---	11.24	37.20	42.70	C-1	11	30.68	---	16.05	46.73	54.60
1-1/16" x 11-1/2" wide																
3' long	Inst	EA	C-1	16	22.28	---	10.53	32.81	37.97	C-1	11	26.33	---	15.05	41.37	48.75
6' long	Inst	EA	C-1	12	47.30	---	14.05	61.35	68.23	C-1	8	55.90	---	20.07	75.97	85.80
For beech treads, ADD	Inst	%	---	---	40.0	---	---	---	---	---	---	40.0	---	---	---	---
For mitered return nosings, ADD	Inst	LF	C-1	60	2.48	---	2.81	5.28	6.66	C-1	42	2.93				

Stairs, shop fabricated. Per flight

Basement stairs, soft wood, open risers

Description	Oper	Unit	Crew Size	Avg Day Prod	Avg Mat'l	Avg Equip	Avg Labor	Avg Total	Avg O&P	Crew Size	Avg Day Prod	Avg Mat'l	Avg Equip	Avg Labor	Avg Total	Avg O&P
3' wide, 8' high	Inst	Flt	C-18	4.0	93.50	---	84.27	177.77	219.07	C-18	2.8	110.50	---	120.39	230.89	289.88
Box stairs, no handrails, 3' wide																
Oak treads																
2' high	Inst	Flt	C-18	5.0	110.00	---	67.42	177.42	210.45	C-18	3.5	130.00	---	96.31	226.31	273.51
4' high	Inst	Flt	C-18	4.0	247.50	---	84.27	331.77	373.07	C-18	2.8	292.50	---	120.39	412.89	471.88
6' high	Inst	Flt	C-18	3.5	401.50	---	96.31	497.81	545.01	C-18	2.5	474.50	---	137.59	612.09	679.51
8' high	Inst	Flt	C-18	3.0	511.50	---	112.37	623.87	678.92	C-18	2.1	604.50	---	160.52	765.02	843.68
Pine treads for carpet																
2' high	Inst	Flt	C-18	5.0	82.50	---	67.42	149.92	182.95	C-18	3.5	97.50	---	96.31	193.81	241.01
4' high	Inst	Flt	C-18	4.0	154.00	---	84.27	238.27	279.57	C-18	2.8	182.00	---	120.39	302.39	361.38
6' high	Inst	Flt	C-18	3.5	242.00	---	96.31	338.31	385.51	C-18	2.5	286.00	---	137.59	423.59	491.01
8' high	Inst	Flt	C-18	3.0	297.00	---	112.37	409.37	464.42	C-18	2.1	351.00	---	160.52	511.52	590.18
Stair rail with balusters, 5 risers	Inst	EA	C-18	15.0	143.00	---	22.47	165.47	176.48	C-18	10.5	169.00	---	32.10	201.10	216.84
For 4' wide stairs, ADD	Inst	%	C-18	---	25.0	5.0	---	---	---	C-18	---	25.0	5.0	---	---	---

			Costs Based On Large Volume							Costs Based On Small Volume					
Description	Oper	Unit	Crew Size	Avg Day Prod	Avg Mat'l Unit Cost	Avg Equip Unit Cost	Avg Labor Unit Cost	Avg Total Unit Cost	Avg Price Incl O&P	Crew Size	Avg Day Prod	Avg Mat'l Unit Cost	Avg Equip Unit Cost	Avg Labor Unit Cost	Avg Total Unit Cost	Avg Price Incl O&P

Stairs (continued)

Open stairs, prefinished, metal stringers, 3'-6" wide treads, no railings

Description	Oper	Unit	Crew Size	Avg Day Prod	Avg Mat'l Unit Cost	Avg Equip Unit Cost	Avg Labor Unit Cost	Avg Total Unit Cost	Avg Price Incl O&P	Crew Size	Avg Day Prod	Avg Mat'l Unit Cost	Avg Equip Unit Cost	Avg Labor Unit Cost	Avg Total Unit Cost	Avg Price Incl O&P
2' high	Inst	Flt	C-18	5.0	346.50	---	67.42	413.92	446.95	C-18	3.5	409.50	---	96.31	505.81	553.01
4' high	Inst	Flt	C-18	4.0	429.00	---	84.27	513.27	554.57	C-18	2.8	507.00	---	120.39	627.39	686.38
6' high	Inst	Flt	C-18	3.5	748.00	---	96.31	844.31	891.51	C-18	2.5	884.00	---	137.59	1021.59	1089.01
8' high	Inst	Flt	C-18	3.0	1166.00	---	112.37	1278.37	1333.42	C-18	2.1	1378.00	---	160.52	1538.52	1617.18

Adjustments

For 3 piece wood railings and balusters, ADD

Description	Oper	Unit	Crew Size	Avg Day Prod	Avg Mat'l Unit Cost	Avg Equip Unit Cost	Avg Labor Unit Cost	Avg Total Unit Cost	Avg Price Incl O&P	Crew Size	Avg Day Prod	Avg Mat'l Unit Cost	Avg Equip Unit Cost	Avg Labor Unit Cost	Avg Total Unit Cost	Avg Price Incl O&P
2' high	Inst	EA	C-18	15.0	104.50	---	22.47	126.97	137.98	C-18	10.5	123.50	---	32.10	155.60	171.34
4' high	Inst	EA	C-18	14.0	137.50	---	24.08	161.58	173.38	C-18	9.8	162.50	---	34.40	196.90	213.75
6' high	Inst	EA	C-18	13.0	214.50	---	25.93	240.43	253.14	C-18	9.1	253.50	---	37.04	290.54	308.70
8' high	Inst	EA	C-18	12.0	269.50	---	28.09	297.59	311.36	C-18	8.4	318.50	---	40.13	358.63	378.29
For 3'-6" x 3'-6" platform ADD	Inst	EA	C-18	4.0	121.00	---	84.27	205.27	246.57	C-18	2.8	143.00	---	120.39	263.39	322.38

Curved stairways, oak, unfinished, curved balustrade

Open one side

Description	Oper	Unit	Crew Size	Avg Day Prod	Avg Mat'l Unit Cost	Avg Equip Unit Cost	Avg Labor Unit Cost	Avg Total Unit Cost	Avg Price Incl O&P	Crew Size	Avg Day Prod	Avg Mat'l Unit Cost	Avg Equip Unit Cost	Avg Labor Unit Cost	Avg Total Unit Cost	Avg Price Incl O&P
9' high	Inst	Flt	C-18	0.70	4840	---	482	5322	5558	C-18	0.49	5720	---	688	6408	6745
10' high	Inst	Flt	C-18	0.70	5280	---	482	5762	5998	C-18	0.49	6240	---	688	6928	7265
Open both sides																
9' high	Inst	Flt	C-18	0.50	8030	---	674	8704	9035	C-18	0.35	9490	---	963	10453	10925
10' high	Inst	Flt	C-18	0.50	8690	---	674	9364	9695	C-18	0.35	10270	---	963	11233	11705

Steel, Reinforcing. See Concrete, Masonry.

Stucco. See Plaster & Stucco.

Studs. See Framing.

Subflooring. See Framing.

Suspended Ceiling Systems

See also Acoustical treatment.

Suspended ceiling panels and grid system

Description	Oper	Unit	Crew Size	Avg Day Prod	Avg Mat'l Unit Cost	Avg Equip Unit Cost	Avg Labor Unit Cost	Avg Total Unit Cost	Avg Price Incl O&P	Crew Size	Avg Day Prod	Avg Mat'l Unit Cost	Avg Equip Unit Cost	Avg Labor Unit Cost	Avg Total Unit Cost	Avg Price Incl O&P
Suspended ceiling panels and grid system	Demo	SF	L-2	1780	---	---	0.14	0.14	0.21	L-2	1424	---	---	0.18	0.18	0.26

Suspended Ceilings

Suspended ceilings consist of a grid of small metal hangers which are hung from the ceiling framing with wire or strap, and drop-in panels sized to fit the grid system. The main advantage to this type of ceiling finish is that it can be adjusted to any height desired. It can cover a lot of flaws and obstructions (uneven plaster, pipes, wiring, and ductwork). With a high ceiling, the suspended ceiling reduces the sound transfer from the floor above and increases the ceiling's insulating value. One great advantage of suspended ceilings is that the area directly above the tiles is readily accessible for repair and maintenance work; the tiles merely have to be lifted out from the grid. This system also eliminates the need for any other ceiling finish.

Installation Procedure:

1. Decide on ceiling height. It must clear the lowest obstacles but remain above windows and doors. Snap a chalk line around walls at the new height. Check it with a carpenter's level.

2. Install wall mold/angle at the marked line; this supports the tiles around the room's perimeter. Check with a level. Use nails (6d) or screws to fasten molding to studs (16" or 24" o.c.); on masonry, use concrete nails 24" o.c. The molding should be flush along the chalk line with the bottom leg of the L-angle facing into the room.

3. Fit wall molding at inside corners by straight cutting (90 degrees) and overlapping; for outside corners, miter cut corner (45 degrees) and butt together. Cut the molding to required lengths with tin snips or a hacksaw.

4. Mark center line of room above wall molding. Decide where first main runner should be to achieve desired border width. Position main runners by running taut string lines across room. Repeat for first row of cross tees closest to one wall. Then attach hanger wires with screw eyes to the joists at 4' intervals.

5. Trim main runners at wall to line up slot with cross tee string. Rest trimmed end on wall molding; insert wire through holes; recheck with level; twist wire several times.

6. Insert cross tee and tabs into main runner's slots; push down to lock. Check with level. Repeat.

7. Drop 2' x 4' panels into place.

Estimating Materials:

1. Measure the ceiling and plot it on graph paper. Mark the direction of ceiling joists. On the ceiling itself, snap chalk lines along the joist lines.

2. Draw ceiling layout for grid system onto graph paper. Plan the ceiling layout, figuring full panels across the main ceiling and evenly trimmed partial panels at the edges. To calculate the width of border panels in each direction, determine the width of the gap left after full panels are placed all across the dimension; then divide by 2.

3. For purposes of ordering material, determine the room size in even number of feet. If the room length or width is not divisible by 2 feet, increase the dimension to the next larger unit divisible by 2 feet. Example: Room size (actual) = 11'-6" x 14'-4" would be considered a 12' x 16' room. This allows for waste and/or breakage. In this example, 192 SF of material would be ordered.

4. Main runners/tees must run perpendicular to ceiling joists. For a 2' x 2' grid, cross and main tees are 2' apart. For a 2' x 4' grid, main tees are 4' apart and 4' cross tees connect the main tees; then 2' cross tees can be used to connect the 4' cross tees. The long panels of the grid are set parallel to the ceiling joists, so the T-shaped main runner is attached perpendicular to joists at 4' o.c.

5. Wall molding/angle is manufactured in 10' lengths. Main runners/tees are manufactured in 12' lengths. Cross tees are either 2' long or 4' long. Hanger wire is 12 gauge, and attached to joists with either screw eyes or hook-and-nail. Drop-in panels are either 8 SF (2' x 4') or 4 SF (2' x 2').

6. To determine the number of drop-in panels required, divide the nominal room size in square feet (e.g., 192 SF in above example) by square foot of panel to be used.

7. The quantity of 12 gauge wire depends on the "drop" distance of the ceiling. Figure one suspension wire and screw eye for each 4' of main runner; if any main run is longer than 12' then splice plates are needed, with a wire and screw eye on each side of the splice. The length of each wire should be at least 2" longer than the "drop" distance, to allow for a twist after passing through runner or tee.

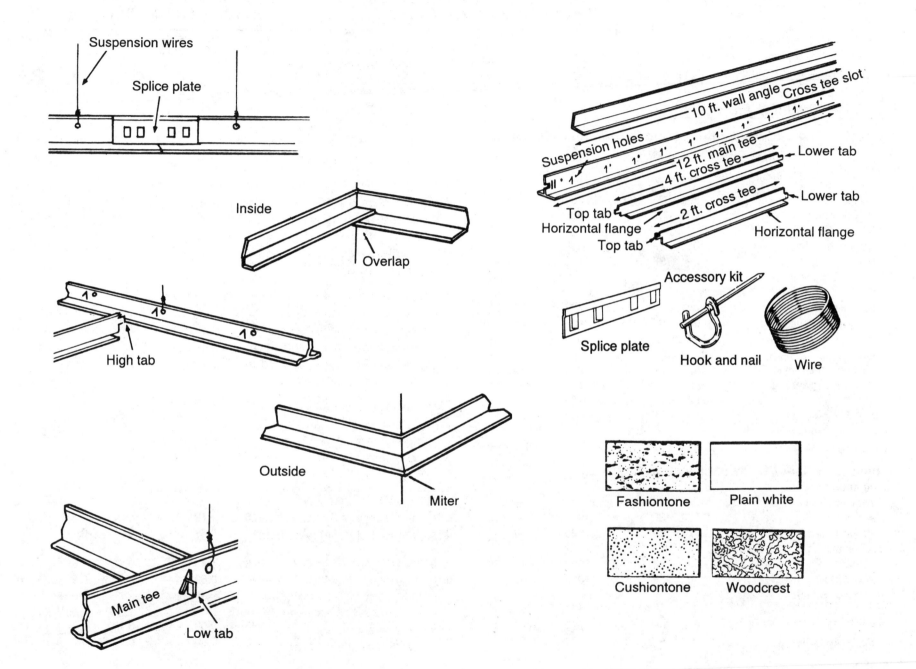

Suspension wires

Splice plate

Inside

Overlap

10 ft. wall angle
Cross tee slot

Suspension holes

1' 1' 1' 1' 1'
1' 1' 1' 1' 1'

12 ft. main tee
4 ft. cross tee
Lower tab

High tab

Top tab
Horizontal flange
Top tab
2 ft. cross tee
Lower tab
Horizontal flange

Accessory kit

Splice plate

Hook and nail

Wire

Outside

Miter

Fashiontone

Plain white

Cushiontone

Woodcrest

Main tee

Low tab

			Costs Based On Large Volume							Costs Based On Small Volume					
Description	Oper	Unit	Crew Size	Avg Day Prod	Avg Mat'l Unit Cost	Avg Equip Unit Cost	Avg Labor Unit Cost	Avg Total Unit Cost	Avg Price Incl O&P	Crew Size	Avg Day Prod	Avg Mat'l Unit Cost	Avg Equip Unit Cost	Avg Labor Unit Cost	Avg Total Unit Cost	Avg Price Incl O&P

Suspended ceilings (continued)

Complete 2' x 4' slide lock grid system (baked enamel). Labor includes installing grid and laying in panels

Luminous panels

Prismatic panels

Acrylic	Inst	SF	C-2	380	1.56	---	0.47	2.03	2.27	C-2	304	1.80	---	0.59	2.39	2.68
Polystyrene	Inst	SF	C-2	380	1.13	---	0.47	1.61	1.84	C-2	304	1.31	---	0.59	1.90	2.19

Ribbed panels

Acrylic	Inst	SF	C-2	380	1.77	---	0.47	2.24	2.48	C-2	304	2.04	---	0.59	2.63	2.92
Polystyrene	Inst	SF	C-2	380	1.13	---	0.47	1.61	1.84	C-2	304	1.31	---	0.59	1.90	2.19

Fiberglass panels

1/2" T

Plain white	Inst	SF	C-2	405	0.99	---	0.45	1.43	1.65	C-2	324	1.14	---	0.56	1.70	1.97
Sculptured white	Inst	SF	C-2	405	1.05	---	0.45	1.50	1.72	C-2	324	1.22	---	0.56	1.77	2.04

5/8" T

Fissured	Inst	SF	C-2	405	1.01	---	0.45	1.46	1.68	C-2	324	1.17	---	0.56	1.73	2.00
Fissured, fire rated	Inst	SF	C-2	405	1.13	---	0.45	1.58	1.79	C-2	324	1.31	---	0.56	1.86	2.13

Mineral fiber panels

1/2" T

1/2" T	Inst	SF	C-2	405	1.09	---	0.45	1.54	1.76	C-2	324	1.26	---	0.56	1.82	2.09
9/16" T	Inst	SF	C-2	405	1.22	---	0.45	1.67	1.89	C-2	324	1.41	---	0.56	1.97	2.24
For wood grained grid components, ADD	Inst	%	---	---	7.0	---	---	---	---	---	---	7.0	---	---	---	---

Panels only (labor to lay panels in grid system)

Luminous panels

Prismatic panels

Acrylic	Inst	SF	C-2	1020	1.03	---	0.18	1.20	1.29	C-2	816	1.19	---	0.22	1.41	1.51
Polystyrene	Inst	SF	C-2	1020	0.60	---	0.18	0.77	0.86	C-2	816	0.69	---	0.22	0.91	1.02

Ribbed panels

Acrylic	Inst	SF	C-2	1020	1.24	---	0.18	1.41	1.50	C-2	816	1.43	---	0.22	1.65	1.75
Polystyrene	Inst	SF	C-2	1020	0.60	---	0.18	0.77	0.86	C-2	816	0.69	---	0.22	0.91	1.02

Fiberglass panels

1/2" T

Plain white	Inst	SF	C-2	1220	0.46	---	0.15	0.60	0.68	C-2	976	0.53	---	0.18	0.71	0.80
Sculptured white	Inst	SF	C-2	1220	0.52	---	0.15	0.67	0.74	C-2	976	0.60	---	0.18	0.78	0.88

Description	Oper	Unit	Costs Based On Large Volume							Costs Based On Small Volume						
			Crew Size	Avg Day Prod	Avg Mat'l Unit Cost	Avg Equip Unit Cost	Avg Labor Unit Cost	Avg Total Unit Cost	Avg Price Incl O&P	Crew Size	Avg Day Prod	Avg Mat'l Unit Cost	Avg Equip Unit Cost	Avg Labor Unit Cost	Avg Total Unit Cost	Avg Price Incl O&P

Suspended ceilings (continued)

5/8" T

Description	Oper	Unit	Crew Size	Avg Day Prod	Avg Mat'l Unit Cost	Avg Equip Unit Cost	Avg Labor Unit Cost	Avg Total Unit Cost	Avg Price Incl O&P	Crew Size	Avg Day Prod	Avg Mat'l Unit Cost	Avg Equip Unit Cost	Avg Labor Unit Cost	Avg Total Unit Cost	Avg Price Incl O&P
Fissured	Inst	SF	C-2	1220	0.48	---	0.15	0.63	0.70	C-2	976	0.56	---	0.18	0.74	0.83
Fissured, fire rated	Inst	SF	C-2	1220	0.60	---	0.15	0.75	0.82	C-2	976	0.69	---	0.18	0.87	0.97
Mineral fiber panels																
1/2" T	Inst	SF	C-2	1220	0.56	---	0.15	0.71	0.78	C-2	976	0.65	---	0.18	0.83	0.92
9/16" T	Inst	SF	C-2	1220	0.69	---	0.15	0.84	0.91	C-2	976	0.80	---	0.18	0.98	1.07

Grid system components

Main runner, 12' L pieces

Description	Oper	Unit	Crew Size	Avg Day Prod	Avg Mat'l Unit Cost	Avg Equip Unit Cost	Avg Labor Unit Cost	Avg Total Unit Cost	Avg Price Incl O&P	Crew Size	Avg Day Prod	Avg Mat'l Unit Cost	Avg Equip Unit Cost	Avg Labor Unit Cost	Avg Total Unit Cost	Avg Price Incl O&P
Enamel	Inst	EA	---	---	5.79	---	---	---	5.79	---	---	6.68	---	---	---	6.68
Wood grain	Inst	EA	---	---	6.63	---	---	---	6.63	---	---	7.65	---	---	---	7.65
Cross tees																
2' L pieces																
Enamel	Inst	EA	---	---	1.04	---	---	---	1.04	---	---	1.20	---	---	---	1.20
Wood grain	Inst	EA	---	---	1.17	---	---	---	1.17	---	---	1.35	---	---	---	1.35
4' L pieces																
Enamel	Inst	EA	---	---	1.95	---	---	---	1.95	---	---	2.25	---	---	---	2.25
Wood grain	Inst	EA	---	---	2.21	---	---	---	2.21	---	---	2.55	---	---	---	2.55
Wall molding, 10' L pieces																
Enamel	Inst	EA	---	---	3.71	---	---	---	3.71	---	---	4.28	---	---	---	4.28
Wood grain	Inst	EA	---	---	4.88	---	---	---	4.88	---	---	5.63	---	---	---	5.63
Main runner hold-down clips																
1/2" T panels (1000/carton)	Inst	Ctn	---	---	10.40	---	---	---	10.40	---	---	12.00	---	---	---	12.00
5/8" T panels (1000/carton)	Inst	Ctn	---	---	14.30	---	---	---	14.30	---	---	16.50	---	---	---	16.50

Telephone prewiring. See Electrical.

Television antenna. See Electrical.

Thermostat. See Electrical.

Tile. See Acoustical treatment, Ceramic, Masonry, Resilient flooring.

			Costs Based On Large Volume							Costs Based On Small Volume						
Description	Oper	Unit	Crew Size	Avg Day Prod	Avg Mat'l Unit Cost	Avg Equip Unit Cost	Avg Labor Unit Cost	Avg Total Unit Cost	Avg Price Incl O&P	Crew Size	Avg Day Prod	Avg Mat'l Unit Cost	Avg Equip Unit Cost	Avg Labor Unit Cost	Avg Total Unit Cost	Avg Price Incl O&P

Toilets. See Plumbing.

Trash Compactors

Includes wiring, connection, and installation in a pre-cutout area.

Description	Oper	Unit	Crew Size	Avg Day Prod	Avg Mat'l Unit Cost	Avg Equip Unit Cost	Avg Labor Unit Cost	Avg Total Unit Cost	Avg Price Incl O&P	Crew Size	Avg Day Prod	Avg Mat'l Unit Cost	Avg Equip Unit Cost	Avg Labor Unit Cost	Avg Total Unit Cost	Avg Price Incl O&P
White	Inst	EA	E-1	4	592.80	---	48.98	641.78	663.83	E-1	3	684.00	---	61.23	745.23	772.78
Colors	Inst	EA	E-1	4	625.30	---	48.98	674.28	696.33	E-1	3	721.50	---	61.23	782.73	810.28
To remove and reset	Reset	EA	E-1	11	---	---	17.81	17.81	25.83	E-1	9	---	---	22.26	22.26	32.28
Material adjustments																
Stainless steel trim kit	---	LS	---	---	33.80	---	---	---	33.80	---	---	39.00	---	---	---	39.00
Stainless steel panel and trim kit	---	LS	---	---	83.20	---	---	---	83.20	---	---	96.00	---	---	---	96.00
"Black Glas" acrylic front panels and trim kit	---	LS	---	---	83.20	---	---	---	83.20	---	---	96.00	---	---	---	96.00
Hardwood top	---	LS	---	---	65.00	---	---	---	65.00	---	---	75.00	---	---	---	75.00

Trim. See Molding and trim.

Trusses. See Framing.

Vanity units. See Cabinetry.

Wallpaper

Butt joint. Includes application of glue sizing. Material prices are not given because they vary radically

Description	Oper	Unit	Crew Size	Avg Day Prod	Avg Mat'l Unit Cost	Avg Equip Unit Cost	Avg Labor Unit Cost	Avg Total Unit Cost	Avg Price Incl O&P	Crew Size	Avg Day Prod	Avg Mat'l Unit Cost	Avg Equip Unit Cost	Avg Labor Unit Cost	Avg Total Unit Cost	Avg Price Incl O&P
Paper, single rolls 36 SF/Roll	Inst	Roll	Q-1	22.4	---	3.91	7.23	11.14	14.53	Q-1	14.6	---	6.01	11.12	17.13	22.36
Vinyl, with fabric backing (prepasted), single rolls 36 SF/Roll	Inst	Roll	Q-1	20.0	---	4.38	8.10	12.47	16.28	Q-1	13.0	---	6.73	12.45	19.19	25.04
Vinyl, with fabric backing (not prepasted), single rolls 36 SF/Roll	Inst	Roll	Q-1	18.4	---	4.76	8.80	13.56	17.69	Q-1	12.0	---	7.32	13.54	20.86	27.22
Paste	Inst	Gal		---	13.00	---	---	---	13.00	---	---	13.00	---	---	---	13.00

Wallpaper

Most wall coverings today are really not wallpapers in the technical sense. Wallpaper is a paper material (which may or not be coated with washable plastic).

The greater part of today's products, though still called wallpaper, are actually composed of a vinyl on a fabric backing — not a paper backing. Some vinyl/fabric coverings come prepasted.

Vinyl coverings with non-woven fabric, woven fabric or synthetic fiber backing are fire, mildew and fade resistant. Also, these coverings can be stripped from plaster walls, whereas those coverings with paper backing must be steamed or scraped from the walls.

Vinyl coverings may also be stripped from gypsum wallboard, but unless the vinyl covering has a synthetic fiber backing, it is likely that the wallboard paper surface may be damaged.

One gallon of paste (approximately two-thirds pound of dry paste and water) should adequately cover 12 rolls with paper backing and 6 rolls with fabric backing. The rougher the texture of the surface to be pasted, the greater the quantity of wet paste needed.

1. **Dimensions.**

   a. Single roll is 36 SF. The paper is 18" wide x 24'-0" long.

   b. Double roll is 72 SF. The paper is 18" wide x 48'-0" long.

2. **Installation.** New paper may be applied over existing paper if the existing paper has butt joints, a smooth surface and is tight to the wall. When new paper is applied direct to plaster or sheetrock, the wall should receive a coat of glue size before the paper is applied.

3. **Estimating technique.** Determine the gross area, deduct openings and other areas not to be papered and add 20% to the net area for waste. The net area plus 20% divided by the number of SF per roll will provide the number of rolls needed.

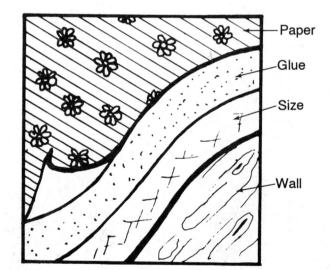

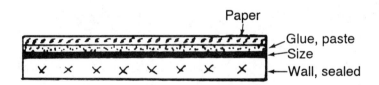

The steps in wallpapering

263

			Costs Based On Large Volume							Costs Based On Small Volume						
Description	Oper	Unit	Crew Size	Avg Day Prod	Avg Mat'l Unit Cost	Avg Equip Unit Cost	Avg Labor Unit Cost	Avg Total Unit Cost	Avg Price Incl O&P	Crew Size	Avg Day Prod	Avg Mat'l Unit Cost	Avg Equip Unit Cost	Avg Labor Unit Cost	Avg Total Unit Cost	Avg Price Incl O&P

Wallpaper (continued)

Labor adjustments

When ceiling is 9'-0" high or more, DECREASE OUTPUT

Description	Oper	Unit	Crew Size	Avg Day Prod	Mat'l	Equip	Labor	Total	O&P	Crew Size	Avg Day Prod	Mat'l	Equip	Labor	Total	O&P
more, DECREASE OUTPUT	Inst	Roll	Q-1	4.0	---	---	---	---	---	Q-1	2.6	---	---	---	---	---

In kitchens, baths, and in other rooms, when floor area is less than 50.0 SF

DECREASE OUTPUT	Inst	Roll	Q-1	6.4	---	---	---	---	---	Q-1	4.2	---	---	---	---	---

Water closets. See Plumbing, Fixtures, Toilets.

Water heater. See Plumbing.

Water softeners. See Plumbing.

Weather vanes. See Cupolas.

Windows

Windows, with related trim and frame

To 12 SF

Description	Oper	Unit	Crew Size	Avg Day Prod	Mat'l	Equip	Labor	Total	O&P	Crew Size	Avg Day Prod	Mat'l	Equip	Labor	Total	O&P
Aluminum	Demo	EA	L-2	21.0	---	---	12.38	12.38	18.33	L-2	14.7	---	---	17.69	17.69	26.19
Wood	Demo	EA	L-2	16.0	---	---	16.26	16.26	24.06	L-2	11.2	---	---	23.22	23.22	34.37

13 SF to 50 SF

Description	Oper	Unit	Crew Size	Avg Day Prod	Mat'l	Equip	Labor	Total	O&P	Crew Size	Avg Day Prod	Mat'l	Equip	Labor	Total	O&P
Aluminum	Demo	EA	L-2	13.0	---	---	20.01	20.01	29.61	L-2	9.1	---	---	28.58	28.58	42.30
Wood	Demo	EA	L-2	10.0	---	---	26.01	26.01	38.49	L-2	7.0	---	---	37.15	37.15	54.99

Aluminum

Vertical slide, includes screen and hardware, with satin anodized finish

Single glazed

Description	Oper	Unit	Crew Size	Avg Day Prod	Mat'l	Equip	Labor	Total	O&P	Crew Size	Avg Day Prod	Mat'l	Equip	Labor	Total	O&P
1'-6" x 3'-0" H	Inst	Set	C-1	7.0	57.20	---	24.08	81.28	93.08	C-1	4.9	66.00	---	34.40	100.40	117.25
2'-0" x 2'-0" H	Inst	Set	C-1	7.0	54.60	---	24.08	78.68	90.48	C-1	4.9	63.00	---	34.40	97.40	114.25
2'-0" x 2'-6" H	Inst	Set	C-1	7.0	59.80	---	24.08	83.88	95.68	C-1	4.9	69.00	---	34.40	103.40	120.25
2'-0" x 3'-0" H	Inst	Set	C-1	6.5	63.70	---	25.93	89.63	102.34	C-1	4.6	73.50	---	37.04	110.54	128.70
2'-0" x 3'-6" H	Inst	Set	C-1	6.5	68.90	---	25.93	94.83	107.54	C-1	4.6	79.50	---	37.04	116.54	134.70
2'-0" x 4'-0" H	Inst	Set	C-1	6.5	72.80	---	25.93	98.73	111.44	C-1	4.6	84.00	---	37.04	121.04	139.20
2'-0" x 4'-6" H	Inst	Set	C-1	6.5	78.00	---	25.93	103.93	116.64	C-1	4.6	90.00	---	37.04	127.04	145.20
2'-0" x 5'-0" H	Inst	Set	C-1	6.0	81.90	---	28.09	109.99	123.76	C-1	4.2	94.50	---	40.13	134.63	154.29
2'-0" x 6'-0" H	Inst	Set	C-1	6.0	91.00	---	28.09	119.09	132.86	C-1	4.2	105.00	---	40.13	145.13	164.79

Windows (continued)

Description	Oper	Unit	Costs Based On Large Volume							Costs Based On Small Volume						
			Crew Size	Avg Day Prod	Avg Mat'l Unit Cost	Avg Equip Unit Cost	Avg Labor Unit Cost	Avg Total Unit Cost	Avg Price Incl O&P	Crew Size	Avg Day Prod	Avg Mat'l Unit Cost	Avg Equip Unit Cost	Avg Labor Unit Cost	Avg Total Unit Cost	Avg Price Incl O&P
2'-6" x 3'-0" H	Inst	Set	C-1	6.5	71.50	---	25.93	97.43	110.14	C-1	4.6	82.50	---	37.04	119.54	137.70
2'-6" x 3'-6" H	Inst	Set	C-1	6.5	75.40	---	25.93	101.33	114.04	C-1	4.6	87.00	---	37.04	124.04	142.20
2'-6" x 4'-0" H	Inst	Set	C-1	6.5	80.60	---	25.93	106.53	119.24	C-1	4.6	93.00	---	37.04	130.04	148.20
2'-6" x 4'-6" H	Inst	Set	C-1	6.5	85.80	---	25.93	111.73	124.44	C-1	4.6	99.00	---	37.04	136.04	154.20
2'-6" x 5'-0" H	Inst	Set	C-1	6.0	89.70	---	28.09	117.79	131.56	C-1	4.2	103.50	---	40.13	143.63	163.29
2'-6" x 6'-0" H	Inst	Set	C-1	6.0	100.10	---	28.09	128.19	141.96	C-1	4.2	115.50	---	40.13	155.63	175.29
3'-0" x 2'-0" H	Inst	Set	C-1	6.5	67.60	---	25.93	93.53	106.24	C-1	4.6	78.00	---	37.04	115.04	133.20
3'-0" x 3'-0" H	Inst	Set	C-1	6.5	78.00	---	25.93	103.93	116.64	C-1	4.6	90.00	---	37.04	127.04	145.20
3'-0" x 3'-6" H	Inst	Set	C-1	6.5	83.20	---	25.93	109.13	121.84	C-1	4.6	96.00	---	37.04	133.04	151.20
3'-0" x 4'-0" H	Inst	Set	C-1	6.5	88.40	---	25.93	114.33	127.04	C-1	4.6	102.00	---	37.04	139.04	157.20
3'-0" x 4'-6" H	Inst	Set	C-1	6.5	93.60	---	25.93	119.53	132.24	C-1	4.6	108.00	---	37.04	145.04	163.20
3'-0" x 5'-0" H	Inst	Set	C-9	5.8	98.80	---	51.48	150.28	175.51	C-9	4.1	114.00	---	73.54	187.54	223.58
3'-0" x 6'-0" H	Inst	Set	C-9	5.2	111.80	---	57.42	169.22	197.36	C-9	3.6	129.00	---	82.03	211.03	251.23
3'-6" x 3'-6" H	Inst	Set	C-1	6.5	91.00	---	25.93	116.93	129.64	C-1	4.6	105.00	---	37.04	142.04	160.20
3'-6" x 4'-0" H	Inst	Set	C-1	5.5	97.50	---	30.65	128.15	143.16	C-1	3.9	112.50	---	43.78	156.28	177.73
3'-6" x 4'-6" H	Inst	Set	C-1	4.5	102.70	---	37.46	140.16	158.51	C-1	3.2	118.50	---	53.51	172.01	198.23
3'-6" x 5'-0" H	Inst	Set	C-9	5.2	107.90	---	57.42	165.32	193.46	C-9	3.6	124.50	---	82.03	206.53	246.73
4'-0" x 3'-0" H	Inst	Set	C-1	6.5	93.60	---	25.93	119.53	132.24	C-1	4.6	108.00	---	37.04	145.04	163.20
4'-0" x 3'-6" H	Inst	Set	C-1	5.5	101.40	---	30.65	132.05	147.06	C-1	3.9	117.00	---	43.78	160.78	182.23
4'-0" x 4'-0" H	Inst	Set	C-9	5.8	105.30	---	51.48	156.78	182.01	C-9	4.1	121.50	---	73.54	195.04	231.08
4'-0" x 4'-6" H	Inst	Set	C-9	5.8	111.80	---	51.48	163.28	188.51	C-9	4.1	129.00	---	73.54	202.54	238.58
4'-0" x 5'-0" H	Inst	Set	C-9	5.2	118.30	---	57.42	175.72	203.86	C-9	3.6	136.50	---	82.03	218.53	258.73
For bronze finish ADD	Inst	%	---	---	10.0	---	---	---	---	---	---	10.0	---	---	---	---
Dual glazed																
1'-6" x 3'-0" H	Inst	Set	C-1	6.8	101.40	---	24.79	126.19	138.33	C-1	4.8	117.00	---	35.41	152.41	169.76
2'-0" x 2'-0" H	Inst	Set	C-1	6.8	96.20	---	24.79	120.99	133.13	C-1	4.8	111.00	---	35.41	146.41	163.76
2'-0" x 2'-6" H	Inst	Set	C-1	6.8	105.30	---	24.79	130.09	142.23	C-1	4.8	121.50	---	35.41	156.91	174.26
2'-0" x 3'-0" H	Inst	Set	C-1	6.3	113.10	---	26.75	139.85	152.96	C-1	4.4	130.50	---	38.22	168.72	187.45
2'-0" x 3'-6" H	Inst	Set	C-1	6.3	120.90	---	26.75	147.65	160.76	C-1	4.4	139.50	---	38.22	177.72	196.45
2'-0" x 4'-0" H	Inst	Set	C-1	6.3	128.70	---	26.75	155.45	168.56	C-1	4.4	148.50	---	38.22	186.72	205.45
2'-0" x 4'-6" H	Inst	Set	C-1	6.3	137.80	---	26.75	164.55	177.66	C-1	4.4	159.00	---	38.22	197.22	215.95
2'-0" x 5'-0" H	Inst	Set	C-1	5.8	145.60	---	29.06	174.66	188.90	C-1	4.1	168.00	---	41.51	209.51	229.86
2'-0" x 6'-0" H	Inst	Set	C-1	5.8	161.20	---	29.06	190.26	204.50	C-1	4.1	186.00	---	41.51	227.51	247.86

Windows (continued)

Description	Oper	Unit	Crew Size	Avg Day Prod	Avg Mat'l Unit Cost	Avg Equip Unit Cost	Avg Labor Unit Cost	Avg Total Unit Cost	Avg Price Incl O&P	Crew Size	Avg Day Prod	Avg Mat'l Unit Cost	Avg Equip Unit Cost	Avg Labor Unit Cost	Avg Total Unit Cost	Avg Price Incl O&P
2'-6" x 3'-0" H	Inst	Set	C-1	6.3	126.10	---	26.75	152.85	165.96	C-1	4.4	145.50	---	38.22	183.72	202.45
2'-6" x 3'-6" H	Inst	Set	C-1	6.3	133.90	---	26.75	160.65	173.76	C-1	4.4	154.50	---	38.22	192.72	211.45
2'-6" x 4'-0" H	Inst	Set	C-1	6.3	143.00	---	26.75	169.75	182.86	C-1	4.4	165.00	---	38.22	203.22	221.95
2'-6" x 4'-6" H	Inst	Set	C-1	6.3	152.10	---	26.75	178.85	191.96	C-1	4.4	175.50	---	38.22	213.72	232.45
2'-6" x 5'-0" H	Inst	Set	C-1	5.8	159.90	---	29.06	188.96	203.20	C-1	4.1	184.50	---	41.51	226.01	246.36
2'-6" x 6'-0" H	Inst	Set	C-1	5.8	185.90	---	29.06	214.96	229.20	C-1	4.1	214.50	---	41.51	256.01	276.36
3'-0" x 2'-0" H	Inst	Set	C-1	6.3	118.30	---	26.75	145.05	158.16	C-1	4.4	136.50	---	38.22	174.72	193.45
3'-0" x 3'-0" H	Inst	Set	C-1	6.3	137.80	---	26.75	164.55	177.66	C-1	4.4	159.00	---	38.22	197.22	215.95
3'-0" x 3'-6" H	Inst	Set	C-1	6.3	146.90	---	26.75	173.65	186.76	C-1	4.4	169.50	---	38.22	207.72	226.45
3'-0" x 4'-0" H	Inst	Set	C-1	6.3	156.00	---	26.75	182.75	195.86	C-1	4.4	180.00	---	38.22	218.22	236.95
3'-0" x 4'-6" H	Inst	Set	C-1	6.3	166.40	---	26.75	193.15	206.26	C-1	4.4	192.00	---	38.22	230.22	248.95
3'-0" x 5'-0" H	Inst	Set	C-9	5.6	175.50	---	53.32	228.82	254.95	C-9	3.9	202.50	---	76.17	278.67	315.99
3'-0" x 6'-0" H	Inst	Set	C-9	5.0	193.70	---	59.72	253.42	282.68	C-9	3.5	223.50	---	85.31	308.81	350.61
3'-6" x 3'-6" H	Inst	Set	C-1	6.3	159.90	---	26.75	186.65	199.76	C-1	4.4	184.50	---	38.22	222.72	241.45
3'-6" x 4'-0" H	Inst	Set	C-1	5.3	171.60	---	31.80	203.40	218.98	C-1	3.7	198.00	---	45.43	243.43	265.69
3'-6" x 4'-6" H	Inst	Set	C-1	4.3	182.00	---	39.20	221.20	240.40	C-1	3.0	210.00	---	56.00	266.00	293.43
3'-6" x 5'-0" H	Inst	Set	C-9	5.0	192.40	---	59.72	252.12	281.38	C-9	3.5	222.00	---	85.31	307.31	349.11
4'-0" x 3'-0" H	Inst	Set	C-1	6.3	162.50	---	26.75	189.25	202.36	C-1	4.4	187.50	---	38.22	225.72	244.45
4'-0" x 3'-6" H	Inst	Set	C-1	5.3	172.90	---	31.80	204.70	220.28	C-1	3.7	199.50	---	45.43	244.93	267.19
4'-0" x 4'-0" H	Inst	Set	C-9	5.6	184.60	---	53.32	237.92	264.05	C-9	3.9	213.00	---	76.17	289.17	326.49
4'-0" x 4'-6" H	Inst	Set	C-9	5.6	197.60	---	53.32	250.92	277.05	C-9	3.9	228.00	---	76.17	304.17	341.49
4'-0" x 5'-0" H	Inst	Set	C-9	5.0	208.00	---	59.72	267.72	296.98	C-9	3.5	240.00	---	85.31	325.31	367.11
For bronze finish ADD	Inst	%	---	---	7.0	---	---	---	---	---	---	7.0	---	---	---	---

Horizontal slide, includes screen and hardware, with satin anodized finish
1 sliding lite, 1 fixed lite
Single glazed

Description	Oper	Unit	Crew Size	Avg Day Prod	Avg Mat'l Unit Cost	Avg Equip Unit Cost	Avg Labor Unit Cost	Avg Total Unit Cost	Avg Price Incl O&P	Crew Size	Avg Day Prod	Avg Mat'l Unit Cost	Avg Equip Unit Cost	Avg Labor Unit Cost	Avg Total Unit Cost	Avg Price Incl O&P
2'-0" x 2'-0" H	Inst	Set	C-1	6.5	46.80	---	25.93	72.73	85.44	C-1	4.6	54.00	---	37.04	91.04	109.20
2'-0" x 3'-0" H	Inst	Set	C-1	6.5	58.50	---	25.93	84.43	97.14	C-1	4.6	67.50	---	37.04	104.54	122.70
2'-6" x 3'-0" H	Inst	Set	C-1	6.5	63.70	---	25.93	89.63	102.34	C-1	4.6	73.50	---	37.04	110.54	128.70
3'-0" x 1'-0" H	Inst	Set	C-1	6.5	42.90	---	25.93	68.83	81.54	C-1	4.6	49.50	---	37.04	86.54	104.70
3'-0" x 1'-6" H	Inst	Set	C-1	6.5	49.40	---	25.93	75.33	88.04	C-1	4.6	57.00	---	37.04	94.04	112.20
3'-0" x 2'-0" H	Inst	Set	C-1	6.5	54.60	---	25.93	80.53	93.24	C-1	4.6	63.00	---	37.04	100.04	118.20
3'-0" x 2'-6" H	Inst	Set	C-1	6.5	61.10	---	25.93	87.03	99.74	C-1	4.6	70.50	---	37.04	107.54	125.70

			Costs Based On Large Volume							Costs Based On Small Volume						
Description	Oper	Unit	Crew Size	Avg Day Prod	Avg Mat'l Unit Cost	Avg Equip Unit Cost	Avg Labor Unit Cost	Avg Total Unit Cost	Avg Price Incl O&P	Crew Size	Avg Day Prod	Avg Mat'l Unit Cost	Avg Equip Unit Cost	Avg Labor Unit Cost	Avg Total Unit Cost	Avg Price Incl O&P

Windows (continued)

Description	Oper	Unit	Crew Size	Avg Day Prod	Avg Mat'l Unit Cost	Avg Equip Unit Cost	Avg Labor Unit Cost	Avg Total Unit Cost	Avg Price Incl O&P	Crew Size	Avg Day Prod	Avg Mat'l Unit Cost	Avg Equip Unit Cost	Avg Labor Unit Cost	Avg Total Unit Cost	Avg Price Incl O&P
3'-0" x 3'-0" H	Inst	Set	C-1	6.5	67.60	---	25.93	93.53	106.24	C-1	4.6	78.00	---	37.04	115.04	133.20
3'-0" x 3'-6" H	Inst	Set	C-1	5.5	74.10	---	30.65	104.75	119.76	C-1	3.9	85.50	---	43.78	129.28	150.73
3'-0" x 4'-0" H	Inst	Set	C-1	4.5	80.60	---	37.46	118.06	136.41	C-1	3.2	93.00	---	53.51	146.51	172.73
3'-0" x 5'-0" H	Inst	Set	C-9	3.9	102.70	---	76.56	179.26	216.78	C-9	2.7	118.50	---	109.37	227.87	281.47
3'-4" x 2'-0" H	Inst	Set	C-1	6.5	57.20	---	25.93	83.13	95.84	C-1	4.6	66.00	---	37.04	103.04	121.20
3'-4" x 2'-6" H	Inst	Set	C-1	6.5	63.70	---	25.93	89.63	102.34	C-1	4.6	73.50	---	37.04	110.54	128.70
3'-4" x 3'-0" H	Inst	Set	C-1	5.5	70.20	---	30.65	100.85	115.86	C-1	3.9	81.00	---	43.78	124.78	146.23
3'-4" x 3'-6" H	Inst	Set	C-1	5.5	78.00	---	30.65	108.65	123.66	C-1	3.9	90.00	---	43.78	133.78	155.23
3'-4" x 4'-0" H	Inst	Set	C-1	4.5	84.50	---	37.46	121.96	140.31	C-1	3.2	97.50	---	53.51	151.01	177.23
4'-0" x 1'-0" H	Inst	Set	C-1	5.5	49.40	---	30.65	80.05	95.06	C-1	3.9	57.00	---	43.78	100.78	122.23
4'-0" x 1'-6" H	Inst	Set	C-1	5.5	55.90	---	30.65	86.55	101.56	C-1	3.9	64.50	---	43.78	108.28	129.73
4'-0" x 2'-0" H	Inst	Set	C-1	4.5	62.40	---	37.46	99.86	118.21	C-1	3.2	72.00	---	53.51	125.51	151.73
4'-0" x 2'-6" H	Inst	Set	C-1	4.5	70.20	---	37.46	107.66	126.01	C-1	3.2	81.00	---	53.51	134.51	160.73
4'-0" x 3'-0" H	Inst	Set	C-1	4.5	76.70	---	37.46	114.16	132.51	C-1	3.2	88.50	---	53.51	142.01	168.23
4'-0" x 3'-6" H	Inst	Set	C-1	4.2	84.50	---	40.13	124.63	144.29	C-1	2.9	97.50	---	57.33	154.83	182.92
4'-0" x 4'-0" H	Inst	Set	C-1	4.2	91.00	---	40.13	131.13	150.79	C-1	2.9	105.00	---	57.33	162.33	190.42
4'-0" x 5'-0" H	Inst	Set	C-9	3.5	114.40	---	85.31	199.71	241.51	C-9	2.5	132.00	---	121.87	253.87	313.59
5'-0" x 2'-0" H	Inst	Set	C-9	4.2	71.50	---	71.09	142.59	177.43	C-9	2.9	82.50	---	101.56	184.06	233.83
5'-0" x 2'-6" H	Inst	Set	C-9	4.2	79.30	---	71.09	150.39	185.23	C-9	2.9	91.50	---	101.56	193.06	242.83
5'-0" x 3'-0" H	Inst	Set	C-9	3.9	87.10	---	76.56	163.66	201.18	C-9	2.7	100.50	---	109.37	209.87	263.47
5'-0" x 3'-6" H	Inst	Set	C-9	3.9	94.90	---	76.56	171.46	208.98	C-9	2.7	109.50	---	109.37	218.87	272.47
5'-0" x 4'-0" H	Inst	Set	C-9	3.5	104.00	---	85.31	189.31	231.11	C-9	2.5	120.00	---	121.87	241.87	301.59
5'-0" x 5'-0" H	Inst	Set	C-9	3.5	135.20	---	85.31	220.51	262.31	C-9	2.5	156.00	---	121.87	277.87	337.59
6'-0" x 2'-0" H	Inst	Set	C-9	4.1	79.30	---	72.83	152.13	187.81	C-9	2.9	91.50	---	104.04	195.54	246.52
6'-0" x 2'-6" H	Inst	Set	C-9	4.1	88.40	---	72.83	161.23	196.91	C-9	2.9	102.00	---	104.04	206.04	257.02
6'-0" x 3'-0" H	Inst	Set	C-9	3.9	96.20	---	76.56	172.76	210.28	C-9	2.7	111.00	---	109.37	220.37	273.97
6'-0" x 3'-6" H	Inst	Set	C-9	3.5	105.30	---	85.31	190.61	232.41	C-9	2.5	121.50	---	121.87	243.37	303.09
6'-0" x 4'-0" H	Inst	Set	C-9	3.5	114.40	---	85.31	199.71	241.51	C-9	2.5	132.00	---	121.87	253.87	313.59
6'-0" x 5'-0" H	Inst	Set	C-9	3.5	149.50	---	85.31	234.81	276.61	C-9	2.5	172.50	---	121.87	294.37	354.09
8'-0" x 2'-0" H	Inst	Set	C-9	3.9	96.20	---	76.56	172.76	210.28	C-9	2.7	111.00	---	109.37	220.37	273.97
8'-0" x 2'-6" H	Inst	Set	C-9	3.9	106.60	---	76.56	183.16	220.68	C-9	2.7	123.00	---	109.37	232.37	285.97

Description	Oper	Unit	Costs Based On Large Volume							Costs Based On Small Volume						
			Crew Size	Avg Day Prod	Avg Mat'l Unit Cost	Avg Equip Unit Cost	Avg Labor Unit Cost	Avg Total Unit Cost	Avg Price Incl O&P	Crew Size	Avg Day Prod	Avg Mat'l Unit Cost	Avg Equip Unit Cost	Avg Labor Unit Cost	Avg Total Unit Cost	Avg Price Incl O&P

Windows (continued)

Description	Oper	Unit	Crew Size	Avg Day Prod	Avg Mat'l Unit Cost	Avg Equip Unit Cost	Avg Labor Unit Cost	Avg Total Unit Cost	Avg Price Incl O&P	Crew Size	Avg Day Prod	Avg Mat'l Unit Cost	Avg Equip Unit Cost	Avg Labor Unit Cost	Avg Total Unit Cost	Avg Price Incl O&P
8'-0" x 3'-0" H	Inst	Set	C-9	3.5	117.00	---	85.31	202.31	244.11	C-9	2.5	135.00	---	121.87	256.87	316.59
8'-0" x 4'-0" H	Inst	Set	C-9	3.2	148.20	---	93.31	241.51	287.23	C-9	2.2	171.00	---	133.30	304.30	369.62
8'-0" x 5'-0" H	Inst	Set	C-9	3.0	178.10	---	99.53	277.63	326.40	C-9	2.1	205.50	---	142.19	347.69	417.36
For bronze finish, ADD	Inst	%	---	---	15.0	---	---	---	---	---	---	15.0	---	---	---	---
Dual glazed																
2'-0" x 2'-0" H	Inst	Set	C-1	6.3	84.50	---	26.75	111.25	124.36	C-1	4.4	97.50	---	38.22	135.72	154.45
2'-0" x 2'-6" H	Inst	Set	C-1	6.3	93.60	---	26.75	120.35	133.46	C-1	4.4	108.00	---	38.22	146.22	164.95
2'-0" x 3'-0" H	Inst	Set	C-1	6.3	102.70	---	26.75	129.45	142.56	C-1	4.4	118.50	---	38.22	156.72	175.45
2'-0" x 3'-6" H	Inst	Set	C-1	6.3	117.00	---	26.75	143.75	156.86	C-1	4.4	135.00	---	38.22	173.22	191.95
2'-0" x 4'-0" H	Inst	Set	C-1	6.3	120.90	---	26.75	147.65	160.76	C-1	4.4	139.50	---	38.22	177.72	196.45
2'-6" x 2'-0" H	Inst	Set	C-1	6.3	91.00	---	26.75	117.75	130.86	C-1	4.4	105.00	---	38.22	143.22	161.95
2'-6" x 2'-6" H	Inst	Set	C-1	6.3	101.40	---	26.75	128.15	141.26	C-1	4.4	117.00	---	38.22	155.22	173.95
2'-6" x 3'-0" H	Inst	Set	C-1	6.1	111.80	---	27.63	139.43	152.97	C-1	4.3	129.00	---	39.47	168.47	187.81
2'-6" x 3'-6" H	Inst	Set	C-1	6.1	120.90	---	27.63	148.53	162.07	C-1	4.3	139.50	---	39.47	178.97	198.31
2'-6" x 4'-0" H	Inst	Set	C-1	5.8	131.30	---	29.06	160.36	174.60	C-1	4.1	151.50	---	41.51	193.01	213.36
3'-0" x 2'-0" H	Inst	Set	C-1	6.3	98.80	---	26.75	125.55	138.66	C-1	4.4	114.00	---	38.22	152.22	170.95
3'-0" x 2'-6" H	Inst	Set	C-1	6.3	106.60	---	26.75	133.35	146.46	C-1	4.4	123.00	---	38.22	161.22	179.95
3'-0" x 3'-0" H	Inst	Set	C-1	6.3	119.60	---	26.75	146.35	159.46	C-1	4.4	138.00	---	38.22	176.22	194.95
3'-0" x 3'-6" H	Inst	Set	C-1	5.3	130.00	---	31.80	161.80	177.38	C-1	3.7	150.00	---	45.43	195.43	217.69
3'-0" x 4'-0" H	Inst	Set	C-1	4.3	141.70	---	39.20	180.90	200.10	C-1	3.0	163.50	---	56.00	219.50	246.93
3'-0" x 5'-0" H	Inst	Set	C-1	3.7	172.90	---	45.55	218.45	240.77	C-1	2.6	199.50	---	65.08	264.58	296.46
3'-4" x 2'-0" H	Inst	Set	C-1	6.3	104.00	---	26.75	130.75	143.86	C-1	4.4	120.00	---	38.22	158.22	176.95
3'-4" x 2'-6" H	Inst	Set	C-1	6.3	114.40	---	26.75	141.15	154.26	C-1	4.4	132.00	---	38.22	170.22	188.95
3'-4" x 3'-0" H	Inst	Set	C-1	5.3	126.10	---	31.80	157.90	173.48	C-1	3.7	145.50	---	45.43	190.93	213.19
3'-4" x 3'-6" H	Inst	Set	C-1	5.3	136.50	---	31.80	168.30	183.88	C-1	3.7	157.50	---	45.43	202.93	225.19
3'-4" x 4'-0" H	Inst	Set	C-1	4.3	148.20	---	39.20	187.40	206.60	C-1	3.0	171.00	---	56.00	227.00	254.43
3'-4" x 5'-0" H	Inst	Set	C-1	4.3	180.70	---	39.20	219.90	239.10	C-1	3.0	208.50	---	56.00	264.50	291.93
4'-0" x 2'-0" H	Inst	Set	C-1	4.3	113.10	---	39.20	152.30	171.50	C-1	3.0	130.50	---	56.00	186.50	213.93
4'-0" x 2'-6" H	Inst	Set	C-1	4.3	124.80	---	39.20	164.00	183.20	C-1	3.0	144.00	---	56.00	200.00	227.43
4'-0" x 3'-0" H	Inst	Set	C-1	4.3	137.80	---	39.20	177.00	196.20	C-1	3.0	159.00	---	56.00	215.00	242.43
4'-0" x 3'-6" H	Inst	Set	C-1	4.0	149.50	---	42.14	191.64	212.28	C-1	2.8	172.50	---	60.20	232.70	262.19
4'-0" x 4'-0" H	Inst	Set	C-1	4.0	161.20	---	42.14	203.34	223.98	C-1	2.8	186.00	---	60.20	246.20	275.69
4'-0" x 5'-0" H	Inst	Set	C-1	3.3	196.30	---	51.08	247.38	272.40	C-1	2.3	226.50	---	72.96	299.46	335.22

			Costs Based On Large Volume							Costs Based On Small Volume						
Description	Oper	Unit	Crew Size	Avg Day Prod	Avg Mat'l Unit Cost	Avg Equip Unit Cost	Avg Labor Unit Cost	Avg Total Unit Cost	Avg Price Incl O&P	Crew Size	Avg Day Prod	Avg Mat'l Unit Cost	Avg Equip Unit Cost	Avg Labor Unit Cost	Avg Total Unit Cost	Avg Price Incl O&P

Windows (continued)

Description	Oper	Unit	Crew Size	Avg Day Prod	Avg Mat'l Unit Cost	Avg Equip Unit Cost	Avg Labor Unit Cost	Avg Total Unit Cost	Avg Price Incl O&P	Crew Size	Avg Day Prod	Avg Mat'l Unit Cost	Avg Equip Unit Cost	Avg Labor Unit Cost	Avg Total Unit Cost	Avg Price Incl O&P
5'-0" x 2'-0" H	Inst	Set	C-1	4.0	128.70	---	42.14	170.84	191.48	C-1	2.8	148.50	---	60.20	208.70	238.19
5'-0" x 2'-6" H	Inst	Set	C-1	4.0	141.70	---	42.14	183.84	204.48	C-1	2.8	163.50	---	60.20	223.70	253.19
5'-0" x 3'-0" H	Inst	Set	C-9	3.7	156.00	---	80.70	236.70	276.24	C-9	2.6	180.00	---	115.29	295.29	351.78
5'-0" x 3'-6" H	Inst	Set	C-9	3.7	167.70	---	80.70	248.40	287.94	C-9	2.6	193.50	---	115.29	308.79	365.28
5'-0" x 4'-0" H	Inst	Set	C-9	3.3	184.60	---	90.48	275.08	319.42	C-9	2.3	213.00	---	129.26	342.26	405.60
5'-0" x 5'-0" H	Inst	Set	C-9	3.3	245.70	---	90.48	336.18	380.52	C-9	2.3	283.50	---	129.26	412.76	476.10
6'-0" x 2'-0" H	Inst	Set	C-9	3.9	144.30	---	76.56	220.86	258.38	C-9	2.7	166.50	---	109.37	275.87	329.47
6'-0" x 2'-6" H	Inst	Set	C-9	3.9	158.60	---	76.56	235.16	272.68	C-9	2.7	183.00	---	109.37	292.37	345.97
6'-0" x 3'-0" H	Inst	Set	C-9	3.7	172.90	---	80.70	253.60	293.14	C-9	2.6	199.50	---	115.29	314.79	371.28
6'-0" x 3'-6" H	Inst	Set	C-9	3.3	189.80	---	90.48	280.28	324.62	C-9	2.3	219.00	---	129.26	348.26	411.60
6'-0" x 4'-0" H	Inst	Set	C-9	3.3	204.10	---	90.48	294.58	338.92	C-9	2.3	235.50	---	129.26	364.76	428.10
6'-0" x 5'-0" H	Inst	Set	C-9	3.3	263.90	---	90.48	354.38	398.72	C-9	2.3	304.50	---	129.26	433.76	497.10
For bronze finish, ADD	Inst	%	---	---	24.0	---	---	---	---	---	---	24.0	---	---	---	---
2 sliding lites, 1 fixed lite																
Single glazed																
6'-0" x 2'-0" H	Inst	Set	C-9	3.7	97.50	---	80.70	178.20	217.74	C-9	2.6	112.50	---	115.29	227.79	284.28
6'-0" x 2'-6" H	Inst	Set	C-9	3.7	106.60	---	80.70	187.30	226.84	C-9	2.6	123.00	---	115.29	238.29	294.78
6'-0" x 3'-0" H	Inst	Set	C-9	3.5	118.30	---	85.31	203.61	245.41	C-9	2.5	136.50	---	121.87	258.37	318.09
6'-0" x 4'-0" H	Inst	Set	C-9	3.5	149.50	---	85.31	234.81	276.61	C-9	2.5	172.50	---	121.87	294.37	354.09
6'-0" x 5'-0" H	Inst	Set	C-9	3.5	188.50	---	85.31	273.81	315.61	C-9	2.5	217.50	---	121.87	339.37	399.09
7'-0" x 2'-0" H	Inst	Set	C-9	3.7	105.30	---	80.70	186.00	225.54	C-9	2.6	121.50	---	115.29	236.79	293.28
7'-0" x 2'-6" H	Inst	Set	C-9	3.7	128.70	---	80.70	209.40	248.94	C-9	2.6	148.50	---	115.29	263.79	320.28
7'-0" x 3'-0" H	Inst	Set	C-9	3.5	148.20	---	85.31	233.51	275.31	C-9	2.5	171.00	---	121.87	292.87	352.59
7'-0" x 4'-0" H	Inst	Set	C-9	3.5	161.20	---	85.31	246.51	288.31	C-9	2.5	186.00	---	121.87	307.87	367.59
7'-0" x 5'-0" H	Inst	Set	C-9	2.8	201.50	---	106.64	308.14	360.39	C-9	2.0	232.50	---	152.34	384.84	459.49
8'-0" x 2'-0" H	Inst	Set	C-9	3.5	113.10	---	85.31	198.41	240.21	C-9	2.5	130.50	---	121.87	252.37	312.09
8'-0" x 2'-6" H	Inst	Set	C-9	3.5	126.10	---	85.31	211.41	253.21	C-9	2.5	145.50	---	121.87	267.37	327.09
8'-0" x 3'-0" H	Inst	Set	C-9	3.5	139.10	---	85.31	224.41	266.21	C-9	2.5	160.50	---	121.87	282.37	342.09
8'-0" x 3'-6" H	Inst	Set	C-9	3.5	159.90	---	85.31	245.21	287.01	C-9	2.5	184.50	---	121.87	306.37	366.09
8'-0" x 4'-0" H	Inst	Set	C-9	2.8	172.90	---	106.64	279.54	331.79	C-9	2.0	199.50	---	152.34	351.84	426.49
8'-0" x 5'-0" H	Inst	Set	C-9	2.8	214.50	---	106.64	321.14	373.39	C-9	2.0	247.50	---	152.34	399.84	474.49
9'-0" x 3'-0" H	Inst	Set	C-9	3.5	156.00	---	85.31	241.31	283.11	C-9	2.5	180.00	---	121.87	301.87	361.59
9'-0" x 3'-6" H	Inst	Set	C-9	3.5	178.10	---	85.31	263.41	305.21	C-9	2.5	205.50	---	121.87	327.37	387.09
9'-0" x 4'-0" H	Inst	Set	C-9	2.8	193.70	---	106.64	300.34	352.59	C-9	2.0	223.50	---	152.34	375.84	450.49

Windows (continued)

Description	Oper	Unit	Costs Based On Large Volume							Costs Based On Small Volume					
			Crew Size	Avg Day Prod	Avg Mat'l Unit Cost	Avg Equip Unit Cost	Avg Labor Unit Cost	Avg Total Unit Cost	Avg Price Incl O&P	Crew Size	Avg Day Prod	Avg Mat'l Unit Cost	Avg Equip Unit Cost	Avg Labor Unit Cost	Avg Total Unit Cost	Avg Price Incl O&P
10'-0" x 2'-0" H	Inst	Set	C-9	3.5	189.80	---	85.31	275.11	316.91	C-9	2.5	219.00	---	121.87	340.87	400.59
10'-0" x 3'-0" H	Inst	Set	C-9	3.5	166.40	---	85.31	251.71	293.51	C-9	2.5	192.00	---	121.87	313.87	373.59
10"-0" x 3'-6" H	Inst	Set	C-9	2.8	191.10	---	106.64	297.74	349.99	C-9	2.0	220.50	---	152.34	372.84	447.49
10'-0" x 4'-0" H	Inst	Set	C-9	2.8	200.20	---	106.64	306.84	359.09	C-9	2.0	231.00	---	152.34	383.34	457.99
10'-0" x 5'-0" H	Inst	Set	C-9	2.8	282.10	---	106.64	388.74	440.99	C-9	2.0	325.50	---	152.34	477.84	552.49
For bronze finish, ADD	Inst	%	---	---	13.5	---	---	---	---	---	---	13.5	---	---	---	---
Dual glazed																
6'-0" x 2'-0" H	Inst	Set	C-9	3.5	167.70	---	85.31	253.01	294.81	C-9	2.5	193.50	---	121.87	315.37	375.09
6'-0" x 2'-6" H	Inst	Set	C-9	3.5	185.90	---	85.31	271.21	313.01	C-9	2.5	214.50	---	121.87	336.37	396.09
6'-0" x 3'-0" H	Inst	Set	C-9	3.3	205.40	---	90.48	295.88	340.22	C-9	2.3	237.00	---	129.26	366.26	429.60
6'-0" x 3'-6" H	Inst	Set	C-9	3.3	228.80	---	90.48	319.28	363.62	C-9	2.3	264.00	---	129.26	393.26	456.60
6'-0" x 4'-0" H	Inst	Set	C-9	3.3	256.10	---	90.48	346.58	390.92	C-9	2.3	295.50	---	129.26	424.76	488.10
6'-0" x 5'-0" H	Inst	Set	C-9	3.3	314.60	---	90.48	405.08	449.42	C-9	2.3	363.00	---	129.26	492.26	555.60
7'-0" x 2'-0" H	Inst	Set	C-9	3.5	184.60	---	85.31	269.91	311.71	C-9	2.5	213.00	---	121.87	334.87	394.59
7'-0" x 2'-6" H	Inst	Set	C-9	3.5	204.10	---	85.31	289.41	331.21	C-9	2.5	235.50	---	121.87	357.37	417.09
7'-0" x 3'-0" H	Inst	Set	C-9	3.3	223.60	---	90.48	314.08	358.42	C-9	2.3	258.00	---	129.26	387.26	450.60
7'-0" x 3'-6" H	Inst	Set	C-9	3.3	256.10	---	90.48	346.58	390.92	C-9	2.3	295.50	---	129.26	424.76	488.10
7'-0" x 4'-0" H	Inst	Set	C-9	3.3	278.20	---	90.48	368.68	413.02	C-9	2.3	321.00	---	129.26	450.26	513.60
7'-0" x 5'-0" H	Inst	Set	C-9	2.6	340.60	---	114.84	455.44	511.72	C-9	1.8	393.00	---	164.06	557.06	637.45
8'-0" x 2'-0" H	Inst	Set	C-9	3.3	198.90	---	90.48	289.38	333.72	C-9	2.3	229.50	---	129.26	358.76	422.10
8'-0" x 2'-6" H	Inst	Set	C-9	3.3	221.00	---	90.48	311.48	355.82	C-9	2.3	255.00	---	129.26	384.26	447.60
8'-0" x 3'-0" H	Inst	Set	C-9	3.3	244.40	---	90.48	334.88	379.22	C-9	2.3	282.00	---	129.26	411.26	474.60
8'-0" x 3'-6" H	Inst	Set	C-9	3.3	276.90	---	90.48	367.38	411.72	C-9	2.3	319.50	---	129.26	448.76	512.10
8'-0" x 4'-0" H	Inst	Set	C-9	2.6	300.30	---	114.84	415.14	471.42	C-9	1.8	346.50	---	164.06	510.56	590.95
8'-0" x 5'-0" H	Inst	Set	C-9	2.6	364.00	---	114.84	478.84	535.12	C-9	1.8	420.00	---	164.06	584.06	664.45
For bronze finish, ADD	Inst	%	---	---	24.0	---	---	---	---	---	---	24.0	---	---	---	---

Wood

Awning windows, pine; exterior treated and primed with natural interior; glazed 1/2" insulating glass; unit includes weatherstripping and exterior trim

Unit is one ventilating lite wide

Description	Oper	Unit	Crew Size	Avg Day Prod	Avg Mat'l Unit Cost	Avg Equip Unit Cost	Avg Labor Unit Cost	Avg Total Unit Cost	Avg Price Incl O&P	Crew Size	Avg Day Prod	Avg Mat'l Unit Cost	Avg Equip Unit Cost	Avg Labor Unit Cost	Avg Total Unit Cost	Avg Price Incl O&P
20", 24", 32", 36" x 32" W	Inst	Set	C-1	8.0	167.70	---	21.07	188.77	199.09	C-1	5.6	193.50	---	30.10	223.60	238.35
20", 24", 32", 36" x 40" W	Inst	Set	C-1	8.0	183.30	---	21.07	204.37	214.69	C-1	5.6	211.50	---	30.10	241.60	256.35
20", 24", 32", 36" x 48" W	Inst	Set	C-1	8.0	195.00	---	21.07	216.07	226.39	C-1	5.6	225.00	---	30.10	255.10	269.85

			Costs Based On Large Volume							Costs Based On Small Volume					
Description	Oper	Unit	Crew Size	Avg Day Prod	Avg Mat'l Unit Cost	Avg Equip Unit Cost	Avg Labor Unit Cost	Avg Total Unit Cost	Avg Price Incl O&P	Crew Size	Avg Day Prod	Avg Mat'l Unit Cost	Avg Equip Unit Cost	Avg Labor Unit Cost	Avg Total Unit Cost	Avg Price Incl O&P

Windows (continued)

Unit is two ventilating lites wide

Description	Oper	Unit	Crew Size	Avg Day Prod	Avg Mat'l Unit Cost	Avg Equip Unit Cost	Avg Labor Unit Cost	Avg Total Unit Cost	Avg Price Incl O&P	Crew Size	Avg Day Prod	Avg Mat'l Unit Cost	Avg Equip Unit Cost	Avg Labor Unit Cost	Avg Total Unit Cost	Avg Price Incl O&P
20", 24", 32", 36" x 64" W	Inst	Set	C-9	6.0	314.60	---	49.77	364.37	388.75	C-9	4.2	363.00	---	71.09	434.09	468.93
20", 24", 32", 36" x 80" W	Inst	Set	C-9	6.0	345.80	---	49.77	395.57	419.95	C-9	4.2	399.00	---	71.09	470.09	504.93
20", 24", 32", 36" x 96" W	Inst	Set	C-9	6.0	369.20	---	49.77	418.97	443.35	C-9	4.2	426.00	---	71.09	497.09	531.93

Unit is three ventilating lites wide

Description	Oper	Unit	Crew Size	Avg Day Prod	Avg Mat'l Unit Cost	Avg Equip Unit Cost	Avg Labor Unit Cost	Avg Total Unit Cost	Avg Price Incl O&P	Crew Size	Avg Day Prod	Avg Mat'l Unit Cost	Avg Equip Unit Cost	Avg Labor Unit Cost	Avg Total Unit Cost	Avg Price Incl O&P
20", 24", 32", 36" x 96" W	Inst	Set	C-9	4.5	470.60	---	66.35	536.95	569.47	C-9	3.2	543.00	---	94.79	637.79	684.24
20", 24", 32", 36" x 120" W	Inst	Set	C-9	4.5	520.00	---	66.35	586.35	618.87	C-9	3.2	600.00	---	94.79	694.79	741.24
20", 24", 32", 36" x 144" W	Inst	Set	C-9	4.5	556.40	---	66.35	622.75	655.27	C-9	3.2	642.00	---	94.79	736.79	783.24
For aluminum clad (baked white enamel exterior, ADD	Inst	%	---	---	12.0	---	---	---	---	---	---	12.0	---	---	---	---

Casement windows, rough opening sizes, glazed 1/2" insulating glass; unit includes hardware drip cap, and weatherstripping; screens must be added

1 ventilating lite, no fixed lites; 20", 24", 28" W

Description	Oper	Unit	Crew Size	Avg Day Prod	Avg Mat'l Unit Cost	Avg Equip Unit Cost	Avg Labor Unit Cost	Avg Total Unit Cost	Avg Price Incl O&P	Crew Size	Avg Day Prod	Avg Mat'l Unit Cost	Avg Equip Unit Cost	Avg Labor Unit Cost	Avg Total Unit Cost	Avg Price Incl O&P
32" H	Inst	Set	C-9	10.0	140.40	---	29.86	170.26	184.89	C-9	7.0	162.00	---	42.66	204.66	225.56
40" H	Inst	Set	C-9	9.5	154.70	---	31.43	186.13	201.53	C-9	6.7	178.50	---	44.90	223.40	245.40
48" H	Inst	Set	C-9	9.0	171.60	---	33.18	204.78	221.03	C-9	6.3	198.00	---	47.40	245.40	268.62
56" H	Inst	Set	C-9	8.5	188.50	---	35.13	223.63	240.84	C-9	6.0	217.50	---	50.18	267.68	292.27
64" H	Inst	Set	C-9	8.0	204.10	---	37.32	241.42	259.71	C-9	5.6	235.50	---	53.32	288.82	314.95
72" H	Inst	Set	C-9	7.5	222.30	---	39.81	262.11	281.62	C-9	5.3	256.50	---	56.87	313.37	341.24

2 ventilating lites, no fixed lites; 40", 48", 56" W

Description	Oper	Unit	Crew Size	Avg Day Prod	Avg Mat'l Unit Cost	Avg Equip Unit Cost	Avg Labor Unit Cost	Avg Total Unit Cost	Avg Price Incl O&P	Crew Size	Avg Day Prod	Avg Mat'l Unit Cost	Avg Equip Unit Cost	Avg Labor Unit Cost	Avg Total Unit Cost	Avg Price Incl O&P
32" H	Inst	Set	C-9	8.0	263.90	---	37.32	301.22	319.51	C-9	5.6	304.50	---	53.32	357.82	383.95
40" H	Inst	Set	C-9	7.5	292.50	---	39.81	332.31	351.82	C-9	5.3	337.50	---	56.87	394.37	422.24
48" H	Inst	Set	C-9	7.0	326.30	---	42.66	368.96	389.86	C-9	4.9	376.50	---	60.94	437.44	467.30
56" H	Inst	Set	C-9	6.5	360.10	---	45.94	406.04	428.55	C-9	4.6	415.50	---	65.62	481.12	513.28
64" H	Inst	Set	C-9	6.0	390.00	---	49.77	439.77	464.15	C-9	4.2	450.00	---	71.09	521.09	555.93
72" H	Inst	Set	C-9	5.5	426.40	---	54.29	480.69	507.29	C-9	3.9	492.00	---	77.56	569.56	607.56

2 ventilating lites, 1 fixed lite 60", 72", 84" W

Description	Oper	Unit	Crew Size	Avg Day Prod	Avg Mat'l Unit Cost	Avg Equip Unit Cost	Avg Labor Unit Cost	Avg Total Unit Cost	Avg Price Incl O&P	Crew Size	Avg Day Prod	Avg Mat'l Unit Cost	Avg Equip Unit Cost	Avg Labor Unit Cost	Avg Total Unit Cost	Avg Price Incl O&P
32" H	Inst	Set	C-9	7.0	362.70	---	42.66	405.36	426.26	C-9	4.9	418.50	---	60.94	479.44	509.30
40" H	Inst	Set	C-9	6.5	401.70	---	45.94	447.64	470.15	C-9	4.6	463.50	---	65.62	529.12	561.28
48" H	Inst	Set	C-9	6.0	451.10	---	49.77	500.87	525.25	C-9	4.2	520.50	---	71.09	591.59	626.43
56" H	Inst	Set	C-9	5.5	497.90	---	54.29	552.19	578.79	C-9	3.9	574.50	---	77.56	652.06	690.06
64" H	Inst	Set	C-9	5.0	542.10	---	59.72	601.82	631.08	C-9	3.5	625.50	---	85.31	710.81	752.61
72" H	Inst	Set	C-9	4.5	594.10	---	66.35	660.45	692.97	C-9	3.2	685.50	---	94.79	780.29	826.74

			Costs Based On Large Volume							Costs Based On Small Volume					
Description	Oper	Unit	Crew Size	Avg Day Prod	Avg Mat'l Unit Cost	Avg Equip Unit Cost	Avg Labor Unit Cost	Avg Total Unit Cost	Avg Price Incl O&P	Crew Size	Avg Day Prod	Avg Mat'l Unit Cost	Avg Equip Unit Cost	Avg Labor Unit Cost	Avg Total Unit Cost	Avg Price Incl O&P

Windows (continued)

2 ventilating lites, 2 fixed lites, 80", 96", 112" W

32" H	Inst	Set	C-9	6.0	464.10	---	49.77	513.87	538.25	C-9	4.2	535.50	---	71.09	606.59	641.43
40" H	Inst	Set	C-9	5.5	514.80	---	54.29	569.09	595.69	C-9	3.9	594.00	---	77.56	671.56	709.56
48" H	Inst	Set	C-9	5.0	578.50	---	59.72	638.22	667.48	C-9	3.5	667.50	---	85.31	752.81	794.61
56" H	Inst	Set	C-9	4.5	638.30	---	66.35	704.65	737.17	C-9	3.2	736.50	---	94.79	831.29	877.74
64" H	Inst	Set	C-9	4.0	696.80	---	74.65	771.45	808.03	C-9	2.8	804.00	---	106.64	910.64	962.89
72" H	Inst	Set	C-9	3.5	764.40	---	85.31	849.71	891.51	C-9	2.5	882.00	---	121.87	1003.87	1063.59

3 ventilating lites, 2 fixed lites; 100", 120", 140" W

32" H	Inst	Set	C-9	5.0	588.90	---	59.72	648.62	677.88	C-9	3.5	679.50	---	85.31	764.81	806.61
40" H	Inst	Set	C-9	4.5	653.90	---	66.35	720.25	752.77	C-9	3.2	754.50	---	94.79	849.29	895.74
48" H	Inst	Set	C-9	4.0	734.50	---	74.65	809.15	845.73	C-9	2.8	847.50	---	106.64	954.14	1006.39
56" H	Inst	Set	C-9	3.5	812.50	---	85.31	897.81	939.61	C-9	2.5	937.50	---	121.87	1059.37	1119.09
64" H	Inst	Set	C-9	3.0	884.00	---	99.53	983.53	1032.30	C-9	2.1	1020.00	---	142.19	1162.19	1231.86
72" H	Inst	Set	C-9	2.5	969.80	---	119.44	1089.24	1147.76	C-9	1.8	1119.00	---	170.62	1289.62	1373.23

For aluminum clad (baked white enamel) exterior, ADD	Inst	%	---	---	20.0	---	---	---	---	---	---	20.0	---	---	---	---

Double hung windows, glazed 7/16" insulating glass, rough opening sizes, two lites per unit; includes screens and weatherstripping

1'-10" x 4'-8"	Inst	Set	C-1	4.0	122.20	---	42.14	164.34	184.98	C-1	2.8	141.00	---	60.20	201.20	230.69
2'-2" x 3'-4", 4'-0", 4'-8"	Inst	Set	C-1	4.0	120.90	---	42.14	163.04	183.68	C-1	2.8	139.50	---	60.20	199.70	229.19
2'-6" x 3'-4", 4'-0", 4'-8"	Inst	Set	C-1	4.0	109.20	---	42.14	151.34	171.98	C-1	2.8	126.00	---	60.20	186.20	215.69
2'-10" x 3'-4", 4'-0", 4'-8"	Inst	Set	C-1	4.0	111.80	---	42.14	153.94	174.58	C-1	2.8	129.00	---	60.20	189.20	218.69
2'-10" x 5'-4", 5'-8"	Inst	Set	C-1	4.0	132.60	---	42.14	174.74	195.38	C-1	2.8	153.00	---	60.20	213.20	242.69
3'-2" x 3'-4", 4'-0", 4'-8"	Inst	Set	C-1	4.0	122.20	---	42.14	164.34	184.98	C-1	2.8	141.00	---	60.20	201.20	230.69
3'-2" x 5'-4", 5'-8"	Inst	Set	C-1	4.0	143.00	---	42.14	185.14	205.78	C-1	2.8	165.00	---	60.20	225.20	254.69
3'-6" x 3'-4", 4'-0", 4'-8"	Inst	Set	C-1	4.0	140.40	---	42.14	182.54	203.18	C-1	2.8	162.00	---	60.20	222.20	251.69
3'-6" x 5'-4", 5'-8"	Inst	Set	C-1	4.0	143.00	---	42.14	185.14	205.78	C-1	2.8	165.00	---	60.20	225.20	254.69

For aluminum clad (baked white enamel) exterior, ADD	Inst	%	---	---	50.0	---	---	---	---	---	---	50.0	---	---	---	---

			Costs Based On Large Volume						Costs Based On Small Volume						
Description	Oper	Unit	Crew Size	Avg Day Prod	Avg Mat'l Unit Cost	Avg Equip Unit Cost	Avg Labor Unit Cost	Avg Total Unit Cost	Avg Price Incl O&P	Crew Size	Avg Day Prod	Avg Mat'l Unit Cost	Avg Equip Unit Cost	Avg Labor Unit Cost	Avg Total Unit Cost	Avg Price Incl O&P

Windows (continued)

Picture Windows

Unit with one fixed lite, 3/4" insulating glass, frame size

Description	Oper	Unit	Crew Size	Avg Day Prod	Avg Mat'l Unit Cost	Avg Equip Unit Cost	Avg Labor Unit Cost	Avg Total Unit Cost	Avg Price Incl O&P	Crew Size	Avg Day Prod	Avg Mat'l Unit Cost	Avg Equip Unit Cost	Avg Labor Unit Cost	Avg Total Unit Cost	Avg Price Incl O&P
4'-0" W x 3'-6" H; 3'-6" W x 4' 0" H	Inst	EA	C-9	11.0	230.10	---	27.14	257.24	270.55	C-9	7.7	265.50	---	38.78	304.28	323.28
4'-8" W x 3'-6" H; 4'-0" W x 4'-0" H; 3'-6" W x 4'-8"H	Inst	EA	C-9	11.0	250.90	---	27.14	278.04	291.35	C-9	7.7	289.50	---	38.78	328.28	347.28
4'-8" W x 4'-0" H; 4'-0" W x 4'-10" H	Inst	EA	C-9	11.0	269.10	---	27.14	296.24	309.55	C-9	7.7	310.50	---	38.78	349.28	368.28
4'-8" W x 4'-8" H	Inst	EA	C-9	10.0	288.60	---	29.86	318.46	333.09	C-9	7.0	333.00	---	42.66	375.66	396.56
4'-8" W x 5'-6" H; 5'-4" W x 4'-8" H	Inst	EA	C-9	10.0	327.60	---	29.86	357.46	372.09	C-9	7.0	378.00	---	42.66	420.66	441.56
5'-4" W x 5'-6" H	Inst	EA	C-9	10.0	360.10	---	29.86	389.96	404.59	C-9	7.0	415.50	---	42.66	458.16	479.06

Unit with one 3/4" fixed insulated lite and two glazed S.S.B. casement flankers; rough opening, sizes; unit includes hardware, casing and molding; 15", 19", 23" flankers

Description	Oper	Unit	Crew Size	Avg Day Prod	Avg Mat'l Unit Cost	Avg Equip Unit Cost	Avg Labor Unit Cost	Avg Total Unit Cost	Avg Price Incl O&P	Crew Size	Avg Day Prod	Avg Mat'l Unit Cost	Avg Equip Unit Cost	Avg Labor Unit Cost	Avg Total Unit Cost	Avg Price Incl O&P
40" H																
88" W	Inst	Set	C-9	8.0	533.00	---	37.32	570.32	588.61	C-9	5.6	615.00	---	53.32	668.32	694.45
96" W	Inst	Set	C-9	8.0	551.20	---	37.32	588.52	606.81	C-9	5.6	636.00	---	53.32	689.32	715.45
104" W	Inst	Set	C-9	8.0	562.90	---	37.32	600.22	618.51	C-9	5.6	649.50	---	53.32	702.82	728.95
112" W	Inst	Set	C-9	8.0	579.80	---	37.32	617.12	635.41	C-9	5.6	669.00	---	53.32	722.32	748.45
48" H																
80" W	Inst	Set	C-9	8.0	566.80	---	37.32	604.12	622.41	C-9	5.6	654.00	---	53.32	707.32	733.45
88" W	Inst	Set	C-9	8.0	586.30	---	37.32	623.62	641.91	C-9	5.6	676.50	---	53.32	729.82	755.95
96" W	Inst	Set	C-9	7.0	603.20	---	42.66	645.86	666.76	C-9	4.9	696.00	---	60.94	756.94	786.80
104" W	Inst	Set	C-9	7.0	616.20	---	42.66	658.86	679.76	C-9	4.9	711.00	---	60.94	771.94	801.80
112" W	Inst	Set	C-9	7.0	618.80	---	42.66	661.46	682.36	C-9	4.9	714.00	---	60.94	774.94	804.80
56" H																
80" W	Inst	Set	C-9	8.0	612.30	---	37.32	649.62	667.91	C-9	5.6	706.50	---	53.32	759.82	785.95
88" W	Inst	Set	C-9	7.0	630.50	---	42.66	673.16	694.06	C-9	4.9	727.50	---	60.94	788.44	818.30
96" W	Inst	Set	C-9	7.0	651.30	---	42.66	693.96	714.86	C-9	4.9	751.50	---	60.94	812.44	842.30
104" W	Inst	Set	C-9	6.0	690.30	---	49.77	740.07	764.45	C-9	4.2	796.50	---	71.09	867.59	902.43
112" W	Inst	Set	C-9	6.0	707.20	---	49.77	756.97	781.35	C-9	4.2	816.00	---	71.09	887.09	921.93
120" W	Inst	Set	C-9	6.0	712.40	---	49.77	762.17	786.55	C-9	4.2	822.00	---	71.09	893.09	927.93

Description	Oper	Unit	Costs Based On Large Volume							Costs Based On Small Volume						
			Crew Size	Avg Day Prod	Avg Mat'l Unit Cost	Avg Equip Unit Cost	Avg Labor Unit Cost	Avg Total Unit Cost	Avg Price Incl O&P	Crew Size	Avg Day Prod	Avg Mat'l Unit Cost	Avg Equip Unit Cost	Avg Labor Unit Cost	Avg Total Unit Cost	Avg Price Incl O&P

Windows (continued)

64" H

Description	Oper	Unit	Crew Size	Avg Day Prod	Avg Mat'l Unit Cost	Avg Equip Unit Cost	Avg Labor Unit Cost	Avg Total Unit Cost	Avg Price Incl O&P	Crew Size	Avg Day Prod	Avg Mat'l Unit Cost	Avg Equip Unit Cost	Avg Labor Unit Cost	Avg Total Unit Cost	Avg Price Incl O&P
96" W	Inst	Set	C-9	6.0	720.20	---	49.77	769.97	794.35	C-9	4.2	831.00	---	71.09	902.09	936.93
104" W	Inst	Set	C-9	6.0	752.70	---	49.77	802.47	826.85	C-9	4.2	868.50	---	71.09	939.59	974.43
112" W	Inst	Set	C-9	6.0	768.30	---	49.77	818.07	842.45	C-9	4.2	886.50	---	71.09	957.59	992.43
120" W	Inst	Set	C-9	6.0	783.90	---	49.77	833.67	858.05	C-9	4.2	904.50	---	71.09	975.59	1010.43
Material adjustments																
For glazed 1/2" insulated																
flankers, ADD	Inst	%	---	---	15.0	---	---	---	---	---	---	15.0	---	---	---	---
For screen per flanker, ADD	Inst	EA	C-9	---	6.50	---	---	---	6.50	C-9	---	7.50	---	---	---	7.50

Garden Windows, unit extends out 13-1/4" from wall, bronze tempered glass in top panels of all units, wood shelves included

1 fixed lite

Single glazed

Description	Oper	Unit	Crew Size	Avg Day Prod	Avg Mat'l Unit Cost	Avg Equip Unit Cost	Avg Labor Unit Cost	Avg Total Unit Cost	Avg Price Incl O&P	Crew Size	Avg Day Prod	Avg Mat'l Unit Cost	Avg Equip Unit Cost	Avg Labor Unit Cost	Avg Total Unit Cost	Avg Price Incl O&P
3'-0" x 3'-0" H	Inst	EA	C-9	4.5	401.70	---	66.35	468.05	500.57	C-9	3.2	463.50	---	94.79	558.29	604.74
3'-0" x 4'-0" H	Inst	EA	C-9	4.5	438.10	---	66.35	504.45	536.97	C-9	3.2	505.50	---	94.79	600.29	646.74
4'-0" x 3'-0" H	Inst	EA	C-9	4.5	435.50	---	66.35	501.85	534.37	C-9	3.2	502.50	---	94.79	597.29	643.74
4'-0" x 4'-0" H	Inst	EA	C-9	4.0	468.00	---	74.65	542.65	579.23	C-9	2.8	540.00	---	106.64	646.64	698.89
4'-0" x 5'-0" H	Inst	EA	C-9	3.5	501.80	---	85.31	587.11	628.91	C-9	2.5	579.00	---	121.87	700.87	760.59
5'-0" x 3'-0" H	Inst	EA	C-9	4.0	465.40	---	74.65	540.05	576.63	C-9	2.8	537.00	---	106.64	643.64	695.89
5'-0" x 3'-8" H	Inst	EA	C-9	3.5	492.70	---	85.31	578.01	619.81	C-9	2.5	568.50	---	121.87	690.37	750.09
5'-0" x 4'-0" H	Inst	EA	C-9	3.5	501.80	---	85.31	587.11	628.91	C-9	2.5	579.00	---	121.87	700.87	760.59
5'-0" x 5'-0" H	Inst	EA	C-9	3.0	546.00	---	99.53	645.53	694.30	C-9	2.1	630.00	---	142.19	772.19	841.86
6'-0" x 3'-0" H	Inst	EA	C-9	3.5	496.60	---	85.31	581.91	623.71	C-9	2.5	573.00	---	121.87	694.87	754.59
6'-0" x 3'-8" H	Inst	EA	C-9	3.2	520.00	---	93.31	613.31	659.03	C-9	2.2	600.00	---	133.30	733.30	798.62
6'-0" x 4'-0" H	Inst	EA	C-9	3.0	540.80	---	99.53	640.33	689.10	C-9	2.1	624.00	---	142.19	766.19	835.86
Dual glazed																
3'-0" x 3'-0" H	Inst	EA	C-9	4.5	551.20	---	66.35	617.55	650.07	C-9	3.2	636.00	---	94.79	730.79	777.24
3'-0" x 4'-0" H	Inst	EA	C-9	4.5	607.10	---	66.35	673.45	705.97	C-9	3.2	700.50	---	94.79	795.29	841.74
4'-0" x 3'-0" H	Inst	EA	C-9	4.5	601.90	---	66.35	668.25	700.77	C-9	3.2	694.50	---	94.79	789.29	835.74
4'-0" x 4'-0" H	Inst	EA	C-9	4.0	652.60	---	74.65	727.25	763.83	C-9	2.8	753.00	---	106.64	859.64	911.89
4'-0" x 5'-0" H	Inst	EA	C-9	3.5	716.30	---	85.31	801.61	843.41	C-9	2.5	826.50	---	121.87	948.37	1008.09

Description	Oper	Unit	Costs Based On Large Volume							Costs Based On Small Volume						
			Crew Size	Avg Day Prod	Avg Mat'l Unit Cost	Avg Equip Unit Cost	Avg Labor Unit Cost	Avg Total Unit Cost	Avg Price Incl O&P	Crew Size	Avg Day Prod	Avg Mat'l Unit Cost	Avg Equip Unit Cost	Avg Labor Unit Cost	Avg Total Unit Cost	Avg Price Incl O&P

Windows (continued)

Description	Oper	Unit	Crew Size	Avg Day Prod	Avg Mat'l Unit Cost	Avg Equip Unit Cost	Avg Labor Unit Cost	Avg Total Unit Cost	Avg Price Incl O&P	Crew Size	Avg Day Prod	Avg Mat'l Unit Cost	Avg Equip Unit Cost	Avg Labor Unit Cost	Avg Total Unit Cost	Avg Price Incl O&P
5'-0" x 3'-0" H	Inst	EA	C-9	4.0	646.10	---	74.65	720.75	757.33	C-9	2.8	745.50	---	106.64	852.14	904.39
5'-0" x 3'-8" H	Inst	EA	C-9	3.5	685.10	---	85.31	770.41	812.21	C-9	2.5	790.50	---	121.87	912.37	972.09
5'-0" x 4'-0" H	Inst	EA	C-9	3.5	703.30	---	85.31	788.61	830.41	C-9	2.5	811.50	---	121.87	933.37	993.09
5'-0" x 5'-0" H	Inst	EA	C-9	3.0	783.90	---	99.53	883.43	932.20	C-9	2.1	904.50	---	142.19	1046.69	1116.36
6'-0" x 3'-0" H	Inst	EA	C-9	3.5	703.30	---	85.31	788.61	830.41	C-9	2.5	811.50	---	121.87	933.37	993.09
6'-0" x 3'-8" H	Inst	EA	C-9	3.2	743.60	---	93.31	836.91	882.63	C-9	2.2	858.00	---	133.30	991.30	1056.62
6'-0" x 4'-0" H	Inst	EA	C-9	3.0	781.30	---	99.53	880.83	929.60	C-9	2.1	901.50	---	142.19	1043.69	1113.36
1 fixed lite, 1 sliding lite																
Single glazed																
3'-0" x 3'-0" H	Inst	EA	C-9	4.5	513.50	---	66.35	579.85	612.37	C-9	3.2	592.50	---	94.79	687.29	733.74
3'-0" x 4'-0" H	Inst	EA	C-9	4.5	564.20	---	66.35	630.55	663.07	C-9	3.2	651.00	---	94.79	745.79	792.24
4'-0" x 3'-0" H	Inst	EA	C-9	4.5	551.20	---	66.35	617.55	650.07	C-9	3.2	636.00	---	94.79	730.79	777.24
4'-0" x 4'-0" H	Inst	EA	C-9	4.0	598.00	---	74.65	672.65	709.23	C-9	2.8	690.00	---	106.64	796.64	848.89
4'-0" x 5'-0" H	Inst	EA	C-9	3.5	656.50	---	85.31	741.81	783.61	C-9	2.5	757.50	---	121.87	879.37	939.09
5'-0" x 3'-0" H	Inst	EA	C-9	4.0	590.20	---	74.65	664.85	701.43	C-9	2.8	681.00	---	106.64	787.64	839.89
5'-0" x 3'-8" H	Inst	EA	C-9	3.5	631.80	---	85.31	717.11	758.91	C-9	2.5	729.00	---	121.87	850.87	910.59
5'-0" x 4'-0" H	Inst	EA	C-9	3.5	638.30	---	85.31	723.61	765.41	C-9	2.5	736.50	---	121.87	858.37	918.09
5'-0" x 5'-0" H	Inst	EA	C-9	3.0	717.60	---	99.53	817.13	865.90	C-9	2.1	828.00	---	142.19	970.19	1039.86
6'-0" x 3'-0" H	Inst	EA	C-9	3.5	618.80	---	85.31	704.11	745.91	C-9	2.5	714.00	---	121.87	835.87	895.59
6'-0" x 3'-8" H	Inst	EA	C-9	3.2	656.50	---	93.31	749.81	795.53	C-9	2.2	757.50	---	133.30	890.80	956.12
6'-0" x 4'-0" H	Inst	EA	C-9	3.0	673.40	---	99.53	772.93	821.70	C-9	2.1	777.00	---	142.19	919.19	988.86

Wiring. See Electrical.

Index

Practical References for Builders

Audio: Estimating Remodeling

Listen to the "hands-on" estimating instruction in this popular remodeling seminar. Make your own unit price estimate based on the prints enclosed. Then check your completed estimate with those prepared in the actual seminar. After listening to these tapes you will know how to establish an operating budget for your business, determine indirect costs and profit, and estimate remodeling with the unit cost method. **Includes seminar workbook, project survey and unit price estimating form, and six 20-minute cassettes, $65.00**

Manual of Professional Remodeling

This is the practical manual of professional remodeling written by an experienced and successful remodeling contractor. Shows how to evaluate a job to avoid 30-minute jobs that take all day, what to fix and what to leave alone, and what to watch for in dealing with subcontractors. Includes chapters on calculating space requirements, repairing structural defects, remodeling kitchens, baths, walls and ceilings, doors and windows, floors, roofs, installing fireplaces and chimneys (including built-ins), skylights, and exterior siding. Includes blank forms, checklists, sample contracts, and proposals you can copy and use. **400 pages, 8½ x 11, $19.75**

Remodeler's Handbook

The complete manual of home improvement contracting: Planning the job, estimating costs, doing the work, running your company and making profits. Pages of sample forms, contracts, documents, clear illustrations and examples. Chapters on evaluating the work, rehabilitation, kitchens, bathrooms, adding living area, re-flooring, re-siding, re-roofing, replacing windows and doors, installing new wall and ceiling cover, repainting, upgrading insulation, combating moisture damage, estimating, selling your services, and bookkeeping for remodelers. **416 pages, 8½ x 11, $23.00**

Remodeling Kitchens & Baths

This book is your guide to succeeding in a very lucrative area of the remodeling market: how to repair and replace damaged floors; how to redo walls, ceilings, and plumbing; and how to modernize the home wiring system to accommodate today's heavy electrical demands. Show how to install new sinks and countertops, ceramic tile, sunken tubs, whirlpool baths, luminous ceilings, skylights, and even special lighting effects. Completely illustrated, with manhour tables for figuring your labor costs. **384 pages, 8½ x 11, $26.25**

Running Your Remodeling Business

Everything you need to know about operating a remodeling business, from making your first sale to insuring your profits: how to advertise, write up a contract, estimate, schedule your jobs, arrange financing (for both you and your customers), and when and how to expand your business. Explains what you need to know about insurance, bonds, and liens, and how to collect the moeny you've earned. Includes sample business forms for your use. **272 pages, 8½ x 11, $21.00**

Profits in Buying & Renovating Homes

Step-by-step instructions for selecting, repairing, improving, and selling highly profitable "fixer-uppers." Shows which price ranges offer the highest profit-to-investment ratios, which neighborhoods offer the best return, practical directions for repairs, and tips on dealing with buyers, sellers, and real estate agents. Shows you how to determine your profit before you buy, what bargains to avoid, and simple, inexpensive upgrades that will charm your buyers and ensure your profits. **304 pages, 8½ x 11, $19.75**

Electrical Construction Estimator

If you estimate electrical jobs, this is your guide to current material costs, reliable manhour estimates per unit, and the total installed cost for all common electrical work: conduit, wire, boxes, fixtures, switches, outlets, loadcenters, panelboards, raceway, duct, signal systems, and more. Explains what every estimator should know before estimating each part of an electrical system. **416 pages, 8½ x 11, $25.00. Revised annually**

Estimating Electrical Construction

A practical approach to estimating materials and labor for residential and commercial electrical construction. Written by the A.S.P.E. National Estimator of the Year, it explains how to use labor units, the plan take-off and the bid summary to establish an accurate estimate. Covers dealing with suppliers, pricing sheets, and how to modify labor units. Provides extensive labor unit tables, and blank forms for use in estimating your next electrical job. **272 pages, 8½ x 11, $19.00**

Audio Estimating Electrical Work

Listen to Trade Service's two-day seminar and study electrical estimating at your own speed and at a fraction of the cost of attending the actual seminar. You'll learn what to expect from specifications, how to adjust labor units from a price book to your job, how to make an accurate take-off from the plans, and how to spot hidden costs that other estimators may miss. **Includes six 30-minute tapes, a workbook including price sheets, specification sheet, bid summary, estimate recap sheet, blueprints used in the actual seminar, and blank forms for your own use. $65**

Residential Electrician's Handbook

Explains what every builder needs to know about designing electrical systems for residential construction. Shows how to draw up an electrical plan from the blueprints, including the service entrance, grounding, lighting requirements for kitchen, bedroom and bath and how to lay them out. Explains how to plan electrical heating systems and what equipment you'll need, how to plan outdoor lighting, and much more. If you are a builder who ever has to plan an electrical system, you should have this book. **194 pages, 8½ x 11, $16.75**

Remodeling Contractor's Handbook

Everything you need to know to make a remodeling business grow: Identifying a market for your business, inexpensive sales and advertising techniques that work, and how to prepare accurate estimates. Also explains building a positive company image, training effective sales people, placing loans for customers, and bringing in profitable work to keep your company growing. **304 pages, 8½ x 11, $18.25**

National Construction Estimator

Current building costs in dollars and cents for residential, commercial, and industrial construction. Estimated prices for every commonly used building material. The manhours, recommended crew and labor cost for installation. Includes Estimate Writer, an electronic version of the book on computer disk - at no extra cost on 5½" high density (1.2Mb) disk. The 1991 National Construction Estimator and Estimate Writer on 1.2Mb disk cost **$22.50** (Add $10 if you want Estimate Writer on 5½" double density 360K disks or 3½" 720K disks.) **576 pages, 8½ x 11, $22.50. Revised annually**

Bookkeeping for Builders

This book will show you simple, practical instructions for setting up and keeping accurate records — with a minimum of effort and frustration. Shows how to set up the essentials of a record-keeping system: the payment journal, income journal, general journal, records for fixed assets, accounts receivable, payables and purchases, petty cash, and job costs. You'll be able to keep the records required by the I.R.S., as well as accurate and organized business records for your own use. **208 pages, 8½ x 11, $19.75**

Spec Builder's Guide

Explains how to plan and build a home, control your construction costs, and then sell the house at a price that earns a decent return on the time and money you've invested. Includes professional tips on the time and money you've invested. Includes professional tips to ensure success as a spec builder: how government statistics help you judge the housing market, cutting costs at every opportunity without sacrificing quality, and taking advantage of construction cycles. Every chapter includes checklists, diagrams, charts, figures, and estimating tables. **448 pages, 8½ x 11, $27.00**

Contractor's Survival Manual

How to survive hard times in construction and take full advantage of the profitable cycles. Shows what to do when the bills can't be paid, finding money and buying time, transferring debt, and all the alternatives to bankruptcy. Explains how to build profits, avoid problems in zoning and permits, taxes, time-keeping, and payroll. Unconventional advice includes how to invest in inflation, get high appraisals, trade and postpone income, and how to stay hip-deep in profitable work. **160 pages, 8½ x 11, $16.75**

Video: Stair Framing

Shows how to use a calculator to figure the rise and run of each step, the height of each riser, the number of treads, and the tread depths. Then watch how to take these measurements to construct an actual set of stairs. You'll see how to mark and cut your carriages, treads, and risers, and install a stairway that fits your calculations for the perfect set of stairs. **60 minutes, VHS, $24.75**

Roof Framing

Frame any type of roof in common use today, even if you've never framed a roof before. Shows how to use a pocket calculator to figure any common, hip, valley, and jack rafter length in seconds. Over 400 illustrations take you through every measurement and every cut on each type of roof: gable, hip, Dutch, Tudor, gambrel, shed, gazebo and more. **480 pages, 5½ x 8½, $22.00**

Contractor's Guide to the Building Code Revised

This completely revised edition explains in plain English exactly what the Uniform Code requires and shows how to design and construct residential and light commercial buildings that will pass inspection the first time. Suggests how to work with the inspector to minimize construction costs, what common building shortcuts are likely to be cited, and where exceptions are granted. **544 pages, 5½ x 8½, $24.25**

Drywall Contracting

How to do professional quality drywall work, how to plan and estimate each job, and how to start and keep your drywall business thriving. Covers the eight essential steps in making any drywall estimate, how to achieve the six most commonly-used surface treatments, how to work with metal studs, and how to solve and prevent most common drywall problems. **288 pages, 5½ x 8½, $18.25**

Construction Estimating Reference Data

Collected in this single volume are the building estimator's 300 most useful estimating reference tables. Labor requirements for nearly every type of construction are included: site work, concrete work, masonry, steel, carpentry, thermal & moisture protection, doors and windows, finishes, mechanical and electrical. Each section explains in detail the work being estimated and gives the appropriate crew size and equipment needed. **368 pages, 8½ x 11, $26.00**

Building Cost Manual

Square foot costs for residential, commercial, industrial, and farm buildings. In a few minutes you work up a reliable budget estimate based on the actual materials and design features, area, shape, wall height, number of floors and support requirements. Most important, you include all the important variables that can make any building unique from a cost standpoint. **240 pages, 8½ x 11, $14.00. Revised annually**

Estimating Tables for Home Building

Produce accurate estimates in minutes for nearly any home or multi-family dwelling. This handy manual has the tables you need to find the quantity of materials and labor for most residential construction. Includes overhead and profit, how to develop unit costs for labor and materials, and how to be sure you've considered every cost in the job. **336 pages, 8½ x 11, $21.50**

Estimating Home Building Costs

Estimate every phase of residential construction from site costs to the profit margin you should include in your bid. Shows how to keep track of manhours and make accurate labor cost estimates for footings, foundations, framing and sheathing finishes, electrical, plumbing and more. Explains the work being estimated and provides sample cost estimate worksheets with complete instructions for each job phase. **320 pages, 5½ x 8½, $17.00**

Cost Records for Construction Estimating

How to organize and use cost information from jobs just completed to make more accurate estimates in the future. Explains how to keep the cost records you need to reflect the time spent on each part of the job. Shows the best way to track costs for sitework, footing, foundations, framing, interior finish, siding and trim, masonry, and subcontract expense. Provides sample forms. **208 pages, 8½ x 11, $15.75**

Carpentry Estimating

Simple, clear instructions show you how to take off quantities and figure costs for all rough and finish carpentry. Shows how much overhead and profit to include, how to convert piece prices to MBF prices or linear foot prices, and how to use the tables included to quickly estimate manhours. All carpentry is covered; floor joists, exterior and interior walls and finishes, ceiling joists and rafters, stairs, trim, windows, doors, and much more. Includes sample forms, checklists, and the author's factor worksheets to save you time and help prevent errors. **320 pages, 8½ x 11, $25.50**

Wood-Frame House Construction

From the layout of the outer walls, excavation and formwork, to finish carpentry and painting, every step of construction is covered in detail, with clear illustrations and explanations. Everything the builder needs to know about framing, roofing, siding, insulation and vapor barrier, interior finishing, floor coverings, and stairs — complete step-by-step "how to" information on what goes into building a frame house. **240 pages, 8½ x 11, $14.25. Revised edition**

Rough Carpentry

All rough carpentry is covered in detail: sills, girders, columns, joists, sheathing, ceiling, roof and wall framing, roof trusses, dormers, bay windows, furring and grounds, stairs and insulation. Many of the 24 chapters explain practical code approved methods for saving lumber and time without sacrificing quality. Chapters on columns, headers, rafters, joists and girders show how to use simple engineering principles to selected the right lumber dimension for whatever species and grade you are using. **288 pages, 8½ x 11, $17.00**

Carpentry Layout

Explains the easy way to figure: cuts for stair carriages, treads and risers; lengths for common, hip and jack rafters; spacing for joists, studs, rafters and pickets; layout for rake and bearing walls. Shows how to set foundation corner stakes, even for a complex home on a hillside. Practical examples show how to use a hand-held calculator as a powerful layout tool. Written in simple language any carpenter can understand. **240 pages, 5½ x 8½, $16.25**

Construction Surveying & Layout

A practical guide to simplified construction surveying: How land is divided, how to use a transit and tape to find a known point, how to draw an accurate survey map from your field notes, how to use topographic surveys, and the right way to level and set grade. You'll learn how to make a survey for any residential or commercial lot, driveway, road, or bridge — including how to figure cuts and fills and calculate excavation quantities. If you've been wanting to make your own surveys, or just read and verify the accuracy of surveys made by others, you should have this guide. **256 pages, 5½ x 8½, $19.25**

Paint Contractor's Manual

How to start and run a profitable paint contracting company: getting set up and organized to handle volume work, avoiding the mistakes most painters make, getting top production from your crews and the most value from your advertising dollar. Shows how to estimate all prep and painting. Loaded with manhour estimates, sample forms, contracts, charts, tables and examples you can use. **224 pages, 8½ x 11, $19.25**

Painter's Handbook

Loaded with "how-to" information you'll use every day to get professional results on any job: the best way to prepare a surface for painting or repainting; selecting and using the right materials and tools (including airless spray); tips for repainting kitchens, bathrooms, cabinets, eaves and porches; how to match and blend colors; why coatings fail and what to do about it. Thirty profitable specialties that could be your gravy train in the painting business. Every professional painter needs this practical handbook. **320 pages, 8½ x 11, $21.25**

Painting Cost Guide

A complete guide for estimating painting costs for just about any type of residential, commercial, or industrial painting, whether by brush, spray, or roller. Provides typical costs and bid prices for fast, medium and slow work, including: material costs per gallon, square feet covered per gallon, square feet covered per manhour, labor cost per 100 square feet, material cost per 100 square feet, overhead and taxes per 100 square feet, and how much to add for profit. **448 pages, 8½ x 11, $27.50. Revised annually**

Estimating Painting Costs

Here is an accurate step-by-step estimating system, based on a set of easy-to-use manhour tables that anyone can use for estimating painting costs: from simple residential repaints to complicated commercial jobs — even heavy industrial and government work. Explains taking field measurements, doing take-offs from plans and specs, predicting productivity, figuring labor, material costs, overhead and profit. Includes manhour and material tables, plus samples, forms, and checklists for your use. **448 pages, 8½ x 11, $28.00**

Home Wiring: Improvement, Extension, Repairs

How to repair electrical wiring in older homes, extend or expand an existing electrical system in homes being remodeled, and bring the electrical system up to modern standards in any residence. Shows how to use the anticipated loads and demand factors to figure the amperage and number of new circuits needed, and how to size and install wiring, conduit, switches, and auxiliary panels and fixtures. Explains how to test and troubleshoot fixtures, circuit wiring, and switches, as well as how to service or replace low voltage systems. **224 pages, 5½ x 8½, $15.00**

Order Form 2 (identical to Order Form 1)

Book list (column 1):

- ☐ 22.50 National Const. Estimator w/ free Estimate Writer on 5¼" (1.2Mb) disk — *Add $10 for Estimate Writer on either* ☐ 5¼" (360K) disk or ☐ 3½" (720K) disk.
- ☐ 26.50 Nat. Repair & Remodel Est.
- ☐ 19.25 Paint Contractor's Manual
- ☐ 21.25 Painter's Handbook
- ☐ 27.50 Painting Cost Guide
- ☐ 13.00 Plan & Design Plumb System
- ☐ 19.25 Planning Drain, Waste & Vent
- ☐ 21.00 Plumber's Exam Prep. Guide
- ☐ 18.00 Plumber's Handbook Rev.
- ☐ 19.75 Profits Buy & Renov Homes
- ☐ 14.25 Rafter Length Manual
- ☐ 23.00 Remodeler's Handbook
- ☐ 18.25 Remodeling Contr. Handbook
- ☐ 26.25 Remodeling Kitchens & Baths
- ☐ 11.50 Residential Electrical Design
- ☐ 16.75 Residential Electr. Hndbk.
- ☐ 18.25 Residential Wiring
- ☐ 22.00 Roof Framing
- ☐ 14.00 Roofers Handbook
- ☐ 17.00 Rough Carpentry
- ☐ 21.00 Run. Your Remodeling Bus.
- ☐ 27.00 Spec Builder's Guide
- ☐ 15.50 Stair Builder's Handbook
- ☐ 14.25 Wood-Frame House Const.

Book list (column 2):

- ☐ 16.75 Contractor's Survival Manual
- ☐ 16.50 Cont. Year-Rd Tax Guide
- ☐ 15.75 Cost Records for Const. Est.
- ☐ 9.50 Dial-A-Length Rafterule
- ☐ 18.25 Drywall Contracting
- ☐ 13.75 Electrical Blueprint Reading
- ☐ 25.00 Electrical Const. Estimator
- ☐ 19.00 Estimating Electrical Const.
- ☐ 17.00 Estimat. Home Blding Costs
- ☐ 28.00 Estimating Painting Costs
- ☐ 17.25 Estimating Plumbing Costs
- ☐ 21.50 Est. Tables for Home Building
- ☐ 22.75 Excav. & Grading Hndbk Rev.
- ☐ 9.25 E-Z Square
- ☐ 23.25 Fences & Retaining Walls
- ☐ 15.25 Finish Carpentry
- ☐ 24.75 Hdbk of Const. Contr. Vol. 1
- ☐ 24.75 Hdbk of Const. Contr. Vol. 2
- ☐ 15.00 Home Wiring: Improv. Ext. Repair
- ☐ 17.50 How to Sell Remodeling
- ☐ 19.50 How to Succ'd w/Own Const Bus.
- ☐ 24.50 HVAC Contracting
- ☐ 24.00 Illustrated Guide to NE Code
- ☐ 20.25 Manual of Electrical Contr.
- ☐ 19.75 Manual of Prof. Remodeling
- ☐ 17.25 Masonry & Concrete Const.
- ☐ 26.50 Masonry Estimating

Book list (column 3):

- ☐ 95.00 Audio: Const. Field Super.
- ☐ 65.00 Audio: Estimating Electrical
- ☐ 65.00 Audio: Estimating Remodel
- ☐ 19.95 Audio: Plumbers Examination
- ☐ 22.00 Basic Plumbing with Illust.
- ☐ 30.00 Berger Building Cost File
- ☐ 11.25 Bluprt Read for Bldg Trades
- ☐ 19.75 Bookkeeping for Builders
- ☐ 24.95 Bider's Comp. Dictionary
- ☐ 20.00 Bider's Guide to Account Rev.
- ☐ 15.25 Bider's Guide to Const. Fin.
- ☐ 15.50 Bider's Office Manual Revised
- ☐ 14.00 Building Cost Manual
- ☐ 11.75 Building Layout
- ☐ 22.00 Cabinetmaking: Design-Finish
- ☐ 25.50 Carpentry Estimating
- ☐ 19.75 Carpentry for Resid. Const.
- ☐ 19.00 Carpentry in Com. Const.
- ☐ 16.25 Carpentry Layout
- ☐ 17.75 Computers: Bider's New Tool
- ☐ 14.50 Concrete and Formwork
- ☐ 20.50 Concrete Const. & Estimating
- ☐ 26.00 Const. Estimating Refer. Data
- ☐ 22.00 Construction Superintending
- ☐ 19.25 Const. Surveying & Layout
- ☐ 19.00 Cont. Growth & Profit Guide
- ☐ 24.25 Cont Guide to Bldg Code Rev

10 Day Money Back Guarantee

Craftsman Book Co
6058 Corte del Cedro
Carlsbad, CA 92009

☎ **In a hurry?**
We accept phone orders charged to your MasterCard, Visa or American Express
Call 1-800-829-8123
FAX (619) 438-0398

Name _____
Company _____
Address _____
City/State/Zip _____
Total enclosed _____ (In Calif. add 6% tax)
☐ Visa ☐ MasterCard or ☐ AmEx
Card # _____
Exp. date _____

Order Form 3 (identical to Order Form 1)

Book list (column 1):

- ☐ 22.50 National Const. Estimator w/ free Estimate Writer on 5¼" (1.2Mb) disk — *Add $10 for Estimate Writer on either* ☐ 5¼" (360K) disk or ☐ 3½" (720K) disk.
- ☐ 26.50 Nat. Repair & Remodel Est.
- ☐ 19.25 Paint Contractor's Manual
- ☐ 21.25 Painter's Handbook
- ☐ 27.50 Painting Cost Guide
- ☐ 13.00 Plan & Design Plumb System
- ☐ 19.25 Planning Drain, Waste & Vent
- ☐ 21.00 Plumber's Exam Prep. Guide
- ☐ 18.00 Plumber's Handbook Rev.
- ☐ 19.75 Profits Buy & Renov Homes
- ☐ 14.25 Rafter Length Manual
- ☐ 23.00 Remodeler's Handbook
- ☐ 18.25 Remodeling Contr. Handbook
- ☐ 26.25 Remodeling Kitchens & Baths
- ☐ 11.50 Residential Electrical Design
- ☐ 16.75 Residential Electr. Hndbk.
- ☐ 18.25 Residential Wiring
- ☐ 22.00 Roof Framing
- ☐ 14.00 Roofers Handbook
- ☐ 17.00 Rough Carpentry
- ☐ 21.00 Run. Your Remodeling Bus.
- ☐ 27.00 Spec Builder's Guide
- ☐ 15.50 Stair Builder's Handbook
- ☐ 14.25 Wood-Frame House Const.

Book list (column 2):

- ☐ 16.75 Contractor's Survival Manual
- ☐ 16.50 Cont. Year-Rd Tax Guide
- ☐ 15.75 Cost Records for Const. Est.
- ☐ 9.50 Dial-A-Length Rafterule
- ☐ 18.25 Drywall Contracting
- ☐ 13.75 Electrical Blueprint Reading
- ☐ 25.00 Electrical Const. Estimator
- ☐ 19.00 Estimating Electrical Const.
- ☐ 17.00 Estimat. Home Blding Costs
- ☐ 28.00 Estimating Painting Costs
- ☐ 17.25 Estimating Plumbing Costs
- ☐ 21.50 Est. Tables for Home Building
- ☐ 22.75 Excav. & Grading Hndbk Rev.
- ☐ 9.25 E-Z Square
- ☐ 23.25 Fences & Retaining Walls
- ☐ 15.25 Finish Carpentry
- ☐ 24.75 Hdbk of Const. Contr. Vol. 1
- ☐ 24.75 Hdbk of Const. Contr. Vol. 2
- ☐ 15.00 Home Wiring: Improv. Ext. Repair
- ☐ 17.50 How to Sell Remodeling
- ☐ 19.50 How to Succ'd w/Own Const Bus.
- ☐ 24.50 HVAC Contracting
- ☐ 24.00 Illustrated Guide to NE Code
- ☐ 20.25 Manual of Electrical Contr.
- ☐ 19.75 Manual of Prof. Remodeling
- ☐ 17.25 Masonry & Concrete Const.
- ☐ 26.50 Masonry Estimating

Book list (column 3):

- ☐ 95.00 Audio: Const. Field Super.
- ☐ 65.00 Audio: Estimating Electrical
- ☐ 65.00 Audio: Estimating Remodel
- ☐ 19.95 Audio: Plumbers Examination
- ☐ 22.00 Basic Plumbing with Illust.
- ☐ 30.00 Berger Building Cost File
- ☐ 11.25 Bluprt Read for Bldg Trades
- ☐ 19.75 Bookkeeping for Builders
- ☐ 24.95 Bider's Comp. Dictionary
- ☐ 20.00 Bider's Guide to Account Rev.
- ☐ 15.25 Bider's Guide to Const. Fin.
- ☐ 15.50 Bider's Office Manual Revised
- ☐ 14.00 Building Cost Manual
- ☐ 11.75 Building Layout
- ☐ 22.00 Cabinetmaking: Design-Finish
- ☐ 25.50 Carpentry Estimating
- ☐ 19.75 Carpentry for Resid. Const.
- ☐ 19.00 Carpentry in Com. Const.
- ☐ 16.25 Carpentry Layout
- ☐ 17.75 Computers: Bider's New Tool
- ☐ 14.50 Concrete and Formwork
- ☐ 20.50 Concrete Const. & Estimating
- ☐ 26.00 Const. Estimating Refer. Data
- ☐ 22.00 Construction Superintending
- ☐ 19.25 Const. Surveying & Layout
- ☐ 19.00 Cont. Growth & Profit Guide
- ☐ 24.25 Cont Guide to Bldg Code Rev

10 Day Money Back Guarantee

Craftsman Book Co
6058 Corte del Cedro
Carlsbad, CA 92009

☎ **In a hurry?**
We accept phone orders charged to your MasterCard, Visa or American Express
Call 1-800-829-8123
FAX (619) 438-0398

Name _____
Company _____
Address _____
City/State/Zip _____
Total enclosed _____ (In Calif. add 6% tax)
☐ Visa ☐ MasterCard or ☐ AmEx
Card # _____
Exp. date _____

BUSINESS REPLY MAIL

FIRST CLASS   PERMIT NO. 271   CARLSBAD, CA

POSTAGE WILL BE PAID BY ADDRESSEE

Craftsman Book Company
6058 Corte Del Cedro
P. O. Box 6500
Carlsbad, CA 92008—0992

NO POSTAGE
NECESSARY
IF MAILED
IN THE
UNITED STATES

BUSINESS REPLY MAIL

FIRST CLASS   PERMIT NO. 271   CARLSBAD, CA

POSTAGE WILL BE PAID BY ADDRESSEE

Craftsman Book Company
6058 Corte Del Cedro
P. O. Box 6500
Carlsbad, CA 92008—0992

NO POSTAGE
NECESSARY
IF MAILED
IN THE
UNITED STATES

BUSINESS REPLY MAIL

FIRST CLASS   PERMIT NO. 271   CARLSBAD, CA

POSTAGE WILL BE PAID BY ADDRESSEE

Craftsman Book Company
6058 Corte Del Cedro
P. O. Box 6500
Carlsbad, CA 92008—0992

NO POSTAGE
NECESSARY
IF MAILED
IN THE
UNITED STATES

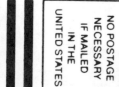